T0316409

Plasma Modeling

Methods and Applications

IOP Plasma Physics Series

Series Editors

Richard Dendy
Culham Centre for Fusion Energy and the University of Warwick, UK

Uwe Czarnetzki
Ruhr-University Bochum, Germany

About the Series

The IOP Plasma Physics ebook series aims at comprehensive coverage of the physics and applications of natural and laboratory plasmas, across all temperature regimes. Books in the series range from graduate and upper-level undergraduate textbooks, research monographs and reviews.

The conceptual areas of plasma physics addressed in the series include:
- Equilibrium, stability and control
- Waves: fundamental properties, emission, and absorption
- Nonlinear phenomena and turbulence
- Transport theory and phenomenology
- Laser-plasma interactions
- Non-thermal and suprathermal particle populations
- Beams and non-neutral plasmas
- High energy density physics
- Plasma-solid interactions, dusty, complex and non-ideal plasmas
- Diagnostic measurements and techniques for data analysis

The fields of application include:
- Nuclear fusion through magnetic and inertial confinement
- Solar-terrestrial and astrophysical plasma environments and phenomena
- Advanced radiation sources
- Materials processing and functionalisation
- Propulsion, combustion and bulk materials management
- Interaction of plasma with living matter and liquids
- Biological, medical and environmental systems
- Low temperature plasmas, glow discharges and vacuum arcs
- Plasma chemistry and reaction mechanisms
- Plasma production by novel means

Plasma Modeling

Methods and Applications

Gianpiero Colonna

Consiglio Nazionale delle Ricerche (CNR), PLASMI Lab at NANOTEC, Bari, Italy

Antonio D'Angola

*Scuola di Ingegneria, Università della Basilicata, Potenza, Italy and
Consiglio Nazionale delle Ricerche (CNR), PLASMI Lab at NANOTEC, Bari, Italy*

IOP Publishing, Bristol, UK

© IOP Publishing Ltd 2016

All rights reserved. No part of this publication may be reproduced, stored in a retrieval system or transmitted in any form or by any means, electronic, mechanical, photocopying, recording or otherwise, without the prior permission of the publisher, or as expressly permitted by law or under terms agreed with the appropriate rights organization. Multiple copying is permitted in accordance with the terms of licences issued by the Copyright Licensing Agency, the Copyright Clearance Centre and other reproduction rights organisations.

Permission to make use of IOP Publishing content other than as set out above may be sought at permissions@iop.org.

Gianpiero Colonna and Antonio D'Angola have asserted his right to be identified as the authors of this work in accordance with sections 77 and 78 of the Copyright, Designs and Patents Act 1988.

ISBN   978-0-7503-1200-4 (ebook)
ISBN   978-0-7503-1201-1 (print)
ISBN   978-0-7503-1202-8 (mobi)

DOI   10.1088/978-0-7503-1200-4

Version: 20161201

IOP Expanding Physics
ISSN 2053-2563 (online)
ISSN 2054-7315 (print)

British Library Cataloguing-in-Publication Data: A catalogue record for this book is available from the British Library.

Published by IOP Publishing, wholly owned by The Institute of Physics, London

IOP Publishing, Temple Circus, Temple Way, Bristol, BS1 6HG, UK

US Office: IOP Publishing, Inc., 190 North Independence Mall West, Suite 601, Philadelphia, PA 19106, USA

To Mariachiara, Gerardina, Michelangelo and Maria Teresa, our future.

Contents

Part II   Fluid and Hybrid Models

Preface

In the last few years, research activity in the fields of plasma physics and chemistry has grown exponentially, not only focusing on fundamental aspects, but also on industrial applications. Due to its inherent complexity, both experiments and numerical modeling are necessary to fully characterize plasma systems. The most commonly used numerical simulation techniques for plasma modeling include fluid dynamics, kinetic, and hybrid models. These simulation models differ significantly in terms of their principles, strengths, and limitations. Kinetic models, such as particle-in-cell with Monte Carlo collisions, are used for non-equilibrium systems, and fluid models are employed for faster computations, reducing the accuracy. Hybrid models provide balance between precision and efficiency.

The golden age of plasma modeling occurred in the 1980s–90s, when several research groups across the world developed theoretical and numerical approaches to investigate plasma behaviors. Currently, plasma modeling mainly addresses techno-logical applications and, due to the complexity of systems, commercial multi-purpose codes are often used. These numerical codes are based on the fluid dynamics approach, describing realistic discharge geometries. *COMSOL* Multiphysics and *ANSYS Fluent* are examples of very popular commercial general-purpose software platforms, based on advanced numerical methods, for modeling and simulating physics-based prob-lems. They are suitable for the design and development of optimized products by researchers and engineers working for leading technical enterprises, research laboratories, and universities. In strong non-equilibrium cases it is necessary to calculate the energy distribution function of particles. Kinetic modeling is recurrent in gas discharge investigations due to the non-equilibrium energy distribution of charged particles, which are not representable by the Maxwell distribution function, with abrupt changes in the slope, dumps and bumps, affecting the macroscopic properties of the plasma. In this context, a critical aspect is to determine the actual distribution function of free electrons, solving the Boltzmann equation. The electron energy distribution directly affects the plasma composition, also inducing non-equilibrium in internal states which, in turn, influences the electron gas, creating a strong synergy between the different plasma components. *BOLSIG+*, originally developed in the LAPLACE laboratory in Toulouse, is an example a freeware non-commercial code for the numerical solution of the Boltzmann equation for free electrons in weakly ionized gases in the framework of the two-term approximation.

Several non-commercial codes have been developed by research groups to simulate one- or two-dimensional problems and suitable plasma applications. It is worth mentioning *nonPDPSIM*, developed by the Kushner research group, which is is a two-dimensional computational platform that solves, using the hybrid approach, transport equations for charged and neutral species, Poisson's equation for the electric potential, the electron energy conservation equation for the electron temper-ature, radiation transport and Navier–Stokes equations for the neutral gas flow, coupled with a Monte Carlo solver for the electron energy distribution function. The two-dimensional platform has three major modules addressing plasma dynamics, fluid dynamics, and surface kinetics, and is implemented on an unstructured mesh to enable resolution of a large dynamic range in space.

The aim of this book is to present a review of the different approaches that can be adopted to model a plasma, giving a comprehensive bibliography for students and researchers who want to start a project in plasma modeling, orienting the reader toward the choice of a suitable numerical method. The chapters in this book have been written by experienced researchers in the field, who have developed advanced theoretical models and high-performance numerical codes. The theoretical models used in commercial or freeware codes are described throughout this book, guiding students, researchers, and engineers in choosing the proper approach for the problem under investigation. This book describes the theoretical and numerical approaches that the plasma community currently uses to investigate plasmas, and thus can help the reader in creating a customized in-house code, particularly when commercial or freeware codes cannot be used.

The book is divided into three parts: part 1, Kinetic theory; part 2, Fluid and hybrid models; and part 3, Applications. In part 1, the first chapter provides a general overview of the Boltzmann equation, methods of solution, and applications. The second and third chapters present the solution of the Boltzmann equation for free electrons in two-term and multi-term approximations. Chapter 4 describes the particle-in-cell algorithm. Chapter 5 presents the ergodic method, which is suitable for multi-scale problems in plasma physics.

In part 2, chapter 6 develops the theory of the fluid model for a plasma, while chapter 7 focuses on magnetohydrodynamics. Chapter 8 describes the self-consistent model, which couples state-to-state kinetics with free electron kinetics. Chapter 9 discusses hybrid models, considering electrons as a fluid and following ions as particles.

Finally, some applications are presented in part 3. In particular, chapter 10 shows the results for conventional and innovative plasma torches under local thermodynamic equilibrium, while chapter 11 focuses on non-equilibrium in high-pressure discharges, also considering the problem of fast gas heating. Chapter 12 discusses the role of radiation heating for different space vehicles entering planetary atmospheres. Chapter 13 treats the complex problem of dust formation in low pressure plasmas and chapter 14 examines the problem of the validation and verification of numerical codes, a fundamental aspect for modeling activities.

In each chapter, the authors have tried to find the right compromise in synthesizing a subject that would be worthy of its own book. Due to limitations of space, the editors were not able to dedicate discussion to some theoretical approaches, such as Monte Carlo and molecular dynamics, and some applications, such as waves in plasmas, self-organized plasmas, probe theory, and so on. Nevertheless, the theoretical and numerical approaches described in the book can be applied to any specific application. The extended reference section in each chapter will help in finding additional material for more in-depth analysis.

Gianpiero Colonna and Antonio D'Angola

Acknowledgements

The Editors want to thank all the contributing authors for their efforts and invaluable work. Without their co-operation this book would not have been published.

We want to thank Professor Mario Capitelli and Professor Gianni Coppa for their precious guidance over the last 25 years.

For chapter 1:

The author wants to thank Annarita Laricchiuta for useful suggestions, in particular on the section devoted to cross section.

For chapter 3:

I would like to thank my colleagues Dr Markus M Becker and Dr Florian Sigeneger for their helpful discussions and kind assistance in preparing this paper.

For chapter 4:

The work presented in Chapter 4 is supported by the European Commission via the Deep-ER project (http://www.deep-er.eu), by the Onderzoekfonds KU Leuven (Research Fund KU Leuven, GOA scheme and Space Weaves RUN project), by the US Air Force EOARD Project FA2386-14-1-0052 and by the Interuniversity Attraction Poles Programme of the Belgian Science Policy Oce (IAPP7/08 CHARM).

For chapter 12:

This work was performed in the framework of the Program of Fundamental Research of the Russian Academy of Sciences and was supported by the Russian Foundation of Basic Research grant #16-01-00379.

For chapter 13:

The authors wish to thank S Ratynskaia, M Rosenberg and P Tolias for useful comments on the manuscript.

This work was funded by the US Department of Energy Office of Science, Office of Fusion Energy Sciences, under the auspices of the National Nuclear Security Administration of the U.S. Department of Energy by Los Alamos National Laboratory, operated by Los Alamos National Security LLC under Contract No. DE-AC52-06NA25396.

For chapter 14:

The simulations presented herein were carried out at the Swiss National Supercomputing Centre (CSCS), under project ID s346, and at the Computational Simulation Centre of the International Fusion Energy Research Centre (IFERCCSC), Aomori, Japan, under the Broader Approach collaboration between

Euratom and Japan, implemented by Fusion for Energy and JAEA. This work was carried out within the framework of the EUROfusion Consortium. It was supported by the Swiss National Science Foundation and received funding from the European Unions Horizon 2020 research and innovation programme under grant agreement number 633053. The views and opinions expressed herein do not necessarily reflect those of the European Commission

About the editors

Gianpiero Colonna

Gianpiero Colonna graduated in physics in 1991 with a thesis on the stationary solution of the Boltzmann equation. He obtained a PhD in chemistry in 1995 with a thesis on the state-to-state kinetics in the boundary layer of hypersonic flows, under the supervision of Mario Capitelli, and held a post-doctoral position at the University of British Columbia in Vancouver, under the supervision of Bernie Shizgal. After a period as a Researcher at the Pharmacy Faculty of the University of Bari, he became a Researcher at the National Research Council (CNR). He is currently a Senior Researcher at the same institution. His research activities have focused on plasma modeling, state-to-state self-consistent kinetics in gas discharges and hypersonic flows, the thermodynamic and transport properties of plasmas, and modeling the plasma plume produced by nanosecond laser pulses. He is the author of about two hundred papers and conference proceedings. He has also co-authored two books of the series *Fundamental Aspects of Plasma Chemical Physics*.

Antonio D'Angola

Antonio D'Angola received his master's degree in nuclear engineering in 1996 and his PhD in energetics in 2000, both at Politecnico di Torino, Italy. Since February 2004 he has been an Assistant Professor at the University of Basilicata in nuclear reactor physics. In 2000 he was a Visiting Researcher at the Plasma Technology Research Centre (CRTP) of Sherbrooke, Quebec. His scientific activities and areas of interest are: numerical methods for the simulation of plasmas by using particle-in-cell and Monte Carlo codes, the calculation of the thermodynamic and transport properties of ionized plasmas for industrial and aerospace applications and the investigation of non-neutral plasmas for ultra-high vacuum systems and laser–plasma interactions for medical applications. He is Associate Researcher at the National Research Council (CNR) and Associate Editor of the journal *Frontiers in Plasma Physics*. He is the author of about one hundred papers and conference proceedings. He has also co-authored a book of the series *Fundamental Aspects of Plasma Chemical Physics*.

Contributors

Jayr Amorim
jayr.de.amorim@gmail.com Departamento de Física,
Instituto Tecnológico de Aeronáutica, 12228-900
Sáo José dos Campos, Brazil

Davide Bernardi
davide.bernardi@enea.it
ENEA C.R. Brasimone
Camugnano (BO), Italy

Gianpiero Colonna
gianpiero.colonna@cnr.it
PLASMI Lab @ CNR-NANOTEC
Bari, Italy

Gianni Coppa
gianni.coppa@polito.it
Dipartimento Energia
Politecnico di Torino
Turin, Italy

Andrea Cristofolini
andrea.cristofolini@unibo.it
Department of Electrical, Electronic, and Information Engineering 'Guglielmo Marconi'
ALMA MATER STUDIORUM – Università di Bologna
Bologna, Italy

Giuliano D'Ammando
giuliano.dammando@cnr.it
PLASMI Lab @ CNR-NANOTEC
Bari, Italy

Antonio D'Angola
antonio.dangola@unibas.it
Scuola di Ingegneria, Università della Basilicata
Potenza, Italy

Gian Luca Delzanno
delzanno@lanl.gov
Los Alamos National Laboratory
Los Alamos, NM, USA

Ambrogio Fasoli
ambrogio.fasoli@epfl.ch
École Polytechnique Fédérale de Lausanne, Swiss Plasma Center
Lausanne, Switzerland

Ivo Furno
ivo.furno@epfl.ch
École Polytechnique Fédérale de Lausanne, Swiss Plasma Center
Lausanne, Switzerland

Daniela Grasso
daniela.grasso@infm.polito.it
ISC-CNR and Politecnico di Torino
Dipartimento Energia Turin, Italy

Vasco Guerra
vguerra@tecnico.ulisboa.pt
Instituto de Plasmas e Fusão Nuclear, Instituto Superior Técnico
Lisbon, Portugal

Philippe Guittienne
philippe.guittienne@epfl.ch
Helyssen Sàrl
Route de la Louche 31, CH-1092
Belmont, Switzerland

Alan Howling
alan.howling@epfl.ch
École Polytechnique Fédérale de Lausanne, Swiss Plasma Center
Lausanne, Switzerland

Remy Jacquier
remy.jacquier@epfl.ch
École Polytechnique Fédérale de Lausanne, Swiss Plasma Center
Lausanne, Switzerland

Giovanni Lapenta
giovanni.lapenta@wis.kuleuven.be
KU Leuven
Departement Wiskunde
Afdeling Plasma-astrofysica
Leuven, Belgium

Detlef Loffhagen
loffhagen@inp-greifswald.de
Department of Plasma Modelling
Leibniz Institute for Plasma Science and Technology (INP Greifswald)
Greifswald, Germany

Pierpaolo Minelli
pierpaolo.minelli@cnr.it
PLASMI Lab @ CNR-NANOTEC
Bari, Italy

Fabio Peano
fabio.peano@gmail.com
Italy

Federico Peinetti
federico.peinetti@gmail.com
Italy

Lucia Daniela Pietanza
luciadaniela.pietanza@cnr.it
PLASMI Lab @ CNR-NANOTEC
Bari, Italy

Paolo Ricci
paolo.ricci@epfl.ch
École Polytechnique Fédérale de Lausanne, Swiss Plasma Center
CH
Lausanne, Switzerland

Marco A. Ridenti
marcoridenti@gmail.com
Departamento de Física, Instituto Tecnológico de Aeronáutica, 12228-900
Sáo José dos Campos, Brazil

Fabio Riva
fabio.riva@epfl.ch
École Polytechnique Fédérale de Lausanne, Swiss Plasma Center
Lausanne, Switzerland

Luis O. Silva
luis.silva@tecnico.ulisboa.pt
GoLP, Instituto Superior Técnico, Universidade de Lisboa,
Lisbon, Portugal

Sergey T. Surzhikov
surg@ipmnet.ru
Institute for Problems in Mechanics, Russian Academy of Sciences
Moscow, Russia

Francesco Taccogna
francesco.taccogna@cnr.it
PLASMI Lab @ CNR-NANOTEC
Bari, Italy

Xianzhu Tang
xtang@lanl.gov
Los Alamos National Laboratory
Los Alamos, NM, USA

Christian Theiler
christian.theiler@epfl.ch
École Polytechnique Fédérale de Lausanne, Swiss Plasma Center
Lausanne, Switzerland

Renato Zaffina
renato.zaffina@unibas.it
Scuola di Ingegneria, Università della Basilicata
Potenza, Italy

Part I

Kinetic Theory

IOP Publishing

Plasma Modeling
Methods and Applications
Gianpiero Colonna

Chapter 1

Boltzmann and Vlasov equations in plasma physics

If you ask researchers in physics and chemistry what the scientific discovery at the origin of modern science is, most of the answers will be Einstein's relativity or quantum physics. In our opinion, this role should be assigned to the *Boltzmann kinetic equation* and its application to the kinetic theory of gases. The same opinion was expressed by Max Planck in his Nobel lecture [1]:

> … my trying to give you the story of the origin of the quantum theory in broad outlines and to couple with this, a picture in a small frame, of the development of this theory up to now, and its present-day significance for physics. … For this reason, I busied myself, from then on, that is, from the day of its establishment, with the task of elucidating a true physical character for the formula, and this problem led me automatically to a consideration of the connection between entropy and probability, that is, Boltzmann's trend of ideas; until after some weeks of the most strenuous work of my life, light came into the darkness, and a new undreamed-of perspective opened up before me.

The Boltzmann equation introduced into physics the idea of probability, which was then used some years later in quantum physics. Also the concept of time changed, relating its direction to that of entropy exchange [2], breaking the symmetry between the past and the future [3], even if the laws of mechanics are invariant for time inversion. Despite its importance, students in physics and chemistry often confuse the Boltzmann equation with the Boltzmann distribution.

The Boltzmann equation has many applications in modern science [4, 5], from fluid dynamics to material science:

- The Chapmann–Enskog expansion is a method for calculating the transport properties of gases [6].

doi:10.1088/978-0-7503-1200-4ch1
© IOP Publishing Ltd 2016

- Numerical methods, such as the Bhatnagar–Gross–Krook (BGK) [7] approximation or *lattice Boltzmann method* (LBM) [8, 9], are commonly used to determine the flow field of complex fluids.
- The spectroscopic signature of gases as well as radiative heating are estimated by solving the Boltzmann equation for photons (see chapter 12).
- The thermal and electrical properties of metals and semiconductor materials can be estimated by solving the Boltzmann equation for electrons and phonons, adapted to account for quantum statistics [10].
- Mathematical aspects, such as the existence and unicity of solutions and the accuracy of approximate methods are also determined using the Boltzmann equation [11].

The Boltzmann equation is one of the most powerful tools for investigating the plasma state, from the electron kinetics in weakly ionized gases [12] to fusion [13] and astrophysical [14, 15] plasmas. In this chapter we present the theoretical foundation of the Boltzmann and Vlasov equations, giving an overview of their applications.

1.1 Fundamentals

Let us consider the single-particle phase space Φ, characterized by the sextuplet $(\mathbf{r}, \mathbf{v})$ of particle position and velocity. The state of the particle corresponds to a single point in this space, while a system with $\mathcal{N}$ particles is represented by a set of $\mathcal{N}$ points. In order to provide a pictorial view, we limit ourselves to a one-dimensional (1D) space coordinate and one velocity component. This representation can also be considered as the projection of the entire volume in the two-dimensional (2D) plane (x, v_x). To determine the time evolution of the volume in the Φ space, the motion of each particle, according to the classical mechanic laws, must be followed:

$$\frac{\mathrm{d}\mathbf{r}_i}{\mathrm{d}t} = \mathbf{v}_i$$
$$\frac{\mathrm{d}\mathbf{v}_i}{\mathrm{d}t} = \mathbf{a}_i.$$

$$(1.1)$$

The ensemble of points changes its shape, as can be observed in figure 1.1 (left) for a 2D example, because each particle moves independently of the others. The acceleration $\mathbf{a}_i$ originates from two contributions

$$\mathbf{a}_i = \mathbf{a}_{\mathrm{ext}}(\mathbf{r}_i) + \sum_{k \neq i}^{\mathcal{N}} \frac{\mathbf{F}_{ik}}{m_i},$$

$$(1.2)$$

one, $\mathbf{a}_{\mathrm{ext}}$, is due to an external field, and the other, $\mathbf{F}_{ik}/m_i$ where m_i is the particle mass, is due to intermolecular forces $\mathbf{F}_{ik}$. This approach, successfully applied to predict the orbit of planets in the solar system, in principle allows one to determine exactly the evolution of any N-particle system, such as a matter in the gas phase. In practice, the number of molecules in a macroscopic system ($\mathcal{N} \approx 10^{23}$) and the

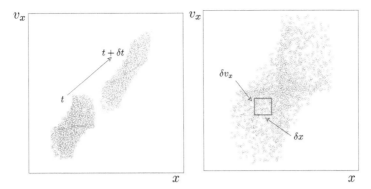

Figure 1.1. Projection of the phase space $(\mathbf{r}, \mathbf{v})$ on the 2D plane (x, v_x) of an $\mathcal{N}$-particle system. Evolution of the sampling points with $a_x(x) = a_{x0} + a_{x1}x^2$ (left) and the starting distribution with the projection of a sampling volume $(\delta\mathbf{x}, \delta\mathbf{v}_x)$ (right).

uncertainty in the initial conditions[1] make such a deterministic procedure unfeasible for any numerical simulation, even for modern high-performance computers.

An alternative approach is to describe the evolution of an N-particle system by using the laws of statistics, developing the theory of *physical kinetics*. Let us consider the following assumptions:

1. A infinitesimal sampling volume, such as that depicted in figure 1.1 (right), contains a very large number of particles, so that a *macroscopic* infinitesimal volume still contains a large number of particles: the gas can be considered as a continuous medium.

2. The mean particle distance, $\bar{d} = \sqrt[3]{\frac{V}{\mathcal{N}}}$, with V the physical volume, is larger than the thermal de Broglie length, $\lambda_{\text{th},i} = \sqrt{\frac{2\pi\hbar}{m_i kT}}$: quantum effects are negligible and the classical statistics and laws of motion can be used.

3. The interparticle interaction distance $\bar{r}_i$ is much smaller than the mean-free-path λ_{mfp}: particle interactions can be described as binary collisions.

As a first step, let us limit ourselves to a system of free particles moving in an external field, without collisions. In a infinitesimal hypercube identified by the vertices $(\mathbf{r} - \frac{1}{2}\delta\mathbf{r}, \mathbf{v} - \frac{1}{2}\delta\mathbf{v})$ and $(\mathbf{r} + \frac{1}{2}\delta\mathbf{r}, \mathbf{v} + \frac{1}{2}\delta\mathbf{v})$, see figure 1.1 (right), which has a volume in the phase space of

$$\delta\Gamma = \delta x \delta y \delta z \delta v_x \delta v_y \delta v_z = \delta V \delta^3 v, \tag{1.3}$$

where δV is the physical volume, the number of particles in the infinitesimal hypercube can be written as

---

[1] The uncertainty of initial position and velocity is assured by *Heisenberg's principle*.

$$\delta \mathcal{N} = f(\mathbf{r}, \mathbf{v}, t)\delta\Gamma, \tag{1.4}$$

which defines the *distribution function* as the number of particles per unit volume in the phase space. For real systems, n_s different species can be present, each one with its own internal structure which, in the framework of quantum physics, can be represented by discrete levels, characterized by a set of quantum numbers ℓ.[2] This consideration brings us to a generalized definition of the distribution function, considering equation (1.4) for particles of the sth species in the ℓth level, i.e.

$$\delta \mathcal{N}_{s,\ell}(t) = f_s(\mathbf{r}, \mathbf{v}, \ell, t)\delta\Gamma. \tag{1.5}$$

1.1.1 The convection operator

Let us now evaluate the rate at which the number of particles inside the sampling volume changes. It corresponds to the flux of particles crossing the surface of the hypercube. For the sake of simplicity, the discussion is restricted to the 2D problem depicted in figure 1.1 (right), with the square sampling volume, where the particle flux is the sum of four contributions

$$\frac{\partial \delta \mathcal{N}_{s,\ell}}{\partial t} = \Phi = \Phi_{\text{left}} + \Phi_{\text{right}} + \Phi_{\text{bottom}} + \Phi_{\text{top}}, \tag{1.6}$$

two through the vertical sides ('left' and 'right'), due to particles with the fixed velocity changing position

$$\Phi_{\text{left}} = + \int_{v_x - \frac{1}{2}\delta v_x}^{v_x + \frac{1}{2}\delta v_x} w_x f_s\left(x - \frac{1}{2}\delta x, w_x, \ell, t\right)\mathrm{d}w_x$$

$$\approx v_x f_s\left(x - \frac{1}{2}\delta x, v_x, \ell, t\right)\delta v_x$$

$$\Phi_{\text{right}} = - \int_{v_x - \frac{1}{2}\delta v_x}^{v_x + \frac{1}{2}\delta v_x} w_x f_s\left(x + \frac{1}{2}\delta x, w_x, \ell, t\right)\mathrm{d}w_x \tag{1.7}$$

$$\approx -v_x f_s\left(x + \frac{1}{2}\delta x, v_x, \ell, t\right)\delta v_x,$$

and two through the horizontal sides ('bottom' and 'top'), due to particles accelerated in the same positions

---

[2] As an example, for atomic species we have the quadruplet (n, l, m, s) corresponding to the principal, angular, magnetic, and spin quantum numbers, and for molecules (n, J, v, s) corresponding to the electronic state, and rotational, vibrational, and total spin quantum numbers. However, internal levels can be ordered by energy and identified by a single integer number.

$$\Phi_{\text{bottom}} = + \int_{x-\frac{1}{2}\delta x}^{x+\frac{1}{2}\delta x} a_x(x')f_s\left(x', v_x - \frac{1}{2}\delta v_x, \ell, t\right)dx'$$

$$\approx a_x(x)f_s\left(x, v_x - \frac{1}{2}\delta v_x, \ell, t\right)\delta x$$

$$\Phi_{\text{top}} = - \int_{x-\frac{1}{2}\delta x}^{x+\frac{1}{2}\delta x} a_x(x')f_s\left(x', v_x + \frac{1}{2}\delta v_x, \ell, t\right)dx' \tag{1.8}$$

$$\approx -a_x(x)f_s\left(x, v_x + \frac{1}{2}\delta v_x, \ell, t\right)\delta x.$$

Combining the terms in equations (1.7) and (1.8), we obtain

$$\Phi_{\text{left}} + \Phi_{\text{right}} = v_x\delta v_x\left[f_s\left(x - \frac{1}{2}\delta x, v_x, \ell, t\right) - f_s\left(x + \frac{1}{2}\delta x, v_x, \ell, t\right)\right]$$

$$\Phi_{\text{bottom}} + \Phi_{\text{top}} = a_x(x)\delta x\left[f_s\left(x, v_x - \frac{1}{2}\delta v_x, \ell, t\right) - f_s\left(x, v_x + \frac{1}{2}\delta v_x, \ell, t\right)\right]. \tag{1.9}$$

Considering that the sampling volume in the phase space is small, the parentheses on the right-hand side in equation (1.9) can be approximated with the derivatives

$$f_s\left(x - \frac{1}{2}\delta x, v_x, \ell, t\right) - f_s\left(x + \frac{1}{2}\delta x, v_x, \ell\right) \approx -\frac{\partial f_s(x, v_x, \ell, t)}{\partial x}\delta x$$

$$f_s\left(x, v_x - \frac{1}{2}\delta v_x, \ell, t\right) - f_s\left(x, v_x + \frac{1}{2}\delta v_x, \ell, t\right) \approx -\frac{\partial f_s(x, v_x, \ell, t)}{\partial v_x}\delta v_x \tag{1.10}$$

and equations (1.6)–(1.8) give the total flux as

$$\frac{\partial \delta \mathcal{N}_{s,\ell}}{\partial t} = \Phi = -\delta x\delta v_x\left[v_x\frac{\partial f_s(x, v_x, \ell, t)}{\partial x} + a_x(x)\frac{\partial f_s(x, v_x, \ell, t)}{\partial v_x}\right]. \tag{1.11}$$

To extend this result to the whole phase space, we should note that in multi-dimensional space:
- The integrals in equations (1.7) and (1.8) must be performed over all the variables apart from the one perpendicular to the surface, i.e. $\delta x\delta v_x \rightarrow \delta\Gamma$.
- Two contributions must be considered for each independent direction, i.e. equation (1.9) should also be written for the y- and z-directions.
- Acceleration depends on the position in the phase space, i.e. $a_x(x) \rightarrow [a_x(\mathbf{r}), a_y(\mathbf{r}), a_z(\mathbf{r})]$ [3]

---

[3] The acceleration can also depend on the velocity, as happens in the presence of a magnetic field for charged species, an aspect discussed in chapter 2.

These assumptions, together with equation (1.5), give

$$\frac{\partial \delta \mathcal{N}_{s,\ell}}{\partial t} = \delta \Gamma \frac{\partial f_s(\mathbf{r}, \mathbf{v}, \ell, t)}{\partial t} = -\delta \Gamma \left[\mathbf{v} \cdot \nabla_r + \mathbf{a}(\mathbf{r}) \cdot \nabla_v \right] f_s(\mathbf{r}, \mathbf{v}, \ell, t), \qquad (1.12)$$

where

$$\mathbf{v} \cdot \nabla_r \equiv v_x \frac{\partial}{\partial x} + v_y \frac{\partial}{\partial y} + v_z \frac{\partial}{\partial z}$$

$$\mathbf{a}(\mathbf{r}) \cdot \nabla_v \equiv a_x(\mathbf{r}) \frac{\partial}{\partial v_x} + a_y(\mathbf{r}) \frac{\partial}{\partial v_y} + a_z(\mathbf{r}) \frac{\partial}{\partial v_z}. \qquad (1.13)$$

The square parentheses in equation (1.12) can be considered as a convection operator, because they describes the evolution of the distribution function due to the free motion of particles subjected to an external force.

1.1.2 The collisional operator

Let us now describe the contribution of collisions to the evolution of the distribution in the phase space. If the gas density is sufficiently small only binary encounters can be considered. If the mean free path is much larger than the interaction distance, these encounters can be treated as collisions. In a collision at given particle velocities, the final state is completely determined by the deflection angle and, in the case of internal excitation and chemical reactions, by the energy necessary to activate the considered process. The uncertainty in assigning position and speed to the colliding particles makes it necessary to describe the collisions statistically. The cross section of a process is the measure of the probability of a collision output. Let us define the following quantities:

- The differential cross section $\frac{d\sigma}{d\Omega}$ is the ratio between the number of particles deflected in the unit time in a given solid angle $d\Omega$ and the incoming particle flux per unit time and per unit surface. It is a function of the modulus of the relative velocity $|\mathbf{v}_A - \mathbf{v}_B|$, the initial and final levels and the solid angle.
- The total cross section σ is the integral of the differential cross section over the whole solid angle. For elastic collisions this quantity is not relevant, including also the contribution of collisions with very small deflection angles. On the other hand, it is often used in inelastic collisions, whose angular dependence on differential cross section is assumed to be isotropic.

Including the collisional operator, equation (1.14) becomes the Boltzmann equation in the usual form

$$\frac{\partial f_s(\mathbf{r}, \mathbf{v}, \ell, t)}{\partial t} + \left[\mathbf{v} \cdot \nabla_r + \mathbf{a}(\mathbf{r}) \cdot \nabla_v \right] f_s(\mathbf{r}, \mathbf{v}, \ell, t) = \left(\frac{\delta f_s}{\delta t} \right)_{coll}. \qquad (1.14)$$

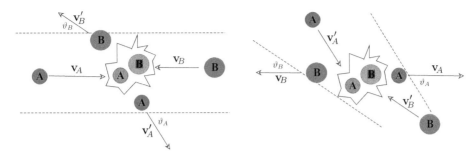

Figure 1.2. Scheme of a binary elastic collision (left) with the deflection angle and a picture of the reverse process (right) obtained by reversing time and space. The dashed lines are the directions of the velocities before the collisions and θ are the corresponding deflection angles.

Let us construct the collisional operator analyzing the binary encounter sketched in figure 1.2. In a sampling volume given by equation (1.3), the number of particles A with velocity $\mathbf{v}_A$ flowing through a unitary surfaces moving jointly with B is given by[4]

$$\Phi_A(\mathbf{v}_A,\,\mathbf{v}_B) = f_A|\mathbf{v}_A - \mathbf{v}_B|\delta^3 v_A = f_A v_{AB}\delta^3 v_A, \tag{1.15}$$

where $v_{AB} = |\mathbf{v}_A - \mathbf{v}_B|$ is the relative velocity and, from the definition of the differential cross section,[5] the number of particles A deflected by the angle ϑ_A in the unit time by a single particle B (see figure 1.3) is

$$\left[\frac{\mathrm{d}\delta\mathcal{N}_A(\vartheta_A)}{\mathrm{d}t}\right]_1 = \Phi_A(\mathbf{v}_A,\,\mathbf{v}_B)\frac{\mathrm{d}\sigma_{AB}}{\mathrm{d}\Omega}(\vartheta_A)\delta\Omega_A$$

$$= f_A v_{AB}\delta^3 v_A\frac{\mathrm{d}\sigma_{AB}}{\mathrm{d}\Omega}(\vartheta_A)\mathrm{d}\Omega_A, \tag{1.16}$$

with

$$\mathrm{d}\Omega_A = \sin\vartheta_A\,\mathrm{d}\vartheta_A\,\mathrm{d}\phi, \tag{1.17}$$

with ϕ the angle of the deflection plane.[6] The velocities after the collision $\mathbf{v}'_A$ and $\mathbf{v}'_B$, as well as ϑ_B, are univocally determined by ϑ_A and ϕ, because energy and momentum conservation laws can be used to determine all the other components. Equation (1.16) gives the number of particles undergoing the considered collision with a single B particle. To have the total number of particles deflected in the given direction we have to multiply this quantity by the number of scattering centers, i.e.

$$\frac{\partial\delta\mathcal{N}_A(\vartheta_A)}{\partial t} = \left[\frac{\partial\delta\mathcal{N}_A(\vartheta_A)}{\partial t}\right]_1 f_B\delta^3 v_B\delta V = f_A f_B v_{AB}\frac{\mathrm{d}\sigma_{AB}}{\mathrm{d}\Omega}(\vartheta_A)\delta^3 v_A\delta^3 v_B\delta V\,\mathrm{d}\Omega_A. \tag{1.18}$$

---

[4] For sake of simplicity, let us consider implicitly the variables of the distribution substituting $f_{A/B}(\mathbf{r},\,\mathbf{v}_{A/B},\,\ell_{A/B},\,t) \to f_{A/B}$ and $f_{A/B}(\mathbf{r},\,\mathbf{v}'_{A/B},\,\ell'_{A/B},\,t) \to f'_{A/B}$.
[5] Let us write explicitly only the dependence of the cross section on the deflection angle, using the symbol $\frac{\mathrm{d}\sigma'}{\mathrm{d}\Omega}$ for the reverse process.
[6] The directions of the velocities after the collision, if not parallel, univocally define the deflection plane. Therefore the same angle ϕ can be used for both particles.

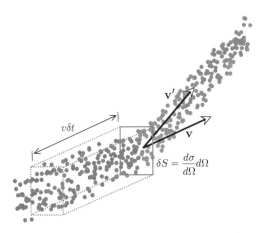

Figure 1.3. The number of particles, all with the same velocity **v** relative to the scattering center (blue dot), flowing through the surface δS in the time interval δt are those contained in the volume $\delta V = v\,\delta t\delta S$. Considering δS as the differential cross section, all these particles are deflected in the same solid angle $d\Omega$ with velocity **v'**.

It should be noted that equation (1.18) must be symmetric with respect to exchange between A and B, because this expression gives the same rate for the number of particles B deflected by the angle θ_B, therefore

$$\frac{d\sigma_{AB}}{d\Omega}(\vartheta_A)d\Omega_A = \frac{d\sigma_{BA}}{d\Omega}(\vartheta_B)d\Omega_B. \tag{1.19}$$

The variation expressed by equation (1.18) is the loss rate for particles A and B in the state before the collision. At the same time, the gain rates coming from the reverse process must be considered, as illustrated in figure 1.2 (right), given by

$$\left[\frac{\partial \delta \mathcal{N}_A(\vartheta_A)}{\partial t}\right]_r = f'_A f'_B\, v'_{AB}\delta^3 v'_A \delta^3 v'_B \frac{d\sigma'_{AB}}{d\Omega}(\vartheta_A)d\Omega_A \delta V, \tag{1.20}$$

noting that the deflection angles are the same as in the direct process. However, to maintain parity with respect to space–time inversion, the probability of a process must be equal to that of its reverse, i.e.

$$v'_{AB}\frac{d\sigma'_{AB}}{d\Omega}(\vartheta_A) = v_{AB}\frac{d\sigma_{AB}}{d\Omega}(\vartheta_A). \tag{1.21}$$

This equation is known as a *micro-reversibility* principle. This assumption can be demonstrated for collisions with spherical symmetry but it cannot be proved in general. For strange geometries, such as those of large organic molecules, the micro-reversibility principle is not valid. Nonetheless, in this book the micro-reversibility principle is considered as valid. The micro-reversibility principle has a corresponding law from the macroscopic point of view, the *detailed balance* principle. The global variation of the distribution function of f_A is obtained by summing over all the collision partners B, integrating over all its initial states and over all the final states of A, thus obtaining the final expression for the collisional operator

$$\left(\frac{\mathrm{d}f_A}{\mathrm{d}t}\right)_{\mathrm{coll}} = \sum_{\ell_A'} \sum_{B,\ell_B,\ell_B'} \int v_{AB}\frac{\mathrm{d}\sigma_{AB}}{\mathrm{d}\Omega}(\vartheta_A)\left(f_A'f_B' - f_Af_B\right)\mathrm{d}^3v_B \; \mathrm{d}\Omega_A, \tag{1.22}$$

that has been divided by the volume $\delta\Gamma_A = \delta V \delta^3 v_A$ to obtain the equation for the distribution function.

1.1.3 Boltzmann's *H*-theorem

It is straightforward to demonstrate that in equilibrium, i.e. in the presence of Boltzmann and Maxwell distributions, at the same temperature, the collisional operator is null, reproducing an important result already obtained in the framework of statistical physics. Another important result, obtained for the first time by Boltzmann, is the demonstration of the growth of entropy for an isolated rarefied gas, namely, the *H-theorem*. Let us define the functional

$$H(t) = \iiint \mathrm{d}^3v f \ln f \tag{1.23}$$

and its time derivative[7] given by

$$\frac{\partial H(t)}{\partial t} = \int \mathrm{d}^3v \frac{\partial f}{\partial t}(\ln f + 1). \tag{1.24}$$

Substituting the Boltzmann equation (see equation (1.14)) in this expression and noting that the convection operator in equation (1.12) gives a null integral[8], only the contribution of the collision integral remains. Distinguishing the colliders as in the previous section we obtain

$$\frac{\partial H(t)}{\partial t} = \int \mathrm{d}^3v_A(\ln f_A + 1)v_{AB}\frac{\mathrm{d}\sigma_{AB}}{\mathrm{d}\Omega}(\vartheta_A)\left(f_A'f_B' - f_Af_B\right)\mathrm{d}^3v_A\mathrm{d}^3v_B\mathrm{d}\Omega_A. \tag{1.25}$$

The argument of the integral in equation (1.25) is invariant by exchanging A and B and averaging the two expressions we obtain

$$\frac{\partial H(t)}{\partial t} = \frac{1}{2}\int \mathrm{d}^3v_A\left[\ln\left(f_Af_B\right) + 2\right]v_{AB}\frac{\mathrm{d}\sigma_{AB}}{\mathrm{d}\Omega}(\vartheta_A)\left(f_A'f_B' - f_Af_B\right)\mathrm{d}^3v_A\mathrm{d}^3v_B\mathrm{d}\Omega_A. \tag{1.26}$$

As a consequence of the micro-reversibility principle, the integral must be the same when the initial and final states are exchanged

$$\frac{\partial H(t)}{\partial t} = \frac{1}{2}\int \mathrm{d}^3v_A\left[\ln\left(f_A'f_B'\right) + 2\right]v_{AB}\frac{\mathrm{d}\sigma_{AB}}{\mathrm{d}\Omega}(\vartheta_A)\left(f_Af_B - f_A'f_B'\right)\mathrm{d}^3v_A\mathrm{d}^3v_B\mathrm{d}\Omega_A, \tag{1.27}$$

---

[7] It should be noted [16, 17] that the entropy is defined as the integral of $(f - f \ln f)$. The following demonstration can be applied in a straightforward manner to entropy [16], showing that the time derivative of S has the opposite sign with respect to that of H.

[8] This assertion can be easily demonstrated, by applying Gauss's theorem and transforming the volume integral of the derivatives in a surface integral of the function. At infinity the distribution function is null, therefore the surface integral gives a null contribution [16].

and averaging equations (1.26) and (1.27) we have the final expression

$$\frac{\partial H(t)}{\partial t} = \frac{1}{4} \int d^3v_A \left[\ln\left(f_A f_B\right) - \ln\left(f'_A f'_B\right) \right] v_{AB}$$
$$\times \frac{d\sigma_{AB}}{d\Omega}(\vartheta_A)\left(f'_A f'_B - f_A f_B\right) d^3v_A d^3v_B d\Omega_A. \tag{1.28}$$

The sign of the integrand depends on the product

$$\left[\ln\left(f_A f_B\right) - \ln\left(f'_A f'_B\right) \right]\left(f'_A f'_B - f_A f_B\right), \tag{1.29}$$

which is always negative or null when the initial and final distributions are the same, conditions verified at the equilibrium. The integral of a negative function is always negative, it is

$$\frac{\partial H(t)}{\partial t} \leqslant 0, \tag{1.30}$$

demonstrating the *H*-theorem.[9]

1.1.4 The Vlasov equation

The above theory cannot be applied for ionized gases, as they involve the Coulomb interaction, a long-range force. As a consequence charge–charge interaction cannot be treated as a collision. If the gas is weakly ionized, the charge–charge interactions are negligible with respect to collisions with neutrals, but for highly ionized gases the Coulomb forces provide the dominant contribution, and some adjustments are necessary for the Boltzmann equation to be applied to plasmas.

Let us consider a test particle s with charge q_s, position $\mathbf{r}_s$ in a plasma in stationary conditions. The s particle will feel the electric field produced by all the other charges

$$\mathbf{E}(\mathbf{r}_s, t) = \frac{1}{4\pi\epsilon_0} \sum_{i \neq s}^{N} \frac{q_i \mathbf{r}_{s,i}(t)}{|\mathbf{r}_{s,i}(t)|^3}, \tag{1.31}$$

where $\mathbf{r}_{s,i} = \mathbf{r}_i(t) - \mathbf{r}_s$ is the position of i relative to s. We can separate two contributions to $\mathbf{E}(\mathbf{r}_s, t)$, a constant mean field

$$\mathbf{E}_0(\mathbf{r}_s) = \frac{1}{\delta t} \int_t^{t+\delta t} \mathbf{E}(\mathbf{r}_s, t) dt, \tag{1.32}$$

where δt is a time interval sufficiently long, and $\mathbf{E}_r(\mathbf{r}_s, t)$, due to fluctuations, is a function of time with null mean

$$\frac{1}{\delta t} \int_t^{t+\delta t} \mathbf{E}_r(\mathbf{r}_s, t) = \frac{1}{\delta t} \int_t^{t+\delta t} \left[\mathbf{E}(\mathbf{r}_s, t) - \mathbf{E}_0(\mathbf{r}_s) \right] dt \approx \mathbf{0}. \tag{1.33}$$

---

[9] The demonstration given here refers to homogeneous systems. For non-homogeneous systems see [5].

The time interval δt must be a macroscopic time, i.e. $\delta t \gg 1/\tau_c$ where τ_c is the mean time between collisions. In principle, $\mathbf{E}_r(\mathbf{r}_s, t)$ is a continuous function of time, with rapid oscillations with mean time amplitude of the order of τ_c. This means that if we sample this field periodically, at successive equal time intervals of amplitude δt, much larger than τ_c, i.e. $\mathbf{E}_r(\mathbf{r}_s, t)$, $\mathbf{E}_r(\mathbf{r}_s, t + 2\delta t)$, ..., $\mathbf{E}_r(\mathbf{r}_s, t + n\delta t)$, we will have a random series. If the system is not in a stationary condition, the above assumptions remain valid if the sampling time δt is small with the respect to the evolution time of whole system, i.e. all the macroscopic quantities are almost constant in this time interval. Similar arguments can be used for the magnetic field. This property of the local field suggests that the mean field contributes to the acceleration in the Boltzmann equation, while $\mathbf{E}_r$ must contribute to the collision integral, due its random evolution. The mean field does not produce entropy, a consequence of the H-theorem, differently from the random field.

In fully ionized gases it could be demonstrated that the entropy variation due to collisions is much slower than other macroscopic quantities such as charge density ρ_c,

$$\left| \frac{1}{S} \frac{\partial S}{\partial t} \right| \ll \left| \frac{1}{\rho_c} \frac{\partial \rho_c}{\partial t} \right|, \tag{1.34}$$

therefore the collision integral can be considered null. The Boltzmann equation in equation (1.14) becomes

$$\frac{\partial f_s}{\partial t} + \left[\mathbf{v} \cdot \nabla_r + \frac{q_s}{m_s} \left(\mathbf{E}_0 + \mathbf{v} \wedge \mathbf{B}_0 \right) \cdot \nabla_v \right] f_s = 0, \tag{1.35}$$

where the local fields are related only to local macroscopic quantities such as charge density and current,

$$\begin{cases} \rho_c = \sum_s q_s \int f_s \, \mathrm{d}^3 v \\ \mathbf{j}_c = \sum_s q_s \int \mathbf{v} f_s \, \mathrm{d}^3 v, \end{cases} \tag{1.36}$$

through the Maxwell equations

$$\begin{cases} \nabla \cdot \mathbf{E}_0 = \dfrac{\rho_c}{\epsilon_0} \\ \nabla \cdot \mathbf{B}_0 = 0 \\ \nabla \wedge \mathbf{E}_0 = -\dfrac{\partial \mathbf{B}_0}{\partial t} \\ \nabla \wedge \mathbf{B}_0 = \dfrac{\mathbf{j}_c}{\epsilon_0 c^2} + \dfrac{1}{c^2} \dfrac{\partial \mathbf{E}_0}{\partial t}. \end{cases} \tag{1.37}$$

This system is known as the *Vlasov equation*. In most practical applications the time derivatives of the fields in the Maxwell equations are negligible, and the quasi-static fields are considered. The validity of the Vlasov equation depends on two conditions:

- The mean particle distance must be smaller than the Debye length

$$\lambda_D = \sqrt{\frac{\epsilon_0 k T V}{\sum_s q_s^2 \mathcal{N}_s}}, \tag{1.38}$$

i.e. the number of particles inside the Debye sphere must be statistically significant [16].

- The collision frequency is much smaller than the plasma frequency

$$\omega_p = \sqrt{\sum_s \frac{q_s^2 \mathcal{N}_s}{\epsilon_0 m_s V}} \tag{1.39}$$

in order to neglect the collision integral.

1.2 Cross sections

The collisional term in the Boltzmann equation results from the dynamics of binary collisions in the plasma promoting elastic, inelastic, and reactive processes that govern the temporal evolution of the kinetics at a microscopic level. The construction of a reliable database for the elementary probabilities represents a challenge even today, due to the requirements of accuracy, consistency, and completeness, issues that significantly affect the predictive capability of kinetic models, not only in the estimation of macroscopic plasma parameters, but also in shedding light on relevant collisional mechanisms [18]. Modern approaches in quantum molecular dynamics are fully exploited to aim for the highest accuracy. However, the wide spectrum of theoretical methods, including semiclassical, quasi-classical, classical, and semi-empirical, provides methods that are suitable for specific classes of processes, offering a favorable balance between the reliability of dynamics information and the computational load characterizing the derivation of a complete dataset. This is, in particular, the case for state-to-state kinetics (see section 8.1), where the dependence of the probability on the quantum state of colliders is also considered.

Elastic collisions usually dominate the kinetics and fully determine the transport properties of thermal equilibrium plasmas [6]. Transport cross sections, i.e. the momentum transfer and higher ℓ moments of an elastic cross section, and their thermal average, i.e. the collision integrals $\Omega_{ij}^{\ell,s}(T)$ of the Chapman–Enskog theory [19], can be obtained by quantum [20] or semiclassical Wentzel-Kramers-Brillouin (WKB) estimation of phase-shifts, or by direct classical trajectory integration of the interaction potential [21]. The interaction potential from accurate electronic structure calculations, or modeled with either full-range analytical forms [6] or a phenomenological potential [22], is assumed to be isotropic, so is dependent only on the approaching line of colliding partners, also for interactions involving molecules. Some examples do exist in the literature of classical trajectory calculations on *ab initio* surfaces for molecular collisions [23], considering anisotropic potentials, however, for small rapidly rotating molecules an isotropic angle-averaged potential can be assumed [24]. Resonant charge exchange in ion–parent-atom interactions is the only non-elastic process considered in transport calculations, due to the significant contribution in the ℓ-odd collision integrals [6]. A rigorous quantum coupled-channel approach [20] is required; however, it is worth mentioning the asymptotic approach [25, 26], based on the idea of expanding the cross section in terms of small parameters connected to the geometry of the collision.

In the case of electron elastic scattering, the quantum nature of the interactions is not negligible and manifests itself in the features of the energy profile of the cross section, as the low-energy peaks due to resonances, or the *Ramsauer minimum* [27]. Therefore a quantum approach is required and the collision integrals are evaluated by integration of theoretical [28] or experimental differential elastic electron-scattering cross sections.

Inelastic and reactive channels are considered negligible in transport, except for quantum effects at low temperatures. However, they play a key role in non-equilibrium plasmas, chemical species undergoing excitation, and dissociation and ionization processes induced by electron impact or collisions with heavy particles with increasing threshold energies.

Quantum approaches (the R-matrix [29], close-coupling, the Schwinger multi-channel [30], local complex potential [31]) represent the best theoretical framework for the investigation of the dynamics in electron–molecule scattering. They involve non-adiabatic effects as the resonant electron capture with the formation of long-living negative ion intermediates or vibrionic coupling. However, other methods have been successfully used in the past: first-order approaches, such as the Born, Born–Bethe, and Born–Ochkur approaches [32], which are accurate for high collision energies; semiclassical methods [33–35] that assume that the impinging electron moves classically in linear trajectories and the quantum molecular target; and the classical Gryzinskii method [36]. More recently, novel simplified approaches have been proposed in the literature, such as the BEf-scaling method [37], based on a rescaling procedure of the Born cross section for dipole-allowed vibrionic excitations with the experimental optical oscillator strengths, and the similarity approach [38] or the binary encounter dipole (BED) model for ionization [39, 40]. Simplified approaches are still attractive, demonstrating in many cases striking agreement with experiments and representing very promising tools for more complex molecular systems.

For the dynamics of heavy-particle collisions, leading to energy transfer among internal and translation degrees of freedom of colliding partners, rearrangement reactions, or dissociation, the quasi-classical trajectory (QCT) method [18, 41, 42] has been well assessed. It has proved to compare well with and complement the quantum approaches [43]. The QCT method provides accurate potential energy surface (PES) vales and robust trajectory statistics, allowing the derivation of state-specific rate coefficients through the pseudo-quantization of the energy in reactant and product channels, and is able to fully exploit the advantages of high-performance code parallelization [44]. Furthermore, the QCT approach can follow the dynamics in collisional systems characterized by non-adiabatic coupling among different electronic states, with trajectories hopping [45] from ground to excited PESs. The mechanism of vibrational energy transfer in atom–molecule and molecule–molecule collisions has also been investigated in the framework of the semiclassical approach [46], the forced-harmonic oscillator [47], and the widely used SSH method [48].

A number of different open-access web-databases are available [49–52], where dynamical information has been collected, validated, and shared with the modeling community. These also include the state-specific cross sections and rate coefficients of elementary processes involving ground and excited chemical species, with resolution on the electronic, vibrational, and rotational degrees of freedom [53, 54].

1.3 Solution of the Boltzmann equation

The first approach for an analytical solution of the Boltzmann equation was the expansion series called *Chapman–Enskog* theory [55], which writes the distribution function as

$$f = f_0 + \sum_{n=1}^{\infty} \epsilon^n f_n, \tag{1.40}$$

where ϵ is a small parameter, and stops the series at the first order. The first term f_0 is the Maxwell distribution, which gives a null collision integral. Therefore the transport coefficients are related to the perturbation f_1. This approach allows one to calculate the transport properties relative to the specific cross sections. Also, the perturbation is expanded in a polynomial series [6], allowing different degrees of approximation. The Chapman–Enskog theory is valid for small deviations from the Maxwell distributions. The Chapman–Enskog theory leads to the Navier–Stokes equations. However, in extremely rarefied gases, Navier–Stokes equations fail and more accurate approaches are required. An improved model was introduced by Grad [56] extending the moment equations to thirteen terms.

Even if it is very difficult solve the complete Boltzmann equation due to the large dimension of the phase space and the non-linearity of the collisional operator, different numerical methods have aspired to this. In recent years the lattice Boltzmann method (LBM) [8, 57] has been the subject of growing interest to simulate complex fluids. This method writes the Boltzmann equation in a discrete space $\mathcal{L}$ with a finite number of possible velocities (see figure 1.4). The time must be discrete as well in order to obtain that any advancement in the position is still a point of the lattice, i.e.

$$x_i \in \mathcal{L} \Rightarrow x_i + \mathbf{v}_j \delta t \in \mathcal{L}. \tag{1.41}$$

An interesting feature of the LBM is that it is a solver of the Navier–Stokes equations [8, 57], a characteristic inherited from the original Boltzmann equation. In contrast to the Navier–Stokes equations, the LBM has a linear convective operator. In most cases, the collisional operator is also linearized, using the BGK approach [9]. This method is practical and suitable for investigating complex systems, such as incompressible flows through porous media [58] and multi-phase/multi-component flows [59].

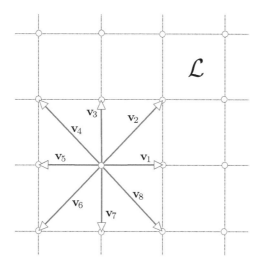

Figure 1.4. Example of a 2D lattice with discrete velocities.

Even if the LBM can be derived directly from the Boltzmann equation in the BGK approximation [7] under small Mach number expansion [60], the two equations are substantially different. Whereas the Boltzmann equation is valid for rarefied gases, the LBM is mainly used for incompressible flows and dense matter. Moreover, the *H*-theorem is not applicable in LBM.

Another interesting approach based on the Boltzmann equation is the *direct simulation Monte Carlo* (DSMC), developed by G Bird in the early 1960s [61, 62]. The method has been developed to investigate rarefied flows in very low density, where the Navier–Stokes equations fail, and is based on the idea that the free motion and the collisions can be decoupled. The DSMC considers the atomic structure of the matter. The gas is represented by a large number of pseudo-particles, each one representing a large number of real particles in the same position and with the same velocity. In the convection phase all the particles move according to the classical laws of motion for a given time interval, δt, which is small with respect to the mean collision time. Figure 1.1 (left) can also be considered a pictorial view of this system. The computational domain is divided into a grid (see figure 1.1 (right) for an example of grid elements) with the purpose of calculating statistical means, such as density, velocity, temperature, and so on. In each grid element, according to the total cross section, the number of collisions is determined and a Monte Carlo approach is used to sample the collisions. Random numbers are generated to decide which particles will collide and the result of the collision, changing the state of the particles accordingly. Convective and collision phases are repeated one after the other. With this method it is possible to calculate the distribution function directly, with a sensitivity depending on the number of particles. DSMC has been used to model hypersonic entry into a planetary atmosphere [63], and also for real spacecraft, such as the space shuttle [64, 65].

For fluid dynamics applications, simplified cross-section models are used, such as hard spheres [66], variable hard spheres [67], and variable soft spheres [68], since the collision probabilities are available in closed form. More advanced models include cross sections obtained from quantum or semiclassical calculations (see section 1.2), also including state-to-state kinetics [69–71], as described in chapter 8. The inclusion of detailed cross-section datasets in DSMC calculations is a complex problem, not only due to the large amount memory required to store the data and the time needed to calculate the probability of each event, but also because these data are not easily available. A novel approach, called DSMC-QCT [72, 73], does not go through the cross-section evaluation, but instead follows the motion of particles on a multi-dimensional PES. Once the colliding particles have been chosen, random numbers are generated to select the collisional parameters. The interaction is then followed in the center-of-mass frame of reference, moving the particles on the given PES, until the interparticle distance is large enough. The final state obtained in this way determines the new states of the colliders.

DSMC is not adequate for modeling plasma, although some attempts have been made in the literature using different time-steps for heavy particles and electrons [74]. Other approaches entail the coupling of particle-in-cell (PIC) (see chapter 4) with DSMC to model plasmas in rarefied flows [75].

The Boltzmann equation has also been applied to calculate the evolution of non-equilibrium distributions of electrons and phonons in metals heated by femtosecond laser pulses [10, 76–82]. The laser pulse is absorbed by the electron gas and the energy is redistributed by electron–electron collisions. At longer times the energy is transferred to the solid lattice with the production of phonons.

In a metal, electrons partially occupy the conduction band, following the Fermi statistic. In this case we generalize the definition of distribution as

$$n_s = \frac{\mathcal{N}_s}{V} = \int_0^\infty g_s f_s^{\mathrm{FD}}(\varepsilon) \mathcal{D}_s(\varepsilon) \mathrm{d}\varepsilon, \qquad (1.42)$$

where g_s is the spin multiplicity of the particle s (= 2 for electrons) and $\mathcal{D}_s$ is the density of states. The equilibrium state is described by the Fermi–Dirac distribution

$$f_s^{\mathrm{FD}}(\varepsilon) = \left(e^{\beta(\varepsilon - E_{\mathrm{F}})} + 1 \right)^{-1}, \qquad (1.43)$$

where E_{F} is the Fermi energy,[10] defined as

$$\mathcal{N}_s = \int_0^{E_{\mathrm{F}}} \mathcal{D}_s(\varepsilon) \mathrm{d}\varepsilon. \qquad (1.44)$$

Even at absolute temperature $T = 0$ the electrons are not in the same quantum state because of the Pauli exclusion principle, but all allowed states below the Fermi energy are fully occupied and the distribution function is a step function

$$\begin{aligned} \varepsilon \leqslant E_{\mathrm{F}} &\Rightarrow f_s^{\mathrm{FD}}(\varepsilon) = 1 \\ \varepsilon > E_{\mathrm{F}} &\Rightarrow f_s^{\mathrm{FD}}(\varepsilon) = 0. \end{aligned} \qquad (1.45)$$

---

[10] Equation (1.43) is only an approximation and is valid only at low temperatures.

This character of the Fermi distribution should be reflected in the Boltzmann equation and, in particular, to account for the Pauli exclusion principle the product of the distributions in the collision integral (see equation (1.22)) becomes

$$f_A f_B \rightarrow f_A(1 - f_A)f_B(1 - f_B), \tag{1.46}$$

which assures that the distribution does not become larger than one.

1.4 Plasma modeling numerical codes

To conclude this chapter, we will give a brief description of the numerical codes commonly used by researchers for modeling plasmas, using the theoretical approaches described in the following chapters. The most used numerical simulation techniques for plasma modeling include fluid dynamics, kinetic, and hybrid models. These simulation models are significantly different in terms of their principles, strengths, and limitations. Kinetic models, such as PIC with Monte Carlo collisions (see chapter 4), are used for non-equilibrium systems, and fluid models are employed for faster computation with the cost of reducing the accuracy. Hybrid models provide balance between precision and efficiency.

COMSOL Multiphysics [83] is a general-purpose software platform, based on advanced numerical methods, for modeling and simulating physics-based problems. It is suitable for the design and development of optimized products by researchers and engineers working for leading technical enterprises, research laboratories, and universities. The *COMSOL* platform is characterized by an environment designed for cross-disciplinary product development with a unified workflow for electrical, mechanical, fluid, and chemical applications. The 'Plasma' module, available in the 'Electrical' suite, is a partial differential equation solver including drift–diffusion approximations for ions, the quasi-neutrality assumption for electron movements, reduced Maxwell equations for electromagnetic fields, electron energy equations for electron temperatures, and the Navier–Stokes equation for neutral background gas. Users can conveniently formulate physical equations according to their require-ments and do not have to spend much time coding numerical algorithms, particularly for in complex geometries. As an example, by using the 'Plasma' module, low-temperature plasma sources and systems (direct current discharges, inductively coupled plasmas, and microwave plasmas) can be simulated and scientists can gain insight into the physics of discharges and gauge the performance of existing or future designs. The only drawback of the solver is the lack of kinetic information. In fact, *COMSOL* does not include the Boltzmann equation solver and in many applications kinetic effects must be considered.

ANSYS Fluent [84] is a computational fluid dynamics (CFD) software tool used by researchers and engineers to optimize a product's performance. *ANSYS Fluent* includes well-validated physical modeling capabilities to deliver fast and accurate results across the widest range of CFD and Multiphysics applications. *ANSYS Fluent* offers many great advantages in the modeling of plasmas. In fact, *ANSYS Fluent* can solve very complex fluid dynamic problems by using a large set of numerical algorithms to solve energy, momentum, mass, and many other kinds of equations. Moreover, chemical reactions, continuity of species, drift–diffusion models, and Maxwell equations can be coupled and solved by taking advantage of a very important capability of such software: the user-defined scalar (UDS) variables. *ANSYS Fluent* can solve the transport equation for an arbitrary UDS in the same way that it solves the transport equation for a scalar such as species mass fraction. Extra scalar transport equations may be needed in certain types of combustion applications or, for example, in plasma applications. *ANSYS Fluent* allows one to define additional scalar transport equations in a model.

By using this capability, for an arbitrary scalar Φ_k, *ANSYS Fluent* solves the equation

$$\frac{\partial \rho \Phi_k}{\partial t} + \frac{\partial}{\partial x_i}\left(\rho u_i \Phi_k - \Gamma_k \frac{\partial \Phi_k}{\partial x_i}\right) = S_{\Phi_k} \qquad k = 1, ..., N, \qquad (1.47)$$

where Γ_k and S_{Φ_k} are the diffusion coefficient and source term supplied by the user for each of the N scalar equations, u_i represents the velocity field, and x_i the spatial coordinates.

In the case of a steady-state problem and in the absence of convective fluxes $\frac{\partial \rho \Phi_k}{\partial t} = 0$ and $u_i = 0$. Furthermore, imposing $\Gamma_k = 0$, equation (1.47) becomes

$$-\nabla^2 \Phi_k = S_{\Phi_k} \qquad k = 1, ..., N. \qquad (1.48)$$

ANSYS Fluent can solve UDS equations in both the fluid and solid regions and is able to impose a continuity condition along a boundary that separates a fluid zone from a solid one. This capability allows it to solve a large number of equations.

Both *COMSOL* and *ANSYS Fluent* are CFD codes based on the theoretical general models described in detail in chapters 6 and 7, and are applied in inductively coupled plasma torches (chapter 10) and aerospace applications (chapter 12).

BOLSIG+ [85] is a non-commercial code for the numerical solution of the Boltzmann equation for free electrons in weakly ionized gases in the framework of the two-term approximation (see chapter 2). Developed by Hagelaar [86] of the LAPLACE laboratory in Toulouse in France, *BOLSIG+* was first released in 2005 as a replacement for the earlier *BOLSIG* solver developed by Pitchford, Boeuf, and Morgan, based on the old code *ELENDIF* by Morgan and Penetrante [87].

BOLSIG+ uses data from the web-accessed database LXcat [49], a selection of complete sets of electron–neutral cross-section sets (see section 1.2) for a wide range of gases, compiled by different contributors and with varying levels of detail.

The main application of *BOLSIG+* is to obtain electron transport coefficients and rates for electron-induced processes starting from cross-section data. These quantities, which can be used as input for fluid models (see chapters 6, 7, and 12), are obtained from the non-Maxwellian electron energy distribution calculated for a given gas composition and under stationary conditions.[11]

---

[11] This procedure, known also as cold gas approximation, should be compared with the self-consistent approach (see chapter 8).

BOLSIG+ is solved under the following assumptions:
- The electric field and collision probabilities are constant in space and time and there are no boundaries for the system.
- A constant gas composition.
- The electron production/loss due to ionization/attachment is assumed to result in exponential growth/decay of the electron number density.

Under the above assumptions, the Boltzmann equation reduces to a convection–diffusion equation with non-local source term in energy space, which is then discretized by an exponential scheme and solved for the electron energy distribution function by a standard matrix inversion technique (see section 2.7).

There are other non-commercial codes. These codes have been developed by research groups and generalized for a category of plasma applications. Typically these codes can simulate only 1D or 2D problems for a limited range of plasma applications. Among them, one of the best and most popular codes is *nonPDPSIM* [88], developed by Kushner and his research group. *nonPDPSIM* is a 2D computational platform that solves transport equations for charged and neutral species, Poisson's equation for the electric potential, the electron energy conservation equation for the electron temperature, and radiation transport and Navier–Stokes equations for the neutral gas flow. The 2D platform has three major modules addressing plasma dynamics, fluid dynamics, and surface kinetics, and is implemented on an unstructured mesh to enable resolution of a large dynamic range in space. *CFD-GEOM* or *SKYMESH2*, a mesh generation program, has been incorporated into the program to allow for unstructured meshes. The model is coupled with a thermodynamics module providing local thermodynamic equilibrium properties. To speed up the simulation, a fully implicit scheme and sparse matrix techniques are used. The source function for electron impact reactions and secondary electron emission, and the electron transport coefficients needed by this segment of the code, are computed assuming the local field approximation, solving the electron energy equation and/or using a Monte Carlo simulation. The continuity equations for neutral species are solved explicitly.

Bibliography

[1] Planck M 1967 The Nobel lecture *Nobel Lectures, Physics 1901–1921* (Amsterdam: Elsevier) http://www.nobelprize.org/nobel_prizes/physics/laureates/1918/planck-lecture.html

[2] Bobylev A V and Cercignani C 1999 On the rate of entropy production for the Boltzmann equation *J. Stat. Phys.* **94** 603–18

[3] Eddington A 2012 *The Nature of the Physical World: Gifford Lectures (1927)* (Cambridge: Cambridge University Press)

[4] Cercignani C 1988 *The Boltzmann equation and Its Applications* (*Applied Mathematical Sciences* vol 67) (New York: Springer)

[5] Cercignani C 1995 Aerodynamical applications of the Boltzmann equation *Riv. Nuovo Cimento* **18** 1–40

[6] Capitelli M, Bruno D and Laricchiuta A 2013 *Fundamental Aspects of Plasma Chemical Physics: Transport* (*Springer Series on Atomic, Optical, and Plasma Physics* vol 74) (Berlin: Springer)

[7] Bhatnagar P L, Gross E P and Krook M 1954 A model for collision processes in gases. I. Small amplitude processes in charged and neutral one-component systems *Phys. Rev.* **94** 511

[8] Chen S and Doolen G D 1998 Lattice Boltzmann method for fluid flows *Annu. Rev. Fluid Mech.* **30** 329–64

[9] Qian Y, d'Humières D and Lallemand P 1992 Lattice BGK models for Navier–Stokes equation *Europhys. Lett.* **17** 479

[10] Greenwood D 1958 The Boltzmann equation in the theory of electrical conduction in metals *Proc. Phys. Soc.* **71** 585

[11] Villani C 2002 A review of mathematical topics in collisional kinetic theory *Handbook Math. Fluid Dyn.* **1** 71–305

[12] Rockwood S D 1973 Elastic and inelastic cross sections for electron-Hg scattering from Hg transport data *Phys. Rev.* A **8** 2348–60

[13] Brizard A J 1995 Nonlinear gyrokinetic Vlasov equation for toroidally rotating axisymmetric tokamaks *Phys. Plasmas* **2** 459–71

[14] Hénon M 1982 Vlasov equation? *Astron. Astrophys.* **114** 211

[15] Maruca B A, Kasper J C and Gary S P 2012 Instability-driven limits on helium temperature anisotropy in the solar wind: observations and linear Vlasov analysis *Astrophys. J.* **748** 137

[16] Landau L D, Lifshitz E and Pitaevskii L 1993 *Physical Kinetics* (*Course of Theoretical Physics* vol 10) (Oxford: Pergamon)

[17] Capitelli M, Colonna G and D'Angola A 2011 *Fundamental Aspects of Plasma Chemical Physics: Thermodynamics* (*Springer Series on Atomic, Optical, and Plasma Physics* vol 66) (New York: Springer)

[18] Capitelli M, Celiberto R, Colonna G, Esposito F, Gorse C, Hassouni K, Laricchiuta A and Longo S 2016 *Fundamental Aspects of Plasma Chemical Physics: Kinetics* (*Springer Series on Atomic, Optical, and Plasma Physics* vol 86) (New York: Springer)

[19] Hirschfelder J O, Curtiss C F and Bird R B 1966 *Molecular Theory of Gases and Liquids* (New York: Wiley)

[20] Krstić P S and Schultz D R 1999 Elastic scattering and charge transfer in slow collisions: isotopes of H and H$^+$ colliding with isotopes of H and with He *J. Phys. B: At. Mol. Phys.* **32** 3485–509

[21] Colonna G and Laricchiuta A 2008 General numerical algorithm for classical collision integral calculation *Comput. Phys. Commun.* **178** 809–16

[22] Capitelli M, Cappelletti D, Colonna G, Gorse C, Laricchiuta A, Liuti G, Longo S and Pirani F 2007 On the possibility of using model potentials for collision integral calculations of interest for planetary atmospheres *Chem. Phys.* **338** 62–8

[23] Viehland L A, Dickinson A S and Maclagan R G A R 1996 Transport coefficients for NO$^+$ ions in helium gas: a test of the NO$^+$–He interaction potential *Chem. Phys.* **211** 1–5

[24] Stallcop J R, Partridge H and Levin E 2002 Effective potential energies and transport cross sections for atom–molecule interactions of nitrogen and oxygen *Phys. Rev.* A **64** 042722

[25] Nikitin E E and Smirnov B M 1978 Quasiresonant processes in slow collisions *Sov. Phys. Usp* **21** 95–116

[26] Smirnov B M 2001 Atomic structure and the resonant charge exchange process *Phys. Usp* **44** 221–53

[27] Gryzinski M 1970 Ramsauer effect as a result of the dynamic structure of the atomic shell *Phys. Rev. Lett.* **24** 45–7

[28] Bray I and Stelbovics A T 2003 Convergent close-coupling approach to electron–atom collisions *Many-Particle Quantum Dynamics in Atomic and Molecular Fragmentation* ed J Ullrich and V Shevelko (*Springer Series on Atomic, Optical, and Plasma Physics* vol 35) (New York: Springer) pp 121–35

[29] Tennyson J 2010 Electron–molecule collision calculations using the R-matrix method *Phys. Rep.* **491** 29–76

[30] Gianturco F A and Huo W M (eds) 1995 *Computational Methods for Electron–Molecule Collisions* (*Springer Series on Atomic Optical, and Plasma Physics*) (New York: Springer)

[31] Wadehra J M 1986 Vibrational excitation and dissociative attachment *Nonequilibrium Vibrational Kinetics* (*Topics in Current Physics* vol 39) (New York: Springer) pp 191–232

[32] Chung S and Lin C C 1972 Excitation of the electronic states of the nitrogen molecule by electron impact *Phys. Rev.* A **6** 988–1002

[33] Hazi A U 1981 Impact-parameter method for electronic excitation of molecules by electron impact *Phys. Rev.* A **23** 2232–40

[34] Redmon M J, Garrett B C, Redmon L T and McCurdy C W 1985 Improved impact-parameter method for electronic excitation and dissociation of diatomic molecules by electron impact *Phys. Rev.* A **32** 3354–65

[35] Celiberto R and Rescigno T N 1993 Dependence of electron-impact excitation cross sections on the initial vibrational quantum number in H_2 and D_2 molecules: $X^1\Sigma_g^+ \rightarrow B^1\Sigma_u^+$ and $^1\Sigma_g^+ \rightarrow C^1\Pi_u$ transitions *Phys. Rev.* A **47** 1939–45

[36] Gryziński M 1965 Classical theory of atomic collisions. I. Theory of inelastic collisions *Phys. Rev.* A **138** 336–58

[37] Kim Y-K 2007 Scaled Born cross sections for excitations of H_2 by electron impact *J. Chem. Phys.* **126** 064305

[38] Adamson S, Astapenko V, Deminskii M, Eletskii A, Potapkin B, Sukhanov L and Zaitsevskii A 2007 Electron impact excitation of molecules: calculation of the cross section using the similarity function method and *ab initio* data for electronic structure *Chem. Phys. Lett.* **436** 308–13

[39] Kim Y-K and Rudd M E 1994 Binary-counter-dipole model for electron-impact ionization *Phys. Rev.* A **50** 3954–67

[40] Huo W M 2001 Convergent series representation for the generalized oscillator strength of electron-impact ionization and an improved binary-encounter-dipole model *Phys. Rev.* A **64** 042719

[41] Blais N C and Truhlar D G 1976 Monte Carlo trajectory study of $Ar+H_2$ collisions. I. Potential energy surface and cross sections for dissociation, recombination, and inelastic scattering *J. Chem. Phys.* **65** 5335–56

[42] Blais N C, Zhao M, Mladenovic M, Truhlar D G, Schwenke D W, Sun Y and Kouri D J 1989 Comparison of quasiclassical trajectory calculations to accurate quantum mechanics for state-to-state partial cross sections at low total angular momentum for the reaction $D + H^2 \rightarrow HD + H$ *J. Chem. Phys.* **91** 1038–42

[43] Esposito F, Coppola C M and Fazio D D 2015 Complementarity between quantum and classical mechanics in chemical modeling. The $H + HeH^+ \rightarrow H_2^+ + He$ reaction: a rigorous test for reaction dynamics methods *J. Phys. Chem.* A **119** 12615–26

[44] Lombardi A, Faginas-Lago N and Laganà A 2014 Grid calculation tools for massive applications of collision dynamics simulations: carbon dioxide energy transfer *Lecture Notes in Computer Science* (New York: Springer) pp 627–39

[45] Tully J C 1976 Nonadiabatic processes in molecular collisions *Dynamics of Molecular Collisions, Part B* ed W H Miller (New York: Plenum)

[46] Billing G 1987 Rate constants for vibrational transitions in diatom–diatom collisions *Comput. Phys. Commun.* **44** 121–36

[47] Adamovich I V, Macheret S O, Rich J W and Treanor C E 1998 Vibrational energy transfer rates using a forced harmonic oscillator model *J. Thermophys. Heat Transfer* **12** 57–65

[48] Schwartz R N, Slawsky Z I and Herzfeld K F 1952 Calculation of vibrational relaxation times in gases *J. Chem. Phys.* **20** 1591–9

[49] Plasma Data Exchange Project 2016 LXcat database URL http://fr.lxcat.net/home/

[50] IAEA Nuclear Data section A+M Data Unit 2016 ALADDIN database URL https://www-amdis.iaea.org/ALADDIN/

[51] GASPAR 2016 Gas and Plasma Radiation Database URL http://esther.ist.utl.pt/gaspar/

[52] NIFS 2016 National Institute for Fusion Science database URL http://dbshino.nifs.ac.jp/

[53] Phys4Entry 2016 http://phys4entrydb.ba.imip.cnr.it/Phys4EntryDB/

[54] STELLAR 2016 http://esther.ist.utl.pt/pages/stellar.html

[55] Chapman S and Cowling T G 1970 *The Mathematical Theory of Non-uniform Gases: An account of the kinetic theory of viscosity, thermal conduction and diffusion in gases* (Cambridge: Cambridge University Press)

[56] Grad H 1949 On the kinetic theory of rarefied gases *Commun. Pure Appl. Math* **2** 331–407

[57] Succi S 2001 *The Lattice Boltzmann Equation: For fluid dynamics and beyond* (Oxford: Oxford University Press)

[58] Guo Z and Zhao T 2002 Lattice Boltzmann model for incompressible flows through porous media *Phys. Rev.* E **66** 036304

[59] Shan X and Chen H 1993 Lattice Boltzmann model for simulating flows with multiple phases and components *Phys. Rev.* E **47** 1815

[60] Abe T 1997 Derivation of the lattice Boltzmann method by means of the discrete ordinate method for the Boltzmann equation *J. Comput. Phys.* **131** 241–6

[61] Bird G A 1963 Approach to translational equilibrium in a rigid sphere gas *Phys. Fluids* **6** 1518

[62] Bird G A 1994 *Molecular Gas Dynamics and the Direct Simulation of Gas Flows* (Oxford: Clarendon)

[63] Bondar Y A, Markelov G, Gimelshein S and Ivanov M 2005 DSMC study of shock-detachment process in hypersonic chemically reacting flow *24th Int. Symp. on Rarefied Gas Dynamics, AIP Conf. Proc.* **762** 523–8

[64] Bird G 1990 Application of the direct simulation Monte Carlo method to the full shuttle geometry *AIAA Paper* 90–1692

[65] Gallis M A 2005 DSMC simulations in support of the STS-107 accident investigation 24th *Int. Symp. on Rarefied Gas Dynamics, AIP Conf. Proc.* **762** 523–8

[66] Ohwada T 1993 Structure of normal shock waves: direct numerical analysis of the Boltzmann equation for hard-sphere molecules *Phys. Fluids* A **5** 217–34

[67] Hash D, Hassan H and Moss J N 1994 Direct simulation of diatomic gases using the generalized hard sphere model *J. Thermophys. Heat Transfer* **8** 758–64

[68] Fan J 2002 A generalized soft-sphere model for Monte Carlo simulation *Phys. Fluids* **14** 4399–405

[69] Willauer D L and Varghese P L 1993 Direct simulation of rotational relaxation using state-to-state cross sections *J. Thermophys. Heat Transfer* **7** 49–54

[70] Bruno D, Capitelli M, Esposito F, Longo S and Minelli P 2002 Direct simulation of non-equilibrium kinetics under shock conditions in nitrogen *Chem. Phys. Lett.* **360** 31–7

[71] Kim J G and Boyd I D 2014 Monte Carlo simulation of nitrogen dissociation based on state-resolved cross sections *Phys. Fluids* **26** 012006

[72] Fujita K 2007 Assessment of molecular internal relaxation and dissociation by DSMC-QCT analysis *AIAA Paper* 2007–4345

[73] Valentini P, Schwartzentruber T E, Bender J D, Nompelis I and Candler G V 2015 Direct simulation of rovibrational excitation and dissociation in molecular nitrogen using an *ab initio* potential energy surface *AIAA Paper* 2015–474

[74] Economou D J, Bartel T J, Wise R S and Lymberopoulos D P 1995 Two-dimensional direct simulation Monte Carlo (DSMC) of reactive neutral and ion flow in a high density plasma reactor *IEEE Trans. Plasma Sci.* **23** 581–90

[75] Serikov V V, Kawamoto S and Nanbu K 1999 Particle-in-cell plus direct simulation Monte Carlo (PIC-DSMC) approach for self-consistent plasma-gas simulations *IEEE Trans. Plasma Sci.* **27** 1389–98

[76] Pietanza L, Colonna G and Capitelli M 2005 Electron and phonon dynamics in laser short pulses-heated metals *Appl. Surf. Sci.* **248** 103–7

[77] Pietanza L, Colonna G, Longo S and Capitelli M 2004 Electron and phonon distribution relaxation in metal films under a femtosecond laser pulse *Thin Solid Films* **453** 506–12

[78] Pietanza L, Colonna G, Longo S and Capitelli M 2004 Electron and phonon relaxation in metal films perturbed by a femtosecond laser pulse *Appl. Phys.* A **79** 1047–50

[79] Pietanza L D, Colonna G, Longo S and Capitelli M 2007 Non-equilibrium electron and phonon dynamics in metals under femtosecond laser pulse *Eur. Phys. J.* D **45** 369–89

[80] Shivanian E, Abbasbandy S and Alhuthali M S 2014 Exact analytical solution to the Poisson-Boltzmann equation for semiconductor devices *Eur. Phys. J. Plus.* **129** 1–8

[81] Majorana A 1991 Space homogeneous solutions of the Boltzmann equation describing electron-phonon interactions in semiconductors *Transp. Theory Stat. Phys.* **20** 261–79

[82] Apostolova T, Perlado J and Rivera A 2015 Femtosecond laser irradiation induced-high electronic excitation in band gap materials: A quantum-kinetic model based on Boltzmann equation *Nucl. Instrum. Methods B* **352** 167–70

[83] COMSOL Multiphysics https://www.comsol.it/plasma-module

[84] ANSYS Fluent http://www.ansys.com/Products/Fluids/ANSYS-Fluent

[85] Hagelaar G and Pitchford L 2016 BOSLIG+ Electron Boltzmann equation solver http://www.bolsig.laplace.univ-tlse.fr/

[86] Hagelaar G and Pitchford L 2005 Solving the Boltzmann equation to obtain electron transport coefficients and rate coefficients for fluid models *Plasma Sources Sci. Technol.* **14** 722

[87] Morgan W and Penetrante B 1990 ELENDIF: a time-dependent Boltzmann solver for partially ionized plasmas *Comput. Phys. Commun.* **58** 127–52

[88] Bhoj A N and Kushner M J 2007 Continuous processing of polymers in repetitively pulsed atmospheric pressure discharges with moving surfaces and gas flow *J. Phys. D: Appl. Phys.* **40** 6953

IOP Publishing

Plasma Modeling
Methods and Applications
Gianpiero Colonna and Antonio D'Angola

Chapter 2

The two-term Boltzmann equation

Modeling gas discharges and non-equilibrium plasmas usually requires knowledge of the rate coefficients of the processes induced by electron impact and the transport properties of free electrons. In collision-dominated plasmas, usually known as cold plasmas, the Boltzmann distribution is not adequate to describe the statistic of free electrons [1]. A gas subjected to the action of an electric field is an open system, where the energy is injected into the gas by means of electrons. The mechanisms of energy transfer from free electrons to atoms and molecules are very complex, determining the shape of the electron distribution, usually departing from a Maxwellian distribution. In order to determine the electron energy distribution function (EEDF), the Boltzmann equation (see equation (1.14)) must be solved.

If the applied field is not too large, the two-term, or $P1$, approximation is commonly considered, usually written in the following form:

$$\frac{\partial n_0(\varepsilon)}{\partial t} = -\frac{\partial J_{el}(\varepsilon)}{\partial \varepsilon} - \frac{\partial J_f(\varepsilon)}{\partial \varepsilon} - \frac{\partial J_{ee}(\varepsilon)}{\partial \varepsilon} - \frac{\partial J_{ne}(\varepsilon)}{\partial \varepsilon}$$
$$+ S_{in}(\varepsilon) + S_{sup}(\varepsilon) + S_{ion}(\varepsilon) + S_{rec}(\varepsilon) + S_{chem}(\varepsilon),$$

(2.1)

corresponding to the first-order expansion in spherical harmonics, where J are fluxes in the energy space corresponding to elastic collisions ('el'), external fields ('f'), electron–electron collisions ('ee'), and the cooling term due to nozzle expansion ('ne'), and S are source terms due to inelastic ('in'), superelastic ('sup'), ionization ('ion'), and chemical processes in general ('chem'). For large values of the electric fields further terms are necessary in the spherical-harmonic expansion [2–8] (see chapter 3 for details). An alternative approach is the *test particle Monte Carlo* (TP-MC) method [9], where collisions of a single particle in a bath of heavy species are randomly sampled. It can be demonstrated that the TP-MC method is equivalent to solving the stationary, homogeneous Boltzmann equation with expansion in spherical harmonics, including infinite terms.

doi:10.1088/978-0-7503-1200-4ch2
© IOP Publishing Ltd 2016

The two-term approximation is a good compromise between computational time and accuracy, so it is often used to model the electron energy distribution in stationary [1, 10, 11] and time-dependent [12–16] approaches. Another important application is the self-consistent coupling of the Boltzmann equation with state-to state chemical kinetics [12, 17, 18] (see chapter 8), determining at the same time the distribution function of the internal states of atoms and molecules, and gas composition.

For demonstrating the effect of elementary processes, such as superelastic [11, 19] and electron–electron [20, 21] collisions, on the shape of the EEDF, the *P*1 approximation is one of the most commonly used approaches. It simultaneously provides the fractional power going into different channels [22]. The *P*1 approximation has also been applied to calculating the EEDF in non-homogeneous conditions [23–27] and investigating magnetohydrodynamics (MHD) effects [28, 29].

The macroscopic properties of the electron gas are determined by the effective shape of the EEDF, which is strictly related to the energy dependence of the inelastic cross sections. This behavior was the origin of the swarm data approach [30–32] to determine the cross sections of electron collisions with different species [26, 27, 33–40].

It is worth noting that a on-line solver called *BOLSIG+* [41] can be used free of charge through the LXCat website [42], which has a large cross section database.

In this chapter we will present the two-term approximation of the Boltzmann equation in the general form, including MHD effects, space gradients, and flow conditions.

2.1 Two-term distribution

The two-term approximation decomposes the distribution function as

$$f(\mathbf{r}, \mathbf{v}, t) = f_0(\mathbf{r}, v, t) + \frac{\mathbf{v}}{v} \cdot \mathbf{f}_1(\mathbf{r}, v, t), \tag{2.2}$$

where f_0 and $\mathbf{f}_1$ represent the isotropic and anisotropic contributions to the velocity distribution. Let us take as angle variables the director cosines

$$\varphi_i = \frac{v_i}{v} = \cos \vartheta_i. \tag{2.3}$$

For functions of the type

$$F(\mathbf{v}) = G(v)\Phi(\boldsymbol{\varphi}) \tag{2.4}$$

the integrals in the velocity space can be transformed as

$$\iiint_{-\infty}^{\infty} F(\mathbf{v}) \mathrm{d}^3 v = \frac{\pi}{2} \int_0^{\infty} v^2 \mathrm{d}v G(v) \iiint_{-1}^{1} \Phi(\boldsymbol{\varphi}) \mathrm{d}^3 \varphi. \tag{2.5}$$

It should be noted that equation (2.5) is not general, but it is valid in the remainder of this chapter as demonstrated in appendix 2.8. The integrals over the φ variables are of the following type:

$$\int_{-1}^{1} \varphi_i^{a-1} d\varphi_i = 2\frac{a|2}{a}, \tag{2.6}$$

where the symbol $a|2$ is the modulus function (0 for even numbers and 1 for odd numbers). The isotropic (f_0) and the anisotropic ($\mathbf{f}_1$) parts of the distribution are related to the exact distribution

$$f_0(\mathbf{r}, v, t) = \frac{1}{8} \iiint_{-1}^{1} f(\mathbf{r}, \mathbf{v}, t) d^3\varphi \tag{2.7}$$

$$\mathbf{f}_1(\mathbf{r}, v, t) = \frac{3}{8} \iiint_{-1}^{1} \boldsymbol{\varphi} f(\mathbf{r}, \mathbf{v}, t) d^3\varphi. \tag{2.8}$$

The first three moments of the distribution (particle density, mean velocity, and mean energy in electronvolts) in the two-term approximation are calculated as

$$N_e = \iiint_{v} f(\mathbf{r}, \mathbf{v}, t) d^3v = 4\pi \int_{0}^{\infty} v^2 f_0(\mathbf{r}, v, t) dv \tag{2.9}$$

$$\mathbf{v}_d = \frac{1}{N_e} \iiint_{v} \mathbf{v} f(\mathbf{r}, \mathbf{v}, t) d^3v = \frac{4\pi}{3N_e} \int_{0}^{\infty} v^3 \mathbf{f}_1(\mathbf{r}, v, t) dv \tag{2.10}$$

$$\varepsilon_{\text{med}} = \frac{m_e}{2eN_e} \iiint_{v} v^2 f(\mathbf{r}, \mathbf{v}, t) d^3v = \frac{4\pi m_e}{eN_e} \int_{0}^{\infty} v^4 f_0(\mathbf{r}, v, t) dv. \tag{2.11}$$

It should be noted that the scalar properties, such as density and mean energy, are obtained from f_0 while the vectorial properties, such as the mean velocity, are calculated from $\mathbf{f}_1$.

2.2 Differential equations

For free electrons in magnetized plasma, the acceleration $\mathbf{a}$ in equation (1.14) is given by

$$\mathbf{a} = \mathcal{A} + \mathbf{v} \wedge \mathcal{R}, \tag{2.12}$$

where

$$\mathcal{A} = -\frac{e}{m_e}\mathbf{E} + \mathbf{g} \tag{2.13}$$

$$\mathcal{R} = -\frac{e}{m_e}\mathbf{B}, \tag{2.14}$$

with $\mathbf{g}$ the gravity acceleration, which is usually negligible in weakly ionized gas applications.

To determine the differential equation for f_0 we must integrate all the terms in equation (1.14) over the φ variables and, substituting f with equation (2.2), we obtain

$$\iiint_{-1}^{1} \frac{\partial f}{\partial t} d^3\varphi = 8\frac{\partial f_0}{\partial t} \tag{2.15}$$

$$\nabla_v f = \frac{\mathbf{v}}{v}\frac{\partial f_0}{\partial v} + \frac{\mathbf{f}_1}{v} + \frac{\mathbf{v}}{v}\mathbf{v} \cdot \frac{\partial}{\partial v}\frac{\mathbf{f}_1}{v} = \boldsymbol{\varphi}\frac{\partial f_0}{\partial v} + \frac{\mathbf{f}_1}{v} + \boldsymbol{\varphi} v \boldsymbol{\varphi} \cdot \frac{\partial}{\partial v}\frac{\mathbf{f}_1}{v} \tag{2.16}$$

$$\iiint_{-1}^{1} \mathbf{v} \cdot \nabla_r f d^3\varphi = \frac{8v}{3}\nabla_r \cdot \mathbf{f}_1 \tag{2.17}$$

$$\iiint_{-1}^{1} \mathcal{A} \cdot \nabla_v f d^3\varphi = \mathcal{A} \cdot \left(8\frac{\mathbf{f}_1}{v} + \frac{8v}{3}\frac{\partial}{\partial v}\frac{\mathbf{f}_1}{v}\right) = \frac{8\mathcal{A}}{3v^2} \cdot \frac{\partial v^2 \mathbf{f}_1}{\partial v} \tag{2.18}$$

$$\iiint_{-1}^{1} \mathbf{v} \wedge \mathcal{R} \cdot \nabla_v f d^3\varphi = 0 \tag{2.19}$$

$$S_0 = \frac{1}{8} \iiint_{-1}^{1} \left(\frac{\delta f}{\delta t}\right)_{coll} d^3\varphi. \tag{2.20}$$

The calculation of S_0 is described in detail in section 2.6 (see also [43–45]).

The equation for $\mathbf{f}_1$ comes from integration of equation (1.14) multiplied by $\boldsymbol{\varphi}$:

$$\iiint_{-1}^{1} \boldsymbol{\varphi}\frac{\partial f}{\partial t} d^3\varphi = \frac{8}{3}\frac{\partial \mathbf{f}_1}{\partial t} \tag{2.21}$$

$$\iiint_{-1}^{1} \boldsymbol{\varphi}\mathbf{v} \cdot \nabla_r f d^3\varphi = \frac{8}{3}v\nabla_r f_0 \tag{2.22}$$

$$\iiint_{-1}^{1} \boldsymbol{\varphi}\mathcal{A} \cdot \nabla_v f d^3\varphi = \frac{8}{3}\mathcal{A}\frac{\partial f_0}{\partial v} \tag{2.23}$$

$$\iiint_{-1}^{1} \boldsymbol{\varphi}\mathbf{v} \wedge \mathcal{R} \cdot \nabla_v f d^3\varphi = \frac{8}{3}\mathcal{R} \wedge \mathbf{f}_1 \tag{2.24}$$

$$\mathbf{S}_1 = \frac{3}{8} \iiint_{-1}^{1} \boldsymbol{\varphi}\left(\frac{\delta f}{\delta t}\right)_{coll} d^3\varphi. \tag{2.25}$$

The term $\mathbf{S}_1$ is given by (see section 2.6.1)

$$\mathbf{S}_1 = -\nu_e \mathbf{f}_1, \tag{2.26}$$

where ν_e is the total elastic collision frequency given in equation (2.78). Other processes, such as inelastic, superelastic, and electron–electron collisions must be taken into account in the collision frequency (see [43]), but their contribution is usually negligible.

From basic algebra, the vector product can be written in matrix form as

$$\mathbf{a} \wedge \mathbf{b} = \begin{pmatrix} 0 & -a_z & a_y \\ a_z & 0 & -a_x \\ -a_y & a_x & 0 \end{pmatrix} \begin{pmatrix} b_x \\ b_y \\ b_z \end{pmatrix}. \tag{2.27}$$

Combining the magnetic field (equation (2.24)) and elastic collision (equation (2.26)) contributions,

$$\mathcal{R} \wedge \mathbf{f_1} + \nu_e \, \mathbf{f_1} = \begin{pmatrix} \nu_e & -\mathcal{R}_z & \mathcal{R}_y \\ \mathcal{R}_z & \nu_e & -\mathcal{R}_x \\ -\mathcal{R}_y & \mathcal{R}_x & \nu_e \end{pmatrix} \begin{pmatrix} f_{1x} \\ f_{1y} \\ f_{1z} \end{pmatrix} = \hat{\Omega} \mathbf{f_1}, \tag{2.28}$$

and the space gradient (equation (2.22)) and electric field (equation (2.23)) terms,

$$\mathbf{Q_0} = -\left(v \nabla_r f_0 + \mathcal{A} \frac{\partial f_0}{\partial v} \right), \tag{2.29}$$

depending only on the isotropic distribution, we obtain the equation for the anisotropic part. The final result is the following system of equations:

$$\begin{cases} \dfrac{\partial f_0}{\partial t} + \dfrac{v}{3} \nabla_r \cdot \mathbf{f_1} + \dfrac{\mathcal{A}}{3v^2} \cdot \dfrac{\partial v^2 \mathbf{f_1}}{\partial v} = S_0 \\ \dfrac{\partial \mathbf{f_1}}{\partial t} + \hat{\Omega} \mathbf{f_1} = \mathbf{Q_0}. \end{cases} \tag{2.30}$$

In the equation for the anisotropic distribution, $\hat{\Omega} \mathbf{f_1}$ is the dissipation term while $\mathbf{Q_0}$ is the driving force.

2.3 Quasi-stationary approximation

Except in cases where high-frequency electric fields are applied, $\mathbf{f_1}$ relaxes more rapidly than the f_0, and the quasi-stationary approximation ($\frac{\partial \mathbf{f_1}}{\partial t} \approx 0$) can be used, calculating the anisotropic distribution by solving the algebraic equation

$$\hat{\Omega} \mathbf{f_1} = \mathbf{Q_0}, \tag{2.31}$$

with the analytical solution

$$\mathbf{f_1} = \hat{\Omega}^{-1} \mathbf{Q_0} = \hat{\omega} \mathbf{Q_0}, \tag{2.32}$$

where

$$\hat{\omega} = \frac{1}{\nu_e\left(\nu_e^2 + \mathcal{R}^2\right)} \begin{pmatrix} \nu_e^2 + \mathcal{R}_x^2 & \mathcal{R}_x\mathcal{R}_y + \nu_e\mathcal{R}_z & \mathcal{R}_x\mathcal{R}_z - \nu_e\mathcal{R}_y \\ \mathcal{R}_x\mathcal{R}_y - \nu_e\mathcal{R}_z & \nu_e^2 + \mathcal{R}_y^2 & \mathcal{R}_y\mathcal{R}_z + \nu_e\mathcal{R}_x \\ \mathcal{R}_x\mathcal{R}_z + \nu_e\mathcal{R}_y & \mathcal{R}_y\mathcal{R}_z - \nu_e\mathcal{R}_x & \nu_e^2 + \mathcal{R}_z^2 \end{pmatrix}. \tag{2.33}$$

Substituting equation (2.32) in equation (2.30) we have

$$\frac{\partial f_0}{\partial t} + \frac{v}{3}\nabla_r \cdot \hat{\omega}\mathbf{Q}_0 + \frac{\mathcal{A}}{3v^2} \cdot \frac{\partial v^2 \hat{\omega}\mathbf{Q}_0}{\partial v} = S_0. \tag{2.34}$$

For homogeneous systems all the space gradients vanish, giving the equation commonly found for the two-term approximation:

$$\frac{\partial f_0}{\partial t} - \frac{1}{3v^2}\frac{\partial}{\partial v}\mathcal{A} \cdot \hat{\omega}\mathcal{A}v^2\frac{\partial f_0}{\partial v} = S_0. \tag{2.35}$$

2.4 Electrons in flow

The differential equations in the previous section have been obtained for a gas at rest. To determine the corrections due to flowing gas, we can suppose that equations (2.34) and (2.35) are valid in a reference frame that moves together with the flow at speed $\mathbf{u}$ in the laboratory. The particle velocity in this new frame ($\mathbf{v}'$) is related to the velocity in the laboratory system ($\mathbf{v}$) by $\mathbf{v}' = \mathbf{v} - \mathbf{u}$. The distribution function in the two reference systems must be equal,

$$f(\mathbf{v}) = f'(\mathbf{v}') = f_0(\mathbf{v}' + \mathbf{u}) + \frac{\mathbf{v}' + \mathbf{u}}{|\mathbf{v}' + \mathbf{u}|} \cdot \mathbf{f}_1(\mathbf{v}' + \mathbf{u}), \tag{2.36}$$

and assuming that the flow speed is much lower than the thermal velocity ($\mathbf{v}' \approx \mathbf{v}$), we can expand the terms in equation (2.36) in a Taylor series truncated at the first order,

$$f_0(\mathbf{v}' + \mathbf{u}) \approx f_0(v') + \mathbf{u} \cdot \boldsymbol{\varphi}'\frac{\partial f_0(v')}{\partial v}$$

$$\frac{\mathbf{v}' + \mathbf{u}}{|\mathbf{v}' + \mathbf{u}|} \approx \boldsymbol{\varphi}' \tag{2.37}$$

$$\mathbf{f}_1(\mathbf{v}' + \mathbf{u}) \approx \mathbf{f}_1(v') + \mathbf{u} \cdot \boldsymbol{\varphi}'\frac{\partial \mathbf{f}_1(v')}{\partial v},$$

which, applied to the definitions of f_0 and $\mathbf{f}_1$, give

$$f_0'(v') = \iiint_{-1}^{1} f'(\mathbf{v}')\mathrm{d}^3\varphi' = f_0(v) + \frac{1}{3}\mathbf{u} \cdot \frac{\partial \mathbf{f}_1(v)}{\partial v}$$

$$\mathbf{f}_1'(v') = \iiint_{-1}^{1} \boldsymbol{\varphi}'f'(\mathbf{v}')\mathrm{d}^3\varphi' = \mathbf{f}_1(v) + \mathbf{u}\frac{\partial f_0(v)}{\partial v}. \tag{2.38}$$

These equations can be used to determine the corrections to the Boltzmann equation for a flowing gas. It must be pointed out that equation (2.25) for $\mathbf{S}_1$, is valid in the reference frame where the gas is at rest, therefore

$$\mathbf{S}_1 = -\nu_e\,\mathbf{f}_1' = -\nu_e\mathbf{u}\frac{\partial f_0(v)}{\partial v} - \nu_e\,\mathbf{f}_1, \tag{2.39}$$

from which it follows that equation (2.34) is still valid if $\mathbf{Q}_0$ is given by

$$\mathbf{Q}_0 = -\left[v\nabla_{\!r}f_0 + (\mathcal{A} + \nu_e\mathbf{u})\frac{\partial f_0}{\partial v}\right] \tag{2.40}$$

and equation (2.35) becomes

$$\frac{\partial f_0}{\partial t} - \frac{1}{3v^2}\frac{\partial}{\partial v}\mathcal{A}\cdot\hat{\omega}(\mathcal{A} + \nu_e\mathbf{u})v^2\frac{\partial f_0}{\partial v} = S_0. \tag{2.41}$$

It should be noted that the term S_0 must also be calculated considering $f'(\mathbf{v}')$. Neglecting electron–electron collisions, the collision integral is a linear function of the isotropic distribution

$$\begin{aligned} S_0 &= \hat{S}_{\mathrm{LC}}\big[f_0'(v')\big] = \hat{S}_{\mathrm{LC}}\left[f_0(v) + \frac{1}{3}\mathbf{u}\cdot\frac{\partial\mathbf{f}_1(v)}{\partial v}\right]\\ &= \hat{S}_{\mathrm{LC}}\left[f_0(v) + \frac{1}{3}\mathbf{u}\cdot\frac{\partial\hat{\omega}\mathbf{Q}_0}{\partial v}\right]. \end{aligned} \tag{2.42}$$

A similar approach can be used to calculate the contribution of electron–electron collisions to $\mathbf{S}_1$ [43], considering the transformation in a reference frame that moves with the electron current ($\mathbf{f}_1' = 0$). In any case, the contribution of the flow to the collision integrals S_0 and S_1 is a second order correction.

To couple electron kinetics with a fluid dynamic model it is necessary to determine the interaction between the two fluids. Essentially, the two systems exchange with each other both energy and momentum. The energy balance is already considered in S_0, providing one includes the corresponding terms in the fluid dynamics equations. On the other hand, the momentum exchange is usually neglected. To estimate the momentum exchange between electrons and heavy particles we can start from the relaxation equation in the absence of fields and gradients:

$$\frac{\partial\mathbf{f}_1}{\partial t} = -\nu_e\mathbf{u}\frac{\partial f_0}{\partial v} - \nu_e\,\mathbf{f}_1, \tag{2.43}$$

obtaining the source of free electron momentum due to drag from heavy particles

$$\begin{aligned} \left(\frac{\partial\rho_e\mathbf{u}_e}{\partial t}\right)_{\mathrm{drag}} &= N_e m_e\frac{\partial\mathbf{v}_d}{\partial t} = \frac{4\pi m_e}{3}\int_0^\infty v^3\frac{\partial\mathbf{f}_1}{\partial t}\mathrm{d}v\\ &= -\frac{4\pi m_e}{3}\int_0^\infty v^3\nu_e\left(\mathbf{u}\frac{\partial f_0}{\partial v} + \mathbf{f}_1\right)\mathrm{d}v. \end{aligned} \tag{2.44}$$

2.5 Electron energy distribution

The two-term Boltzmann equation is usually written for the electron energy distribution defined as

$$n_0(\varepsilon) = 4\pi \sqrt{\frac{2e^3\varepsilon}{m_e^3}} f_0(v) \qquad (2.45)$$

$$\mathbf{n}_1(\varepsilon) = \frac{8\pi e^2\varepsilon}{3m_e^2} \mathbf{f}_1(v), \qquad (2.46)$$

where

$$\varepsilon = \frac{m_e}{2e} v^2$$

is the electron energy in electronvolts. This change of variable transforms the velocity derivative and differential as

$$\frac{\partial}{\partial v} = \sqrt{\frac{2m_e\varepsilon}{e}} \frac{\partial}{\partial \varepsilon} \qquad (2.47)$$

$$dv = \sqrt{\frac{e}{2m_e\varepsilon}} d\varepsilon, \qquad (2.48)$$

obtaining for equations (2.9)–(2.11)

$$N_e = \int_0^\infty n_0(\varepsilon) d\varepsilon \qquad (2.49)$$

$$\mathbf{v}_d = \frac{1}{N_e} \int_0^\infty \mathbf{n}_1(\varepsilon) d\varepsilon \qquad (2.50)$$

$$\varepsilon_{\text{med}} = \frac{1}{N_e} \int_0^\infty n_0(\varepsilon)\varepsilon \, d\varepsilon. \qquad (2.51)$$

The resulting differential equations (see equation (2.30)) become

$$\begin{cases} \dfrac{\partial n_0}{\partial t} + \nabla_r \cdot \mathbf{n}_1 + \dfrac{m_e\mathcal{A}}{e} \cdot \dfrac{\partial \mathbf{n}_1}{\partial \varepsilon} = S_0' \\ \dfrac{\partial \mathbf{n}_1}{\partial t} + \dfrac{2ee}{3m_e}\nabla_r n_0 + \dfrac{2\varepsilon}{3}(\mathcal{A} + \nu_e\mathbf{u})\left(\dfrac{\partial n_0}{\partial \varepsilon} - \dfrac{n_0}{2\varepsilon}\right) + \hat{\Omega}\mathbf{n}_1 = 0, \end{cases} \qquad (2.52)$$

where

$$S_0' = 4\pi \sqrt{\frac{2e^3\varepsilon}{m_e^3}} S_0 \qquad (2.53)$$

and in the quasi-stationary approximation we have

$$\mathbf{n}_1 = -\frac{2e\varepsilon}{3m_e}\hat{\omega}\left[\nabla_r n_0 + \frac{m_e}{e}\left(\mathcal{A} + \nu_e\mathbf{u}\right)\left(\frac{\partial n_0}{\partial\varepsilon} - \frac{n_0}{2\varepsilon}\right)\right]. \qquad (2.54)$$

2.5.1 Current anisotropy

To clarify the role of the matrix $\hat{\omega}$ let us consider the configuration with the magnetic field oriented in the z-direction, $\mathbf{B} = (0, 0, B)$. For this condition, and defining

$$\beta = \frac{eB}{\nu_e m_e}$$

$$\alpha = \frac{\beta}{\nu_e\left(1 + \beta^2\right)} \qquad (2.55)$$

$$\gamma = \frac{1}{\nu_e\left(1 + \beta^2\right)},$$

we have

$$\hat{\omega} = \frac{1}{\nu_e\left(1 + \beta^2\right)}\begin{pmatrix} 1 & \beta & 0 \\ -\beta & 1 & 0 \\ 0 & 0 & 1 + \beta^2 \end{pmatrix} = \begin{pmatrix} \gamma & \alpha & 0 \\ -\alpha & \gamma & 0 \\ 0 & 0 & 1 \end{pmatrix}. \qquad (2.56)$$

Referring to the expression of $\mathbf{n}_1$ in equation (2.54) and considering that the current density is

$$\mathbf{j} = N_e e\mathbf{v}_d = N_e e \int_0^\infty \mathbf{n}_1 \, d\varepsilon \qquad (2.57)$$

for $\mathbf{E} = (0, E, 0)$, the Hall current is oriented in the x-direction and its value is given by[1]

$$j_{\text{Hall}} = -\frac{2N_e e^2}{3m_e}\int_0^\infty \varepsilon\sqrt{\varepsilon}\frac{\partial}{\partial\varepsilon}\frac{n_0}{\sqrt{\varepsilon}}\alpha E \, d\varepsilon \qquad (2.58)$$

2.5.2 Transport properties

The transport properties of the electron gas can be calculated directly from the EEDF starting from equation (2.54), remembering that the mean electron velocity $\mathbf{v}_d$ is calculated from the anisotropic distribution $\mathbf{n}_1$ (see equation (2.50)). We can distinguish three contributions, due to diffusion, drift, and drag:[1]

$$\mathbf{v}_{\text{diff}} = \frac{1}{N_e}\hat{D}_e\nabla_r N_e = -\frac{2e}{3m_e N_e}\int_0^\infty \varepsilon\hat{\omega}n_0 d\varepsilon\nabla_r N_e \qquad (2.59)$$

---

[1] It is straightforward to show that $\frac{\partial}{\partial\varepsilon}\frac{n_0}{\sqrt{\varepsilon}} = \left(\frac{\partial n_0}{\partial\varepsilon} - \frac{n_0}{2\varepsilon}\right)$.

$$\mathbf{v}_{\text{drift}} = \hat{\mu}_e \mathbf{E} = -\frac{2e}{3m_e N_e} \mathbf{E} \int_0^\infty \varepsilon \hat{\omega} \frac{\partial}{\partial \varepsilon} \frac{n_0}{\sqrt{\varepsilon}} d\varepsilon \tag{2.60}$$

$$\mathbf{v}_{\text{drag}} = \hat{C}_{\text{drag}} \mathbf{u} = -\frac{2}{3N_e} \mathbf{u} \int_0^\infty \nu_e \varepsilon \hat{\omega} \frac{\partial}{\partial \varepsilon} \frac{n_0}{\sqrt{\varepsilon}} d\varepsilon \tag{2.61}$$

giving the transport coefficients[2]

$$\hat{D}_e = -\frac{2e}{3m_e N_e} \int_0^\infty \varepsilon \hat{\omega} n_0 d\varepsilon \tag{2.62}$$

$$\hat{\mu}_e = -\frac{2e}{3m_e N_e} \int_0^\infty \varepsilon \hat{\omega} \frac{\partial}{\partial \varepsilon} \frac{n_0}{\sqrt{\varepsilon}} d\varepsilon \tag{2.63}$$

$$\hat{C}_{\text{drag}} = -\frac{2}{3N_e} \int_0^\infty \nu_e \varepsilon \hat{\omega} \frac{\partial}{\partial \varepsilon} \frac{n_0}{\sqrt{\varepsilon}} d\varepsilon. \tag{2.64}$$

In the case of an equilibrium distribution, the diffusion coefficient and electron mobility are related through the Einstein relation to the electron temperature:

$$\frac{kT_e}{e} = \frac{D_e}{\mu_e}. \tag{2.65}$$

In the case of a non-Maxwellian distribution, the electron temperature is not defined and equation (2.65) is used to define the *characteristic energy*, $\varepsilon_{\text{car}} = D_e/\mu_e$, often assumed to be a measure of the electron temperature. For a Maxwellian distribution, in general, we have $\frac{3}{2}\varepsilon_{\text{car}} = \varepsilon_{\text{med}}$, a relation not valid in non-equilibrium.

2.5.3 Nozzle flow

To obtain the two-term approximation of the Boltzmann equation for quasi-1D nozzle flows, analogously to Euler equations, the space gradients must be substituted with

$$\nabla_r f \equiv \begin{pmatrix} \frac{1}{A}\frac{\partial Af}{\partial x} \\ 0 \\ 0 \end{pmatrix}, \tag{2.66}$$

which makes the Boltzmann equation very complex, containing mixed derivatives in space and energy. Moreover it has been observed that, during the expansion, the electron distribution cools down mainly due to collisions, and therefore the

---

[2] In the absence of a magnetic field, the transport coefficients become scalar quantities, since for a generic tensorial property for $\mathbf{B} = 0$ it is $\hat{G} = G_0 \hat{I}$, where $\hat{I}$ is the identity matrix.

contribution of the expansion is usually negligible. However, a correction term can be added estimating the energy variation related to the increase of the mean velocity due to the expansion. It can be easily verified that the mean electron energy variation corresponds to that of the electron kinematic energy. In fact

$$\delta\varepsilon_{\mathrm{med}} = \delta \int_0^\infty \varepsilon n_0(\varepsilon)\mathrm{d}\varepsilon = \int_0^\infty \varepsilon\delta n_0(\varepsilon)\mathrm{d}\varepsilon = \frac{m_e}{e}u\delta u \int_0^\infty \varepsilon\frac{\partial n_0}{\partial\varepsilon}\mathrm{d}\varepsilon \qquad (2.67)$$

and integrating by parts

$$\int_0^\infty \varepsilon\frac{\partial n_0}{\partial\varepsilon}\mathrm{d}\varepsilon = \varepsilon n_0(\varepsilon)|_0^\infty - \int_0^\infty n_0(\varepsilon)\mathrm{d}\varepsilon = 0 - 0 - 1 = -1 \qquad (2.68)$$

we have

$$\delta\varepsilon_{\mathrm{med}} = -\frac{m_e}{e}u\delta u, \qquad (2.69)$$

which is the variation of electron kinematic energy per particle. As a consequence, the correction term introduced in [29] to account for the mean energy variation by expansion or compression is $\frac{\partial J_{\mathrm{ne}}}{\partial\varepsilon}$ where

$$J_{\mathrm{ne}}(\varepsilon) = \frac{m_e}{e}n_0(\varepsilon)u\frac{\partial u}{\partial t}. \qquad (2.70)$$

2.6 The collision integral

Let us now derive the expression of the collision integral for the two-term Boltzmann equation starting from the general definition in equation (1.22). A heavy species s, in collisions with electrons, does not change its velocity, due to the large mass difference. Therefore the integral with respect to the velocity of the s particle can be calculated separately, resulting in its density N_s. Only the dependence on the internal distribution $f_{s,\ell}$ will be considered,[3] giving

$$\left(\frac{\delta f}{\delta t}\right)_{\mathrm{coll}} = \sum_{s,\ell,\ell'} N_s v \int \left[f(\mathbf{v}', \mathbf{r}, t)f_{s,\ell'} - f(\mathbf{v}, \mathbf{r}, t)f_{s,\ell}\right]\frac{\mathrm{d}\sigma_p}{\mathrm{d}\Omega}\mathrm{d}\Omega. \qquad (2.71)$$

Three types of processes can be distinguished:
1. elastic collisions, with $\ell = \ell'$, where the internal state s does not change;
2. inelastic collisions, with $\ell' > \ell$, where the final state has higher energy than the initial one; and
3. superelastic, or second collisions, with $\ell > \ell'$, where the electron gains the energy lost by s.

Considering the two-term approximation we can separate the collision integral into two contributions

---

[3] Using this notation $\sum_\ell f_{s,\ell} = 1$.

$$\left(\frac{\delta f}{\delta t}\right)_{coll} = \delta f_0(v) + \delta f_1(v), \tag{2.72}$$

each one calculated using equation (2.71) with the corresponding term. Trivially, for the properties of integrals, $S_0 \propto \delta f_0(v)$ and $S_1 \propto \delta f_1(v)$, in fact, from equations (2.20) and (2.25) we have

$$S_0 = \sum_{s,\ell,\ell'} N_s v \int \left[f_0(v', \mathbf{r}, t) f_{s,\ell'} - f_0(v, \mathbf{r}, t) f_{s,\ell} \right] \frac{d\sigma_p}{d\Omega} d\Omega \tag{2.73}$$

$$\mathbf{S}_1 = \frac{3}{8} \iiint_{-1}^{1} d^3\varphi\boldsymbol{\varphi} \sum_{s,\ell,\ell'} N_s v \int \left[\boldsymbol{\varphi}' \cdot \mathbf{f}_1(v', \mathbf{r}, t) f_{s,\ell'} - \boldsymbol{\varphi} \cdot \mathbf{f}_1(v, \mathbf{r}, t) f_{s,\ell} \right] \frac{d\sigma_p}{d\Omega} d\Omega, \tag{2.74}$$

where $\boldsymbol{\varphi}'$ differs from φ by a quantity that depends on the deflection angle and plane. Under the assumption that all the deflection planes are equivalent[4] we can limit the calculation in the z-direction considering that $\varphi_z = \cos \vartheta$ and $\varphi_z' = \cos(\vartheta + \vartheta_d) = \cos \vartheta \cos \vartheta_d - \sin \vartheta \sin \vartheta_d$. In equation (2.74) the integrals with mixed terms in different directions as well as those containing the $\sin \vartheta$ are null, therefore only the terms with φ_z^2 give an effective contribution, and therefore

$$\mathbf{S}_1 = \sum_{s,\ell,\ell'} N_s v \int \left[\cos \vartheta_d \, \mathbf{f}_1(v', \mathbf{r}, t) f_{s,\ell'} - \mathbf{f}_1(v, \mathbf{r}, t) f_{s,\ell} \right] \frac{d\sigma_p}{d\Omega} d\Omega. \tag{2.75}$$

In contrast to S_0, only elastic collisions contribute significantly to $\mathbf{S}_1$ (see equation (2.26)), since inelastic and superelastic processes have smaller cross sections.

2.6.1 Elastic collisions with heavy species

From equation (2.75) it is very simple to obtain equation (2.26). Considering that in elastic collisions $v' \approx v$ (only the direction changes) and that the internal distribution is a common factor[5] we obtain

$$\begin{aligned}
\mathbf{S}_1 &= -\sum_s N_s v \mathbf{f}_1(v, \mathbf{r}, t) \int_{\Omega} (1 - \cos \vartheta_d) \frac{d\sigma_p}{d\Omega} d\Omega \\
&= -\sum_s N_s v \mathbf{f}_1(v, \mathbf{r}, t) \sigma_s^{mt}(v) = -\nu_e \, \mathbf{f}_1(v, \mathbf{r}, t),
\end{aligned} \tag{2.76}$$

---

[4] This assumption is not valid only for very large molecules with complex geometry.

[5] This assumption is not really valid, because the elastic cross sections depend on the atomic or molecular level [46, 47]. However, because of the small population of excited levels, the major contribution comes from the ground state. Only in extreme conditions does the dependence of elastic cross sections on the excited state give an appreciable contribution.

where

$$\sigma_s^{mt}(v) = \int_\Omega (1 - \cos \vartheta_d) \frac{\mathrm{d}\sigma_p}{\mathrm{d}\Omega} \mathrm{d}\Omega \tag{2.77}$$

is the *momentum transfer* cross section and

$$\nu_e = \sum_s N_s v \sigma_s^{mt}(v) \tag{2.78}$$

is the elastic collision frequency. When evaluating S_1, the small variation in the velocity module is negligible with respect to the change of direction. On the other hand, the energy lost by an electron in an elastic collision must be considered, because assuming $v' = v$ gives a null contribution to the S_0. To solve this problem we have to obtain a Fokker–Plank equation by expanding in series the distribution with respect to a weakly varying variable. The distribution of the heavy species cannot be neglected here, therefore we will insert f_s in equation (2.73) again, which is considered to be isotropic and not affected by the collision, giving[6]

$$S_{0,s}^{el} = N_s \int_0^{4\pi} v \frac{\mathrm{d}\sigma_p}{\mathrm{d}\Omega} \mathrm{d}\Omega \int_0^\infty f_s(v_s, \ell) [f_0(v') - f_0(v)] \mathrm{d}^3 v_s \tag{2.79}$$

and expanding in series around v and performing the integrations

$$S_{0,s,\ell}^{el} = N_s \frac{\partial}{\partial v} \left(A_{s,\ell} f_0(v) + B_{s,\ell} \frac{\partial f_0(v)}{\partial v} \right). \tag{2.80}$$

It must be noted that, as it is the distribution of a function of the module of the velocity, the probability should be expressed in terms of $f_0 v^2$ [48] (see equation (2.9)), therefore the velocity derivative in the Fokker–Plank equation must be substituted as $\frac{\partial}{\partial v} \to \frac{1}{v^2} \frac{\partial v^2}{\partial v}$. The coefficients $A_{s,\ell}$ and $B_{s,\ell}$, representing the integrals, are not independent and can be related considering the equilibrium condition, i.e. a Boltzmann distribution, giving $A_{s,\ell} = B_{s,\ell} m_e v / kT$. Following these considerations equation (2.80) becomes

$$S_{0,s,\ell}^{el} = \frac{N_s}{v^2} \frac{\partial}{\partial v} v^2 B_{s,\ell} \left(\frac{m_e v}{kT} f_0(v) + \frac{\partial f_0(v)}{\partial v} \right). \tag{2.81}$$

Let us now calculate explicitly the coefficient $B_{\ell,s}$

$$B_{s,\ell} = \frac{1}{2} \int_0^{4\pi} v \frac{\mathrm{d}\sigma_p}{\mathrm{d}\Omega} \mathrm{d}\Omega \int_0^\infty f_s(v_s, \ell) [(\Delta v)^2 \mathrm{d}^3 v_s, \tag{2.82}$$

---

[6] For the sake of brevity, from now on we will consider implicitly the dependence of the distribution on position and time.

where $\Delta v = v' - v$. This quantity can be calculated considering that in elastic collisions, the amplitude of the relative velocity is the same before and after the collision giving

$$(\mathbf{v} - \mathbf{v}_s)^2 = \left(\mathbf{v}' - \mathbf{v}'_s\right)^2 \Rightarrow v^2 + v_s^2 - 2\mathbf{v}\cdot\mathbf{v}_s = v'^2 + v_s'^2 - 2\mathbf{v}'\cdot\mathbf{v}'_s$$

$$\overset{v_s\approx v'_s}{\Longrightarrow}v'^2 - v^2 \overset{v\approx v'}{\approx} 2v\Delta v \approx \mathbf{v}_s\cdot(\mathbf{v} - \mathbf{v}') \Rightarrow \Delta v \approx \mathbf{v}_s\cdot(\mathbf{v} - \mathbf{v}')/v.$$

(2.83)

$(\Delta v)^2$ contains the terms $\propto v_{s,i}v_{s,j}$ and, as the distribution is isotropic, only those terms with $i = j$ give a mean different from zero, i.e. $\left|v_{s,i}^2\right| = \left|v_s^2\right|/3 = kT/m_s$, and equation (2.82) becomes

$$B_{s,\ell} = \frac{kT}{2m_s}f_{s,\ell}\int_0^{4\pi} v\frac{d\sigma_p}{d\Omega}d\Omega(v^2 + v'^2 - 2\mathbf{v}\cdot\mathbf{v}')/v^2$$

$$\overset{v\approx v'}{\approx}\frac{kT}{m_s}f_{s,\ell}\int_0^{4\pi} v\frac{d\sigma_p}{d\Omega}d\Omega(1 - \cos\vartheta_d) = \frac{kT}{m_s}f_{s,\ell}v\sigma_{s,\ell}^{mt} = \frac{kT}{m_s}f_{s,\ell}v_s^{mt}.$$

(2.84)

If we assume that the momentum transfer is almost independent of the internal level we can eliminate in equation (2.84) the internal distribution considering a single coefficient for the whole species.

To calculate the source term for the electron energy distribution $n_0(\varepsilon)$ in equation (2.52) we have to consider the equations in section 2.5 giving

$$S_{0,s}^{\prime el} = 4\pi\sqrt{\frac{2e^3\varepsilon}{m_e^3}}\frac{N_s}{v^2}\sqrt{\frac{2m_e\varepsilon}{e}}\frac{\partial}{\partial\varepsilon}v^2 B_s$$

$$\times\left(\frac{m_e}{kT}\sqrt{\frac{2e\varepsilon}{m_e}}\frac{1}{4\pi}\sqrt{\frac{m_e^3}{2e^3\varepsilon}}n_0(\varepsilon) + \sqrt{\frac{2m_e\varepsilon}{e}}\frac{\partial}{\partial\varepsilon}\frac{1}{4\pi}\sqrt{\frac{m_e^3}{2e^3\varepsilon}}n_0(\varepsilon)\right)$$

$$= \frac{N_s}{\varepsilon}\frac{m_e}{2e}\frac{2e\varepsilon}{m_e}\frac{\partial}{\partial\varepsilon}\varepsilon\frac{2e}{m_e}B_s\frac{m_e^2}{e^2}\left(\frac{e}{kT}n_0(\varepsilon) + \sqrt{\varepsilon}\frac{\partial}{\partial\varepsilon}\frac{n_0(\varepsilon)}{\sqrt{\varepsilon}}\right)$$

(2.85)

$$= N_s\frac{2m_e}{e}\frac{\partial}{\partial\varepsilon}\frac{kT}{m_s}v_s^{mt}\varepsilon\sqrt{\varepsilon}\left(\frac{e}{kT}\frac{n_0(\varepsilon)}{\sqrt{\varepsilon}} + \frac{\partial}{\partial\varepsilon}\frac{n_0(\varepsilon)}{\sqrt{\varepsilon}}\right)$$

$$= 2N_s\frac{m_e}{m_se}\frac{\partial}{\partial\varepsilon}v_s^{mt}\left(ee n_0(\varepsilon) + kT\varepsilon\sqrt{\varepsilon}\frac{\partial}{\partial\varepsilon}\frac{n_0(\varepsilon)}{\sqrt{\varepsilon}}\right)$$

$$= 2N_s\frac{m_e}{m_se}\frac{\partial}{\partial\varepsilon}v_s^{mt}(\varepsilon)\left[n_0(\varepsilon)\left(ee - \frac{kT}{2}\right) + \varepsilon kT\frac{\partial n_0(\varepsilon)}{\partial\varepsilon}\right],$$

therefore, in equation (2.1) we have that the flux due to elastic collision is given by

$$J_{el} = -2\frac{m_e}{e}\bar{\nu}\left[n_0(\varepsilon)\left(\varepsilon\varepsilon - \frac{kT}{2}\right) + \varepsilon kT\frac{\partial n_0(\varepsilon)}{\partial\varepsilon}\right], \tag{2.86}$$

where $\bar{\nu} = \sum_s N_s \nu_s^{mt}/m_s$.

2.6.2 Electron–electron collisions

The procedure to derive the Fokker–Plank equation and the collisional operator in equation (2.86) for electron–molecule elastic collisions can be adapted to electron–electron collisions obtaining for the corresponding flux term J_{ee}

$$J_{ee} = \alpha_{ee}\left\{P(\varepsilon)\left[\frac{n(\varepsilon)}{\varepsilon} - \frac{\partial n(\varepsilon)}{\partial\varepsilon}\right] - Q(\varepsilon)n(\varepsilon)\right\}, \tag{2.87}$$

where

$$\alpha_{ee} = \frac{2\pi}{3}(K_e e)^2\sqrt{\frac{2q_e}{m_e}}N_e \ln\Lambda_{ee} = 2.5821 \cdot 10^{-12}N_e \ln\Lambda_{ee}, \tag{2.88}$$

where

$$\Lambda_{ee} = \sqrt{\frac{\varepsilon_{car}}{4\pi K_e e N_e}} \cdot \frac{\varepsilon_{med}}{K_e e} = 5.1527 \cdot 10^{12}\sqrt{\frac{\varepsilon_{car}\,\varepsilon_{med}^2}{N_e}}. \tag{2.89}$$

$K_e = 1/(4\pi\varepsilon_0)$ is the Coulomb constant in which ε_0 is the vacuum dielectric constant. The numerical constants are obtained for all the quantities in IS units, apart from for the electron energy which is in electronvolts.

The coefficients P and Q are given by

$$P(\varepsilon) = \frac{2}{\sqrt{\varepsilon}}\int_0^\varepsilon \xi n_0(\xi)d\xi + 2\varepsilon\int_\varepsilon^\infty \frac{n_0(\xi)}{\sqrt{\xi}}d\xi = P^a(\varepsilon) + P^b(\varepsilon) \tag{2.90}$$

$$Q(\varepsilon) = \frac{3}{\sqrt{\varepsilon}}\int_0^\varepsilon n_0(\xi)d\xi. \tag{2.91}$$

The following recursion rules can easily be obtained

$$P^a(\varepsilon) = \sqrt{\frac{\varepsilon - \Delta}{\varepsilon}}P^a(\varepsilon - \Delta) + \frac{2}{\sqrt{\varepsilon}}\int_{\varepsilon-\Delta}^\varepsilon \xi n_0(\xi)d\xi \tag{2.92}$$

$$P^b(\varepsilon) = \frac{\varepsilon}{\varepsilon + \Delta}P^b(\varepsilon + \Delta) + 2\varepsilon\int_\varepsilon^{\varepsilon+\Delta} \frac{n_0(\xi)}{\sqrt{\xi}}d\xi \tag{2.93}$$

$$Q(\varepsilon) = \sqrt{\frac{\varepsilon - \Delta}{\varepsilon}}\, Q(\varepsilon - \Delta) + \frac{3}{\sqrt{\varepsilon}} \int_{\varepsilon - \Delta}^{\varepsilon} n_0(\xi)\mathrm{d}\xi \qquad (2.94)$$

which can help in building the operator, as will be discussed later.

2.6.3 Inelastic and superelastic collisions

Inelastic collisions are endothermal processes where electrons give energy to heavy species, resulting in excitation or chemical reactions. Superelastic collisions are the reverse process, where chemical or internal energy is given back to the electrons. The energy transferred ($\varepsilon^\star$), known as the threshold energy, is constant, independent of the energy of the impinging electron.[7] In the class of inelastic and superelastic processes are included:

$e(\varepsilon) + A(i) \rightleftharpoons e\left(\varepsilon - \varepsilon_{i,j}^\star\right) + A(j > i)$ internal transitions

$e(\varepsilon) + A(i) \rightleftharpoons e\left(\varepsilon - \varepsilon_{i,ion}^\star\right) + A^+$ ionization and three-body recombination

$e(\varepsilon) + M^+ \rightleftharpoons$ products $(+h\nu)$ attachment and radiative recombination

$e(\varepsilon) + A(i) \rightleftharpoons e(\varepsilon - \varepsilon^\star) +$ products $(+h\nu)$ other chemical processes.

The presence of the photons, in practice, means that the reverse process must be neglected.[8] A common assumption for inelastic and superelastic collisions is that their cross sections do not depend on the deflection angle, i.e. all the directions are equiprobable after the collision, i.e.

$$\frac{\mathrm{d}\sigma_p}{\mathrm{d}\Omega} = \frac{\sigma_p}{4\pi}. \qquad (2.95)$$

The contributions of these processes are usually neglected in $\mathbf{S}_1$ even if they should be included. In fact, from equations (2.75) and (2.95) the contribution of a single process is

$$\mathbf{S}_{1,p} = -N_s f_{s,\ell}\, v\sigma_p\, \mathbf{f}_1 = -\nu_p\, \mathbf{f}_1, \qquad (2.96)$$

being the integral of $\cos\vartheta_d$ null in the case of inelastic or superelastic processes. Therefore the collision frequency in equation (2.26) should be the sum of the frequencies of all the processes, including elastic, inelastic, and superelastic collisions.

Let us calculate the source term $S'_{0,p}$[9] for inelastic processes starting from equations (2.45), (2.53), (2.73) and under the hypothesis in equation (2.95) of isotropic collisions

---

[7] In some chemical processes the threshold energy is not constant, depending on the interaction mechanism. Nevertheless, it is usually considered constant, equal to the minimum energy required by the transition.

[8] To properly calculate the rate of the reverse process in the presence of photons a radiation transport equation must be considered as described in chapter 8.

[9] p represent the set of initial and final parameter of the considered process.

$$S'_{0,p}(\varepsilon) = N_s v \left[\sqrt{\frac{\varepsilon}{\varepsilon'}} n_0(\varepsilon') f_{s,\ell'} - n_0(\varepsilon) f_{s,\ell} \right] \sigma_p, \tag{2.97}$$

neglecting the change in velocity of the heavy particle, we can say that $\varepsilon' = \varepsilon - \varepsilon^\star$. The negative term corresponds to the inelastic process

$$S^{\mathrm{in}}_{0,p}(\varepsilon) = -N_s v n_0(\varepsilon) f_{s\ell} \sigma_p, \tag{2.98}$$

i.e. to electrons passing from energy ε to ε' and the positive term corresponds to superelastic collisions

$$S^{\mathrm{sup}}_{0,p}(\varepsilon) = N_s v \sqrt{\frac{\varepsilon}{\varepsilon'}} n_0(\varepsilon') f_{s,\ell'} \sigma_p, \tag{2.99}$$

i.e. to electrons passing from energy[10] ε' to ε. To determine the cross section $\sigma'_p(\varepsilon')$ for the superelastic process let us use the general definition of the source term

$$S^{\mathrm{sup}}_{0,p}(\varepsilon) = N_s v' n_0(\varepsilon') f_{s,\ell'} \sigma'_p. \tag{2.100}$$

Comparing to equation (2.99) it gives

$$v' \sigma'_p = v \sqrt{\frac{\varepsilon}{\varepsilon'}} \sigma_p \Rightarrow \sigma'_p = \frac{\varepsilon}{\varepsilon'} \sigma_p. \tag{2.101}$$

Equation (2.101) is valid only for elementary processes, i.e. between states with the same statistical weight. In the general case we can obtain the cross section of the superelastic process by applying the detailed balance principle, i.e. the collision integral is null for Boltzmann and Maxwell distributions at the same temperature:

$$v' \sigma'_p(\varepsilon') n_0^{\mathrm{eq}}(\varepsilon') f^{\mathrm{eq}}_{s,\ell'} = v n_0^{\mathrm{eq}}(\varepsilon) f^{\mathrm{eq}}_{s,\ell} \sigma_p$$

$$\sqrt{\varepsilon'} \, \sigma'_p(\varepsilon') \sqrt{\varepsilon'} \, \mathrm{e}^{-\varepsilon'/kT} g_{s,\ell'} \mathrm{e}^{-\varepsilon_{s,\ell'}/kT} = \sqrt{\varepsilon} \, \sigma_p(\varepsilon) \sqrt{\varepsilon} \, \mathrm{e}^{-\varepsilon/kT} g_{s,\ell} \mathrm{e}^{-\varepsilon_{s,\ell}/kT} \tag{2.102}$$

$$\varepsilon' \sigma'_p(\varepsilon') g_{s,\ell'} = \varepsilon \sigma_p(\varepsilon) g_{s,\ell} \mathrm{e}^{-\frac{\varepsilon - \varepsilon' + \varepsilon_{s,\ell} - \varepsilon_{s,\ell'}}{kT}},$$

where $g_{s,\ell}$ is the statistical weight of the ℓth level. For the energy conservation we have $\varepsilon - \varepsilon' = \varepsilon^\star = \varepsilon_{s,\ell'} - \varepsilon_{s,\ell}$, obtaining

$$\sigma'_p(\varepsilon - \varepsilon^\star) = \frac{g_{s,\ell} \varepsilon}{g_{s,\ell'} \varepsilon'} \sigma_p(\varepsilon) = \frac{g_{s,\ell} \varepsilon}{g_{s,\ell'} (\varepsilon - \varepsilon^\star)} \sigma_p(\varepsilon). \tag{2.103}$$

---

[10] Obviously, for each of these terms there is another one with the opposite sign, $S^{\mathrm{in/sup}}_{0,p}(\varepsilon')$, assuring the conservation of the total number of electrons.

2.6.4 Chemical processes

The contribution of chemical processes induced by electron impact is given by equation (2.98), as for inelastic processes. For the reverse processes, in principle, the Boltzmann collision integral could not be used because it only accounts for binary collisions. However, using the detailed balance principle we can estimate the *superelastic* source term for these processes as well. The general form for a reaction is given by

$$e(\varepsilon) + X_\ell \rightleftharpoons e(\varepsilon - \varepsilon^\star) + \sum_s c_s P_{s,\ell} + c_e e(\varepsilon_e). \tag{2.104}$$

The stoichiometric coefficient c_e of the electron can also be negative to include the attachment. Now we need to define some quantities derived from the thermodynamics:

$$\Delta H_r = \sum_s c_s H_s^f - H_X^f \qquad \text{reaction enthalpy}$$

$$\mathcal{Q}_s = \sum_\ell g_{s,\ell} e^{-\varepsilon_{s,\ell}/kT} \qquad \text{partition function of } s\text{th species}$$

$$\mathcal{Q}_e = \int_0^\infty \sqrt{\varepsilon} e^{-\varepsilon/kT} d\varepsilon \qquad \text{partition function of electrons} \tag{2.105}$$

$$K_{\text{eq}} = N_{X,\text{eq}}^{-1} N_{e,\text{eq}}^{c_e-1} \prod_s N_{s_{\text{eq}}}^{c_s} \qquad \text{global equilibrium constant.}$$

Usually, we can calculate the threshold energy by applying the energy conservation law

$$\varepsilon_t^\star = \Delta H_r + \sum_s c_s \varepsilon_{s,\ell} - \varepsilon_{X,\ell} + \max(0, c_e)\varepsilon_e, \tag{2.106}$$

where the 'max' function excludes the electron in the case of attachment. However, depending on the mechanism of reaction, sometimes the threshold energy is higher than the value in equation (2.106) and some of the energy is transferred to the translation of heavy particles [49], therefore we use the subscript t to indicate that equation (2.106) defines the theoretical threshold.

Now let us apply the detailed balance to the reaction in equation (2.104) considering $\varepsilon' = \varepsilon - \varepsilon^\star$ and $\nu = v(\varepsilon)\sigma$:

$$\frac{N_{e,\text{eq}}}{\mathcal{Q}_e} \sqrt{\varepsilon} e^{-\varepsilon/kT} \frac{N_{X,\text{eq}}}{\mathcal{Q}_X} g_{X,\ell} e^{-\varepsilon_{X,\ell}/kT} \nu$$

$$= \frac{N_{e,\text{eq}}}{\mathcal{Q}_e} \sqrt{\varepsilon'} e^{-\varepsilon'/kT} \frac{N_{e,\text{eq}}^{c_e}}{\mathcal{Q}_e^{c_e}} \sqrt{\varepsilon_e}^{c_e} e^{-c_e \varepsilon_e/kT} \prod_s \frac{N_{s,\text{eq}}^{c_s}}{\mathcal{Q}_s^{c_s}} g_{s,\ell} e^{-\varepsilon_{s,\ell}/kT} \nu'$$

$$\nu = \nu'\sqrt{\frac{\varepsilon'}{\varepsilon}}\,\frac{N_{e,\mathrm{eq}}^{c_e}}{N_{X,\mathrm{eq}}}\prod_s N_s^{c_s}\,\frac{Q_X}{Q_e^{c_e}\prod_s Q_s^{c_s}}\,\frac{\sqrt{\varepsilon_e^{c_e}}}{g_{X,\ell}}\prod_s g_{s,\ell}^{c_s}\,\mathrm{e}^{\left(\varepsilon-\varepsilon'+\varepsilon_{X,\ell}-c_e\varepsilon_e-\sum_s\varepsilon_{s,\ell}\right)/kT}$$

$$= \nu'\sqrt{\frac{\varepsilon'}{\varepsilon}}\,\frac{K_{\mathrm{eq}}}{K_Q}\,K_g\,\mathrm{e}^{\left(\varepsilon^\star-\varepsilon_i^\star+\Delta H_r\right)/kT}\quad\text{for all but attachment}\tag{2.107}$$

$$= \nu'\sqrt{\frac{\varepsilon'}{\varepsilon}}\,\frac{K_{\mathrm{eq}}}{K_Q}\,K_g\,\mathrm{e}^{\left(\varepsilon-\varepsilon_i^\star+\Delta H_r\right)/kT}\quad\text{for attachment,}$$

where

$$K_Q = \frac{Q_e^{c_e}}{Q_X}\prod_s Q_s^{c_s}\qquad K_g = \frac{\sqrt{\varepsilon_e^{c_e}}}{g_{X,\ell}}\prod_s g_{s,\ell}^{c_s}.$$

It should be noted that $\sqrt{\varepsilon}$ can be considered the statistical weight for electrons of energy ε.

2.7 The numerical solution

Once all the terms in equation (2.1) have been defined, we can proceed to its numerical solution. The first step is to write the equation in discrete form in the energy space. The first and second derivative terms come from the J. Following Rockwood [45], a finite difference approach was used, with the central difference for the first derivatives. This procedure is unstable when the electric field is very small, a problem solved by Elliot and Greene in [50]. Alternatively upwind schemes can be used, which properly apply forward or backward discretization according to the derivative coefficient, i.e.

$$\frac{\partial y}{\partial t} = u(x,t)\frac{\partial y}{\partial x} \Rightarrow \frac{\partial y_i}{\partial t} = \begin{cases} u\!\left(x_{i+\frac{1}{2}},t\right)\dfrac{y_{i+1}-y_i}{x_{i+1}-x_i} & \text{if } u\!\left(x_{i+\frac{1}{2}},t\right)\geqslant 0 \\[2ex] u\!\left(x_{i-\frac{1}{2}},t\right)\dfrac{y_i-y_{i-1}}{x_i-x_{i-1}} & \text{if } u\!\left(x_{i-\frac{1}{2}},t\right)< 0, \end{cases}\tag{2.108}$$

so that the y_i coefficient is always negative (see figure 2.1 for the symbols).

The discretization of the S terms does not present particular difficulties, except when some cross section contains very sharp peaks, as in resonant vibrational excitation [51]. In this case, sampling the cross sections as $\sigma_i = \sigma(\varepsilon_i)$ gives values strongly dependent on the grid size. On the other hand, taking the mean value inside each mesh interval as (see equation (2.100))

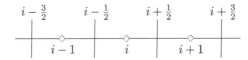

Figure 2.1. Scheme of a uniform grid. The half-integer indexes refer to the interval boundaries.

$$v_i \sigma_i = \frac{1}{\varepsilon_{i+\frac{1}{2}} - \varepsilon_{i-\frac{1}{2}}} \int_{\varepsilon-\frac{1}{2}}^{\varepsilon+\frac{1}{2}} v(\varepsilon)\sigma(\varepsilon)\mathrm{d}\varepsilon \qquad (2.109)$$

reduces the grid dependence.

The discretized equation can be expressed in a general matrix form

$$\frac{\mathrm{d}\mathbf{n}}{\mathrm{d}t} = \left[\hat{C} + \hat{T}(\mathbf{n})\right]\mathbf{n}, \qquad (2.110)$$

where $\mathbf{n} = \{n_0(\varepsilon_i)\}_{i=1, \ldots, N_e}$, or in finite volume discretization, the mean value in the ith interval. The matrix $\hat{C}$ includes all the linear terms and has a diagonal structure

$$\hat{C} = \begin{pmatrix} a_1 & b_1 & 0 & 0 & \cdots & d_1 & 0 & 0 \\ c_2 & a_2 & b_2 & 0 & & 0 & d_2 & 0 \\ 0 & c_3 & a_3 & b_3 & & 0 & 0 & d_3 \\ & \vdots & & & & & \vdots & \\ s_{N_e-3} & 0 & 0 & 0 & \cdots & 0 & 0 & 0 \\ 0 & s_{N_e-2} & 0 & 0 & c_{N_e-2} & a_{N_e-2} & b_{N_e-2} & 0 \\ 0 & 0 & s_{N_e-1} & 0 & 0 & c_{N_e-1} & a_{N_e-1} & b_{N_e-1} \\ 0 & 0 & 0 & s_{N_e} & 0 & 0 & c_{N_e} & a_{N_e} \end{pmatrix}. \qquad (2.111)$$

The first upper and lower diagonals (b and c) come from the flux terms, while each inelastic process contributes to an upper diagonal (d) and the corresponding superelastic process to the symmetric lower diagonal (s). The main diagonal (a) fulfils the conservation condition, i.e.

$$\sum_{j=1}^{N_e} C_{ji} = 0. \qquad (2.112)$$

The non-linear matrix $\hat{T}$, due to electron–electron collisions, has a tridiagonal structure, being generated by the derivative of J_{ee}, and depends linearly on the distribution. Also matrix $\hat{T}$ must fulfil the conservation of total electrons in equation (2.112).

To calculate the stationary solution, the following non-linear equation must be solved:

$$\left[\hat{C} + \hat{T}(\mathbf{n})\right]\mathbf{n} = 0, \qquad (2.113)$$

through iterative methods such as Newton's algorithm, which require the Jacobian matrix, or solving the equation staring from the solution of the linear problem

$$\hat{C}\mathbf{n}^0 = 0 \qquad (2.114)$$

and the successive approximations as

$$\left[\hat{C} + \hat{T}(\mathbf{n}^p) \right]\mathbf{n}^{p+1} = 0. \tag{2.115}$$

Equations (2.114) and (2.115) have ∞^1 solutions, because the determinant is null. This problem can be easily solved, substituting the first equation with the conservation law

$$\sum_{i=1}^{N_e} \Delta_i n_i = N_e, \tag{2.116}$$

with Δ_i the amplitude of the ith interval.

The stationary solution is commonly used coupled with the chemical kinetics, to reduce computational time, as the electron evolution is usually faster than the level and species concentrations [10]. It should be noted that the EEDF is also a function of the composition and of internal level distributions.

Some general features of EEDF can be seen in figure 2.2 where the stationary EEDF is shown for different conditions, as listed in table 2.1, to explore the role of inelastic, superelastic, and electron–electron collisions. Inelastic collisions (curves a, b) induce an abrupt change in the slope at about 20 eV, the threshold energy of the excitation of the metastable state. When superelastic collisions are activated by the He metastable state (curves c, d), some plateaux appear. Electron–electron collisions (curves b, d) reshape the distribution towards a Maxwellian, with the net result of increasing the distribution tails (curve b), hiding the plateaux in the presence of superelastic collisions (curve d).

Curve e has been calculated by adding some molecular species to helium, with vibrational structure and electronically excited states, in order to investigate the sensitivity of the EEDF to the mixture. The presence of a vibrational state keeps the electron distribution cooler than in pure helium, even if the electric field is slightly

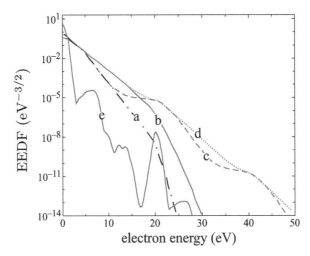

Figure 2.2. EEDF calculated in the conditions listed in table 2.1. The data are from [1, 11].

Table 2.1. Labels for the curves in figure 2.2. The case with ionization degree $N_e/N = 0$ corresponds to the case where electron–electron collisions are neglected. He* and $N_2^\star$ are metastable states, with null values indicating that superelastic collisions are neglected. The conditions are from [1, 11].

Curve	%He	He*/He	%N$_2$	N$_2^\star$/N$_2$	%CO$_2$	%CO	N_e/N	E/N(Td)
a	100	0	0	0	0	0	0	4
b	100	0	0	0	0	0	10^{-3}	4
c	100	10^{-4}	0	0	0	0	0	4
d	100	10^{-4}	0	0	0	0	10^{-3}	4
e	72	10^{-6}	18	10^{-4}	2.5	7.5	0	5

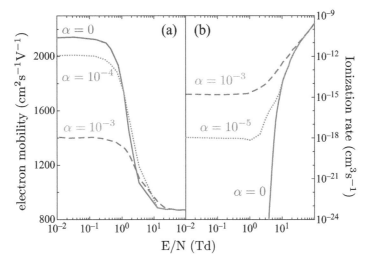

Figure 2.3. Electron mobility (a) and ionization rates (b) in a pure helium mixture, with He*/He = N_e/N = α, as a function of the reduced electric field. The data are from [1].

higher and, due to the lower mean elastic cross section, in place of plateaux we can observe complex peak structures, generated by the combination of inelastic and superelastic collisions of electrons with different species.

The responses of the distribution to changes in external parameters are reflected in the transport properties and global rate coefficients. As an example, in figure 2.3 electron mobility (a) and ionization rate (b) are reported for different excitation and ionization degrees, represented by α, as a function of the reduced electric field. At high field, all the curves converge, independent of α. At low and intermediate fields the differences are relevant, in particular in the case of the ionization degree, where even small excitation increases the ionization rate by orders of magnitude.

In the famous paper [45], Rockwood proposed a time integration based on the Euler method, with the linear part taken implicitly, i.e. with backward discretization, and the electron–electron matrix explicitly, i.e. with forward discretization:

$$(\hat{I} - \Delta t \hat{C})\mathbf{n}(t + \Delta t) = \left\{ \hat{I} + \Delta t \hat{T}[\mathbf{n}(t)] \right\} \mathbf{n}(t). \qquad (2.117)$$

The advantages of equation (2.117) are that LU decomposition can be applied only once at the beginning. Moreover, in this scheme, the electron–electron collision operator implicitly conserves the electron energy. Problems arise when the electron density is high and the time step needed for having a stable explicit integration becomes very small. Taking all the terms implicitly as

$$\left\{\hat{I} - \Delta t \hat{C} - \Delta t \hat{T}[\mathbf{n}(t)]\right\}\mathbf{n}(t + \Delta t) = \mathbf{n}(t) \tag{2.118}$$

makes the procedure stable, but the matrix $\hat{T}$ does not conserve the electron energy. To avoid this problem an iterative procedure can be used to calculate successive approximations of the EEDF, $\mathbf{n}_z(t + \Delta t)$, as

$$\left\{\hat{I} - \Delta t \hat{C} - \Delta t \hat{T}[\mathbf{n}_z(t + \Delta t)]\right\}\mathbf{n}_{z+1}(t + \Delta t) = \mathbf{n}(t). \tag{2.119}$$

To reduce computational time, an *operator-splitting* approach can be used,

$$\left\{\hat{I} - \Delta t \hat{T}[\mathbf{n}(t + \Delta t)]\right\}\left(\hat{I} - \Delta t \hat{C}\right)\mathbf{n}(t + \Delta t) = \mathbf{n}(t), \tag{2.120}$$

first solving the linear part, calculating the zeroth order approximation

$$\left(\hat{I} - \Delta t \hat{C}\right)\mathbf{n}_0 = \mathbf{n}(t) \tag{2.121}$$

and then iterating the non-linear matrix for calculating the successive approximations

$$[\hat{I} - \Delta t \hat{T}(\mathbf{n}_{z-1})]\mathbf{n}_z = \mathbf{n}_0 \tag{2.122}$$

until the convergence criterion based on energy conservation by electron–electron collisions

$$\left| \frac{\mathcal{E}_{z_c} - \mathcal{E}_0}{\mathcal{E}_0} \right| < \epsilon_{\max} \tag{2.123}$$

is fulfilled, where $\mathcal{E}_z$ is the mean energy of the zth approximation, and the new distribution is given by $\mathbf{n}(t + \Delta t) = \mathbf{n}_{z_c}$. This procedure assures the energy conservation for electron–electron collisions, with good efficiency,[11] applying LU decomposition to $\hat{C}$ only once, while $\hat{T}$, being a tridiagonal matrix, can be solved very rapidly, with a CPU time $\propto N_e$. Therefore the critical aspect is the calculation of the $\hat{T}$ matrix.

---

[11] There is another aspect that should be considered for the solution efficiency: the system of equations is very stiff, presenting timescales ranging over many orders of magnitude. A step adapting procedure, such as that proposed in [52], must be used to have reasonable total computational time. Alternatively, higher-order methods for stiff differential equations, such as implicit Runge–Kutta [53], can be adopted.

The traditional approach proposed by Rockwood [45], discretizes the derivative of J_{ee} in the ith interval as

$$\left(\frac{dn_i}{dt}\right)_{ee} = -\frac{J_{ee}^{+}(i) - J_{ee}^{-}(i)}{\Delta\varepsilon}, \tag{2.124}$$

where $J^{+}(i)$ and $J^{-}(i)$ are evaluated at the upper $(i + \frac{1}{2})$ and lower $(i - \frac{1}{2})$ limits of the ith interval, with the trivial relation

$$J_{ee}^{-}(i) = J_{ee}^{+}(i - 1). \tag{2.125}$$

The explicit expression of $J_{ee}^{+}(i)$ is given by

$$J_{ee}^{+}(i) = \alpha\left[\left(\frac{n_{i+\frac{1}{2}}}{4\varepsilon_{i+\frac{1}{2}}} - \frac{n_{i+1} - n_i}{\Delta\varepsilon}\right)P_{i+\frac{1}{2}} - Q_{i+\frac{1}{2}}n_{i+\frac{1}{2}}\right], \tag{2.126}$$

where the population distribution at the upper limit of the ith interval is calculated as the average of the distribution at i and $i + 1$,

$$n_{i+\frac{1}{2}} = \frac{n_i + n_{i+1}}{2}. \tag{2.127}$$

As a consequence, the right side of equation (2.124) depends only on n_{i-1}, n_i, and n_{i+1},

$$\left(\frac{dn_i}{dt}\right)_{ee} = a_{i-1}n_{i-1} - (a_i + b_i)n_i + b_{i+1}n_{i+1}, \tag{2.128}$$

confirming that $\hat{T}$ is a tridiagonal matrix. Comparing equations (2.124) and (2.126), a_i and b_i are given by

$$a_i = \frac{\alpha}{\Delta\varepsilon}\left[u_i^{+}P_{i+\frac{1}{2}} - \frac{Q_{i+\frac{1}{2}}}{2}\right] \tag{2.129}$$

$$b_{i+1} = \frac{\alpha}{\Delta\varepsilon}\left[u_i^{-}P_{i+\frac{1}{2}} - \frac{Q_{i+\frac{1}{2}}}{2}\right], \tag{2.130}$$

where

$$u_i^{\pm} = \left(\frac{1}{\Delta\varepsilon} \pm \frac{1}{4\varepsilon_{i+\frac{1}{2}}}\right). \tag{2.131}$$

From the definitions of P and Q, it follows that a_i and b_i depend linearly on the EEDF

$$\mathbf{a} = \hat{A} \cdot \mathbf{n} \tag{2.132}$$

$$\mathbf{b} = \hat{B} \cdot \mathbf{n}. \tag{2.133}$$

The matrices $\hat{A}$ and $\hat{B}$ are not independent because electron–electron collisions must also conserve the electron energy and, from equation (2.51), the following relation holds:

$$
\begin{aligned}
\frac{d\varepsilon_{\mathrm{med}}}{dt} &= \sum_{i=1}^{N_e} \varepsilon_i \left(\frac{dn_i}{dt} \right)_{\mathrm{ee}} \Delta\varepsilon = \sum_{i=1}^{N_e} \varepsilon_i [a_i - b_i] n_n \Delta\varepsilon \\
&= \sum_{i=1}^{N_e} \sum_{k=1}^{N_e} \big[A_{i,k} - B_{i,k} \big] n_i n_k \varepsilon_i \Delta\varepsilon = 0.
\end{aligned}
\tag{2.134}
$$

With $\mathbf{a} \neq \mathbf{b}$, equation (2.134) is verified if the matrix $\hat{A} - \hat{B}$ is antisymmetric, i.e.

$$\hat{B} = \hat{A}^+, \tag{2.135}$$

where $\hat{A}^+$ is the transpose of $\hat{A}$.

Generally, vanishing fluxes J are imposed as boundary conditions at both sides of the energy interval

$$J_{\mathrm{ee}}^-(1) = J_{\mathrm{ee}}^+(N_\varepsilon) = 0, \tag{2.136}$$

which means

$$b_1 = a_{N_e} = 0 \tag{2.137}$$

obtaining

$$A_{N_e k} = B_{1k} = A_{1k} = B_{1N_e} = 0 \qquad (k = 1, \ldots, N_\varepsilon). \tag{2.138}$$

To calculate $P_{i+\frac{1}{2}}$ and $Q_{i+\frac{1}{2}}$, Rockwood [45] uses the expressions

$$P_{i+\frac{1}{2}} = \frac{P_i + P_{i+1}}{2}; \qquad Q_{i+\frac{1}{2}} = \frac{Q_i + Q_{i+1}}{2}. \tag{2.139}$$

From definitions of P and Q, it seems that the direct evaluation of these quantities at the boundaries of the domain is more straightforward than equation (2.139). However, evaluating $\hat{T}$ using equations (2.132) and (2.133) requires a CPU time $\propto K^2$, with $\hat{A}$ and $\hat{B}$ being full matrices.

An alternative efficient algorithm to calculate the matrix $\hat{T}$ [54] avoids the explicit calculations of matrices $\hat{A}$ and $\hat{B}$ (see equations (2.132) and (2.133)), reducing the computational time, to calculate a_k and b_k. Starting from the recursive relations in equations (2.92)–(2.94)

$$P_{i+1}^a = \sqrt{\frac{\varepsilon_i}{\varepsilon_{i+1}}} P_i^a + \frac{\Delta\varepsilon}{\sqrt{\varepsilon i + 1}}(\varepsilon_i n_i + \varepsilon_{i+1} n_{i+1})$$

$$P_{i+1}^b = \frac{\varepsilon_{i+1}}{\varepsilon_i} P_i^b - \varepsilon_{i+1}\Delta\varepsilon\left(\frac{n_i}{\sqrt{\varepsilon_i}} + \frac{n_{i+1}}{\sqrt{\varepsilon_{i+1}}}\right) \tag{2.140}$$

$$Q_{i+1} = \sqrt{\frac{\varepsilon_i}{\varepsilon_{i+1}}} Q_i + \frac{3\Delta\varepsilon}{2\sqrt{\varepsilon_{i+1}}}(n_i + n_{i+1}),$$

saving the computational time required to recalculate the terms with $k < i$ for each energy mesh point, as done in equations (2.132) and (2.133).

Equation (2.140) can be written in compact form as

$$\hat{A}_a \mathbf{P}^a = \hat{L}_a \mathbf{n}$$

$$\hat{A}_b \mathbf{P}^b = \hat{L}_b \mathbf{n} \tag{2.141}$$

$$\hat{A}_Q \mathbf{Q} = \hat{L}_Q \mathbf{n},$$

where $\hat{A}_Q = \hat{A}_a$. The matrice's $\hat{A}$s and $\hat{L}_Q$ and $\hat{L}_a$ are bidiagonal matrices given by

$$\hat{A}_{Q,i,i} = \hat{A}_{b,i,i} = 1$$

$$\hat{A}_{Q,i,i-1} = \sqrt{\frac{i-1}{i}} \tag{2.142}$$

$$\hat{A}_{b,i,i-1} = \frac{i}{i-1}$$

$$\hat{L}_{Q,i,i} = \hat{L}_{Q,i,i-1} = \frac{3}{2}\left(\frac{\Delta\varepsilon}{i}\right)^{\frac{1}{2}}$$

$$\hat{L}_{a,i,i} = \Delta\varepsilon^{\frac{3}{2}}\sqrt{i} \tag{2.143}$$

$$\hat{L}_{a,i,i-1} = \Delta\varepsilon^{\frac{3}{2}}\frac{i-1}{\sqrt{i}},$$

while $\hat{L}_b$ also has a first row different from zero:

$$\hat{L}_{b,i,i} = -\sqrt{i}\Delta\varepsilon^{\frac{3}{2}} \qquad i > 1$$

$$\hat{L}_{b,i,i-1} = -\Delta\varepsilon^{\frac{3}{2}}\frac{i}{\sqrt{i-1}} \tag{2.144}$$

$$\hat{L}_{b,1,i} = 2\Delta\varepsilon^{\frac{3}{2}}\frac{1}{\sqrt{i}}.$$

The elements a_k of the matrix $\hat{T}$ (see equation (2.128)) are calculated as functions of $\mathbf{P}$ and $\mathbf{Q}$ by considering equation (2.129)

$$\mathbf{a} = -\frac{\alpha}{2\Delta\varepsilon}\left[\hat{M}_P\mathbf{P} + \hat{M}_Q\mathbf{Q}\right] = \hat{A}\mathbf{n}, \tag{2.145}$$

where $\hat{M}_P$ and $\hat{M}_Q$ are defined as follows (see equation (2.131)):

$$\hat{M}_{P,i,i} = \hat{M}_{P,i,i+1} = -u_i^+$$
$$\hat{M}_{Q,i,i} = \hat{M}_{Q,i,i+1} = \frac{1}{2}. \tag{2.146}$$

All the matrices are bidiagonal (except for $\hat{L}_b$ which also has a first row different from zero). In the proposed numerical scheme, matrices $\hat{A}$ and $\hat{B}$ are not explicitly calculated and vectors $\mathbf{a}$ and $\mathbf{b}$ are obtained directly following equations (2.141)–(2.148). The boundary conditions represented by equations (2.137) and (2.138) are introduced in the numerical procedure requiring that

$$\hat{M}_{P,N_e,N_e} = \hat{M}_{Q,N_e,N_e} = 0$$
$$\hat{L}_{Q,l,1} = \hat{L}_{a,l,1} = \hat{L}_{b,l,1} = 0 \qquad l = 1, \dots, N_e. \tag{2.147}$$

To calculate the b_i we have considered the relation (see equation (2.135)) that imposes explicitly the energy conservation. The resulting equations consider the transpose of the matrices already defined (see equations (2.142)–(2.144) and (2.146))

$$\mathbf{b} = -\frac{\alpha}{2\Delta\varepsilon}\left[\hat{L}_a^+\mathbf{x}_a + \hat{L}_b^+\mathbf{x}_b + \hat{L}_Q^+\mathbf{x}_Q\right] = \hat{B}\mathbf{n}, \tag{2.148}$$

where vectors $\mathbf{x}_a$, $\mathbf{x}_b$, and $\mathbf{x}_Q$ are obtained solving the following linear systems:

$$\hat{A}_a^+\mathbf{x}_a = \hat{M}_P^+\mathbf{n}$$
$$\hat{A}_b^+\mathbf{x}_b = \hat{M}_P^+\mathbf{n} \tag{2.149}$$
$$\hat{A}_Q^+\mathbf{x}_Q = \hat{M}_Q^+\mathbf{n}.$$

By using equations (2.141) and (2.145), the matrices $\hat{A}$ and $\hat{B}$ can be formally written as a function of the sparse matrices

$$\hat{A} = -\frac{\alpha}{2\Delta\varepsilon}\left[\hat{M}_P\left(\hat{A}_a^{-1}\hat{L}_a + \hat{A}_b^{-1}\hat{L}_b\right) + \hat{M}_Q\hat{A}_Q^{-1}\hat{L}_Q\right] \tag{2.150}$$

$$\hat{B} = \hat{A}^+ = -\frac{\alpha}{2\Delta\varepsilon}\left[\left(\hat{L}_a^+\left(\hat{A}_a^{-1}\right)^+ + \hat{L}_b^+\left(\hat{A}_b^{-1}\right)^+\right)\hat{M}_P^+ + \hat{L}_Q^+\left(\hat{A}_Q^{-1}\right)^+\hat{M}_Q^+\right]. \tag{2.151}$$

There is another possible approach to calculate the terms b_i starting from equation (2.130). In this way, the vectors $\mathbf{x}$ in equation (2.149) are not calculated, with $\mathbf{b}$ being directly related to $\mathbf{P}_a$, $\mathbf{P}_b$, and $\mathbf{Q}$. One new matrix, $\hat{M}_P^\star$ must be defined to be used in an equation for $\mathbf{b}$ analogous to equation (2.145). This results in a more efficient algorithm, reducing the time needed to solve the linear sparse system in equation (2.149). However, the algorithm based on equation (2.149), imposing explicitly the relation of energy conservation in equation (2.135), needs many fewer iterations to converge, and therefore it is globally more efficient.

Numerical tests have been realized in order to compare the performance of the proposed numerical scheme to the method developed by Rockwood [45]. Four procedures have been verified, which are different in the algorithm used to calculate the electron–electron collision matrices and in performing internal iteration:

RCK0 The matrices $\hat{A}$ and $\hat{B}$ are explicitly calculated following the Rockwood algorithm. No internal iteration is performed.

SPR0 The coefficients a_k and b_k are calculated following equations (2.141)–(2.149). No internal iteration is performed.

RCK1 The same as **RCK0** but with internal iterations up to convergence in equation (2.123).

SPR1 The same as **SPR0** but with internal iterations following the procedure in equation (2.123).

When only electron–electron collisions are considered, the calculations are performed by varying the time step dt (10^{-15}–10^{-11} s), the tolerance ϵ_{max} for the energy conservation (10^{-10}–10^{-3}), and the number of energy groups N_ε (200–2000). The energy domain ($\varepsilon_{max} = N_\varepsilon \Delta\varepsilon = 20$ eV) is the same for all the test cases while the initial is a step functions with a variable width ε_Δ as

$$n(\varepsilon, t = 0) = \begin{cases} \dfrac{n_e}{\varepsilon_\Delta} & \text{if} \quad \varepsilon \leqslant \varepsilon_\Delta \\ 0 & \text{if} \quad \varepsilon > \varepsilon_\Delta, \end{cases} \tag{2.152}$$

where $K_\Delta = \frac{\varepsilon_\Delta}{N_\varepsilon \Delta\varepsilon}$ assumes one of the values 0.1, 0.25, 0.5 0.75, 0.9. Different initial EEDFs correspond to different mean electron energies related to ε_Δ by

$$\bar{\varepsilon}_e = \frac{1}{n_e} \sum_{i=1}^{N_\varepsilon} n_i \varepsilon_i \Delta\varepsilon = \frac{1}{n_e} \sum_{i=1}^{K_\Delta} \frac{n_e}{\varepsilon_\Delta} \Delta\varepsilon (i - 0.5) \Delta\varepsilon = \frac{\varepsilon_{max} K_\Delta}{2} \tag{2.153}$$

When only electron–electron collisions are considered, the stationary solution is a Maxwellian distribution function whose temperature is related to the initial mean energy in equation (2.153). Simulations have been performed in order to verify that the numerical scheme proposed is much more efficient than standard numerical techniques and the energy conservation in electron–electron collisions can be fulfiled numerically under the required tolerance. To evaluate the efficiency of the sparse matrix algorithm with respect to Rockwood we report the ratio

$$\tau_{a,b} = \frac{\text{CPU time for } \mathbf{RCKa}}{\text{CPU time for } \mathbf{SPRb}}. \tag{2.154}$$

Figure 2.4 shows the ratio $\tau_{0,0}$ of **RCK0** and **SPR0** algorithms as a function of the number of energy groups N_ε. The ratio τ has been evaluated when the EEDF reaches the stationary (Maxwellian) distribution. The simulations, performed with

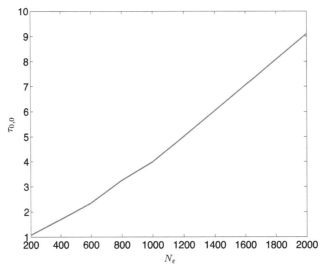

Figure 2.4. Ratio between the CPU times of the **RCK0** and **SPR0** algorithms as a function of the number of energy groups N_ε for $dt = 10^{-12}$ s and $K_\Delta = 0.75$.

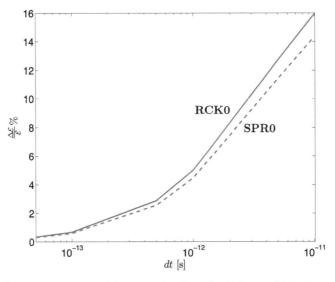

Figure 2.5. Relative percentage errors of the energy density of the stationary distributions calculated by the **RCK0** and **SPR0** algorithms as a function of the time step, for $N_\varepsilon = 200$ and $K_\Delta = 0.1$.

$dt = 10^{-12}$ s and $K_\Delta = 0.75$, show that **SPR0** efficiency grows almost linearly with the mesh size, as expected, because the time needed to calculate the $\hat{T}$ matrix with **RCK0** is proportional to N_ε^2 while with the sparse matrix algorithm it is proportional to N_ε.

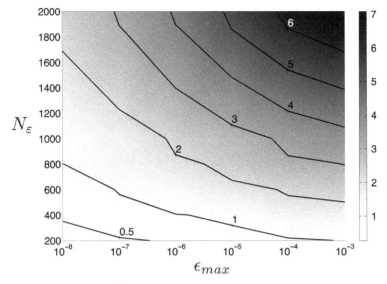

Figure 2.6. Ratio $\tau_{0,1}$ with $dt = 10^{-12}$ s and $K_\Delta = 0.25$. The ratio is evaluated when stationary solutions are reached after $3 \cdot 10^{-10}$ s.

In figure 2.5 the relative percentage errors of the electron mean energy for the stationary solutions the **RCK0** and **SPR0** algorithms are represented for different time steps with $N_\varepsilon = 200$ and $K_\Delta = 0.1$. The relative errors can reach up to 16% for higher time steps. Even if the stationary solutions are Maxwellian, as expected, the calculated electron temperatures do not respect energy conservation. The **SPR0** algorithm gives a smaller error than **RCK0**, a hint that the former should manifest slightly better convergence properties. The difference is so small that we can safely assume that $\tau_{1,1} \approx \tau_{0,0}$.

The sparse matrix approach in evaluating the electron–electron contribution, when energy density conservation in electron–electron collision is required (**SPR1**), can be even faster than the traditional Rockwood algorithm (**RCK0**), as can be observed in figure 2.6. The CPU time needed by the **RCK0** algorithm to reach the stationary Maxwellian distribution can be up to six times higher than **SPR1** in the case of large matrices and small tolerance. The sparse matrix procedure has also been tested [54] in a self-consistent code (see chapter 8) for a helium plasma. The new algorithm has proved to be very efficient and accurate, reducing the computational time dedicated to the electron–electron calculations to a small percentage of the total CPU time.

2.8 Appendix: angle integrals

We will demonstrate that a function of the type

$$\Phi(\boldsymbol{\varphi}) = \begin{cases} \varphi_i^{\,p} & \text{type (a)} \\ \varphi_i^{\,p}\varphi_{j\neq i} & \text{type (b)} \end{cases} \tag{2.155}$$

can be used in the derivation of the two-term Boltzmann equation, demonstrating the expressions in equations (2.5) and (2.6).

The three director cosines are not independent and, to be rigorous, they must be expressed in terms of polar coordinates (ϑ, φ) as

$$\begin{aligned} \varphi_x &= \sin \vartheta \cos \varphi \\ \varphi_y &= \sin \vartheta \sin \varphi \\ \varphi_z &= \cos \vartheta, \end{aligned} \qquad (2.156)$$

giving for the general expression of the integrals in angle coordinates (see equation (2.5))

$$\iiint_{-\infty}^{\infty} F(\mathbf{v})\mathrm{d}^3v = \frac{\pi}{2} \int_0^{\infty} v^2 \mathrm{d}v F'(\mathbf{v}) \int_{-1}^{1} \int_0^{2\pi} \Phi(\boldsymbol{\varphi})\mathrm{d}\cos\vartheta \; \mathrm{d}\varphi. \qquad (2.157)$$

Logically, the value of the integral is obviously independent of which φ_i is considered, because the choice of the reference axis z is arbitrary. Therefore, for the sake of simplicity, we will demonstrate only the case for φ_z, leaving to the reader the verification in the other components.

Type (a)

$$\int_{-1}^{1} \int_0^{2\pi} \varphi_z^{a-1} \mathrm{d}\cos\vartheta \; \mathrm{d}\varphi \; \underline{\underline{\varphi_z = \cos\vartheta}} \int_{-1}^{1} \varphi_z^{a-1} \mathrm{d}\varphi_z \int_0^{2\pi} \mathrm{d}\varphi \\ = 2\pi \frac{1^a - (-1)^a}{a} = \frac{2\pi}{a} a|2. \qquad (2.158)$$

Type (b)

$$\int_{-1}^{1} \int_0^{2\pi} \varphi_z^a \varphi_{x/y} \mathrm{d}\cos\vartheta \; \mathrm{d}\varphi \; \underline{\underline{\text{equation (2.156)}}} \int_0^{\pi} \cos^a \vartheta \sin^2 \vartheta \; \mathrm{d}\vartheta \\ \times \int_0^{2\pi} f_{x/y}(\varphi)\mathrm{d}\varphi = 0, \qquad (2.159)$$

being the function $f_{x/y}(\varphi)$ equal to $\cos\varphi$ or $\sin\varphi$ (see equation (2.156)) whose integral on a period is null.

Bibliography

[1] Capriati G, Colonna G, Gorse C and Capitelli M 1992 A parametric study of electron energy distribution functions and rate and transport coefficients in nonequilibrium helium plasmas *Plasma Chem. Plasma Process.* **12** 237–60
[2] Yachi S, Date H, Kitamori K and Tagashira H 1991 A multi-term Boltzmann equation analysis of electron swarms in gases-the time-of-flight parameters *J. Phys. D: Appl. Phys.* **24** 573

[3] Yachi S, Kitamura Y, Kitamori K and Tagashira H 1988 A multi-term Boltzmann equation analysis of electron swarms in gases *J. Phys. D: Appl. Phys.* **21** 914

[4] Ness K 1994 Multi-term solution of the Boltzmann equation for electron swarms in crossed electric and magnetic fields *J. Phys. D: Appl. Phys.* **27** 1848

[5] Maeda K and Makabe T 1994 Radiofrequency electron swarm transport in reactive gases and plasmas *Phys. Scri.* **T53** 61

[6] Loffhagen D, Winkler R and Braglia G 1996 Two-term and multi-term approximation of the nonstationary electron velocity distribution in an electric field in a gas *Plasma Chem. Plasma Process.* **16** 287–300

[7] Leyh H, Loffhagen D and Winkler R 1998 A new multi-term solution technique for the electron Boltzmann equation of weakly ionized steady-state plasmas *Comput. Phys. Commun.* **113** 33–48

[8] Braglia G, Wilhelm J and Winkler E 1985 Multi-term solutions of Boltzmann's equation for electrons in the real gases Ar, CH_4 and CO_2 *Lett. Nuovo Cimento* **44** 365–78

[9] Dujko S, White R, Petrović L and Robson R 2011 A multi-term solution of the nonconservative Boltzmann equation for the analysis of temporal and spatial non-local effects in charged-particle swarms in electric and magnetic fields *Plasma Sources Sci. Technol.* **20** 024013

[10] Colonna G and Capitelli M 1996 Electron and vibrational kinetics in the boundary layer of hypersonic flow *J. Thermophys. Heat Transfer* **10** 406–12

[11] Colonna G, Capitelli M, DeBenedictis S, Gorse C and Paniccia F 1991 Electron energy distribution functions in CO_2, laser mixture: the effects of second kind collisions from metastable electronic states *Contrib. Plasma Phys.* **31** 575–9

[12] Capitelli M, Colonna G, D'Ammando G, Laporta V and Laricchiuta A 2013 The role of electron scattering with vibrationally excited nitrogen molecules on non-equilibrium plasma kinetics *Phys. Plasmas* **20** 101609

[13] Capitelli M, Colonna G, D'Ammando G, Laporta V and Laricchiuta A 2014 Nonequilibrium dissociation mechanisms in low temperature nitrogen and carbon monoxide plasmas *Chem. Phys.* **438** 31–6

[14] Colonna G, D'Ammando G, Pietanza L and Capitelli M 2014 Excited-state kinetics and radiation transport in low-temperature plasmas *Plasma Phys. Control. Fusion* **57** 014009

[15] Pietanza L, Colonna G, D'Ammando G, Laricchiuta A and Capitelli M 2015 Vibrational excitation and dissociation mechanisms of CO_2 under non-equilibrium discharge and post-discharge conditions *Plasma Sources Sci. Technol.* **24** 042002

[16] Pietanza L D, Colonna G, Laporta V, Celiberto R, D'Ammando G, Laricchiuta A and Capitelli M 2016 Influence of electron molecule resonant vibrational collisions over the symmetric mode and direct excitation-dissociation cross sections of CO_2 on the electron energy distribution function and dissociation mechanisms in cold pure CO_2 plasmas *J. Phys. Chem. A* **120** 2614–28

[17] Colonna G, Pietanza L and d'Ammando G 2012 Self-consistent collisional-radiative model for hydrogen atoms: atom–atom interaction and radiation transport *Chem. Phys.* **398** 37–45

[18] Capitelli M, Celiberto R, Colonna G, Esposito F, Gorse C, Hassouni K, Laricchiuta A and Longo S 2015 *Fundamental Aspects of Plasma Chemical Physics: Kinetics* vol 85 (New York: Springer)

[19] Colonna G, Laporta V, Celiberto R, Capitelli M and Tennyson J 2015 Non-equilibrium vibrational and electron energy distributions functions in atmospheric nitrogen ns pulsed

discharges and μs post-discharges: the role of electron molecule vibrational excitation scaling-laws *Plasma Sources Sci. Technol.* **24** 035004

[20] Colonna G, Gorse C, Capitelli M, Winkler R and Wilhelm J 1993 The influence of electron-electron collisions on electron energy distribution functions in N_2 post discharge *Chem. Phys. Lett.* **213** 5–9

[21] Capitelli M, Colonna G, Gicquel A, Gorse C, Hassouni K and Longo S 1996 Maxwell and non-Maxwell behavior of electron energy distribution function under expanding plasma jet conditions: the role of electron–electron, electron–ion, and superelastic electronic collisions under stationary and time-dependent conditions *Phys. Rev.* E **54** 1843

[22] Pietanza L, Colonna G, D'Ammando G, Laricchiuta A and Capitelli M 2016 Electron energy distribution functions and fractional power transfer in 'cold' and excited CO_2 discharge and post discharge conditions *Phys. Plasmas* **23** 013515

[23] Kortshagen U, Parker G and Lawler J 1996 Comparison of Monte Carlo simulations and nonlocal calculations of the electron distribution function in a positive column plasma *Phys. Rev.* E **54** 6746

[24] Kortshagen U 1993 A non-local kinetic model applied to microwave produced plasmas in cylindrical geometry *J. Phys. D: Appl. Phys.* **26** 1691

[25] Busch C and Kortshagen U 1995 Numerical solution of the spatially inhomogeneous Boltzmann equation and verification of the nonlocal approach for an argon plasma *Phys. Rev.* E **51** 280

[26] Uhrlandt D and Winkler R 1996 Radially inhomogeneous electron kinetics in the DC column plasma *J. Phys. D: Appl. Phys.* **29** 115

[27] Tsendin L 1995 Electron kinetics in non-uniform glow discharge plasmas *Plasma Sources Sci. Technol.* **4** 200

[28] Colonna G and Capitelli M 2008 Boltzmann and master equations for MHD in weakly ionized gases *J. Thermophys. Heat Transfer* **22** 414–23

[29] Colonna G and Capitelli M 2003 The effects of electric and magnetic fields on high enthalpy plasma flows *AIAA paper* 2003–4036

[30] Frost L and Phelps A 1964 Momentum-transfer cross sections for slow electrons in He, Ar, Kr, and Xe from transport coefficients *Phys. Rev.* A **136** 1538

[31] Frost L and Phelps A 1962 Rotational excitation and momentum transfer cross sections for electrons in H_2 and N_2 from transport coefficients *Phys. Rev.* **127** 1621

[32] Engelhardt A and Phelps A 1963 Elastic and inelastic collision cross sections in hydrogen and deuterium from transport coefficients *Phys. Rev.* **131** 2115

[33] Lowke J J, Phelps A V and Irwin B W 1973 Predicted electron transport coefficients and operating characteristics of CO_2-N_2-He laser mixtures *J. Appl. Phys.* **44** 4664–71

[34] Pitchford L and Phelps A 1982 Comparative calculations of electron-swarm properties in N_2 at moderate E/N values *Phys. Rev.* A **25** 540

[35] Phelps A V and Pitchford L C 1985 Anisotropic scattering of electrons by N_2 and its effects on electron transport: tabulation of cross sections and results *JILA Report* vol 23 University of Colorado, Boulder

[36] Phelps A 1991 Cross sections and swarm coefficients for nitrogen ions and neutrals in N_2 and argon ions and neutrals in Ar for energies from 0.1 eV to 10 keV *J. Phys. Chem. Ref. Data* **20** 557–73

[37] Morgan W L 1993 Test of a numerical optimization algorithm for obtaining cross sections for multiple collision processes from electron swarm data *J. Phys. D: Appl. Phys.* **26** 209

[38] Morgan W 1992 A critical evaluation of low-energy electron impact cross sections for plasma processing modeling. I: Cl_2, F_2, and HCl *Plasma Chem. Plasma Process.* **12** 449–76

[39] Morgan W L 1991 Use of numerical optimization algorithms to obtain cross sections from electron swarm data *Phys. Rev. A* **44** 1677

[40] Petrović Z L, Dujko S, Marić D, Malović G, Nikitović Ž, Šaši O, Jovanović J, Stojanović V and Radmilović-Radenović M 2009 Measurement and interpretation of swarm parameters and their application in plasma modelling *J. Phys. D: Appl. Phys.* **42** 194002

[41] Hagelaar G and Pitchford L 2005 Solving the Boltzmann equation to obtain electron transport coefficients and rate coefficients for fluid models *Plasma Sources Sci. Technol.* **14** 722

[42] LXcat 2016 Plasma Data Exchange Project database http://fr.lxcat.net/home

[43] Mitchner M and Kruger C H J 1973 *Partially Ionized Gases* (New York: Wiley)

[44] Golant V E, Zilinskij A P and Sacharov I E 1980 *Fundamentals of Plasma Physics* (New York: Wiley)

[45] Rockwood S D 1973 Elastic and inelastic cross sections for electron-Hg scattering from Hg transport data *Phys. Rev. A* **8** 2348–60

[46] Celiberto R *et al* 2016 Atomic and molecular data for spacecraft re-entry plasmas *Plasma Sources Sci. Technol.* **25** 033004

[47] Gianturco F A and Huo W M (ed) 1995 *Computational Methods for Electron-Molecule Collisions (Springer Series on Atomic, Optical, and Plasma Physics)* (New York: Springer)

[48] Landau L D and Lifschitz E 1981 *Physical Kinetics* (Oxford: Pergamon)

[49] Starikovskiy A and Aleksandrov N 2015 Plasma assisted combustion mechanism for hydrogen and small hydrocarbons *AIAA Paper* 2015–0158

[50] Elliot C J and Greene A E 1976 Electron energy distribution in e-beam generated Xe and Ar plasmas *J. Appl. Phys.* **47** 2946–53

[51] Laporta V, Celiberto R and Tennyson J 2013 Resonant vibrational-excitation cross sections and rate constants for low-energy electron scattering by molecular oxygen *Plasma Sources Sci. Technol.* **22** 025001

[52] Colonna G 1998 Step adaptive method for vibrational kinetics and other initial value problems *Rend. Circ. Mat. Palermo.* Suppl **57** 159–63

[53] Kaps P and Rentrop P 1979 Generalized Runge–Kutta methods of order four with stepsize control for stiff ordinary differential equations *Numer. Math.* **33** 55–68

[54] D'Angola A, Coppa G, Capitelli M, Gorse C and Colonna G 2010 An efficient energy-conserving numerical model for the electron energy distribution function in the presence of electron-electron collisions *Comput. Phys. Commun.* **181** 1204–11

Chapter 3

Multi-term and non-local electron Boltzmann equation

3.1 Introduction

The electron component plays an essential role in the physical understanding of the temporal and spatial behavior of weakly ionized plasmas, because the electrons supply, energetically, the other species of the plasmas. The power supply to the plasma is usually provided by an external electric field which acts on the charged particles and thus maintains the plasma. The power coupled into the plasma is subsequently transferred to the heavy particles by elastic and inelastic electron collision processes. These collision processes lead to considerable changes of the concentrations of the various heavy particle components of the plasma and to the generation of different kinds of excited and charged particles.

Because the mass of the electrons is small compared to that of the heavy particles, the energy exchange between electrons and heavy particles due to elastic collisions is much smaller than the energy exchange between heavy particles with approximately equal masses. This results in a weak energetic contact of the electrons with the heavy particle components of the plasma. Acted upon by the electric field, the electrons are consequently able to attain values of mean kinetic energy which are much larger than the mean energies of the neutral particles and ions of the plasma so that a non-thermal plasma exists.

As a result of the interaction of power gain and power loss through collisions, the electrons are generally not in thermodynamic equilibrium with each other and with the other components of the plasma. This non-equilibrium behavior of the electron component in non-thermal plasmas cannot be described by the known methods of equilibrium thermodynamics, but requires an appropriate microphysical treatment to determine the established non-equilibrium state of the electrons, or to describe their temporal and spatial development in non-thermal plasmas [1, 2].

In order to perform microphysical investigations of the behavior of the electrons, two quite different approaches with specific advantages and disadvantages are available: the solution of the Boltzmann equation of the electrons and particle simulation techniques [3]. The Boltzmann equation method consists of the formulation of an appropriate kinetic equation of the electrons and its solution. Here, the approach for solving the resulting kinetic equation is quite different for investigations of local and non-local plasmas, where the non-locality can take place in time and space. Despite the larger mathematical and numerical effort, the strict kinetic treatment of the electrons using efficient numerical solution techniques has become possible for steady-state, time-dependent, spatially one- and two-dimensional, as well as time- and space-dependent plasma conditions [4–6].

In contrast to the solution of the electron Boltzmann equation, particle simulation methods are characterized by their simpler applicability to more complex geometries. However, they are generally more time-consuming, because a larger number of electrons has to be treated in any simulation to obtain sufficient accuracy for the determination of their distribution function in the entire range of kinetic energies and, in particular, at higher energies. The treatment of the kinetic equation of the electrons usually does not have such limitation and is generally characterized by less expenditure of computing time.

The present contribution focuses on methods for solving the electron Boltzmann equation, where special focus is on techniques to determine the electron velocity distribution function (EVDF) and related macroscopic properties in higher-order accuracy. Such a so-called multi-term treatment includes in its lowest order the solution of the electron kinetic equation in the well-known two-term approximation of the EVDF [4–7], which is represented in chapter 2.

Note that the solution of the electron Boltzmann equation has not only been applied to analyze specific problems related to the kinetic behavior of the electrons. It has also been integrated into a self-consistent description of different non-thermal plasmas by means of hybrid modeling [8–12], where the two-term approximation of the EVDF was used, at most.

3.2 Basic relations

The kinetic description of the electron component starts from the velocity distribution function $f_e(\mathbf{r}, \mathbf{v}, t)$ of the electrons, where $dn_e(\mathbf{r}, t) = f_e(\mathbf{r}, \mathbf{v}, t)d\mathbf{v}$ represents the number of electrons in the velocity space interval $[\mathbf{v}, \mathbf{v} + d\mathbf{v}]$ with $d\mathbf{v} \equiv d^3v$ at the coordinate space position $\mathbf{r}$ and at time t. If this velocity distribution $f_e(\mathbf{r}, \mathbf{v}, t)$ is known, all relevant macroscopic properties of the electrons, such as the electron density $n_e(\mathbf{r}, t)$, their particle flux $\mathbf{j}_e(\mathbf{r}, t)$, and the mean electron energy density $u_e(\mathbf{r}, t)$, can be determined by appropriate averaging of the EVDF over the velocity space. These three macroscopic properties mentioned are given by the averages [13–15]

$$n_e(\mathbf{r}, t) = \int f_e(\mathbf{r}, \mathbf{v}, t)d\mathbf{v}, \tag{3.1}$$

$$\mathbf{j}_e(\mathbf{r},\, t) = \int \mathbf{v} f_e(\mathbf{r},\, \mathbf{v},\, t) d\mathbf{v}, \tag{3.2}$$

$$u_e(\mathbf{r},\, t) = \int \frac{1}{2} m_e v^2 f_e(\mathbf{r},\, \mathbf{v},\, t) d\mathbf{v}, \tag{3.3}$$

where m_e denotes the electron mass and $v = |\mathbf{v}|$.

3.2.1 Boltzmann equation of the electrons

The electrons in weakly ionized plasmas are subject to different interactions such as the action of non-collisional forces and different binary collision processes. These interactions cause the electrons to be redistributed in the six-dimensional phase space consisting of the configuration and velocity spaces. In order to describe this complex redistribution of electrons, an appropriate balance equation for the velocity distribution function $f_e(\mathbf{r},\, \mathbf{v},\, t)$ has to be formulated and solved. According to the concept of binary electron collisions with short-range interaction, this balance equation is the Boltzmann equation [13–15]

$$\frac{\partial f_e}{\partial t} + \mathbf{v} \cdot \nabla_{\mathbf{r}} f_e + \mathbf{a} \cdot \nabla_{\mathbf{v}} f_e = \sum_i \left(C_i^{\mathrm{el}}(f_e) + \sum_p C_{i,p}^{\mathrm{in}}(f_e) \right). \tag{3.4}$$

Here $\mathbf{a}$ represents the acceleration due to non-collisional forces such as electro-magnetic and gravitational forces. Focusing on the action of an electric field $\mathbf{E}$ in the following, the acceleration is given by

$$\mathbf{a} = -\frac{e_0}{m_e} \mathbf{E}, \tag{3.5}$$

where $-e_0$ denotes the charge of the electron. The right-hand side of equation (3.4) describes the change of the EVDF due to binary collisions of electrons with the neutral heavy particle components of the plasma, i.e. atoms and molecules in their ground state as well as in excited states. $C_i^{\mathrm{el}}(f_e)$ and $C_{i,p}^{\mathrm{in}}(f_e)$ are the collision integrals for elastic (el) and various inelastic (in) collisions of electrons with the heavy particle component i. In addition to elastic collisions, the kinetic treatment of the electrons has to include all inelastic electron collision processes, which are important for the plasma under consideration, such as electron impact excitation and de-excitation, dissociation, ionization, and attachment.

In the derivation of the kinetic equation (3.4), collisions of electrons with heavy particle species i are assumed to be instantaneous events taking place at a certain point in space and not affected by any external forces. Thus, the properties of the colliding particles before and after the collision event can be expressed in terms of the momentum and energy conservation laws without considering the specific type of interaction forces. In the case of binary electron collisions in which the nature of the heavy particle species i with mass M_i remains unchanged, the conservation of momentum and energy in each binary collision is given by the relations [14, 15]

$$m_e\mathbf{v}_e + M_i\mathbf{V}_i = m_e\mathbf{v}'_e + M_i\mathbf{V}'_i, \tag{3.6}$$

$$\frac{m_e v_e^2}{2} + \frac{M_i V_i^2}{2} = \frac{m_e v_e'^2}{2} + \frac{M_i V_i'^2}{2} + \Delta E_{\text{int}}, \tag{3.7}$$

where $\mathbf{v}_e$ and $\mathbf{v}'_e$ are the velocity vectors of the electron before and after the collision event, $\mathbf{V}_i$ and $\mathbf{V}'_i$ denote the corresponding velocity vectors of the heavy particle, v_e, v'_e, V_i, and V'_i are the respective absolute values of the velocities, and $\Delta E_{\text{int}} = E'_{\text{int}} - E_{\text{int}}$ is the net change in the internal energy of the heavy particle as a result of the collision process. For elastic collisions, ΔE_{int} is obviously equal to zero because they do not influence the internal state of the heavy particle. Among the various inelastic collision processes, collisions of the first kind with $\Delta E_{\text{int}} > 0$, such as exciting collisions, are distinguished from collisions of the second kind with $\Delta E_{\text{int}} < 0$. Collisions causing atoms or molecules to shift from an excited state to the ground state are an example of the latter.

The change of momentum and kinetic energy of the electrons in binary collisions with heavy particles can be deduced from the conservation equations (3.6) and (3.7). The momentum change [15]

$$m_e\mathbf{v}'_e - m_e\mathbf{v}_e = -\frac{m_e M_i}{m_e + M_i}(1 - \cos\theta)(\mathbf{v}_e - \mathbf{V}_i) \approx -m_e\mathbf{v}_e(1 - \cos\theta) \tag{3.8}$$

is proportional to the relative velocity of the colliding particles, where θ is the scattering angle related to the collision. Because the heavy particle velocity is usually much less than the electron velocity and $m_e \ll M_i$, it can be approximated by the right-hand side expression of equation (3.8). The corresponding change of kinetic energy due to elastic collisions is given by [15]

$$\frac{m_e v_e'^2}{2} - \frac{m_e v_e^2}{2} = -\left(\frac{m_e v_e^2}{2} - \frac{M_i V_i^2}{2}\right)2\frac{m_e}{M_i}(1 - \cos\theta), \tag{3.9}$$

and that due to inelastic collisions takes the form

$$\frac{m_e v_e'^2}{2} - \frac{m_e v_e^2}{2} = -\Delta E_{\text{int}}. \tag{3.10}$$

Similar relations can directly be derived for binary electron collisions in which the nature of the heavy particle changes, such as electron impact dissociation, ionization, or electron attachment.

3.2.2 Expansion of the velocity distribution

The electron Boltzmann equation (3.4) is very complex and comprises a wide range of specific problems of the kinetics of the electrons. Because a general solution of this kinetic equation is difficult to find, various solution methods and numerical techniques have been developed and utilized to analyze different conditions of plasmas, such as steady-state conditions, solely time-dependent or space-dependent plasmas, and eventime- and space-dependent problems [4, 5, 7, 16].

In order to find the solution of the kinetic equation, a decomposition of the velocity distribution function in terms of orthogonal functions in the directions $\mathbf{v}/v$ of the velocity $\mathbf{v}$ is generally applied. Spherical harmonics are such functions [14]. Whether such complex expansion with respect to both angular coordinates of $\mathbf{v}/v$ is required or a reduced expansion concerning only one angular coordinate of the velocity can be used, depends on the structure of the electric field, the anticipated inhomogeneity of the plasma under consideration, and the boundary conditions imposed on the velocity distribution [17–21]. Here, it is generally assumed that the electric field and additionally a potential inhomogeneity in the plasma are parallel to one preferred direction in space only, taken to be the z-direction. For aspects related to the multi-term treatment of the electrons in cylindrical and spherical geometries, the reader is referred to, e.g. [19–21].

Because the electric field and the inhomogeneity in the plasma are parallel to the z-axis of a given coordinate system, the EVDF $f_e(\mathbf{r}, \mathbf{v}, t)$ becomes symmetrical around the field $\mathbf{E}(z, t) = E(z, t)\mathbf{e}_z$, with $\mathbf{e}_z$ being the unit vector in the z-direction. Thus, it possesses the reduced dependence $f_e(z, v, v_z/v, t)$, where v_z is the z component of $\mathbf{v}$, and can be expanded with respect to $v_z/v \equiv \cos\theta$ in Legendre polynomials $P_n(v_z/v)$ according to [14, 15]

$$f_e\left(z, v, \frac{v_z}{v}, t\right) = \sum_{n=0}^{\infty} \hat{f}_n(z, v, t) P_n\left(\frac{v_z}{v}\right). \tag{3.11}$$

The usual procedure to obtain the velocity distribution function f_e is to substitute the expansion equation (3.11) into the electron Boltzmann equation (3.4) and to equate the coefficients of the Legendre polynomials. When employing this expansion and using the orthogonality relation of the Legendre polynomials, a chain of equations for the expansion coefficients $\hat{f}_n$ results. In order to obtain a closed system of equations and its approximate solution, the infinite set of equations is truncated. When replacing the velocity magnitude v by the kinetic energy $U = m_e v^2/2$ of the electrons as usual, the hierarchy of partial differential equations [17, 22]

$$\left(\frac{m_e}{2}\right)^{1/2} U^{1/2} \frac{\partial}{\partial t} f_0(z, U, t) + \frac{U}{3} \frac{\partial}{\partial z} f_1(z, U, t) - \frac{e_0}{3} E(z, t) \frac{\partial}{\partial U}\big(U f_1(z, U, t)\big)$$

$$- \frac{\partial}{\partial U} \left(\sum_i 2\frac{m_e}{M_i} U^2 N_i(z, t)\big(Q_i^{\mathrm{el},0}(U) - Q_i^{\mathrm{el},1}(U)\big) f_0(z, U, t)\right)$$

$$+ \sum_i \sum_p U N_i(z, t) Q_{i,p}^{\mathrm{in},0}(U) f_0(z, U, t) \tag{3.12}$$

$$= \sum_i \sum_p \beta_p^2 \tilde{U}_{i,p} N_i(z, t) Q_{i,p}^{\mathrm{in},0}\big(\tilde{U}_{i,p}\big) f_0\big(z, \tilde{U}_{i,p}, t\big), \quad \tilde{U}_{i,p} = \beta_p U + \rho_p U_{i,p}^{\mathrm{in}},$$

$$\left(\frac{m_e}{2}\right)^{1/2} U^{1/2} \frac{\partial}{\partial t} f_n(z, U, t) + \frac{n}{2n-1} U \frac{\partial}{\partial z} f_{n-1}(z, U, t)$$

$$+ \frac{n+1}{2n+3} U \frac{\partial}{\partial z} f_{n+1}(z, U, t)$$

$$- e_0 E(z, t) \frac{n}{2n-1} \left(U \frac{\partial}{\partial U} f_{n-1}(z, U, t) - \frac{n-1}{2} f_{n-1}(z, U, t) \right)$$

$$- e_0 E(z, t) \frac{n+1}{2n+3} \left(U \frac{\partial}{\partial U} f_{n+1}(z, U, t) + \frac{n+2}{2} f_{n+1}(z, U, t) \right) \tag{3.13}$$

$$+ \sum_i U N_i(z, t) \left(\left(Q_i^{\mathrm{el},0}(U) - Q_i^{\mathrm{el},n}(U) \right) + \sum_p Q_{i,p}^{\mathrm{in},0}(U) \right) f_n(z, U, t)$$

$$= \sum_i \sum_p \beta_p^2 \tilde{U}_{i,p} N_i(z, t) Q_{i,p}^{\mathrm{in},n}\left(\tilde{U}_{i,p} \right) f_n\left(z, \tilde{U}_{i,p}, t \right),$$

$$1 \leqslant n \leqslant l-1, \quad l \geqslant 2, \quad f_l(z, U, t) \equiv 0$$

is obtained in the framework of the multi-term approximation with l coefficients. This hierarchy describes the time- and space-dependent behavior of the isotropic ($n = 0$) and the anisotropic ($n = 1, \ldots, l - 1$) components

$$f_n(z, U, t) = 2\pi \left(\frac{2}{m_e}\right)^{3/2} \hat{f}_n(z, v(U), t) \tag{3.14}$$

of the velocity distribution function $f_e(z, v, v_z/v, t)$.

In order to derive the set of equations (3.12) and (3.13), further assumptions regarding the collision integrals for elastic and inelastic collisions of electrons with the heavy particle species i are required [23, 24]. As usual and sufficient, an additional expansion with respect to the ratio m_e/M_i of electron mass to heavy particle mass is applied and only the leading terms regarding m_e/M_i are considered in each equation of the hierarchy (3.12) and (3.13). That is, the terms of zeroth-order in m_e/M_i are taken into account in the collision integrals relevant to inelastic collisions and the first-order terms are included in the collision integral part relevant to elastic collisions in equation (3.12). Furthermore, the heavy gas particles are assumed to be at rest before the collision events.

The terms with the shifted energy arguments $\tilde{U}_{i,p} = \beta_p U + \rho_p U_{i,p}^{\mathrm{in}}$ occurring on the right-hand sides of equations (3.12) and (3.13) describe the in-scattering of electrons at the kinetic energy U, which have carried out the inelastic collision process p with the heavy particle component i at the energy $\beta_p U + \rho_p U_{i,p}^{\mathrm{in}}$ and are scattered to the energy U as a result of the change of kinetic energy $\rho_p U_{i,p}^{\mathrm{in}}$ in this process with positive value $U_{i,p}^{\mathrm{in}}$. Here, $\rho_p = 1$ for exciting, dissociating, and ionizing electron collisions and $\rho_p = -1$ for de-exciting collision processes. Furthermore, $\beta_p = 1$ for the excitation, de-excitation, and the dissociation process. Regarding the ionization process, it has been supposed that the remaining kinetic energy of the colliding electron is equally shared between the two out-coming electrons after each ionizing event, i.e. $\beta_p = 2$ [25]. Different assumptions of the energy sharing in ionizing electron collision processes are discussed, e.g. in [26, 27]. Moreover, electrons that are attached to a molecule in the collision event are completely lost, so there is no in-scattering term in this case, i.e. $\beta_p = \rho_p = 0$ for electron attachment processes.

The substitution of the Legendre polynomial expansion (3.11) into the electron Boltzmann equation (3.4) also involves the expansion of the collision integrals on the right-hand side of equation (3.4). As a result of that expansion, the sequences of generalized total collision cross sections

$$Q_i^{\mathrm{el},n}(U) = \int \sigma_i^{\mathrm{el}}(U, \cos\theta) P_n(\cos\theta) \sin\theta \, d\theta \, d\varphi, \tag{3.15}$$

$$Q_{i,p}^{\mathrm{in},n}(U) = \int \sigma_{i,p}^{\mathrm{in},n}(U, \cos\theta) P_n(\cos\theta) \sin\theta \, d\theta \, d\varphi \tag{3.16}$$

with $n = 0, \ldots, l-1$ arise in the hierarchy equations (3.12) and (3.13). These total collision cross sections result from the integration of the product of the differential collision cross section $\sigma_i^{\mathrm{el}}(U, \cos\theta)$ and $\sigma_{i,p}^{\mathrm{in},n}(U, \cos\theta)$, respectively, and the Legendre polynomial $P_n(\cos\theta)$ over the entire solid scattering angle $\sin\theta \, d\theta \, d\varphi$. The series of equations (3.15) and (3.16) describe the impact of anisotropic electron–heavy-particle scattering due to elastic and the various inelastic collision processes, respectively. In particular, the difference

$$Q_i^d(U) = Q_i^{\mathrm{el},0}(U) - Q_i^{\mathrm{el},1}(U) \tag{3.17}$$

occurring in equation (3.12) and in equation (3.13) for $n = 1$ represents the well-known momentum transfer cross section for elastic collisions. Because the differential collision cross-section data available is quite limited, isotropic scattering is usually assumed for the inelastic collision processes. In that case, the relations $Q_{i,p}^{\mathrm{in},n}(U) = 0$ for $n \geqslant 1$ are found from equation (3.16) due to the orthogonality of the Legendre polynomials, and the associated terms on the right-hand side of equation (3.13) are zero as well. With regard to the elastic collision processes, data for the total (or integral) collision cross section $Q_i^{\mathrm{el},0}$ for elastic collisions as well as for the elastic momentum transfer cross section Q_i^d (3.17) are known for a number of atoms and molecules. Thus, anisotropic scattering in elastic collisions is taken into consideration in the framework of the two-term approximation ($l = 2$) of the hierarchy equations (3.12) and (3.13) when using the data for Q_i^d. In the multi-term approximation the corresponding differences $Q_i^{\mathrm{el},0}(U) - Q_i^{\mathrm{el},n}(U)$ with $2 \leqslant n \leqslant l-1$ occurring in equation (3.13) are often approximated by the cross section Q_i^d for elastic momentum transfer.

The set (3.12) and (3.13) of partial differential equations forms the basis for the studies and numerical methods discussed below.

3.2.3 Macroscopic balances

The equation system (3.12) and (3.13) provides a microscopic description of the behavior of the electrons. The consistent macroscopic balance equations of the electrons are obtained by appropriate integration of the hierarchy equations (3.12) and (3.13) with $n = 1$ with respect to the kinetic energy U of the electrons. The resulting particle, momentum, and power balance equations have the form

$$\frac{\partial}{\partial t}n_e(z, t) + \frac{\partial}{\partial z}j_z(z, t) = \sum_i \sum_p n_e(z, t)v_{i,p}^{io}(z, t) - \sum_i \sum_p n_e(z, t)v_{i,p}^{at}(z, t), \quad (3.18)$$

$$\frac{\partial}{\partial t}j_z(z, t) + \frac{\partial}{\partial z}\Pi_{zz}(z, t) = -\frac{e_0}{m_e}n_e(z, t)E(z, t) - \sum_i I_i^{el}(z, t) - \sum_{i,p} I_{i,p}^{in}(z, t), \quad (3.19)$$

$$\frac{\partial}{\partial t}u_e(z, t) + \frac{\partial}{\partial z}Q_z(z, t) = -e_0 j_z(z, t)E(z, t) - \sum_i P_i^{el}(z, t) - \sum_{i,p} P_{i,p}^{in}(z, t). \quad (3.20)$$

The particle density $n_e(z, t)$, the particle flux $\mathbf{j}_e(z, t) = j_z(z, t)\mathbf{e}_z$, the zz-component of the momentum flux tensor $\Pi_{zz}(z, t)$, the mean energy density $u_e(z, t)$, and the energy flux $\mathbf{Q}_e(z, t) = Q_z(z, t)\mathbf{e}_z$ of the electrons are given by the integral expressions

$$n_e(z, t) = \int_0^\infty U^{1/2}f_0(z, U, t)\mathrm{d}U, \quad (3.21)$$

$$j_z(z, t) = \frac{1}{3}\left(\frac{2}{m_e}\right)^{1/2}\int_0^\infty Uf_1(z, U, t)\mathrm{d}U, \quad (3.22)$$

$$\Pi_{zz}(z, t) = \frac{2}{3m_e}\int_0^\infty U^{3/2}\left(f_0(z, U, t) + \frac{2}{5}f_2(z, U, t)\right)\mathrm{d}U, \quad (3.23)$$

$$u_e(z, t) = \int_0^\infty U^{3/2}f_0(z, U, t)\mathrm{d}U, \quad (3.24)$$

$$Q_z(z, t) = \frac{1}{3}\left(\frac{2}{m_e}\right)^{1/2}\int_0^\infty U^2 f_1(z, U, t)\mathrm{d}U. \quad (3.25)$$

Note that the particle and energy fluxes $\mathbf{j}_e$ and $\mathbf{Q}_e$ have only a z-component because the velocity distribution function f_e is rotationally symmetric around the direction of $\mathbf{e}_z$ owing to the assumption that the electric field action and the possible inhomogeneity in the plasma take place parallel to this direction only.

The right-hand side terms of equations (3.18), (3.19), and (3.20) describe the gain and loss of electrons, momentum, and power due to collision processes and the action of the electric field. According to the particle balance equation (3.18), the temporal variation of the electron density $n_e(z, t)$ at each z-position results from the net particle flux into or out of the respective plasma volume, represented by the divergence of the electron particle flux $j_z(z, t)$, as well as from the electron gain due to ionization and electron loss due to attachment. The corresponding mean collision frequencies $\nu_{i,p}^{in}(z, t)$ for individual inelastic collision processes such as ionization (io) or attachment (at) are given by the representation

$$\nu_{i,p}^{in}(z, t) = \frac{1}{n_e(z, t)}\left(\frac{2}{m_e}\right)^{1/2} \int_0^\infty U N_i(z, t) Q_{i,p}^{in,0}(U) f_0(z, U, t) dU. \qquad (3.26)$$

The momentum balance equation (3.19) states that the temporal change of the particle flux $j_z(z, t)$ is determined at each position by the net momentum flux into or out of the plasma volume considered as described by the spatial derivative of $\Pi_{zz}(z, t)$, the momentum gain from the electric field $I^f(z, t) = -e_0 n_e(z, t) E(z, t)/m_e$, and the momentum dissipation due to elastic and inelastic collisions, where

$$I_i^{el}(z, t) = \frac{2}{3m_e} \int_0^\infty U^{3/2} N_i(z, t) Q_i^d(U) f_1(z, U, t) dU, \qquad (3.27)$$

$$I_{i,p}^{in}(z, t) = \frac{2}{3m_e} \int_0^\infty U\Big[U^{1/2} N_i(z, t) Q_{i,p}^{in,0}(U) \\ - \beta_p^{1/2}\left(U - \rho_p U_{i,p}^{in}\right)^{1/2} N_i(z, t) Q_{i,p}^{in,1}(U)\Big] f_1(z, U, t) dU. \qquad (3.28)$$

The power balance equation (3.20) has a similar structure to the momentum balance. It shows that the temporal change of the mean energy density $u_e(z, t)$ is caused at the position z by the net power transfer in the macroscopic electron transport, as given by the divergence of the electron energy flux $Q_z(z, t)$, the power gain from the electric field $P^f(z, t) = -e_0 j_z(z, t) E(z, t)$, and the power loss by elastic and inelastic collision processes, where

$$P_i^{el}(z, t) = 2\frac{m_e}{M_i}\left(\frac{2}{m_e}\right)^{1/2} \int_0^\infty U^2 N_i(z, t) Q_i^d(U) f_0(z, U, t) dU, \qquad (3.29)$$

$$P_{i,p}^{in}(z, t) = \rho_p U_{i,p}^{in} n_e(z, t) \nu_{i,p}^{in}(z, t) \quad \text{for} \quad \text{in} \neq \text{at}, \qquad (3.30)$$

$$P_{i,p}^{at}(z, t) = \left(\frac{2}{m_e}\right)^{1/2} \int_0^\infty U^2 N_i(z, t) Q_{i,p}^{at,0}(z, t) f_0(z, U, t) dU. \qquad (3.31)$$

It should be mentioned that all macroscopic properties of the electrons of interest are obtained by an energy space averaging over the isotropic distribution f_0 as well as of the first (f_1) and second (f_2) contribution to the distribution anisotropy. This characteristic feature of the macroscopic properties is an immediate consequence of the orthogonality properties of the expansion (3.11) applied to the velocity distribution with respect to the angular component v_z/v.

3.3 Numerical treatment

In this section details about the numerical solution of the coupled set of equations (3.12) and (3.13) and results for selected conditions are given. In order to solve this time- and space-dependent hierarchy of equations for f_n with $n = 0, ..., (l - 1)$ or the corresponding system of equations related to purely time-dependent, solely space-dependent, and eventime- and space-dependent problems, quite different solution and numerical techniques have been developed and employed. These techniques are most commonly based on the finite difference method.

When using the finite difference method to solve a partial differential equation, completed by appropriate boundary conditions and, if necessary, initial values, the solution domain of interest spanned by the independent variables is first divided into a discrete set of grid points. For example, the space coordinate z, the kinetic energy U, and the time t are independent variables in the hierarchy of equations (3.12) and (3.13), and the respective grid spacings are Δz, ΔU, and Δt. When using the subscripts j, k, and m, the grid points (z_j, U_k, t_m) may have the coordinates $j\Delta z$, $k\Delta U$, and $m\Delta t$. Then, the partial differential equations are discretized at the grid points or at points between the grid points such as the centered points, where the derivatives of f_n are approximated by suitable finite difference expressions including function values at neighboring grid points and the grid spacings. This procedure leads to a system of algebraic equations for the discretized distributions $f_n^{j,k,m}$ at the grid points, whose values are then determined by means of appropriate numerical algorithms. By making the grid spacings sufficiently small, the result for the discretized distributions $f_n^{j,k,m}$ are expected to be a sufficiently close approximation to the distributions f_n at any grid point.

Independent of the problem considered, it is always desirable to have a fast numerical solution. However, aspects regarding, e.g., the stability of the numerical scheme and the convergence and accuracy of the result obtained have to be kept in mind [28–30]. In order to evaluate the accuracy of the solution of system (3.12) and (3.13), the simultaneous fulfilment of the relevant macroscopic balance equations, (3.18), (3.19), and (3.20), can be checked [17], similar to the method of manufactured solutions. Therefore, all macroscopic properties involved are calculated according to equations (3.21)–(3.31) using the numerically obtained distributions f_0, f_1, and f_2. By appropriate choice of the respective grid spacings, all macroscopic balances should be simultaneously satisfied up to a certain numerical error.

In the following sections 3.3.1 and 3.3.2 details about the numerical method for a multi-term solution of the respective time-dependent or space-dependent electron kinetic equation resulting from the hierarchy equations (3.12) and (3.13) are

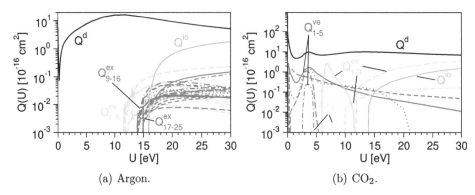

(a) Argon.   (b) CO_2.

Figure 3.1. Collision cross sections for argon and carbon dioxide.

presented. The applicability of the respective methods are demonstrated using results for the temporal and spatial relaxation of plasma electrons acted upon by a constant electric field. In order to illustrate the main aspects of such relaxation behavior of the electron component and to discuss the impact of higher-order terms of the Legendre polynomial expansion on the results, weakly ionized plasmas in argon and carbon dioxide are considered as representatives of atomic and molecular gas plasmas. The total collision cross sections used for the investigations are shown in figure 3.1. The individual collision cross sections for argon and CO_2 have been taken from Hayashi [31, 32]. In addition to the momentum transfer cross section for elastic collisions Q_i^d according to equation (3.17) being used for all differences $Q_i^{el,0} - Q_i^{el,n}$ with $1 \leqslant n \leqslant l - 1$ occurring in equation (3.13), several cross sections for inelastic collisions are contained in the sets of data, where isotropic scattering has been assumed in all inelastic collision processes. With respect to Ar, 25 electronic excitation (ex) cross sections with the energy losses 11.55, 11.62, 11.72, 11.83, 12.91, 13.08, 13.09, 13.15, 13.17, 13.27, 13.30, 13.33, 13.48, 13.84, 13.90, 13.98, 14.01, 14.06, 14.09, 14.15, 14.21, 14.23, 14.30, 14.71, and 15.20 eV, and the ionization (io) cross section with the energy loss 15.76 eV have been taken into account. Regarding CO_2, the cross section set includes five total inelastic collision cross sections for vibrational excitation (ve), three total cross sections for electronic excitation (ex), as well as the total cross sections for electron impact ionization (io) and for dissociative electron attachment (at). The corresponding energy losses are 0.083, 0.172, 0.291, 0.36, and 2.5 eV for the vibrational excitation processes, 5.7, 9.0, and 11.0 eV for the electronic excitation processes, and 13.7 eV for the electron impact ionization.

3.3.1 Solution method for time-dependent conditions

The microphysical treatment of the kinetics of the electron component of time-dependent plasmas takes place primarily by means of solving the non-stationary, spatially homogeneous Boltzmann equation of the electrons. Starting from the set of equations (3.12) and (3.13), the behavior of the EVDF in a purely time-dependent plasma results from the hierarchy of partial differential equations

$$\left(\frac{m_e}{2}\right)^{1/2} U^{1/2} \frac{\partial}{\partial t} f_0(U, t) - \frac{e_0}{3} E(t) \frac{\partial}{\partial U} \big(U f_1(U, t)\big)$$

$$- \frac{\partial}{\partial U}\left(\sum_i 2\frac{m_e}{M_i} U^2 N_i(t) Q_i^d(U) f_0(U, t)\right)$$

$$+ \sum_i \sum_p U N_i(t) Q_{i,p}^{\mathrm{in},0}(U) f_0(U, t)$$

$$= \sum_i \sum_p \beta_p^2 \tilde{U}_{i,p} N_i(t) Q_{i,p}^{\mathrm{in},0}\big(\tilde{U}_{i,p}\big) f_0\big(\tilde{U}_{i,p}, t\big), \quad \tilde{U}_{i,p} = \beta_p U + \rho_p U_{i,p}^{\mathrm{in}},$$

(3.32)

$$\left(\frac{m_e}{2}\right)^{1/2} U^{1/2} \frac{\partial}{\partial t} f_n(U, t) - e_0 E(t)\frac{n}{2n-1}\left(U\frac{\partial}{\partial U} f_{n-1}(U, t) - \frac{n-1}{2} f_{n-1}(U, t)\right)$$

$$- e_0 E(t)\frac{n+1}{2n+3}\left(U\frac{\partial}{\partial U} f_{n+1}(U, t) + \frac{n+2}{2} f_{n+1}(U, t)\right)$$

$$+ \sum_i U N_i(t)\left[\big(Q_i^{\mathrm{el},0}(U) - Q_i^{\mathrm{el},n}(U)\big) + \sum_p Q_{i,p}^{\mathrm{in},0}(U)\right] f_n(U, t)$$

$$= \sum_i \sum_p \beta_p^2 \tilde{U}_{i,p} N_i(t) Q_{i,p}^{\mathrm{in},n}\big(\tilde{U}_{i,p}\big) f_n\big(\tilde{U}_{i,p}, t\big),$$

$$1 \leqslant n \leqslant l-1, \quad l \geqslant 2, \quad f_l(U, t) \equiv 0.$$

(3.33)

This hierarchy describes the temporal evolution of the isotropic distribution $f_0(U, t)$ and the anisotropic components $f_n(U, t)$ with $n \geqslant 1$ of the velocity distribution function due to the action of a time-dependent electric field and the binary electron–heavy particle collisions. Thus, a system of l partial differential equations of first-order including additional terms $f_n(\beta_p U + \rho_p U_{i,p}^{\mathrm{in}}, t)$ with shifted energy arguments has to be solved within the framework of the multi-term approximation.

The system (3.32) and (3.33) of partial differential equations of first-order is solved numerically as an initial-boundary value problem in an appropriate energy region $0 \leqslant U \leqslant U_\infty$ and for times $t \geqslant 0$. Here, the time represents the evolution direction of the kinetic problem and U_∞ denotes a sufficiently large energy, which is used as an upper limit of the kinetic energy in the numerical approach. The initial values for all expansion coefficients $f_n(U, t = 0)$ with $n = 0, \ldots, l-1$ are chosen according to the kinetic problem under consideration. Possible boundary conditions of the equation system for all times $t \geqslant 0$ are [17, 33, 34]

$$f_n(U = 0, t) = 0 \quad \text{for } n \geqslant 1,$$

(3.34)

$$f_n(U \geqslant U_\infty, t) = 0 \quad \text{for } n \geqslant 0.$$

(3.35)

These boundary conditions enforce the suppression of singular contributions to the solution the equation system (3.32) and (3.33).

In order to obtain the numerical solution, a finite difference approximation for the equation system (3.32) and (3.33) is applied using the grid point representation

$$U_k = k\Delta U, \quad \Delta U = U_\infty / n_\infty, \quad 0 \leqslant k \leqslant n_\infty,$$

$$t_m = m\Delta t, \quad m \geqslant 0, \tag{3.36}$$

$$f_n^{k,m} = f_n(U_k, t_m), \quad 0 \leqslant n \leqslant l-1$$

of an equidistant energy grid with the energy boundaries $U_0 = 0$ and $U_{n_\infty} = U_\infty$. In accordance with [17], the partial differential equations are discretized at all centered points $(U_{k+1/2}, t_{m+1/2})$ with $0 \leqslant k \leqslant n_\infty - 1$ and $m \geqslant 0$ using the second-order-correct centered difference expressions

$$f_n^{k+1/2, m+1/2} = \frac{1}{4}\left(f_n^{k+1, m+1} + f_n^{k, m+1} + f_n^{k+1, m} + f_n^{k, m}\right),$$

$$\left(\frac{\partial}{\partial U} f_n(U, t)\right)^{k+1/2, m+1/2} = \frac{1}{2\Delta U}\left(f_n^{k+1, m+1} - f_n^{k, m+1} + f_n^{k+1, m} - f_n^{k, m}\right), \tag{3.37}$$

$$\left(\frac{\partial}{\partial t} f_n(U, t)\right)^{k+1/2, m+1/2} = \frac{1}{2\Delta t}\left(f_n^{k+1, m+1} - f_n^{k+1, m} + f_n^{k, m+1} - f_n^{k, m}\right)$$

for the functions f_n and their partial derivatives. Furthermore, the second-order-correct expression with respect to time

$$f_n\left(\left(\tilde{U}_{i,p}\right)_{k+1/2}, t_{m+1/2}\right) = \frac{1}{2}\left[f_n\left(\left(\tilde{U}_{i,p}\right)_{k+1/2}, t_{m+1}\right) + f_n\left(\left(\tilde{U}_{i,p}\right)_{k+1/2}, t_m\right)\right] \tag{3.38}$$

is used for the treatment of the terms with shifted energy argument on the right-hand sides of equations (3.32) and (3.33). The corresponding representation of $f_n((\tilde{U}_{i,p})_{k+1/2}, t_\mu)$ for $\mu = m$ and $m+1$ on the energy grid is obtained by means of a parabolic interpolation using three neighboring energy grid points. For instance, such an interpolation has the representation [17]

$$f_n\left(U_{k+1/2} + U_{i,p}^{\text{in}}, t_\mu\right) = \frac{1}{2}\alpha_{i,p}^{(-3)}\alpha_{i,p}^{(-1)}f_n^{k+k_{i,p}, \mu} - \alpha_{i,p}^{(-3)}\alpha_{i,p}^{(+1)}f_n^{k+k_{i,p}+1, \mu}$$

$$+ \frac{1}{2}\alpha_{i,p}^{(-1)}\alpha_{i,p}^{(+1)}f_n^{k+k_{i,p}+2, \mu}, \tag{3.39}$$

$$k_{i,p} = \text{int}\left(\frac{U_{i,p}^{\text{in}}}{\Delta U}\right), \quad \alpha_{i,p}^{(\tau)} = \frac{U_{i,p}^{\text{in}}}{\Delta U} - k_{i,p} + \frac{\tau}{2}, \quad \tau = -3, -1, +1$$

for electron impact excitation processes, where int(x) denotes the integer part of x. Such discretization of equations (3.32) and (3.33) finally leads to the finite difference equation system

$$v_n^{k,m} f_{n-1}^{k-1,m+1} + w_n^m f_{n-1}^{k,m+1} + g_n^{k,m} f_n^{k,m+1} + h_n^{k,m} f_n^{k+1,m+1}$$
$$+ r_n^{k,m} f_{n+1}^{k,m+1} + s_n^{k,m} f_{n+1}^{k+1,m+1} = d_n^{k,m} + \tilde{d}_n^{k,m+1} \tag{3.40}$$

for $0 \leqslant n \leqslant l - 1$ with $0 \leqslant k \leqslant n_\infty - 1$ and $m \geqslant 0$, where the coefficients are given by

$$g_0^{k,m} = A^k \frac{\Delta U}{\Delta t} - D^{k,m} + F^{k,m} \frac{\Delta U}{2}, \quad h_0^{k,m} = A^k \frac{\Delta U}{\Delta t} + D^{k,m} + F^{k,m} \frac{\Delta U}{2},$$

$$g_n^{k,m} = A^k \frac{\Delta U}{\Delta t} + H^{k,m} \frac{\Delta U}{2}, \quad h_n^{k,m} = g_n^{k,m}, \quad \text{for } 1 \leqslant n \leqslant l - 1,$$

$$r_n^{k,m} = \frac{n+1}{2n+3} e_0 E(t_{m+1/2}) \left(U_{k+1/2} - \frac{n+2}{2} \frac{\Delta U}{2} \right), \quad \text{for } 0 \leqslant n \leqslant l - 2,$$

$$s_n^{k,m} = -\frac{n+1}{2n+3} e_0 E(t_{m+1/2}) \left(U_{k+1/2} + \frac{n+2}{2} \frac{\Delta U}{2} \right), \quad \text{for } 0 \leqslant n \leqslant l - 2,$$

$$v_n^{k,m} = \frac{n}{2n-1} e_0 E(t_{m+1/2}) \left(U_{k+1/2} + \frac{n-1}{2} \frac{\Delta U}{2} \right), \quad \text{for } 1 \leqslant n \leqslant l - 1,$$

$$w_n^{k,m} = \frac{n}{2n-1} e_0 E(t_{m+1/2}) \left(-U_{k+1/2} + \frac{n-1}{2} \frac{\Delta U}{2} \right), \quad \text{for } 1 \leqslant n \leqslant l - 1,$$

$$v_0^{k,m} = w_0^{k,m} = r_{l-1}^{k,m} = s_{l-1}^{k,m} = 0,$$

$$d_n^{k,m} = -v_n^{k,m} f_{n-1}^{k,m} - w_n^{k,m} f_{n-1}^{k+1,m} + x_n^{k,m} f_n^{k,m} + y_n^{k,m} f_n^{k+1,m} \tag{3.41}$$

$$- r_n^{k,m} f_{n+1}^{k,m} - s_n^{k,m} f_{n+1}^{k+1,m} + \Delta U \sum_i \sum_p I_{i,p,n}^{k,m} \left(\frac{1}{2} \alpha_{i,p}^{(-3)} \alpha_{i,p}^{(-1)} f_n^{k+k_{i,p},m} \right.$$

$$\left. - \alpha_{i,p}^{(-3)} \alpha_{i,p}^{(+1)} f_n^{k+k_{i,p}+1,m} + \frac{1}{2} \alpha_{i,p}^{(-1)} \alpha_{i,p}^{(+1)} f_n^{k+k_{i,p}+2,m} \right)$$

$$x_0^{k,m} = A^k \frac{\Delta U}{\Delta t} + D^{k,m} - F^{k,m} \frac{\Delta U}{2}, \quad y_0^{k,m} = A^k \frac{\Delta U}{\Delta t} - D^{k,m} - F^{k,m} \frac{\Delta U}{2},$$

$$x_n^{k,m} = A^k \frac{\Delta U}{\Delta t} - H^{k,m} \frac{\Delta U}{2}, \quad y_n^{k,m} = x_n^{k,m}, \quad \text{for } 1 \leqslant n \leqslant l - 1,$$

$$\tilde{d}_n^{k,m+1} = \Delta U \sum_i \sum_p I_{i,p,n}^{k,m} \left(\frac{1}{2} \alpha_{i,p}^{(-3)} \alpha_{i,p}^{(-1)} f_n^{k+k_{i,p},m+1} \right.$$

$$\left. - \alpha_{i,p}^{(-3)} \alpha_{i,p}^{(+1)} f_n^{k+k_{i,p}+1,m+1} + \frac{1}{2} \alpha_{i,p}^{(-1)} \alpha_{i,p}^{(+1)} f_n^{k+k_{i,p}+2,m+1} \right)$$

with the further coefficients

$$A^k = \left(\frac{m_e}{2} U_{k+1/2}\right)^{1/2}, \quad D^{k,m} = -\left(U_{k+1/2}\right)^2 \sum_i 2\frac{m_e}{M_i} N_i(t_{m+1/2}) Q_i^d(U_{k+1/2}),$$

$$F^{k,m} = -U_{k+1/2} \sum_i 2\frac{m_e}{M_i} N_i(t_{m+1/2})\left(2Q_i^d(U_{k+1/2}) + U_{k+1/2}\frac{Q_i^d(U_{k+1}) - Q_i^d(U_k)}{\Delta U}\right)$$

$$+ U_{k+1/2} \sum_i \sum_p N_i(t_{m+1/2}) Q_{i,p}^{in,0}(U_{k+1/2}), \tag{3.42}$$

$$H^{k,m} = U_{k+1/2} \sum_i N_i(t_{m+1/2})\left(Q_i^{el,0}(U_{k+1/2}) - Q_i^{el,n}(U_{k+1/2}) + \sum_p Q_{i,p}^{in,0}(U_{k+1/2})\right),$$

$$I_{i,p,n}^{k,m} = \beta_p^2\left(\tilde{U}_{i,p}\right)_{k+1/2} N_i(t_{m+1/2}) Q_{i,p}^{in,n}\left(\left(\tilde{U}_{i,p}\right)_{k+1/2}\right).$$

In combination with the discrete form [17]

$$f_n^{0,m} = 0 \quad \text{for all odd } n \geqslant 1, \quad f_n^{k \geqslant n_\infty, m} = 0 \quad \text{for all even } n \geqslant 0 \tag{3.43}$$

of the selected boundary conditions according to equations (3.34) and (3.35) at $m \geqslant 0$, which has proven to be suitable, equation (3.40) constitutes a complete system of linear equations for the determination of the function values $f_n^{k,m+1}$ with $0 \leqslant n \leqslant l - 1$ for $0 \leqslant k \leqslant n_\infty$ at the time step $m + 1$, if all corresponding function values $f_n^{k,m}$ at the preceding time step m or their initial values at $m = 0$ are known. This linear system is tridiagonal in the l dependent variables f_n with $0 \leqslant n \leqslant l - 1$, where the diagonal terms of the coefficient matrix are $g_n^{k,m}$ for even $n \geqslant 0$ and $h_n^{k,m}$ for odd $n \geqslant 1$. It is converted to the general band form and solved using the general band algorithm given in [35].

On closer inspection, equation (3.40) also shows an additional dependence on function values $f_n^{k,m+1}$ in the formal inhomogeneities $\tilde{d}_n^{k,m+1}$ given in equation (3.41). In order to avoid the numerically much more expensive solution of a non-tridiagonal system, an iterative treatment of the function values $f_n^{k,m+1}$ in $\tilde{d}_n^{k,m+1}$ with $0 \leqslant n \leqslant l - 1$ is applied while using the general band algorithm in each iteration cycle. At each time step t_{m+1} the iterative solution process is finished as soon as the absolute deviation of the ratios $f_n^{k,m+1,J+1}/f_n^{k,m+1,J}$ of two successive discrete function values obtained in the iteration cycles $J + 1$ and J from unity is less than 10^{-5} for all expansion coefficients, i.e. $0 \leqslant n \leqslant l - 1$, at all relevant energy points.

The multi-term method for solving equation system (3.32) and (3.33) has been applied to quite different temporal relaxation processes. Here, results for the temporal relaxation of electrons in argon and carbon dioxide plasmas under the action of a time-independent electric field are presented using the collision cross-section data shown in figure 3.1. The solution procedure started at $t = 0$ from the Gaussian distribution $G_i(U) = c \exp(-(U - U_c)^2/U_w^2)$ with the center energy $U_c = 5$ eV and energy width $U_w = 2$ eV as initial values for the isotropic distribution, i.e. $f_0(U, 0)/n_e(0) = G_f(U)$ with the factor c fixed by the normalization condition equation (3.21). The initial values $f_n(U, 0)$, $n \geqslant 1$, of the anisotropic components were chosen to be zero. The following relaxation results belong to the gas density $N = 3.54 \times 10^{16}$ cm^{-3} assuming a gas pressure of 1 Torr at 0 °C. The electric field was set to $E = 10.62$ V cm^{-1} corresponding to a reduced electric field strength of $E/N = 30$ Td with 1 Td $= 10^{-17}$ V cm^2.

Figure 3.2 shows the temporal relaxation of the isotropic distribution $f_0(U, t)/n_e(t)$ for argon (a) and carbon dioxide (b) calculated using the multi-term method with $l = 8$ expansion coefficients, which was found to yield the converged solution of the electron Boltzmann equation during the entire relaxation course. Starting at $t = 0$ from the same isotropic distribution with a maximum around 5 eV, the temporal evolution finally leads to the establishment of time-independent distributions $f_n(U, t)/n_e(t)$ normalized by the electron density $n_e(t)$. The power input from the field leads to an acceleration of the electrons to larger energies and a widening of $f_0(U, t)/n_e(t)$. Because the electron collision cross sections are very different and cover quite different ranges of kinetic energies (see figure 3.1), the temporal relaxation takes place in different ways in the two gases and leads to very different final distributions.

The corresponding expansion coefficients $f_n(U, t)/n_e(t)$ with $n = 0, 1, 2$ during the early temporal relaxation course at $t = 3$ ns are displayed in figure 3.3. The first two anisotropic components $f_1(U, t)/n_e(t)$ and $f_2(U, t)/n_e(t)$ of the velocity distribution assume positive and negative function values in the kinetic energy range, which are marked by plus and minus signs in parentheses. The opposite signs in the first contribution to the distribution anisotropy $f_1(U, t)/n_e(t)$ are caused by the structure of the isotropic distribution showing one (figure 3.3(a)) or several (figure 3.3(b)) peaks and those in $f_2(U, t)/n_e(t)$ are mainly induced by the structure of $f_1(U, t)/n_e(t)$. The comparison of the magnitude of these distribution anisotropy coefficients with $f_0(U, t)/n_e(t)$ reveals values, in particular for CO_2, which are of the same order. This indicates that higher-order terms can have a remarkable impact on the EVDF not only at steady state but also during the temporal relaxation process.

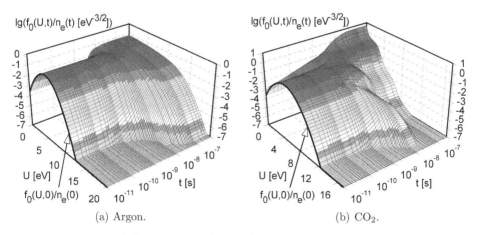

(a) Argon.   (b) CO_2.

Figure 3.2. Temporal relaxation of $f_0(U, t)/n_e(t)$ at $E = 10.62$ V cm^{-1}.

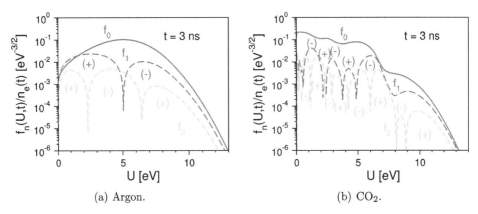

Figure 3.3. Expansion coefficients $f_n(U, t)/n_e(t)$ with $n = 0, 1, 2$ at $t = 3$ ns.

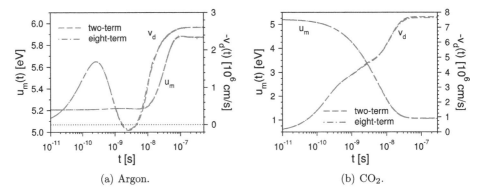

Figure 3.4. Temporal evolution of the mean energy. and drift velocity.

In order to illustrate the impact of higher-order terms of the EVDF expansion on selected macroscopic properties, results for the mean electron energy $u_m(t) = u_e(t)/n_e(t)$ and the electron drift velocity $v_d(t) = j_z(t)/n_e(t) = -\mu_e(t)E(t)$ obtained using time-dependent two-term and eight-term approximation are shown in figure 3.4, where μ_e denotes the electron mobility. Certain differences between two-term and converged eight-term results are found for argon during the temporal relaxation course and for carbon dioxide, in particular when approaching the time-independent state. Furthermore, the results for argon demonstrate a change of sign of the drift velocity during the temporal evolution around $t = 3$ ns which corresponds to a transient state with negative mobility [36]. This behavior was found to depend on the initial conditions chosen. In the time-independent state in pure argon the electron mobility assumes a positive value as expected.

Finally, it should be noted that an adequate method for the estimation of the relaxation time consists in the analysis of the corresponding macroscopic balances resulting from equations (3.18), (3.19), and (3.20). For time-dependent, spatially homogeneous conditions of plasmas of a single background gas component ($i = 1$),

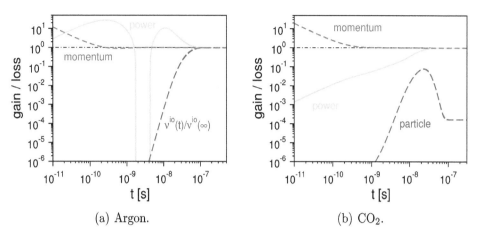

Figure 3.5. Temporal evolution of gain to loss ratios and the ionization frequency.

the establishment of time-independent distributions $f_n(U, t)/n_e(t)$, i.e. of a quasi-stationary or hydrodynamic state with constant $[dn_e(t)/dt]/n_e(t)$, can be generally evaluated by considering the gain and loss terms on the right-hand sides of the macroscopic balances. Figure 3.5 illustrates the temporal behavior of the ratios of the particle gain to loss $\nu^{io}(t)/\nu^{at}(t)$, of the momentum gain to loss $I^f(t)/[I^{el}(t) + \sum_p I_p^{in}(t)]$, and of the power gain to loss $P^f(t)/[P^{el}(t) + \sum_p P_p^{in}(t) + P^{at}(t)]$. Because argon does not attach electrons, here the ionization frequency $\nu^{io}(t)/\nu^{io}(\infty)$ normalized by its time-independent value $\nu^{io}(\infty)$ is shown in figure 3.5(a). Quite different temporal evolutions with changes of several orders of magnitude are found for argon and carbon dioxide until finally the time-independent state is established after some 100 ns at $E = 10.62\,\text{V cm}^{-1}$ and $N = 3.54 \times 10^{16}\,\text{cm}^{-3}$. This relaxation time is mainly determined by the considerably slower establishment of the particle and power gain and loss compared to the much faster relaxation of the momentum gain to loss ratio towards a constant value within about 1 ns. Notice that for argon (figure 3.5(a)) the ratio of power gain to loss assumes negative values around $t = 3$ ns due to the changes in the electron drift direction (see figure 3.4(a)).

3.3.2 Multi-term solution for space-dependent plasmas

The behavior of the electron component in space-dependent plasmas is frequently characterized by the fact that the local gain of power and momentum from the electric field cannot be dissipated at the same position by elastic and inelastic collision processes. The starting point for the following kinetic treatment of the electrons in axially inhomogeneous plasmas is the stationary, spatially one-dimensional electron Boltzmann equation. Beginning with the hierarchy of equations (3.12) and (3.13), the velocity distribution function of the electrons in such a space-dependent plasma can be determined by solving the system of partial differential equations

$$\frac{1}{3}\left[\frac{\partial}{\partial z}\big(Uf_1(z,\,U)\big) - e_0 E(z)\frac{\partial}{\partial U}\big(Uf_1(z,\,U)\big)\right]$$

$$-\frac{\partial}{\partial U}\left(\sum_i 2\frac{m_e}{M_i}U^2 N_i(z)Q_i^d(U)f_0(z,\,U)\right)$$

$$+\sum_i\sum_p UN_i(z)Q_{i,p}^{\mathrm{in},0}(U)f_0(z,\,U)$$

$$=\sum_i\sum_p \beta_p^2 \tilde{U}_{i,p}N_i(z)Q_{i,p}^{\mathrm{in},0}\big(\tilde{U}_{i,p}\big)f_0\big(z,\,\tilde{U}_{i,p}\big),\quad \tilde{U}_{i,p}=\beta_p U + \rho_p U_{i,p}^{\mathrm{in}}, \tag{3.44}$$

$$\frac{n}{2n-1}\left[U\frac{\partial}{\partial z}f_{n-1}(z,\,U) - e_0 E(z)\left(U\frac{\partial}{\partial U}f_{n-1}(z,\,U) - \frac{n-1}{2}f_{n-1}(z,\,U)\right)\right]$$

$$+\frac{n+1}{2n+3}\left[U\frac{\partial}{\partial z}f_{n+1}(z,\,U) - e_0 E(z)\left(U\frac{\partial}{\partial U}f_{n+1}(z,\,U) + \frac{n+2}{2}f_{n+1}(z,\,U)\right)\right]$$

$$+\sum_i UN_i(z)\left(\big(Q_i^{\mathrm{el},0}(U) - Q_i^{\mathrm{el},n}(U)\big) + \sum_p Q_{i,p}^{\mathrm{in},0}(U)\right)f_n(z,\,U) \tag{3.45}$$

$$=\sum_i\sum_p \beta_p^2 \tilde{U}_{i,p}N_i(z)Q_{i,p}^{\mathrm{in},n}\big(\tilde{U}_{i,p}\big)f_n\big(z,\,\tilde{U}_{i,p}\big),$$

$$1\leqslant n\leqslant l-1,\quad l\geqslant 2,\quad f_l(z,\,U)\equiv 0.$$

This equation system describes the development of the isotropic distribution $f_0(z,\,U)$ and the contributions $f_n(z,\,U)$ with $n=1,\,\ldots,\,l-1$ to the anisotropy of the EVDF in the space of the coordinate z and the kinetic energy U of the electrons under the impact of a space-dependent electric field and binary collisions of electrons with heavy particles. Consequently, a system of l first-order partial differential equations comprising additional terms $f_n(z,\,\beta_p U + \rho_p U_{i,p}^{\mathrm{in}})$ with shifted energy argument $(\beta_p U + \rho_p U_{i,p}^{\mathrm{in}})$ has to be solved within the framework of the multi-term approximation.

The direct numerical treatment of the system of equations (3.44) and (3.45) has proven to be hardly practicable so far. This can already be understood on the basis of an analysis of the two-term ($l=2$) approximation with isotropic scattering in inelastic collision processes. Then, a partial differential equation of second-order for the isotropic distribution $f_0(z,\,U)$ results from equations (3.44) and (3.45). This second-order partial differential equation includes cross derivative terms that are known to cause difficulties during the numerical solution.

In order to overcome this problem and to simplify the partial differential equation system (3.44) and (3.45), the kinetic energy U is replaced by the total energy ϵ according to

$$\epsilon = U + W(z), \tag{3.46}$$

i.e. by the sum of kinetic energy U and potential energy $W(z)$ of the electrons in the electric field given by

$$W(z) = e_0 \int_0^z E(\tilde{z}) \mathrm{d}\tilde{z}. \tag{3.47}$$

This transformation leads to the system of first-order partial differential equations

$$\frac{1}{3}\left(U_{z\epsilon}\frac{\partial}{\partial z}\tilde{f}_1(z,\,\epsilon) - e_0 E(z)\tilde{f}_1(z,\,\epsilon) \right) - \frac{\partial}{\partial\epsilon}\left(\sum_i 2\frac{m_e}{M_i} U_{z\epsilon}^2 N_i(z) Q_i^d(U_{z\epsilon})\tilde{f}_0(z,\,\epsilon) \right)$$

$$+ \sum_i\sum_p U_{z\epsilon} N_i(z) Q_{i,p}^{\mathrm{in},0}(U_{z\epsilon})\tilde{f}_0(z,\,\epsilon) = \sum_i\sum_p \beta_p^2 \hat{U}_{i,p}^{z\epsilon} N_i(z) Q_{i,p}^{\mathrm{in},0}\left(\hat{U}_{i,p}^{z\epsilon}\right)\tilde{f}_0\left(z,\,\hat{\epsilon}_{i,p}\right) \tag{3.48}$$

$$\hat{U}_{i,p}^{z\epsilon} = \beta_p U_{z\epsilon} + \rho_p U_{i,p}^{\mathrm{in}}, \; \hat{\epsilon}_{i,p} = \beta_p\epsilon - \left(\beta_p - 1\right)W(z) + \rho_p U_{i,p}^{\mathrm{in}},$$

$$\frac{n}{2n-1}\left(U_{z\epsilon}\frac{\partial}{\partial z}\tilde{f}_{n-1}(z,\,\epsilon) + \frac{n-1}{2}e_0 E(z)\tilde{f}_{n-1}(z,\,\epsilon) \right)$$

$$+ \frac{n+1}{2n+3}\left(U_{z\epsilon}\frac{\partial}{\partial z}\tilde{f}_{n+1}(z,\,\epsilon) - \frac{n+2}{2}e_0 E(z)\tilde{f}_{n+1}(z,\,\epsilon) \right)$$

$$+ \sum_i U_{z\epsilon} N_i(z)\left(\left(Q_i^{\mathrm{el},0}(U_{z\epsilon}) - Q_i^{\mathrm{el},n}(U_{z\epsilon}) \right) + \sum_p Q_{i,p}^{\mathrm{in},0}(U_{z\epsilon}) \right)\tilde{f}_n(z,\,\epsilon) \tag{3.49}$$

$$= \sum_i\sum_p \beta_p^2 \hat{U}_{i,p}^{z\epsilon} N_i(z) Q_{i,p}^{\mathrm{in},n}\left(\hat{U}_{i,p}^{z\epsilon}\right)\tilde{f}_n\left(z,\,\hat{\epsilon}_{i,p}\right),$$

$$1 \leqslant n \leqslant l-1, \quad l \geqslant 2, \quad \tilde{f}_l(z,\,\epsilon) \equiv 0$$

for the transformed expansion coefficients $\tilde{f}_n(z,\,\epsilon) = f_n(z,\,U_{z\epsilon})$ with $U_{z\epsilon} \equiv \epsilon - W(z)$.

This system of l partial differential equations including additional terms $\tilde{f}_n(z,\,\beta_p\epsilon - (\beta_p - 1)W(z) + \rho_p U_{i,p}^{\mathrm{in}})$ with shifted energy arguments is solved numerically as an initial-boundary value problem for given spatial dependences of the electric field and heavy particle densities. The originally rectangular solution domain $(0 \leqslant z \leqslant z_{\max}, 0 \leqslant U \leqslant U_\infty)$ with the spatial margins $z = 0$ and $z = z_{\max}$ changes its shape due to the transformation to the total energy. The resulting transformed solution area $(0 \leqslant z \leqslant z_{\max}, W(z) \leqslant \epsilon \leqslant W(z) + U_\infty)$ is non-rectangular and its boundaries are determined in part by the spatial course of the electric potential.

The total energy ϵ represents the evolution direction of the kinetic problem. The solution of system (3.48) and (3.49) as an initial-boundary value problem takes place from higher to lower total energies. It starts at a sufficiently large total energy $\epsilon = \epsilon_\infty$, where all expansion coefficients can be assumed to be negligibly small in accordance with equation (3.35). Thus, the relation $\tilde{f}_n(z,\,\epsilon \geqslant \epsilon_\infty) = 0$ for all $n = 0, ..., l - 1$ provides appropriate initial values. By analogy with equation (3.34), possible conditions at the total energy boundary $\epsilon = W(z)$ for all z are

$$\tilde{f}_n(z, \epsilon = W(z)) = 0 \quad \text{for } n \geqslant 1. \tag{3.50}$$

Finally, appropriate conditions at the spatial margins of the solution domain, i.e. at $z = 0$ and $z = z_{\max}$, have to be chosen according to the specific problem under consideration. Examples of such conditions can be found, e.g. in [37–39].

In order to numerically solve the equation system (3.48) and (3.49) using the finite difference method, it can be discretized on an equidistant grid in the coordinate z with

$$z_j = j\Delta z, \quad \Delta z = z_{\max}/j_{\max}, \quad 0 \leqslant j \leqslant j_{\max}. \tag{3.51}$$

For example, when studying the spatial relaxation of electrons in an idealized steady-state Townsend (SST) experiment with plane-parallel geometry similar to, e.g. [4, 40], where the electrons start at $z = 0$ and are accelerated in the positive z-direction with $z_{\max} > 0$, the electric field $E(z)$ is assumed to be negative everywhere. Thus, the potential energy $W(z)$ given by equation (3.47) is negative for all $z > 0$ as well. The corresponding total energy grid consists of a part for $\epsilon \geqslant 0$ and a part belonging to negative total energies [37], and it is given by

$$\begin{aligned} \epsilon_k &= k\Delta\epsilon, \quad \Delta\epsilon = U_\infty/n_\infty, \quad 0 \leqslant k \leqslant n_\infty, \quad n_\infty < j_{\max}, \\ \epsilon_k &= W(z_{-k}), \quad k = -1, -2, \ldots, -j_{\max} + n_\infty. \end{aligned} \tag{3.52}$$

The corresponding discretized distributions at the grid points have the representation

$$\tilde{f}_n^{j,k} = \tilde{f}_n(z_j, \epsilon_k), \quad 0 \leqslant n \leqslant l - 1. \tag{3.53}$$

The discretization of the system of partial differential equations (3.48) and (3.49) can be performed at all centered points ($z_{j+1/2}$, $\epsilon_{k+1/2}$) [22, 37] using second-order-correct centered difference expressions for all functions $\tilde{f}_n$ with $0 \leqslant n \leqslant l - 1$ and their first derivatives analogous to equation (3.37), where $j = j_0, \ldots, j_0 + n_\infty$ with $j_0 = -\min(0, k)$ and $k = n_\infty - 1, \ldots, -j_{\max} + n_\infty$. In addition, the second-order-correct expression

$$\tilde{f}_n\left(z_{j+1/2}, (\hat{\epsilon}_{i,p})_{k+1/2}\right) = \frac{1}{2}\left[\tilde{f}_n\left(z_j, (\hat{\epsilon}_{i,p})_{k+1/2}\right) + \tilde{f}_n\left(z_{j+1}, (\hat{\epsilon}_{i,p})_{k+1/2}\right)\right] \tag{3.54}$$

with respect to the space coordinate is applied to treat the terms with shifted total energy argument on the right-hand side of equations (3.48) and (3.49). The transformed expansion coefficients $\tilde{f}_n(z_\xi, (\hat{\epsilon}_{i,p})_{k+1/2})$ with $\xi = j$ and $j + 1$ are represented on the same energy grid by means of a parabolic interpolation using the three neighboring energy grid points. The discrete form of the equation system (3.48) and (3.49) finally reads

$$\begin{aligned} &v_n^{j,k}\tilde{f}_{n-1}^{j,k} + w_n^{j,k}\tilde{f}_{n-1}^{j+1,k} + g_n^{j,k}\tilde{f}_n^{j,k} + g_n^{j,k}\tilde{f}_n^{j+1,k} + r_n^{j,k}\tilde{f}_{n+1}^{j,k} + s_n^{j,k}\tilde{f}_{n+1}^{j+1,k} \\ &= d_n^{j,k+1} + \tilde{d}_n^{j,k} \end{aligned} \tag{3.55}$$

for $0 \leqslant n \leqslant l - 1$. It has the coefficients

$$g_0^{j,k} = -D_0^{j,k}\frac{\Delta z}{\epsilon_{k+1} - \epsilon_k} + F_0^{j,k}\frac{\Delta z}{2}, \quad h_0^{j,k} = -D_0^{j,k}\frac{\Delta z}{\epsilon_{k+1} - \epsilon_k} - F_0^{j,k}\frac{\Delta z}{2},$$

$$g_n^{j,k} = \frac{\Delta z}{2}\sum_i H_{i,n}^{j,k}, \quad h_n^{j,k} = -g_n^{j,k} \quad \text{for } 1 \leqslant n \leqslant l - 1,$$

$$r_n^{j,k} = -A_n^{j,k} + B_n^j\frac{\Delta z}{2}, \quad s_n^{j,k} = A_n^{j,k} + B_n^j\frac{\Delta z}{2} \quad \text{for } 0 \leqslant n \leqslant l - 2,$$

$$v_n^{j,k} = -C_n^{j,k} + G_n^j\frac{\Delta z}{2}, \quad w_n^{j,k} = C_n^{j,k} + G_n^j\frac{\Delta z}{2} \quad \text{for } 1 \leqslant n \leqslant l - 1, \quad (3.56)$$

$$v_0^{j,k} = w_0^{j,k} = r_{l-1}^{j,k} = s_{l-1}^{j,k} = 0,$$

$$d_n^{j,k+1} = -v_n^{j,k}\tilde{f}_{n-1}^{j,k+1} - w_n^{j,k}\tilde{f}_{n-1}^{j+1,k+1} + h_n^{j,k}\tilde{f}_n^{j,k+1} + h_n^{j,k}\tilde{f}_n^{j+1,k+1}$$
$$- r_n^{j,k}\tilde{f}_{n+1}^{j,k+1} - s_n^{j,k}\tilde{f}_{n+1}^{j+1,k+1},$$

$$\tilde{d}_n^{j,k} = \Delta z\sum_i\sum_p \beta_p^2 I_{i,p,n}^{j,k}\Big(\tilde{f}_n\big(z_j, (\hat{\epsilon}_{i,p})_{k+1/2}\big) + \tilde{f}_n\big(z_{j+1}, (\hat{\epsilon}_{i,p})_{k+1/2}\big)\Big)$$

with the further terms

$$D_0^{j,k} = -\sum_i 2\frac{m_e}{M_i}\Big[U_{ze}^2 N_i(z)Q_i^d(U_{ze})\Big]_{j+1/2,k+1/2},$$

$$F_0^{j,k} = -\sum_i 2\frac{m_e}{M_i}\Bigg(\Big[2U_{ze}N_i(z)Q_i^d(U_{ze})\Big]_{j+1/2,k+1/2}$$

$$+ \big(U_{ze}\big)_{j+1/2,k+1/2}^2\frac{\Big[N(z)Q_i^d(U_{ze})\Big]_{j+1,k+1} - \Big[N(z)Q_i^d(U_{ze})\Big]_{j+1,k}}{\epsilon_{k+1} - \epsilon_k}\Bigg)$$

$$+ \sum_i\sum_p\Big[U_{ze}N_i(z)Q_{i,p}^{in,0}(U_{ze})\Big]_{j+1/2,k+1/2}, \quad (3.57)$$

$$H_{i,n}^{j,k} = \Bigg[U_{ze}N_i(z)\bigg(\Big(Q_i^{el,0}(U_{ze}) - Q_i^{el,n}(U_{ze})\Big) + \sum_p Q_{i,p}^{in,0}(U_{ze})\bigg)\Bigg]_{j+1/2,k+1/2},$$

$$A_n^{j,k} = \frac{n+1}{2n+3}\big(U_{ze}\big)_{j+1/2,k+1/2}, \quad B_n^j = -\frac{n+1}{2n+3}\frac{n+2}{2}e_0 E(z_{j+1/2}),$$

$$C_n^{j,k} = \frac{n}{2n-1}\big(U_{ze}\big)_{j+1/2,k+1/2}, \quad G_n^j = \frac{n}{2n-1}\frac{n-1}{2}e_0 E(z_{j+1/2}),$$

$$I_{i,p,n}^{j,k} = \Big[\hat{U}_{i,p}^{ze}N_i(z)Q_{i,p}^{in,n}\big(\hat{U}_{i,p}^{ze}\big)\Big]_{j+1/2,k+1/2}.$$

In order to complete this finite difference equation system, one boundary condition for each of the l expansion coefficients $\tilde{f}_n$ with $0 \leqslant n \leqslant l - 1$ is required. It has been found that the conditions [37]

$$\tilde{f}_n^{j_0=0,k} = \hat{f}_n^k, \ \tilde{f}_n^{j_0>0,k} = 0 \quad \text{for } n = 1, 3, 5, \ldots, \tag{3.58}$$

$$\tilde{f}_n^{j_0+n_\infty,k} = 0 \quad \text{for } n = 0, 2, 4, \ldots \tag{3.59}$$

represent a proper choice for spatial relaxation studies. The equation (3.58) for $j_0 = 0$ fixes the inflow of electrons at the spatial boundary $z = 0$, where the $\hat{f}_n^k$ with odd n are the discrete forms of appropriate functions $\hat{f}_n(\epsilon)$ of the total energy to be specified according to the problem considered. The condition (3.58) for $j_0 > 0$ is the discrete form of the boundary condition equation (3.50) for odd n. At $z_{j_0+n_\infty}$ the kinetic energy values are always equal to or larger than U_∞ so that the condition (3.59) is in accordance with equation (3.35) for even n. This choice of boundary conditions allows one to suppress singular contributions to the solution of physical relevance at large kinetic energies.

The combination of the equation system (3.55) and the boundary conditions (3.58) and (3.59) represents a complete set of linear equations to determine the function values $\tilde{f}_n^{j,k}$ with $0 \leqslant n \leqslant l - 1$ for $j = j_0, \ldots, j_0 + n_\infty$ at the fixed total energy with index k, if all corresponding function values $\tilde{f}_n^{j,k+1}$ at the larger total energy with index $k + 1$ or at their initial values at $k = n_\infty$ are known. This complete linear equation system is again tridiagonal in the l independent variables $\tilde{f}_n$ with $0 \leqslant n \leqslant l - 1$. After its conversion to the general band form, it is resolved by using the general band algorithm given in [35]. Finally, the transformation back from the total energy ϵ to the kinetic energy U yields the expansion coefficients $f_n(z, U)$ with $0 \leqslant n \leqslant l - 1$ of the EVDF and the related macroscopic properties.

A more detailed examination of the formal inhomogeneities $\tilde{d}_n^{j,k}$ in equation (3.55) also makes clear that the progression from larger to lower total energies induces that the discrete function values $\tilde{f}_n(z_j, (\hat{\epsilon}_{i,p})_{k+1/2})$ with shifted energy argument $\hat{\epsilon}_{i,p}$ are already known for all j in the case of exciting, dissociating, and ionizing collision processes and can thus be directly determined. Only de-exciting electron collision processes comprising function values at total energies smaller than the actual one have to be treated iteratively, in a similar way as discussed in section 3.3.1.

Another issue concerns the discretization of the set of equations (3.48) and (3.49). Here, recent studies indicate that a first-order accurate, fully implicit differencing scheme with respect to the total energy ϵ at the points $(z_{j+1/2}, \epsilon_k)$ appears to be preferable to obtain converged results, in particular for the analysis of molecular gases like CO_2.

In the following, results for the spatial relaxation of electrons in argon and CO_2 plasmas with a gas density of $N = 3.54 \times 10^{16} \text{ cm}^{-3}$ acted upon by a space-independent electric field of $E = -10.62 \text{ V cm}^{-1}$ are represented. The results of these studies, resembling an idealized SST experiment with plane-parallel geometry, have been obtained by solving the equation system (3.48) and (3.49) by means of the multi-term method using, typically, $l = 8$ expansion coefficients. In order to initiate the spatial electron relaxation at $z = 0$, a Gaussian-like distribution $G_z(U) = U G_t(U)$ with the center $U_c = 5$ eV and width $U_w = 2$ eV was used to prescribe the first anisotropic component $f_1(z = 0, U)$ of the EVDF and the further anisotropic components $f_n(z = 0, U)$ with odd $n \geqslant 3$ were chosen to be zero. The corresponding boundary values in the transformed space read $\hat{f}_1(\epsilon) = G_z(\epsilon)$ and $\hat{f}_{n=3,5,\ldots}(\epsilon) = 0$. The factor c in G_z is used to normalize all expansion coefficients on the electron density $n_e(0)$ at $z = 0$. Because a negative value is chosen for the constant electric field E, the electrons are accelerated in the positive z-direction and therefore the spatial relaxation of the properties of the electrons happens in that direction as well.

Figure 3.6 illustrates the spatial relaxation of the isotropic distribution $f_0(z, U)/n_e(z)$ normalized by the electron density $n_e(z)$ for argon (a) and carbon dioxide (b). It clearly demonstrates that the spatial relaxation behavior of $f_0(z, U)/n_e(z)$ initiated by the boundary value of the anisotropic distribution $f_1(0, U)/n_e(0)$ is very different for the two gases considered. A strongly damped spatial relaxation with a comparatively small relaxation length takes place in the molecular gas CO_2 (figure 3.6(b)), where the normalized space-independent distributions $f_n(z, U)/n_e(z)$ belonging to the hydrodynamic state with constant $[dn_e(z)/dz]/n_e(z)$ appear already after about 1.5 cm. In contrast, a weakly damped, pronouncedly periodic evolution is found for the rare gas Ar. Therefore, only the initial part of the spatial relaxation of $f_0(z, U)/n_e(z)$ is presented in figure 3.6(a). This periodic behavior results from the interplay of the electron acceleration caused by the electric field action and the energetic backscattering of electrons due to inelastic collision processes. It is more pronounced in the case of argon due to the larger value of the lowest threshold energy of 11.55 eV compared to CO_2.

Figure 3.7 presents the corresponding expansion coefficients $f_n(z, U)/n_e(z)$ with $n = 0, 1, 2$ during the initial part of the spatial relaxation process at $z = 1$ cm. Two peaks are obvious for the expansion coefficients in argon (figure 3.7(a)), which are separated by about 11 eV and are caused by the interaction of the electric field action and the inelastic collision processes mentioned. The impact of expansion coefficients

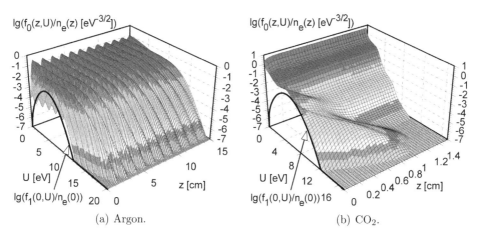

(a) Argon. (b) CO_2.

Figure 3.6. Spatial relaxation of $f_0(z, U)/n_e(z)$ in argon and CO_2 at $E = -10.62$ V cm^{-1}.

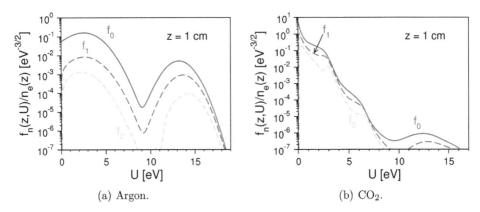

Figure 3.7. Expansion coefficients $f_n(z, U)/n_e(z)$ with $n = 0, 1, 2$ at $z = 1$ cm.

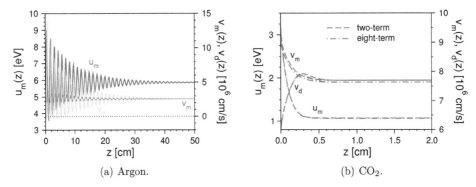

Figure 3.8. Spatial relaxation of mean energy, mean velocity, and drift velocity.

higher than $l = 2$ is found to be generally small during the entire spatial relaxation in argon. In contrast, the anisotropy coefficients $f_n(z, U)/n_e(z)$ with $n = 1$ and 2 reach values of the order of the isotropic distribution $f_0(z, U)/n_e(z)$ for CO_2 (figure 3.7(b)) indicating the impact of higher-order terms on the EVDF already during the course of the spatial relaxation.

The spatial relaxation of the mean energy $u_m(z) = u_e(z)/n_e(z)$ and mean velocity $v_m(z) = j_z(z)/n_e(z)$ of the electrons in argon and carbon dioxide at $E = -10.62$ V cm^{-1} and $N = 3.54 \times 10^{16}$ cm^{-3} is shown in figure 3.8. The mean velocity is composed of the electron drift velocity $v_d(z) = -\mu_e(z)E(z)$ and the diffusion velocity part, where the electron mobility μ_e is given by [41]

$$\mu_e(z) = -\frac{e_0}{3n_e(z)}\left(\frac{2}{m_e}\right)^{1/2}\int_0^\infty \frac{1}{\sum_i N_i(z)\left(Q_i^d(U) + \sum_p Q_{i,p}^{in,0}(U)\right)}$$

$$\times\left[U\left(\frac{\partial}{\partial U}f_0(z, U) + \frac{2}{5}\frac{\partial}{\partial U}f_2(z, U)\right) + \frac{3}{5}f_2(z, U)\right] \qquad (3.60)$$

in the multi-term approximation. Thus, the figure displays the electron drift velocity v_d in addition. It is found for argon (figure 3.8(a)) that the macroscopic properties are subject to larger periodic alterations during the weakly damped relaxation process and run into the hydrodynamic state after more than 50 cm. There the mean velocity is about 0.5% smaller than the drift velocity because of a small electron density increase due to ionization in accordance with [40]. Furthermore, the electron drift velocity v_d changes its sign several times and becomes negative during the early relaxation course up to about $z = 5$ cm so that states with negative mobility appear again. Compared with this, the mean electron velocity v_m always remains positive due to the impact of the diffusive flux.

The strongly damped spatial relaxation process in CO_2 (figure 3.8(b)) does not exhibit any negative mobility. There, the mean velocity v_m is greater than the drift velocity v_d during the early relaxation course. When approaching the hydrodynamic state, v_m and v_d are almost equal because the impact of the dissociative electron attachment is very small. At the same time, certain differences between two-term and converged eight-term results, in particular for the velocities, are obvious during the entire spatial relaxation.

In order to assess the relaxation length, analysis of the corresponding macroscopic balances resulting from equations (3.18), (3.19), and (3.20), similar to that discussed at the end of section 3.3.1, is the most appropriate method. Figure 3.9 represents the spatial relaxation of the corresponding particle, momentum, and power gain to loss ratios $\nu^{io}(z)/\nu^{at}(z)$, $I^f(z)/[I^{el}(z) + \sum_p I_p^{in}(z)]$ and $P^f(z)/[P^{el}(z) + \sum_p P_p^{in}(z) + P^{at}(z)]$, respectively. For argon (figure 3.9(a)), the ionization frequency $\nu^{io}(z)/\nu^{io}(\infty)$ normalized by its space-independent value $\nu^{io}(\infty)$ is shown and a shift by 1 and 2, respectively, was added to the power and momentum gain to loss ratio for a clearer presentation. In contrast to the temporal relaxation, the spatial relaxation process of the power and momentum balances takes place on the same spatial scale in accordance with [4]. However, the relaxation length required to reach hydrodynamic conditions is finally determined by the establishment of a constant particle gain to loss ratio, as figure 3.9(b) clearly shows.

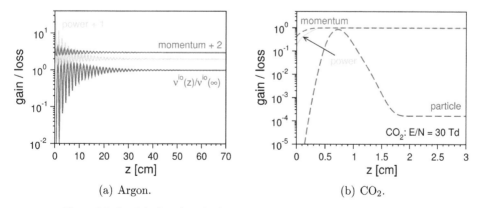

(a) Argon.   (b) CO_2.

Figure 3.9. Spatial relaxation of gain to loss ratios and the ionization frequency.

3.4 Concluding remarks

An introduction to the kinetic treatment of electrons by solving their Boltzmann equation has been given. The focus is on the determination of the velocity distribution function and related macroscopic properties of the electrons in the framework of a generally valid multi-term description. The basic relations of this multi-term approach have been represented and details about the numerical methods for solving both the time-dependent and the spatially one-dimensional electron Boltzmann equation in higher-order accuracy have been discussed.

The corresponding numerical multi-term methods have been applied to study the electron relaxation in a constant electric field in argon and carbon dioxide. The results of the investigations have illustrated the pronounced non-equilibrium behavior of the electrons including similarities and differences for the two gases considered. They have also shown the impact of terms higher than two of the Legendre polynomial expansion of the EVDF. Furthermore, the analysis has made it possible to obtain a clear assessment of the non-local response of the electrons during the temporal and spatial course of the relaxation processes. The resulting relaxation times and relaxation lengths approaching steady states should be estimated on the basis of the macroscopic balances using the method presented.

The spectrum of collision processes considered in the kinetic treatment covers elastic electron collisions as well as several binary inelastic collision processes, such as excitation, de-excitation, dissociation, ionization, and attachment. In order to allow for an analysis of various other aspects which can affect the non-local behavior of the electrons, an extension of the kinetic description and the resulting numerical methods is required. These may include, for example, the effect of the neutral gas temperature in elastic collision processes [14, 22, 42], the sharing of the remaining energy between the two electrons after an ionization event [26, 27], the additional impact of the electron–electron interaction [14, 42], and the influence of further non-conservative electron collision processes such as electron generation due to Penning ionization, chemo-ionization, and collisional and associative detachment, or electron loss due to electron–ion recombination, three-body electron attachment, and recombination [12, 26]. The integration of these aspects into the multi-term treatment can be done immediately and has already been partly realized.

Another important point concerns the integration of both the time-dependent and space-dependent multi-term method into a self-consistent description of non-thermal plasmas by means of hybrid modeling [9]. The realization of this numerical approach is still an open task. While the inclusion of the time-dependent multi-term method in such a more complex plasma description can be realized relatively easily, the corresponding integration of the space-dependent multi-term approach is more challenging. This results, for instance, from the transformation to total energy needed for the solution, which requires special care, in particular for conditions under which electric field reversal occurs [43]. Here, the use of an equidistant grid in coordinates and total energy appears to be necessary [44] to allow for an effective iterative solution approach.

Bibliography

[1] Robson R E, White R D and Petrović Z Lj 2005 Colloquium: physically based fluid modeling of collisionally dominated low-temperature plasmas *Rev. Mod. Phys.* **77** 1303–20

[2] Tsendin L D 2009 Electron kinetics in glows—from Langmuir to the present *Plasma Sources Sci. Technol.* **18** 014020

[3] Donkó Z 2011 Particle simulation methods for studies of low-pressure plasma sources *Plasma Sources Sci. Technol.* **20** 024001

[4] Winkler R, Loffhagen D and Sigeneger F 2002 Temporal and spatial relaxation of electrons in low temperature plasmas *Appl. Surf. Sci.* **192** 50–71

[5] Winkler R, Arndt S, Loffhagen D, Sigeneger F and Uhrlandt D 2004 Progress of the electron kinetics in spatial and spatiotemporal plasma structures *Contrib. Plasmaphys.* **44** 437–49

[6] Loffhagen D, Sigeneger F and Winkler R 2008 Electron kinetics in weakly ionized plasmas *Low Temperature Plasmas. Fundamentals, Technologies and Techniques* ed R Hippler *et al* 2nd ed (New York: Wiley) chapter 2

[7] Hagelaar G J M and Pitchford L C 2005 Solving the Boltzmann equation to obtain electron transport coefficients and rate coefficients for fluid models *Plasma Sources Sci. Technol.* **14** 722–33

[8] Porokhova I A, Golubovskii Yu B, Bretagne J, Tichy M and Behnke J F 2001 Kinetic simulation model of magnetron discharges *Phys. Rev.* E **63** 056408

[9] Loffhagen D and Sigeneger F 2009 Advances in Boltzmann equation based modelling of discharge plasmas *Plasma Sources Sci. Technol.* **18** 034006

[10] Dyatko N A, Ionikh Yu Z, Meshchanov A V, Napartovich A P and Barzilovich K A 2010 Specific features of the current-voltage characteristics of diffuse glow discharges in Ar:N$_2$ mixtures *Plasma Phys. Rep.* **36** 1040–64

[11] Pintassilgo C D, Guerra V, Guaitella O and Rousseau A 2010 Modelling of an afterglow plasma in air produced by a pulsed discharge *Plasma Sources Sci. Technol.* **19** 055001

[12] Hübner M, Gortschakow S, Guaitella O, Marinov D, Rousseau A, Röpcke J and Loffhagen D 2016 Kinetic studies of NO formation in pulsed air-like low-pressure dc plasmas *Plasma Sources Sci. Technol.* **25** 035005

[13] Desloge E A 1966 *Statistical Physics* (New York: Holt, Rinehart and Winston)

[14] Shkarofsky I P, Johnston T W and Bachynski M P 1966 *The Particle Kinetics of Plasmas* (London: Addison-Wesley)

[15] Golant V E, Zhilinsky A P and Sakharov I E 1980 *Fundamental of Plasma Physics* (New York: Wiley)

[16] White R D, Robson R E, Dujko S, Nicoletopoulos P and Li B 2009 Recent advances in the application of Boltzmann equation and fluid equation methods to charged particle transport in non-equilibrium plasmas *J. Phys. D: Appl. Phys.* **42** 194001

[17] Loffhagen D and Winkler R 1996 Time-dependent multi-term approximation of the velocity distribution in the temporal relaxation of plasma electrons *J. Phys. D: Appl. Phys.* **29** 618–27

[18] Winkler R, Petrov G, Sigeneger F and Uhrlandt D 1997 Strict calculation of electron energy distribution functions in inhomogeneous plasmas *Plasma Sources Sci. Technol.* **6** 118–32

[19] Robson R E, Winkler R and Sigeneger F 2002 Multiterm spherical tensor representation of Boltzmann's equation for a nonhydrodynamic weakly ionized plasma *Phys. Rev.* E **65** 056410

[20] Sigeneger F and Winkler R 2002 Study of the electron kinetics in cylindrical hollow cathodes by a multi-term approach *Eur. Phys. J.* **19** 211–23

[21] Porokhova I A, Golubovskii Yu B and Behnke J F 2005 Anisotropy of the electron component in a cylindrical magnetron discharge. I. Theory of the multiterm analysis *Phys. Rev.* E **71** 066406

[22] Hannemann M, Hardt P, Loffhagen D, Schmidt M and Winkler R 2000 The electron kinetics in the cathode region of $H_2/Ar/N_2$ discharges *Plasma Sources Sci. Technol.* **9** 387–99

[23] Wilhelm J and Winkler R 1968 Tensorielle Kugelflächenfunktionsentwicklung in der Boltzmann-Gleichung für das Lorentz-Plasma bzw. -Gas *Beitr. Plasmaphys* **8** 167–88

[24] Winkler R, Braglia G L and Wilhelm J 1988 Impact of nonisotropic elastic and inelastic scattering on the electron velocity distribution in weakly ionized plasmas *Nuovo Cimento* D **10** 1209–34

[25] Margenau H 1948 Theory of high frequency gas discharges. I. Methods for calculating electron distribution functions *Phys. Rev.* **73** 297–308

[26] Loffhagen D and Winkler R 1994 A new nonstationary Boltzmann solver in self-consistent modelling of discharge pumped plasmas for excimer lasers *J. Comput. Phys.* **112** 91–101

[27] Grubert G K and Loffhagen D 2014 Boltzmann equation and Monte Carlo analysis of the electrons in spatially inhomogeneous, bounded plasmas *J. Phys. D: Appl. Phys.* **47** 025204

[28] Richtmyer R D and Morton K W 1967 *Difference Methods for Initial Value Problems* 2nd edn (New York: Wiley)

[29] Thomée V 1969 Stability theory for partial difference operators *SIAM Rev.* **11** 152–95

[30] Thomas J W 1995 *Numerical Partial Differential Equations: Finite Difference Methods* (*Texts in Applied Mathematics* vol 22) (New York: Springer)

[31] Hayashi M 1992 Electron collision cross sections *Plasma Material Science Handbook* ed Japan Society for the Promotion of Science (Tokyo: Ohmsha) pp. 748–66

[32] Hayashi M 2003 Bibliography of electron and photon cross sections with atoms and molecules published in the 20th century: argon *National Institute for Fusion Science Technical Report* NIFS-DATA-72

[33] Winkler R, Braglia G L, Hess A and Wilhelm J 1984 Fundamentals of a technique for determining electron distribution functions by multi-term even-order expansion in Legendre polynomials. 1. Theory, *Beitr. Plasmaphys.* **24** 657–74

[34] Winkler R, Braglia G L and Wilhelm J 1988 Modification of the electron velocity distribution function in weakly ionized plasmas by nonisotropic elastic-scattering processes *Nuovo Cimento* D **10** 1031–60

[35] von Rosenberg D U 1969 *Methods for the Numerical Solution of Partial Differential Equations* (*Modern Analytic and Computational Methods in Science and Mathematics* vol 16) (New York: Elsevier)

[36] Dyatko N A, Kochetov I V and Napartovich A P 2014 Non-thermal plasma instabilities induced by deformation of the electron energy distribution function *Plasma Sources Sci. Technol.* **23** 043001

[37] Petrov G and Winkler R 1997 Multi-term treatment of electron kinetics in inhomogeneous nonequilibrium plasmas *J. Phys. D: Appl. Phys.* **30** 53–66

[38] Loffhagen D, Sigeneger F and Winkler R 2002 Study of the electron kinetics in the anode region of a glow discharge by a multiterm approach and Monte Carlo simulations *J. Phys. D: Appl. Phys.* **35** 1768–76

[39] Becker M M, Grubert G K and Loffhagen D 2010 Boundary conditions for the electron kinetic equation using expansion techniques *Eur. Phys. J. Appl. Phys.* **51** 11001

[40] Dujko S, White R D and Petrović Z Lj 2008 Monte Carlo studies of non-conservative electron transport in the steady-state Townsend experiment *J. Phys. D: Appl. Phys.* **41** 245205

[41] Grubert G K, Becker M M and Loffhagen D 2009 Why the local-mean-energy approximation should be used in hydrodynamic plasma descriptions instead of the local-field approximation *Phys. Rev.* E **80** 036405

[42] Loffhagen D 2005 Impact of electron–electron collisions on the spatial electron relaxation in non-isothermal plasmas *Plasma Chem. Plasma Process.* **25** 519–38

[43] Loffhagen D, Sigeneger F and Winkler R 2004 The effect of a field reversal on the spatial transition of the electrons from an active plasma to a field-free remote plasma *Eur. Phys. J. Appl. Phys.* **25** 45–56

[44] Grubert G K and Loffhagen D 2007 Nonequilibrium properties of electrons in oxygen plasmas *IEEE Trans. Plasma Sci.* **35** 1215–22

IOP Publishing

Plasma Modeling
Methods and Applications
Giovanni Lapenta

Chapter 4

Particle-based simulation of plasmas

4.1 Types of interacting systems

The first step to decide how to model a system of interacting particles is to distinguish between weakly and strongly interacting systems. The distinction has a profound meaning and a direct impact on how to model the system on a computer.

Figure 4.1 summarizes the situation visually. Let us consider a system made of a collection of charged particles in a box with the side of Debye length, λ_D (the box is three-dimensional (3D) but is depicted as two-dimensional (2D) for convenience). We choose the Debye length because a basic property of plasmas is to shield the effects of localized charges over distances exceeding the Debye length. Of course the shielding is exponential and the effect is not totally canceled over one Debye length, but such a length provides a conventional reasonable choice for the interaction range. The electric field in each point of the box is computed by the superposition of the contributions of each particle.

Let us conduct an ideal thought experiment based on using a device that is able to detect the local electric field in one spatial position. We identify in the figure such an ideal measurement device with an icon connected to a point in the system. We try to conduct a thought experiment where no law of physics is violated in any step, but where the difficulties of experimental work are eliminated.

If we consider the configuration in figure 4.1(a), we note that within the domain there are few particles and the measurement obtained by our fantastical electric field meter would be very jumpy. The particles in the box move constantly, interacting with each other and agitated by their thermal motion. As a particle passes by the detector, the measurement detects a jump up and when a particle moves away it detects a jump down. On average at any given time very few particles are near the detector and their specific positions are key in determining the value measured. The effect of a given particle on the electric field at the location of measurement decays very rapidly with distance and only when the particle is nearby the effect is strong.

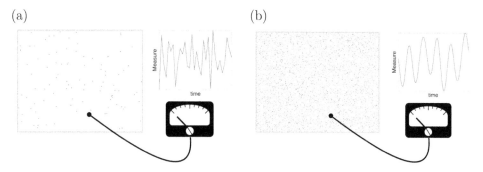

Figure 4.1. Thought experiment for strongly (a) versus weakly (b) coupled systems.

The same effect is detected by each of the particles in the system. The electric field each particle feels is a sum of the contributions of all others, but only when another particle passes by the electric field would a jump be registered: in common terms this event is called a *collision*. The particle trajectories would then be affected by a series of close encounters registered as jumps (collisions) in the trajectory.

The system described goes by the name of a *strongly coupled system* in the language of kinetic theory, a system where the evolution is determined by the close encounters and by the relative configuration of any two pairs of particles. The condition just described is characterized by the presence of few particles in the Debye box.

The opposite situation is that of a weakly coupled system. The corresponding configuration is described in figure 4.1(b).

Now the system is characterized by being composed of an extremely large number of particles in the Debye box. At any given point, the number of particles contributing to the electric field is very large. Regardless of the particle motion, the field is given by the superposition of many contributions. As a consequence, by simple averaging of the effects of all the particles contributing to the measurement, the measurement is smooth and does not jump in time. Similarly the trajectory of a particle is at any time affected by a large number of other particles. The trajectory is smooth and without jumps. These systems are called *weakly coupled*. If in the strongly coupled system the characteristic feature was the presence of a succession of collisions, in the weakly coupled system the characteristic feature is the *mean field* produced by the superposition of contributions from a large number of particles.

4.1.1 Strength of interaction

The discriminant factor in the previous discussion was the number of particles present inside the box under consideration. If we choose the conventional box with sides equal to the Debye length, the number of particles present is

$$N_D = n\lambda_D^3 \qquad (4.1)$$

where n is the plasma density.

A system is considered weakly coupled when N_D is large and strongly coupled when N_D is small.

This concept can be further elaborated by considering the energies of the particles in the system. The particles in the box are distributed in a non-uniform, random way, but on average the volume associated with each particle is simply the volume of the box, λ_D^3, divided by the number of particles in the box, N_D. This volume, $V_p = n^{-1}$, can be used to determine the average interparticle distance, $a = V_p^{1/3} \equiv n^{-1/3}$. This relation provides an average statistical distance. The particles are distributed randomly and their distances are also random, but on average the interparticle distance is a.

The electrostatic potential energy between two particles with separation a is

$$E_{pot} = \frac{q^2}{4\pi\epsilon_0 a},$$ (4.2)

where we have assumed equal charge q for the two particles. Conversely, from statistical physics, the kinetic energy of the particles is of the order of the thermal energy:

$$E_{th} = kT,$$ (4.3)

where k is the Boltzmann constant.

A useful measure of the plasma coupling is given by the so-called *plasma coupling parameter*, Γ, defined as

$$\Gamma = \frac{E_{pot}}{E_{th}} = \frac{q^2}{4\pi\epsilon_0 akT}.$$ (4.4)

Recalling the definition of Debye length ($\lambda_D = (\epsilon_0 kT/nq^2)^{1/2}$) and the value of a obtained above, it follows that

$$\Gamma = \frac{q^2 n^{1/3}}{4\pi\epsilon_0 kT} \equiv \frac{1}{4\pi N_D^{2/3}}.$$ (4.5)

The plasma coupling parameter gives a new physical meaning to the number of particles per Debye cube. When many particles are present in the Debye cube, the coupling parameter is small. In this case, the thermal energy far exceeds the potential energy, thus the trajectory of each particle is little influenced by interactions with the other particles: this is the condition outlined above for the weakly coupled systems. Conversely, when the number of particles per Debye cube is small, the coupling parameter is large, the potential energy dominates, and the trajectories are strongly affected by the near-neighbor interactions: this is the condition typical of strongly coupled systems.

4.2 Computer simulation of interacting systems

A computer simulation of a system of interacting particles can be performed, in principle, by simply following each particle in the system. The so-called *particle–particle (PP)* approach describes the motion of N particles by evolving the equations of Newton for each of the N particles, taking as a force acting on the particle the combined effect of all the other particles in the system.

The evolution is discretized in many temporal steps Δt, each chosen so that the particles move only a small distance. After each move, the force is recomputed and a new move is made for all the particles. The main cost of the effort is the computation of the force, which requires one to sum over all the particles in the system. Once the force is computed the new velocities can be computed. Then the new positions can be computed and the cycle can be repeated indefinitely.

For each particle, the number of terms to sum to compute the force is $N - 1$, and considering that there are N particles, but that each pair needs to be computed only once, the total number of force computations is $N(N - 1)/2$.

For strongly coupled systems, where the number of particles per Debye cube is small, the PP approach is feasible and forms the basis of the very successful *molecular dynamics method* used in condensed matter and in biomolecular studies. We refer the reader to a specific textbooks on molecular dynamics to investigate the approach in more depth [1].

The PP approach is also used in the study of gravitational interactions, for example in the cosmological studies of the formation and distribution of galaxies. In that case, specifically the dark matter is studied with a PP approach. The PP approach can be made more efficient by using the *Barnes–Hut* or *tree algorithm* [2] which can reduce the cost (but not without loss of information [3]) to $O(N \log N)$.

Even with the reduced cost of the tree algorithm, PP methods cannot be practical for weakly coupled systems where the number of particles is very large. As the number of particles increases, the cost scales quadratically (or as $N \log N$) and makes the computational effort unmanageable. In that case, one cannot simply describe every particle in the system and a method must be devised to reduce the description to just a statistical sample of the particles. This is the approach of the *particle-in-cell (PIC)* method (sometimes referred to as particle–mesh).

The key idea behind the simulation of weakly coupled systems is to use as the building blocks of the model not single particles, but rather collective clouds of them: each *computational particle* (referred to sometimes as a super-particle) represents a group of particles and can be visualized as a small piece of phase space. The concept is visualized in figure 4.2.

The fundamental advantage of the finite-size particle approach is that the computational particles, being of finite size, interact more weakly than point particles. When two point particles interact, for example via the Coulomb force, the repulsive or attractive force grows as the particles approach, reaching a singularity at zero separation. Finite-size particles, instead, behave as point particles until their respective domains start to overlap. Once overlap occurs the overlap zone is neutralized, not contributing to the force between the particles. At zero distances when the particles fully overlap (assuming here that all particles have the same size) the force becomes zero. The figure presented is 2D for convenience but the reader is invited to try to visualize this also in 3D.

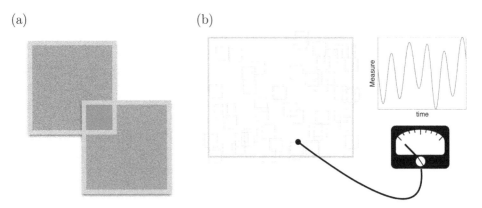

(a)   (b)

Figure 4.2. Finite-size computational particles. (a) shows the overlap of two finite-size particles: the part of the particle clouds that overlaps does not contribute to the force between the two clouds. As two clouds become closer to each other, the force increases only until the overlap starts. After the overlap starts increasing, the force starts to decrease to become zero for complete overlap. (b) shows a Debye cube filled by a set of computational particles to illustrate the idea of the PIC method.

The use of finite-size computational particles allows us to reduce the interaction among particles. A system of finite-size particles behaves as a weakly coupled system. Let us return to our fantastical electric field instrument. The passage of a computational particle is the passage of a cloud of finite size that is affecting the instrument more weakly at a point particle. At any given time multiple clouds are in the vicinity rendering the measurement smooth in time. In the same way, the motion of every cloud is affected by the other clouds in the vicinity. When two clouds begin to overlap their relative force weakens. As a result the motion of each cloud is smooth. Even though we have few computational particles in the Debye cube, the behavior resembles that of weakly coupled systems.

Recalling the definition of the plasma parameter, the use of finite-size particles in practice results in a reduction of the potential energy for the same kinetic energy. The beneficial consequence is that the correct plasma parameter can be achieved by using fewer particles than in the physical system. The conclusion is that the correct coupling parameter is achieved by fewer particles interacting more weakly. The realistic condition is recovered, in essence, by making two mistakes that cancel each other: using fewer particles and using weaker forces. Obviously great mathematical insight has gone into making this rigorous [4, 5]. Let us now see how that can be achieved.

4.3 The particle-in-cell method

The kinetic description of a plasma aims to describe the system via electromagnetic fields and distribution functions. A distribution function expresses the probability of finding a particle with certain properties (e.g. position and velocity). In its nature it is a probabilistic description of nature. It gives up on the hope of following every particle in the system and contents itself by describing the ensemble in statistical terms. A preliminary knowledge of the basic tenets of plasma kinetic theory is assumed in the remainder of this chapter. Classic textbooks provide the necessary preliminary knowledge (see, e.g. [6]).

In extreme summary, the starting point of a kinetic simulation model of plasmas is the Boltzmann equation. The phase-space distribution function $f_s(\mathbf{x}, \mathbf{v}, t)$ for a given species s (electrons or ions) is defined as the number of particles at time t in the infinitesimal element of phase space $\mathbf{dx}$ and $\mathbf{dv}$ around a certain phase-space point $(\mathbf{x}, \mathbf{v})$. The distribution of each species is governed by the Boltzmann equation

$$\frac{\partial f_s}{\partial t} + \mathbf{v} \cdot \frac{\partial f_s}{\partial \mathbf{x}} + \frac{q_s}{m_s}(\mathbf{E} + \mathbf{v} \times \mathbf{B}) \cdot \frac{\partial f_s}{\partial \mathbf{v}} = St(f_s), \tag{4.6}$$

where q_s and m_s are the charge and mass of the species, respectively. The last term is the collision operator that summarizes collisions with a neutral medium (for partially ionized plasmas or for plasmas with immersed dust) or among the charged particles of the plasma. By its definition it is obvious that the integral of f_s over all the phase space is the total number of particles in the system.

The derivatives are written in the standard vector notation in the 3D configuration space $\mathbf{x}$ and 3D velocity space $\mathbf{v}$, forming together the classical phase space.

The electric and magnetic fields are given by the two curl Maxwell's equations

$$\nabla \times \mathbf{E} = -\frac{\partial \mathbf{B}}{\partial t}$$
$$\nabla \times \mathbf{B} = \mu_0 \epsilon_0 \frac{\partial \mathbf{E}}{\partial t} + \mu_0 \mathbf{j} \tag{4.7}$$

supplemented by the two divergence equations

$$\epsilon_0 \nabla \cdot \mathbf{E} = \rho$$
$$\nabla \cdot \mathbf{B} = 0, \tag{4.8}$$

where the net charge density is computed from the distribution functions as

$$\rho(\mathbf{x}, t) = \sum_s q_s \int_{\mathbb{V}} f_s(\mathbf{x}, \mathbf{v}, t) \mathrm{d}\mathbf{v} \tag{4.9}$$

and the current density as

$$\mathbf{j}(\mathbf{x}, t) = \sum_s q_s \int_{\mathbb{V}} \mathbf{v} f_s(\mathbf{x}, \mathbf{v}, t) \mathrm{d}\mathbf{v}, \tag{4.10}$$

where $\mathbb{V}$ is the velocity space.

4.3.1 Mathematical formulation of PIC

The PIC method can be regarded as a representation of the distribution function of each species by a superposition of moving elements, each representing a cloud of physical particles. We follow here the specific approach described in [7].

The mathematical formulation of the PIC method is obtained by assuming that the distribution function of each species is given by the superposition of several elements (called computational particles or superparticles):

$$f_s(\mathbf{x}, \mathbf{v}, t) = \sum_p f_p(\mathbf{x}, \mathbf{v}, t). \tag{4.11}$$

Each element represents a large number of physical particles that are near each other in phase space. For this reason, the choice of elements is made in order to be at the same time physically meaningful (i.e. to represent a group of particles near each other) and mathematically convenient (i.e. to allow the derivation of a manageable set of equations) [8].

The PIC method is based upon assigning to each computational particle a specific functional form for its distribution, a functional form with a number of free parameters whose time evolution will determine the numerical solution of the Vlasov equation [9]. In the standard PIC methods [4, 5], the choice is made to have two free parameters in the functional shape for each spatial dimension. The free parameters will acquire the physical meaning of position and velocity of the computational particle. More advanced methods also take into account the evolving shape of the physical cloud the super-particle is supposed to represent [9].

In the standard PIC model, the functional dependence is assumed to be a tensor product of the shape in each direction of phase space [5]:

$$f_p(\mathbf{x}, \mathbf{v}, t) = N_p S_{\mathbf{x}}(\mathbf{x} - \mathbf{x}_p(t)) S_{\mathbf{v}}(\mathbf{v} - \mathbf{v}_p(t)), \tag{4.12}$$

where $S_{\mathbf{x}}$ and $S_{\mathbf{v}}$ are the *shape functions* for the computational particles and N_p is the number of physical particles that are present in the element of phase space represented by the computational particle.

A number of properties of the shape functions come from their definition:

1. The support of the shape functions is compact in order to describe a small portion of phase space (i.e. it is zero outside a small range).
2. Their integral over the domain is unitary:

$$\int_{V_\xi} S_\xi(\xi - \xi_p) \mathrm{d}\xi = 1, \tag{4.13}$$

   where ξ stands for any coordinate or any velocity direction.
3. While not strictly necessary, Occam's razor suggests choosing symmetric shapes:

$$S_\xi(\xi - \xi_p) = S_\xi(\xi_p - \xi). \tag{4.14}$$

While these definitions still leave very great freedom in choosing the shape functions, traditionally the choices used in practice are very few.

4.3.2 Selection of the particle shapes

A critical choice in the definition of a PIC algorithm is the choice of the shape functions.

For the velocity, S_v, virtually all PIC methods assume a Dirac's delta in each direction:

$$S_v(\mathbf{v} - \mathbf{v}_p) = \delta(v_x - v_{xp})\delta(v_y - v_{yp})\delta(v_z - v_{zp}). \tag{4.15}$$

This choice has the fundamental advantage that if all particles within the element of phase space described by one computational particle have the same speed, they remain closer in phase space during the subsequent evolution.

The original PIC methods developed in the 1950s were based on using a Dirac's delta also as the shape function in space. But now for the spatial shape functions, all commonly used PIC methods are based on the use of the so-called b-splines [10]. The b-spline functions are a series of consecutively higher-order functions obtained from each other by integration. The first b-spline is the flat-top function $b_0(\xi)$ defined as

$$b_0(\xi) = \begin{cases} 1 & \text{if } |\xi| < 1/2 \\ 0 & \text{otherwise.} \end{cases} \tag{4.16}$$

The subsequent b-splines, b_ℓ, are obtained by successive integration via the following generating formula:

$$b_\ell(\xi) = \int_{-\infty}^{\infty} d\xi' b_0(\xi - \xi') b_{\ell-1}(\xi'). \tag{4.17}$$

Figure 4.3 shows the first three b-splines.

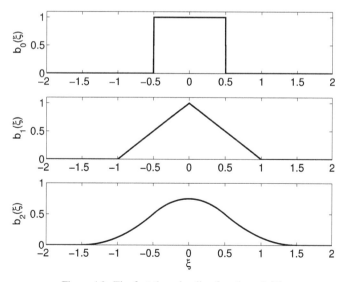

Figure 4.3. The first three b-spline functions, $b_\ell(\xi)$.

Three crucial properties of splines are:

(a) When a b-spline of any order is evaluated on a uniform grid of step 1, the sum over all points of evaluation is unitary, regardless of the central point ξ of the b-spline:

$$\sum_i b_\ell(\xi + i) = 1, \tag{4.18}$$

a property of great convenience in particle interpolation.

(b) The integral of b-splines of any order is unitary:

$$\int_{-\infty}^{\infty} b_\ell(\xi)\mathrm{d}\xi = 1. \tag{4.19}$$

(c) The Dirac's delta $\delta(\xi)$ can be regarded as $b_{-1}(\xi)$.

Based on the b-splines, the spatial shape function of PIC methods is chosen as

$$S_{\mathbf{x}}(\mathbf{x} - \mathbf{x}_p) = \frac{1}{\Delta x_p \Delta y_p \Delta z_p} b_\ell\left(\frac{x - x_p}{\Delta x_p}\right) b_\ell\left(\frac{y - y_p}{\Delta y_p}\right) b_\ell\left(\frac{z - z_p}{\Delta z_p}\right), \tag{4.20}$$

where Δx_p, Δy_p, and Δz_p are the lengths of the supports of the computational particles (i.e. its size) in each spatial dimension.

A common choice in PIC is to use the shape functions equal to the b-splines of order 0, a choice referred to as *cloud-in-cell* because the particle is a uniform square cloud in space with an infinitesimal span in the velocity directions.

4.3.3 Derivation of the equations of motion

Following the procedure in [7], to derive the evolution equations for the free parameters $\mathbf{x}_p$ and $\mathbf{v}_p$, we require the first moments of the Vlasov equation to be exactly satisfied by the functional forms chosen for the elements. This procedure requires some explanation.

First, the Vlasov equation is formally linear in f_s and the equation satisfied by each element is still the same Vlasov equation. The linear superposition of the elements gives the total distribution function and if each element satisfies the Vlasov equation, the superposition does too. A caveat is that the fields really depend on f_s making the Vlasov equation non-linear. As a consequence the fields used in each Vlasov equation for each element must be the total fields due to all elements, the same entering the complete Vlasov equation for f_s. We can then write the Vlasov equation for each element f_p as

$$\frac{\partial f_p}{\partial t} + \mathbf{v} \cdot \frac{\partial f_p}{\partial \mathbf{x}} + \frac{q_s}{m_s}(\mathbf{E} + \mathbf{v} \times \mathbf{B}) \cdot \frac{\partial f_p}{\partial \mathbf{v}} = 0 \tag{4.21}$$

and observe that their sum reproduces correctly the Vlasov equation for the complete distribution function f_s.

Second, the arbitrary functional form chosen for the elements does not satisfy exactly the Vlasov equation. If we substitute the definition (4.12) in equation (4.21), the equation cannot be satisfied exactly at every point. This circumstance is typical of finite element methods [11] and is addressed by requiring that the equations be satisfied only in a weak sense, meaning that some averages are satisfied even though locally point by point the equation is not exactly valid.

In analogy with the usual procedure for finite element methods, we require that the moments of the equation are satisfied. The moments of the Vlasov equation are obtained multiplying by powers of $\mathbf{x}$ and $\mathbf{v}$ and integrating. Using the zeroth-order moment and the six first-order moments (one for each variable in phase space), the following complete set of evolution equations is obtained for the parameters defining the functional dependence of the distribution within each computational particle [7]:

$$\frac{\mathrm{d}N_p}{\mathrm{d}t} = 0$$

$$\frac{\mathrm{d}\mathbf{x}_p}{\mathrm{d}t} = \mathbf{v}_p \tag{4.22}$$

$$\frac{\mathrm{d}\mathbf{v}_p}{\mathrm{d}t} = \frac{q_s}{m_s}\left(\mathbf{E}_p + \mathbf{v}_p \times \mathbf{B}_p\right).$$

A great advantage of the PIC method is that its evolution equations resemble the same Newtonian equations as followed by the regular physical particles. The key difference is that the field is computed as average over the shape function based on the definition of $\mathbf{E}_p$ and $\mathbf{B}_p$ as

$$\mathbf{E}_p = \int S_{\mathbf{x}}(\mathbf{x} - \mathbf{x}_p)\mathbf{E}(\mathbf{x})\mathrm{d}\mathbf{x} \tag{4.23}$$

$$\mathbf{B}_p = \int S_{\mathbf{x}}(\mathbf{x} - \mathbf{x}_p)\mathbf{B}(\mathbf{x})\mathrm{d}\mathbf{x} \tag{4.24}$$

obtained in the course of the procedure for imposing the validity of the moments of the Vlasov equation.

The set of equations above provides a closed description for the Vlasov equation. Once accompanied by an algorithm to solve Maxwell's equations the full Vlasov–Maxwell system can be solved.

In summary, the particle method, as derived above, is a discretization of phase space based on a finite set of computational particles. The continuum distribution function is replaced by a discrete mathematical representation provided by the superposition of moving fixed-shape computational particles of finite size.

The pitfalls of this representation have been studied in depth in the past. A full discussion of this aspect is beyond the scope of the present review. We refer the reader to [4, 5, 9] for additional comments. In extreme summary, a element of phase space initially described correctly by the shape functions chosen for the discretization would be distorted by non-uniform electric and magnetic fields. Instead, the computational particles have a fixed shape and neglect this effect. The Liouville theorem requires the conservation of the phase-space volume of each element, a process that is still correctly described by the discretized system.

Alternatives to PIC have been developed to remove this error by introducing additional degrees of freedom for each computational element (super-particles) [9, 12]. The additional degrees of freedom can describe the particle shape and represent its distortion in phase space. However, the additional complexity of such methods has so far prevented their widespread use.

4.4 Coupling with the field equations: spatial discretization on a grid

So far the attention has focused on the particle discretization of the Vlasov part of the Vlasov–Maxwell system. To complete the simulation method, Maxwell's equations need to be discretized as well.

A wide variety of methods have been developed to do so. Essentially all electromagnetic solvers can be coupled with particle methods [4, 5, 13]. We use a general notation where the continuum operator ∇ is replaced by a discrete operator that approximates it, using the notation ∇_g. The discretized Maxwell's equations become

$$\nabla_g \times \mathbf{E} = -\frac{\partial \mathbf{B}}{\partial t}$$
$$\nabla_g \times \mathbf{B} = \mu_0 \mathbf{J}_g + \mu_0 \epsilon_0 \frac{\partial \mathbf{E}}{\partial t} \tag{4.25}$$

supplemented by the two divergence conditions,

$$\nabla_g \cdot \mathbf{E} = \frac{\rho_g}{\epsilon_0}$$
$$\nabla_g \cdot \mathbf{B} = 0. \tag{4.26}$$

Maxwell's equations require two inputs, the sources ρ_g and $\mathbf{j}_g$ that come from the information carried by the particles. In turn, the outputs are the electric and magnetic fields (or, equivalently the vector and scalar potential). The main concern here is the coupling of Maxwell's equations with the particles and its consequences.

We assume that the fields are known at grid points, not necessarily the same for both fields, and are defined as point values or averages over control volumes: $\mathbf{E}_g$, $\mathbf{B}_g$ where the index g labels the discrete points. Similarly, we assume that the chosen Maxwell scheme needs the sources at discrete grid points defined as averages over control volumes V_g defining each discrete point:

$$\rho_g = \sum_s \frac{q_s}{V_g} \int_\mathbf{v} \int_{V_g} f_s \, d\mathbf{v} d\mathbf{x}$$
$$\mathbf{j}_g = \sum_s \frac{q_s}{V_g} \int_\mathbf{v} \int_{V_g} \mathbf{v} f_s \, d\mathbf{v} d\mathbf{x}. \tag{4.27}$$

In all the formulas above we indicated the grid index as g generically. The cell positions can be chosen differently for different fields and different sources.

Staggered schemes and Yee lattices are commonly used [14, 15]. The considerations below are independent of the details on how space is discretized.

Recalling equations (4.11) and (4.12), the definition of the sources for Maxwell's equations requires a 3D integral of the particle shape over the control volume of each grid point:

$$\rho_g = \sum_p \frac{1}{V_g} \int_{V_g} S_x(\mathbf{x} - \mathbf{x}_p) d\mathbf{x}$$

$$\mathbf{j}_g = \sum_p \mathbf{v}_p \frac{1}{V_g} \int_{V_g} S_x(\mathbf{x} - \mathbf{x}_p) d\mathbf{x}, \tag{4.28}$$

where the integrations in velocity space were done easily by recalling the definition of S_v. The summation is over all particles of all species (for this reason the summation over the species is not indicated explicitly).

The integral in space is straightforward but can involve complicated geometric operations if one allows general shapes for the control volumes. Instead, the use of b-splines introduced above simplifies this operation tremendously. If the particle shapes are chosen as in equation (4.20) and the grid is chosen as uniform, Cartesian, and with cell sizes in each dimension equal to the particle sizes, a simple and elegant reformulation of the integration step can be obtained recalling property (4.17) of b-splines. Each dimension can be treated separately thanks to the choice of a Cartesian uniform grid and the shape of particles as the products of shapes in each direction. If we use the properties of the top hat function (i.e. the b-spline of order 0), the integral needed in each direction can be reformulated as

$$\int_{\Delta x_g} S_x(x - x_p) = \int_{-\infty}^{\infty} S_x(x - x_p) b_0 \left(\frac{x - x_g}{\Delta x_g} \right) dx, \tag{4.29}$$

where Δx_g is the interval in x relative to control volume V_g. Choosing $\Delta x_g = \Delta x_p$ (called simply Δx), recalling the definition of S_x, equation (4.20) and using equation (4.17):

$$\int_{-\infty}^{\infty} S_x(x - x_p) b_0 \left(\frac{x - x_g}{\Delta x} \right) dx = b_{l+1} \left(\frac{x_g - x_p}{\Delta x} \right). \tag{4.30}$$

Collecting the integral in each direction, the so-called interpolation function can be defined as

$$W(\mathbf{x}_g - \mathbf{x}_p) = b_{l+1} \left(\frac{x_g - x_p}{\Delta x} \right) b_{l+1} \left(\frac{y_g - y_p}{\Delta y} \right) b_{l+1} \left(\frac{z_g - z_p}{\Delta z} \right). \tag{4.31}$$

The summation property of the b-splines (4.18) ensures that

$$\sum_g W(\mathbf{x}_g - \mathbf{x}_p) = 1 \tag{4.32}$$

and the sum of the fractional contributions of a particle to the grid is unitary regardless of the position of the particle.

The interpolation function is a direct consequence of the choice made for the shape function. Using b-splines proves a powerful choice as it allows us to write the interpolation function just as the b-spline one order higher than that used for the shape. The calculation of the sources for Maxwell's equations then becomes just the sum of a number of function evaluations without requiring geometrically complex integrals:

$$\rho_g = \frac{1}{V_g} \sum_p q_p W(\mathbf{x}_g - \mathbf{x}_p)$$

$$\mathbf{j}_g = \frac{1}{V_g} \sum_p q_p \mathbf{v}_p W(\mathbf{x}_g - \mathbf{x}_p).$$

(4.33)

An analogous set of operations can be carried out for connecting the output of Maxwell's equations to the particle equations of motion. The needed quantities are the electric and magnetic fields derived above in equations (4.23) and (4.24). The definitions for the particle fields require continuum fields. These can be obtained once again using interpolation:

$$\mathbf{E}(\mathbf{x}) = \sum_g \mathbf{E}_g S_{\mathbf{E}}(\mathbf{x} - \mathbf{x}_g)$$

$$\mathbf{B}(\mathbf{x}) = \sum_g \mathbf{B}_g S_{\mathbf{B}}(\mathbf{x} - \mathbf{x}_g),$$

(4.34)

where $S_{\mathbf{E}}(\mathbf{x} - \mathbf{x}_g)$ and $S_{\mathbf{B}}(\mathbf{x} - \mathbf{x}_g)$ are the fields interpolation functions. Upon substitution in the definition of the particle fields,

$$\mathbf{E}_p = \sum_g \int S_{\mathbf{x}}(\mathbf{x} - \mathbf{x}_p) \mathbf{E}_g S_{\mathbf{E}}(\mathbf{x} - \mathbf{x}_g) d\mathbf{x}$$

$$\mathbf{B}_p = \sum_g \int S_{\mathbf{x}}(\mathbf{x} - \mathbf{x}_p) \mathbf{B}_g S_{\mathbf{B}}(\mathbf{x} - \mathbf{x}_g) d\mathbf{x}.$$

(4.35)

The geometrical complexity of the integrals can be avoided choosing the interpolation functions for the electric and magnetic fields as b-splines of order 0:

$$S_{\mathbf{E},\mathbf{B}}(\mathbf{x} - \mathbf{x}_g) = b_0\left(\frac{x_g - x_p}{\Delta_x}\right) b_0\left(\frac{y_g - y_p}{\Delta_y}\right) b_0\left(\frac{z_g - z_p}{\Delta_z}\right)$$

(4.36)

leading to

$$\mathbf{E}_p = \sum_g \mathbf{E}_g W(\mathbf{x}_g - \mathbf{x}_p) d\mathbf{x}$$

$$\mathbf{B}_p = \sum_g \mathbf{B}_g W(\mathbf{x}_g - \mathbf{x}_p) d\mathbf{x}$$

(4.37)

with the same interpolation function used for the sources.

The majority of PIC codes use b-spline interpolations and can take advantage of the convenient interpolation function defined above. For staggered grid, the interpolation will happen to different grid points for different quantities, but using the same interpolation function. In some implementations, different orders of interpolations can be used for different quantities [16]. A more radical solution is to consider non-uniform [17, 18] or even unstructured grids [19]. In the most complex cases, the exchange of information between particles and cells cannot use the interpolation function derived using b-splines and the integral definitions need to be used directly.

4.5 Temporal discretization of the particle methods

The coupled Vlasov–Maxwell system requires three discretizations. We have first described how to discretize phase space in the Vlasov equation using finite-size computational particles that resemble flat strings of finite size in space but of infinitesimal size in velocity. Next we have given some indications on how to discretize space. Space needs to be discretized on Maxwell's equations and a vast array of possibilities exist in the computational electromagnetism. We focus on how to couple these discretizations with the computational particle information.

The last and most important discretization remaining is time. Again for Maxwell's equations the temporal discretization can be performed in any of the various choices used in computational electromagnetism. What we need to focus on is how to combine the time discretization of the Maxwell equations with that of the particles. Figure 4.4 illustrates the concept. The field equations are solved on a grid where the fields are advanced based on the sources provided by the particles. The particle equations are solved using the fields computed on the grid. The interpolation

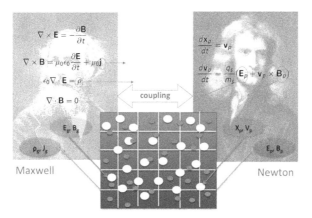

Figure 4.4. Illustration of the coupling between Maxwell's and Newton's equations. Newton's equations need the electric and magnetic fields and Maxwell's equations need the particle positions to compute the sources: current and density. The computational particles are indicated as coloured dots, the larger yellow dots are ions, and the smaller orange dots electrons. The dot size is indicative of the mass of the particles, not their sizes. All computational particles have the same size: their size is identical to the size of the cells but with their center on the center of the particles.

formulas derived above provide the rule to exchange information between the grid and particles.

The question is how to deal with the coupling of Maxwell's equations for the fields and Newton's equations for the computational particles. In principle as the particles move, the fields evolve as well, and any change on one side of figure 4.4 is reflected on the other. The first point to keep in mind is the vastness of the information being exchanged. Modern PIC methods on supercomputers (i.e. parallel computers made of hundreds of thousands of processors) use millions of cells and billions of particles. In fact, the state-of-the-art is starting to reach the trillion particle level. The desire to use more and more particles to cover phase space more accurately suggests keeping the operations per particle as simple as possible.

This guiding principle has long suggested the use of explicit methods, a decision whose wisdom will be questioned below. In explicit methods, the two sets of equations, Maxwell's and Newton's, are solved in a marching order. Using the visualization scheme in figure 4.4, we can assume that each side can be advanced for a small time step, while the other is assumed to be temporarily frozen. We can advance the particles for a small time step in given fields. We then use the new particle positions and speeds to compute the current and density to advance the fields for the same small time step. If the time step is small enough, this procedure has a small error. The method just described is called explicit. The main advantage of the explicit approach is its extreme simplicity, but this is accompanied by the disadvantage that the time step has to be very small.

4.5.1 Explicit temporal discretization of the particle equations

Let us start with discretizing in time Newton's equations for the particles. The simplest and most widely used explicit algorithm is the so-called *leap-frog algorithm* based on staggering the time levels of the velocity and position by a half time step: $x_p(t = n\Delta t) \equiv x_p^n$ and $v_p(t = (n + 1/2)\Delta t) \equiv v_p^{n+1/2}$. The update of position from time level n to time level $n + 1$ uses the velocity at mid-point $v_p^{n+1/2}$, and similarly the update of the velocity from time level $n - 1/2$ to $n + 1/2$ uses the mid-point position x_p^n. This stepping of velocity over position and of position over velocity gives the method its name for its resemblance to the children's game bearing the same name (see figure 4.5).

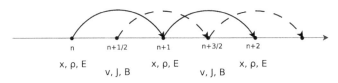

Figure 4.5. Visual representation of the leap-frog algorithm. The time discretization is staggered, with the electric field, charge density, and particle positions at integer times, and the magnetic field, current, and particle velocity at half time steps.

The scheme is summarized by

$$\frac{\mathbf{x}_p^{n+1} - \mathbf{x}_p^n}{\Delta t} = \mathbf{v}_p^{n+1/2}$$

$$\frac{\mathbf{v}_p^{n+1/2} - \mathbf{v}_p^{n-1/2}}{\Delta t} = \frac{q_s}{m_s} \mathbf{E}^n(x_p^n) + \frac{q_s}{m_s} \left(\frac{\mathbf{v}_p^{n+1/2} + \mathbf{v}_p^{n-1/2}}{2} \right) \times \mathbf{B}^n(x_p^n), \tag{4.38}$$

where the use of $\mathbf{E}_p^n$ and $\mathbf{B}_p^n$ implies knowing the solution of Maxwell's equations given the particle information. The first equation is clearly explicit, the second can be formulated explicitly as a roto-translation of the vector $\mathbf{v}_p^{n-1/2}$. The velocity equation can be rewritten in the equivalent explicit form

$$\mathbf{v}_p^{n+1/2} = 2\left(\hat{\mathbf{v}}_p + \beta_s \hat{\mathbf{E}}_p \right) - \mathbf{v}_p^{n-1/2}, \tag{4.39}$$

where hatted quantities have been rotated by the magnetic field:

$$\hat{\mathbf{v}}_p = \alpha_p^n \mathbf{v}_p^{n-1/2}$$

$$\hat{\mathbf{E}}_p = \alpha_p^n \mathbf{E}_p^n \tag{4.40}$$

via a rotation matrix α_p^n defined as

$$\alpha_p^n = \frac{1}{1 + \left(\beta_s B_p^n \right)^2} \left(\mathbb{1} - \beta_s \mathbb{1} \times \mathbf{B}_p^n + \beta_s^2 \mathbf{B}_p^n \mathbf{B}_p^n \right), \tag{4.41}$$

where $\mathbb{1}$ is the dyadic tensor (matrix with diagonal of 1) and $\beta_s = q_p \Delta t / 2 m_p$ (independent of the particle weight and unique to a given species). The shorter notation $\mathbf{E}^n(x_p^n) = \mathbf{E}_p^n$ has been introduced for all fields. An alternative implementation of the rotation procedure above is called the Boris mover [4].

Note that technically the leap-frog algorithm is second-order accurate, when instead the regular explicit Euler scheme is only first-order. Nevertheless, the two differ in practice only for the fact that the velocity is staggered by a half time step. This staggering is achieved by moving the initial velocity of the first time cycle by half a time step using an explicit method:

$$\frac{\mathbf{v}_p^{1/2} - \mathbf{v}_p^0}{\Delta t/2} = \frac{q_s}{m_s} \mathbf{E}_p^0(x_p^0) + \frac{q_s}{m_s} \left(\frac{\mathbf{v}_p^{1/2} + \mathbf{v}_p^0}{2} \right) \times \mathbf{B}_p^0(x_p^0). \tag{4.42}$$

After the first step all subsequent steps would be identical to the Euler explicit scheme.

4.5.2 Explicit PIC cycle

After discretizing the particles' equation of motion, Maxwell's equations also need to be discretized in time. As for space, different methods are also available in time and the interested reader is referred to textbooks on that subject [15]. The simplest approach is to also use the leap-frog algorithm for the fields. The electric field is computed at the integer levels while the magnetic field is computed at the intermediate times (see figure 4.5). Starting form the partially discretized equations (4.25) introduced above, time can be discretized as

$$\nabla_g \times \mathbf{E}^n = \frac{\mathbf{B}^{n+1/2} - \mathbf{B}^{n-1/2}}{\Delta t}$$

$$\nabla_g \times \mathbf{B}^{n+1/2} = \mu_0 \mathbf{J}_g^{n+1/2} + \mu_0 \epsilon_0 \frac{\mathbf{E}^{n+1} - \mathbf{E}^n}{\Delta t}, \tag{4.43}$$

where all differences are time-centered, resulting in a second-order accurate scheme.

The current needs to be accumulated from the particle information at intermediate times $n + 1/2$. Cleverly, the leap-frog algorithm for the particles has the particle velocities at that same time level, but unfortunately the position is staggered by a half time step. A simple choice is to use an averaging method:

$$\mathbf{j}_g^{n+1/2} = \frac{1}{V_g} \sum_p q_p \mathbf{v}_p^{n+1/2} \frac{W(\mathbf{x}_g - \mathbf{x}_p^n) + W(\mathbf{x}_g - \mathbf{x}_p^n)}{2}. \tag{4.44}$$

In a similar manner the magnetic field at integer times needed in the mover is obtained by averaging from the neighboring intermediate times:

$$\mathbf{B}^n = \frac{1}{2}(\mathbf{B}^{n+1/2} + \mathbf{B}^{n-1/2}). \tag{4.45}$$

In the continuum, the divergence conditions are valid at all times if they are valid at the initial time and if the charge continuity equation is satisfied. In PIC the latter condition is not necessarily valid. The divergence of the magnetic field is always zero if the the discretized operator retains the property that the divergence of the curl is zero. This condition is often easily satisfied. But the other divergence condition is a major issue in PIC.

Historically two solutions have been adopted. First, the charge and current interpolation from the particle to the grid can be modified in a manner that ensures the validity of the charge continuity equation: these schemes are called charge conserving [20]. Second, the divergence condition can be applied as a correction to the electric field. At each time step, or at regular intervals, the electric field can be *cleaned* for any component not satisfying the Gauss law. The operation is called divergence cleaning and is numerically rather expensive, as it requires the solving of an elliptic problem with an iterative method. Simplified approaches have been developed [21].

In practice, the two methods can be used in combination because even though a charge-conserving scheme might do a good job at keeping the divergence correct, numerical errors will start to accumulate and a divergence cleaning operation can be judiciously applied regularly after a number of time steps.

The end result is a simple marching algorithm that is illustrated in figure 4.6. Each step in the figure is self-contained and requires only information that is available from the previous steps in the cycle. There is no iteration needed. This is the critical advantage of explicit PIC: the algorithm advances forward without any iterations.

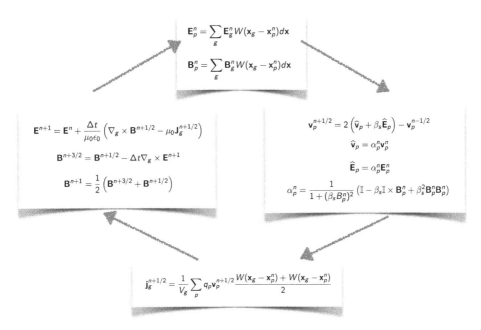

Figure 4.6. Explicit PIC cycle. Four steps are followed in sequence over a time step Δt. Starting, for example, from the top, the electric and magnetic fields from the previous time step are interpolated to the particles. Next the particles are advanced over Δt to find the new position and velocity. The particle information is then interpolated to the grid to obtain the current density in each cell. The fields are then advanced for the same Δt on the grid. A new time step is ready to start.

But this comes at a price. We have assumed in each step that the others remain frozen. During the time step Δt of the cycle, the fields are frozen while the particles move and the particles are frozen while the fields evolve. This is only acceptable if Δt is small. Very small.

4.5.3 Electrostatic explicit methods

In many applications, the currents are sufficiently small to justify an electrostatic approximation. In this case the two curl equations can be ignored and the only remaining equation in the Maxwell system is the Gauss law that can be rewritten in terms of the electrostatic potential as

$$\Delta_g \phi = -\frac{\rho_g}{\epsilon_0},\tag{4.46}$$

where the Laplacian operator is discretized on the grid.

In this case there is no escape form the need of solving the elliptic problem with the curious consequence that explicit electrostatic PIC is in fact more complex and expensive than explicit electromagnetic PIC. But note that using an electromagnetic PIC for an electrostatic case requires one to make sure the divergence cleaning is very accurate. In essence this requires again solving the elliptic Poisson equation and the cost is the same. There is no forgiveness in electromagnetism.

4.5.4 Stability of the explicit PIC method

Explicit PIC is subject to three stability conditions. Exceed them and energy explodes, rendering the simulation meaningless. Users must never exceed these limits.

Stability of the particle mover
The discretization of the equations of motion using the leap-frog algorithm (or other similar explicit time differencing) introduces a stability constraint. As usual stability can be studied with von Neumann analysis [4, 5, 22]: the equations are linearized and their stability is studied by Fourier analysis. Note that the leap-frog algorithm is implicit in the Lorentz force term, but explicit on the electric force. As a consequence only the latter is of primary concern and stability can be studied in the limit of zero magnetic field.

Stability needs to address all temporal scales in the system. Obviously the fastest are the most challenging for an explicit scheme. Among the plasma waves, the fastest electrostatic wave is the Langmuir (i.e. electron plasma) wave.

We proceed using von Neumann analysis [22], meaning we assume the temporal evolution to be harmonic so that all quantities depend on time as $e^{i\omega t}$:

$$\mathbf{x}_p^n = \tilde{\mathbf{x}}_p e^{i\omega n\Delta t}$$
$$\mathbf{v}_p^{n+1/2} = \tilde{\mathbf{v}}_p e^{i\omega(n+1/2)\Delta t},$$
$$\mathbf{E}_p^n = \tilde{\mathbf{E}}_p e^{i\omega n\Delta t} \tag{4.47}$$

where ω is the temporal frequency of the numerical response.

We consider the mover introduced in equation (4.38), repeated here for the case of a Langmuir wave (i.e. without the magnetic field) in a single spatial and single velocity dimension:

$$x_p^{n+1} = x_p^n + \Delta t v_p^{n+1/2}$$
$$v_p^{n+1/2} = v_p^{n-1/2} + \frac{q_p \Delta t}{m_p} E_p^n. \tag{4.48}$$

Substituting in each term the harmonic dependence, we obtain in the Fourier transformed space

$$\tilde{x}_p\left(e^{i\omega\Delta t} - 1\right) = \tilde{v}_p \Delta t e^{i\omega\Delta t/2}$$
$$\tilde{v}_p\left(e^{i\omega\Delta t/2} - e^{-i\omega\Delta t/2}\right) = \frac{q_s \Delta t}{m_s} \tilde{E}_p. \tag{4.49}$$

Recalling now the basic property of cold plasma Langmuir waves that relate the variation of the particle displacement with the electric field in Fourier space as [22]

$$\frac{q_s \tilde{E}}{m_s} = -\omega_{pe}^2 \tilde{x}, \tag{4.50}$$

the final solution is given by the algebraic system

$$\begin{pmatrix} e^{i\omega\Delta t} - 1 & -\Delta t e^{i\omega\Delta t/2} \\ \omega_{pe}^2 \Delta t & e^{i\omega\Delta t/2} - e^{-i\omega\Delta t/2} \end{pmatrix} \begin{pmatrix} \tilde{x}_p \\ \tilde{v}_p \end{pmatrix} = 0. \tag{4.51}$$

The dispersion relation can be computed as the determinant of the system matrix [4, 5]

$$\left(\frac{\omega_{pe}\Delta t}{2} \right)^2 = \sin^2 \left(\frac{\omega\Delta t}{2} \right), \tag{4.52}$$

where the Euler formula for the sin function was used. The dispersion relation (4.52) expresses the frequency of the numerical solution for particle motion in Langmuir waves with physical frequency ω_{pe}. The numerical frequency is not the same as the physically correct frequency.

Notoriously the sin function of real arguments has values only in the $[-1, 1]$ interval. As a consequence, for any values of $\omega_{pe}\Delta t > 2$, von Neumann analysis leads to a complex numerical oscillation frequency, ω. Indeed as the problem is real to begin with, two complex conjugate solutions ω arise, one of the two being damped and the other a growing exponential. The numerical solution becomes unstable, with a numerical growth rate that has no physical equivalence. In practice, the particles heat unboundedly and quickly and the simulation fails within a few time steps.

The first cardinal rule of explicit plasma simulation is that the condition

$$\omega_{pe}\Delta t < 2 \tag{4.53}$$

should never be violated. Indeed, practice advises the use of considerably smaller time steps of order $\omega_{pe}\Delta t = 0.1$, to avoid numerical heating [4].

The temporal stability constraint produces a first great challenge to the application of PIC methods in practice. The range of time scales typical in plasmas covers several orders of magnitude. The electrons are much faster than the ions (due to the mass difference). What the stability constraint imposes is to choose a time step that is able to resolve the fastest frequency in the system: the electron plasma frequency. If we are interested in following longer time scales we still need to use a time step that follows the Langmuir waves, even if our insight tells us there is nothing of interest at those small scales.

Stability due to the explicit time differencing of the field equations
Discretizing Maxwell's equations explicitly leads to the imposition of the Courant–Friedrich–Levy (CFL) condition. This condition arises whenever a hyperbolic equation is discretized explicitly [24]. The CFL condition states that the time step must not exceed the time taken by the characteristic signals of the hyperbolic equation to travel one cell. In the case of Maxwell's equations the characteristic speed of the signal is the speed of light, c. The CFL condition then requires

$$\Delta t < \Delta x/c. \tag{4.54}$$

This condition is in a sense less damaging than the previous one because it can be satisfied by a large Δt if Δx is also large. Often in practice, one needs to resolve long term processes developing over macroscopic scales. In this case both Δt and Δx are large. Additionally it is relatively easy to discretize just Maxwell's equations implicitly removing just this constraint [25]. As we will see below what is difficult is to deal with the coupling of particles and fields implicitly. But just discretizing Maxwell's equations implicitly is a simple task by comparison.

Space resolution and finite grid instability
The introduction of a grid to compute the field characterizes the particle methods for plasma physics as PIC methods. The particles move in a continuum space, but their information is projected to grid points using integration over control volumes. This operation reduces the information, collapsing the continuum particle shapes to discrete contributions in a set of points (the grid values). The loss of information allows one numerical instability to creep in: the finite grid instability.

The mathematical study of the finite grid instability is complex, requiring a careful and complete analysis of all computational steps with the Laplace transformation of time and the Fourier transformation of space. The reader interested in the details is referred to the textbooks on the subject [4, 5] and to the original work cited therein. The summary of the analysis is that to avoid the finite grid instability in explicit particles methods the grid spacing must be chosen to satisfy the constraint

$$\Delta x / \lambda_D < \varsigma, \tag{4.55}$$

where λ_D is the Debye length [4, 5] and the parameter ς is of order one and depends on the details of the implementation. For the widely used choice of shape functions as b-spline of order 0 and consequently interpolation functions of order 1, the scheme referred to as cloud-in-cell, the literature reports $\varsigma \approx \pi$. Using filtering methods and higher-order interpolation this number can be increased significantly allowing the use of larger cells.

A practical consequence of the finite grid instability is a tremendous numerical heating of the plasma characterized by an alternating positive–negative variation of the electric field accompanied by a correlated zig-zag perturbation of the phase space. In a very short time, the energy reaches the bounds of overflow in the machine representation of numbers. The finite grid instability must be avoided, requiring its stability constraint to be respected everywhere in the domain for all directions.

The limit imposed by this stability constraint is devastating. The spatial scale to resolve is the Debye length in the highest density region of the domain where the Debye length is smallest. If using uniform grids, this constraint requires using cell sizes that can be several orders of magnitude smaller than the scales of interest. Note that this consideration has two sides.

First, often the processes of interest do not reach down to the Debye scale. For example, most electromagnetic processes are at their smallest on the electron inertial length, $d_e = c / \omega_{pe}$. This scale can be orders of magnitude larger than the Debye length. We recall that the Debye length is the electron thermal speed divided by the plasma frequency while the electron inertial length is the speed of light divided by the plasma frequency: the inertial length then is obviously larger than the Debye length by the ratio v_{the}/c. Having to resolve the Debye length purely for numerical stability reasons implies a great waste in each direction. When performed in 3D, this waste is compounded. If the time scale constraint encountered before hits the explicit approach once, the spatial stability constraint hits it three times, once per spatial dimension. The combined effect of these two constraint makes it imperative to look beyond explicit particle methods for many practical applications. The ability to relax these two constraints by one order of magnitude, for example, would result in a gain of four orders of magnitude: one for time and three for the three spatial dimensions.

Second, often plasmas present regions of high density in contact with regions of lower density, resulting in a large range of Debye lengths. To avoid the finite grid instability, uniform grids need to resolve the smallest Debye length in the system. Adaptive schemes can be devised to resolve in each region only the local Debye length, resulting in considerable savings [26–28]. However, even in this case, often the local scales of interest are much larger than the Debye length and the local grid resolution would be more desirably chosen according to other considerations than the need to resolve the Debye length.

4.6 Implicit particle methods

An approach to avoid the prohibitively small time steps and grid spacing required by explicit methods is to reduce the physics model used.

For example, the speed of light limitation can be eliminated by retaining only some of the full electromagnetic physics while removing from Maxwell's equations the terms allowing the propagation of light. The approximation is called Darwin [27] and is used when slow-varying magnetic fields are present without any important role in electromagnetic waves.

Other approaches are based on eliminating some of the electron physics to eliminate the need for resolving the Debye length and the plasma frequency. If electrons are treated as fluids and neutrality is enforced, both conditions can be eliminated. The approach is called hybrid and can be implemented in varying degrees of complexity [28]. Another very important approximation is that of averaging over the gyration of particles in large magnetic fields. This approximation is called gyrokinetic and is widely used in fusion research and in many systems where large magnetic fields are present [31].

The other approach to eliminate the stability constraints is to retain the full first-principle description of the electron physics and of the electromagnetic fields, but to rely on more advanced numerical methods that can handle the presence of multiple scales in space and time. To avoid the prohibitively large number of time steps and the exceedingly high resolution needed to run realistic problems with explicit particle methods, implicit methods have been considered for several decades [32–35].

The constraints of the explicit methods come from the artificial freezing of the coupling between fields and particles during the time step. The obvious solution is to reintroduce this coupling and solve the field equations on the grid and the particle equations maintaining their coupling. This coupling of course is non-linear as the particles are the sources of the current and density in Maxwell's equations, and the fields in the particle equations are produced by Maxwell's equations.

In the 1970s, -80s, and -90s, the consensus was that solving the full non-linear coupled system of field equations and particle equations was not feasible. But the situation has changed in recent years. First, computers are of course vastly more powerful. Second and more importantly, the progress made in the domain of Newton–Krylov solvers made this task possible [36, 37].

Two independent teams proposed fully implicit PIC codes [38, 39] capable of solving electrostatic, full electromagnetic [38], fully relativistic [40], or reduced Darwin descriptions of plasmas [41]. The key aspect of these methods is that they completely remove the need to satisfy any of the constraints above. Additionally and equally importantly, these methods are exactly energy conserving.

All previous PIC methods did not conserve energy. Explicit PIC methods can be readily formulated to conserve momentum exactly [42]. With respect to energy conservation, unfortunately the situation is far less rosy. The random aspect linked with the presence of discrete particles induces a tendency for the explicit PIC method to generate numerical heating: random fluctuations in the particle information leads to the generation of electromagnetic waves at small scales (noise). Energy is not only not conserved, but it tends to increase in time [4, 5]. The increase tends to be secular at small time steps but it becomes exponential in the presence of finite-grid instability [43]. The aforementioned stability constraints are then supplemented by the additional requirement of maintaining good energy conservation, typically limiting the time step by an additional order of magnitude to $\omega_{pe}\Delta t \sim 0.1$, but sometimes requiring even smaller time steps to retain an accurate description of particle energization [40].

A class of so-called explicit 'energy-conserving' PIC methods [44] (the quotation marks are used in the literature [45]) has been developed where the energy was conserved only in the limit of a zero time step. Their peculiarity was the use of different interpolation functions for charge and current, and for the magnetic and electric fields, resulting in improved energy conservation. In practice energy conservation was marginally better than in other explicit PICs, but still energy was by no means exactly conserved. For example, the code Celeste belonged to the class of 'energy-conserving' PICs of the old definition [14].

The new fully implicit PICs are exactly energy conserving, the acronym ECPIC (energy-conserving PIC) is then properly deserved. One can prove that if the discretized equations defining the ECPIC are solved exactly, energy is conserved exactly. The ECPIC relies on a Newtonian iteration to solve the model equations. Of course, the Newtonian iteration is stopped at a given tolerance and the energy will then also have the same tolerance in the error. But the method is in principle exactly energy conserving. In practice one can easily achieve relative energy errors below 10^{-9} as in the examples below or any desired accuracy within machine precision. This accuracy is many orders of magnitude better than anything published previously.

The basic common idea of the ECPIC methods is to replace the leap-frog algorithm with an implicit particle mover where the fields appear at the advanced time level, requiring an iteration with the field equations. For example, [38] proposes the θ-scheme

$$\mathbf{x}_p^{n+1} = \mathbf{x}_p^n + \Delta t \mathbf{v}_p^{n+\theta}$$
$$\mathbf{v}_p^{n+1} = \mathbf{v}_p^n + \frac{q_p \Delta t}{m_p}\left(\mathbf{E}_p^{n+\theta}(\bar{\mathbf{x}}_p) + \bar{\mathbf{v}}_p \times \mathbf{B}_p^{n+\theta}(\bar{\mathbf{x}}_p)\right), \tag{4.56}$$

where barred variables are average between time level n and $n + 1$, $\bar{\mathbf{x}} = (\mathbf{x}^{n+1} + \mathbf{x}^n)/2$, and variables of time index $n + \theta$ are defined as $\mathbf{x}^{n+\theta} = \theta \mathbf{x}^{n+1} + (1 - \theta)\mathbf{x}^n$. The same notation applies to all variables, particles, and fields.

The implicit scheme is unconditionally stable. If the same von Neumann analysis performed for the leap-frog scheme is repeated for equation (4.56), the following relation is obtained:

$$\left(\frac{\omega_{\text{pe}} \Delta t}{2}\right)^2 = \tan^2\left(\frac{\omega \Delta t}{2}\right). \tag{4.57}$$

Now the tan function is unbounded and any values of Δt lead to real values for ω without any numerical instability. Figure 4.7 reports the stability analysis for explicit and implicit methods. The frequency of the numerical solution is compared with the

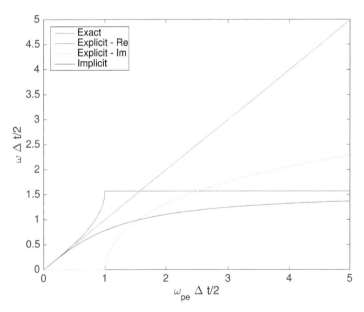

Figure 4.7. Results of the stability analysis of PIC particle movers. The explicit leap-frog mover is compared with the implicit mover in [38]. The explicit method is stable only when $\omega_{pe}\Delta t < 2$. For larger Δt the numerical solution has an imaginary component of frequency that leads to numerical growth. The implicit method is always stable.

actual correct analytical solution in the presence of Langmuir waves. As Δt is increased, the numerical solution of the explicit method increases compared with the correct solution, until the stability limit is reached. When the limit is exceeded, an imaginary part to the the numerical frequency appears, leading to growing, numerically unstable solutions. The implicit method, however, is always stable.

When $\theta = 1/2$, the method is exactly energy conserving [38, 39], provided the field and particle equations are solved to non-linear consistency within a Newtonian iteration.

Of course, accuracy still imposes bounds on the time step. Since the mover uses only one punctual value of the fields, if a particle travels more than one cell per time step the formula becomes inaccurate, leading to severe non-momentum conservation. Even this limitation is removed if sub-cycling is used [39].

As mentioned above, this modern solution was not available until recently. Even today it remains computationally challenging and it requires the deployment of Newton–Krylov non-linear solvers, complicating significantly the optimization of massively parallel PIC codes. The use of non-linear solvers requires global communication and the storage of multiple states of the unknown to form the Krylov space orthogonalization. When applied to a full particle–field coupled system, this leads to extreme memory requirements. This problem can be circumvented with particle enslavement and other preconditions [38, 46], but at the cost of further complexity in code development.

The solution adopted in the earlier days was that of linearizing the particle–field coupling and solving the resulting linear system. This method is semi-implicit in the sense that the formulating equations are still fully coupled and require an iteration, but the iteration is linear and not solved to non-linear consistency. Two approaches have been presented.

The direct implicit method was developed at Lawrence Livermore National Laboratory and is implemented in codes such as AVANTI and the LSP [35, 47–50]. This method is based on replacing the leap-frog algorithm with more advanced implicit movers. For example, the so-called D_1 algorithm (formulated with a recursive filter on the fields [51]) is

$$\mathbf{x}_p^{n+1} = \mathbf{x}_p^n + \Delta t \mathbf{v}_p^{n+1/2}$$
$$\mathbf{v}_p^{n+1/2} = \mathbf{v}_p^{n-1/2} + \frac{q_p}{m_p}\left(\bar{\bar{\mathbf{E}}}_n + \frac{\mathbf{v}_p^{n+1/2} + \mathbf{v}_p^{n-1/2}}{2} \times \mathbf{B}^n\left(\mathbf{x}_p^n\right)\right), \tag{4.58}$$

where the only difference to the standard leap-frog method of equation (4.5) is the calculation of the electric field acceleration as

$$\bar{\bar{\mathbf{E}}}_n = (\mathbf{E}_{n-1}(\mathbf{x}_n) + \mathbf{E}_{n+1}(\mathbf{x}_n))/2. \tag{4.59}$$

A small formal difference but a huge numerical difference. The scheme now requires the new advanced field that can be calculated only after moving the particles. This non-linear iteration is avoided in the direct implicit method by computing the linear response of the plasma.

In the direct implicit method, the plasma response (i.e. the changes in the electric field in response to the particle motion) is obtained by linearizing the particle equations of motion about an estimate (extrapolation) for their unknowns at the new time level $n + 1$. The mathematical procedure used can vary from author to author, but it leads to a form of sensitivity matrix of the linearization that assumes the form of a plasma susceptibility [52]. The field equations can then be formulated using this matrix and solved alone without iterating with the particles. After the new field is computed, the particles can be moved.

An alternative approach, also going back to the late 1970s, is the implicit moment method (IMM), developed at Los Alamos National Laboratory [32, 34, 52]. In this approach, the linearization is obtained by linearizing the moments rather than the particle equations of motion. The method has been used in a family of codes originating with Venus [34], followed by Celeste [16, 53], Parsek2D [54], and now used in iPic3D [55].

The IMM is based on a scheme similar to the θ-scheme described above in the fully implicit ECPIC method:

$$
\begin{aligned}
\mathbf{x}_p^{n+1} &= \mathbf{x}_p^n + \Delta t \mathbf{v}_p^{n+\theta} \\
\mathbf{v}_p^{n+1} &= \mathbf{v}_p^n + \frac{q_p \Delta t}{m_p}\left(\mathbf{E}^{n+\theta}(\overline{\mathbf{x}}_p) + \overline{\mathbf{v}}_p \times \mathbf{B}^n(\overline{\mathbf{x}}_p)\right),
\end{aligned}
\tag{4.60}
$$

differing only in the time level of the magnetic field. As in the case of the leap-frog algorithm (4.39), the velocity equation can be rewritten in the equivalent form

$$
\overline{\mathbf{v}}_p = \hat{\mathbf{v}}_p + \beta_s \hat{\mathbf{E}}_p,
\tag{4.61}
$$

where the same notation is used as in equations (4.40) and (4.41) with the critical difference that now $\hat{\mathbf{E}}_p = \alpha_p^n \mathbf{E}_p^{n+\theta}$ is based on the new advanced electric field that can only be known once the field equations are advanced.

The key defining property of the IMM is that the θ-scheme is now coupled to the Maxwell equations via a linearization procedure. The two curl Maxwell equations are discretized in time with another θ-scheme:

$$
\begin{aligned}
\nabla_g \times \mathbf{E}^{n+\theta} + \frac{1}{c}\frac{\mathbf{B}_g^{n+1} - \mathbf{B}_g^n}{\Delta t} &= 0 \\
\nabla_g \times \mathbf{B}^{n+\theta} - \frac{1}{c}\frac{\mathbf{E}_g^{n+1} - \mathbf{E}_g^n}{\Delta t} &= \frac{4\pi}{c}\overline{\mathbf{J}}_g.
\end{aligned}
\tag{4.62}
$$

The spatial operators in equation (4.62) are discretized on a grid. Consistent with the notation introduced earlier, we index the grid elements generically with the index g. ∇_g is a shorthand for the spatial discretization used. For example in the IMM family of codes the discretization used is different from the Yee scheme [56] and locates on the cell centers for the magnetic field and on the cell vertices for the electric field [7, 14, 55]. However, all the derivations below are not critically dependent on which spatial discretization is used.

The current is computed as an average of the old and new time level, again requiring one to solve the particle equations of motion before the fields can be computed and vice versa. This iteration is avoided by expanding the current in Taylor series truncated at the first order [16]. Computing the current separately for each species s, one obtains

$$\bar{\mathbf{J}}_{sg} = \hat{\mathbf{J}}_{sg} - \frac{\Delta t}{2}\mu_{sg} \cdot \mathbf{E}_\theta - \frac{\Delta t}{2}\nabla \cdot \hat{\Pi}_{sg}, \tag{4.63}$$

where the following expressions were defined:

$$\hat{\mathbf{J}}_{sg} = \sum_p q_p \hat{\mathbf{v}}_p W\left(\mathbf{x}_g - x_p^n\right)$$

$$\hat{\Pi}_{sg} = \sum_p q_p \hat{\mathbf{v}}_p \hat{\mathbf{v}}_p W\left(\mathbf{x}_g - \mathbf{x}_p^n\right), \tag{4.64}$$

with the obvious meaning, respectively, of the current and pressure tensor based on the hatted velocities.

An effective dielectric tensor is defined to express the feedback of the electric field on the plasma current and density:

$$\mu_{sg}^n = -\frac{q_s \rho_s^n}{m_s}, \alpha_{sg}^n \tag{4.65}$$

where the rotation matrix α_s^n is defined in the same manner as that of the particles but based on the local field on the grid rather than the particle field:

$$\alpha_{sg}^n = \frac{1}{1 + \left(\beta_s B_g^n\right)^2}\left(\mathbb{I} - \beta_s \mathbb{I} \times \mathbf{B}_g^n + \beta_s^2 \mathbf{B}_g^n \mathbf{B}_g^n\right). \tag{4.66}$$

With this approach, the current needed for the Maxwell equations can be computed using exclusively particle data from the time level n and no iteration is needed between particles and fields.

Regardless of the use of the direct implicit or IMM, two advantages arise in using semi-implicit methods.

First, these methods avoid any non-linear iteration. The resulting field equations obtained with the procedure described above are linear and can be solved with any linear solver. The three latest implementations of the IMM (Celeste, Parsek, and iPic3D) use the GMRES approach [57, 58]. These methods still require global communication when implemented in parallel, but converge very quickly due to the diagonal dominance of the two curl Maxwell equations. Furthermore these solvers deal only with the grid and the memory requirement is vastly smaller than that of an iteration scheme also dealing with particles [59, 60]. There are typically at least hundreds of particles per cell; the memory requirement of the iteration method is then reduced by the same factor.

Second, the semi-implicit method retains the same structure of the time step loop of the usual explicit PIC (see figure 4.6): the field equations are solved on the grid, the fields are interpolated to the particles, the particles are moved, and then the moments on the grid are computed and the next cycle can continue. The field solver is somewhat more complex and the particle mover is also more complex but the overall scheme is the same as in explicit PIC.

It is important to realize what the added complexity is. The field solver, at first sight, is more complex than the explicit field solver of explicit PIC. In the simplest explicit PIC methods, the fields are marched in a leap-frog manner and require no linear solver. However, even for the so-called charge-conserving schemes the two divergence equations are not satisfied exactly, requiring periodic elliptic solvers to apply so-called divergence cleaning [61]. The need for elliptic solvers is completely eliminated in IMM because of the natural tendency of the IMM to damp any divergence error [58]. As noted above the application of GMRES to the two curl equations is much simpler and converges faster than for elliptic divergence cleaners [60].

The particle mover is also more complex. The mover in equation (4.60) is not explicit in itself. The equation for the velocity requires one to compute the fields at the position $\bar{x}_p$, which is not know until the particle is moved, but to move the particle the velocity needs to be known. This requires a separate iteration for each particle. The iteration is just among the six equations of motion for each particle. All codes in the IMM family use a predictor correct approach that is a Picard-type iteration between the two Newton equations for each particle. The iteration can be truncated at a specified number (typically 3 is used) or until a convergence criterion is satisfied [62, 63]. As we will see below this iteration, even though it is local, is still an expensive operation because it is applied to each particle.

As a rule of thumb, compared with an explicit PIC the IMM is about three times more expensive per time step. The mover requires three steps where the leap-frog algorithm requires only one. The field solver is a small fraction of the cost. The IMM is not bound by the stability constraint and can use larger grid spacing and time steps, winning a large victory in terms of the time needed to complete the simulation of a given box size for a given time interval.

The IMM does not have to satisfy any stability constraints because its model equations are essentially the same as those of the ECPIC. A complete stability analysis of the IMM shows that waves and scales not resolved by a certain simulation are averaged and damped at the sub-grid level [34]. There is, however, a catch. The two great advantages of the semi-implicit methods compared with the ECPIC come at a great loss: energy is no longer conserved. The lack of full non-linear consistency breaks energy conservation and introduces an accuracy constraint. The average particle motion cannot exceed one cell per time step. This limitation comes mathematically from the need for the Taylor series expansion truncated at the first order used above to be a valid approximation. Physically, the requirement is that the moment description introduced by the series expansion can be a good approximation of the particle evolution. The procedure of the IMM can be regarded as a first-order Chapman–Enskog expansion [64]. The IMM in this regards uses a fluid moment description to guess the plasma response to the changes happening to the particles during one time step. This description is not completely consistent with the actual particle evolution. To keep the error under control the time step becomes limited. The limit is usually expressed as $v_{the}\Delta t/\Delta x < 1$ [34]. If this condition is satisfied the method is accurate and the energy error is under control.

A new variant of the IMM has been recently developed to exactly conserve energy [65]. The new method is a variant of the IMM just described, but introduces a new approach to compute the current to enforce exact energy conservation. The advantage is that the finite grid instability is completely eliminated. The method is very new at the moment of writing but it holds promise of being a quantum step forward. More research is needed to verify if this promise is realized in practice.

4.7 Annotated Python code

To make it all more concrete, we now present a complete, functioning PIC code. We choose the simplest case: 1D and electrostatic with electrons being followed as particles but ions forming a fixed background of uniform density. This last assumption is justified by the heavy mass of the ions. Once this code is fully understood going to full 3D electromagnetic is truly a simple step.

We choose the Python language, but on the author's website other versions (classical, relativistic, electrostatic, electromagnetic, implicit, explicit) are available: https://perswww.kuleuven.be/~u0052182. The graphical components to output the data are also available on the website but are not discussed here.

Initialization
As a first step, some Python initializations are needed:

```python
import numpy as np
import pylab as plt
from scipy import sparse
from scipy.sparse import linalg
```

We then need to initialize the definition of the simulation, in terms of domain size and of the grid used to discretize it. Next, the time step is set with the number of cycles to be run. The next step is to define the plasma density. We use normalized units where the plasma frequency is set to be unitary. But physical units can be used.

The ion uniform background is set to exactly balance the charge of the particles (which are all electrons in this simple model):

```python
# Simulation parameters
L = 20*np.pi #20*np.pi # Domain size

NG = 80   # Number of grid cells
N = NG * 200 # Number of particles (200 per cell for example)
dx = L / NG # Cell size

DT = 0.005 # Time step
NT = 50000  # Number of time steps

WP = 1. # Plasma frequency
QM = -1. # Charge/mass ratio
Q = WP**2 / (QM*N/L)  # rho0*L/N: charge carried by a single particle
rho_back = -Q*N/L # Background charge density
```

Particle initialization

We then initialize the electrons. We choose a classic problem of two-stream instability [66] where the initial electrons are subdivided into two equal beams of opposite mean velocity but equal density and thermal speed:

```
#Particle initial properties
V0 = 0.9 # Stream velocity
VT = 0.0000001 # Thermal speed

# perturbation
XP1 = 1.0
mode = 1

# particles (electrons)
xp = np.linspace(0, L-L/N, N).T   # Particle positions
vp = VT * np.random.randn(N) # Particle momentum, initially Maxwellian
pm = np.arange(N)
pm = 1 - 2 * np.mod(pm+1, 2) # Even and odd particles have opposite speed
vp += pm * V0   # Momentum + stream velocity
np.random.shuffle(v) # We reshuffle the indices to avoid any bias

# Add electron perturbation to excite the desired mode
xp += XP1 * (L/N) * np.sin(2 * np.pi * xp / L * mode)
xp[np.where(xp < 0)] += L
xp[np.where(xp >= L)] -= L
```

Grid initialization

In the 1D electrostatic limit, the Maxwell equations reduce to just the Poisson equation:

$$-\epsilon_0 \nabla^2 \phi = e n_i - e n_e, \tag{4.67}$$

where n_i is the uniform background ion density and n_e is the electron density projected to the grid from the particles. The ∇^2 operator is discretized as in the simplest textbook finite difference method [24]:

$$-\epsilon_0 \left(\phi_{i+1} + \phi_{i-1} - 2\phi_i \right) = (e n_i - e n_e) \Delta x^2. \tag{4.68}$$

The potential is computed in the cell centers. The continuum is subdivided in cells indexed by i with $i \in [0, \text{NG} - 1]$ (note that Python as C and C++ counts the indices in vectors from 0, in MATLAB the same index range would be 1 to NG).

Of course, the first cell and the last cell have a problem: we do not have the neighboring cell outside. This challenge is called a *boundary condition* and it is the worst nightmare any computational scientist can experience. Boundary conditions are pure distilled evil. An evil we will avoid with a double trick. First, we use periodic boundary conditions: the left neighbor of the first cell is the last cell and the right neighbor of the last cell is the first cell. Second, the potential is defined minus a constant, and we use this freedom to set the potential in the last cell to 0. The number of unknown potentials then is just NG-1. This tricks works in 1D for periodic boundary conditions, more general boundary conditions in 3D quickly become the aforementioned nightmare.

With these assumptions, the matrix of the discretized Poisson equation is then very simple, it has −2 on the main diagonal and 1 on the two neighboring ones:

```
# Auxiliary vectors
p = np.concatenate([np.arange(N), np.arange(N)])  # Particle indices up to N
Poisson = sparse.spdiags(([1, -2, 1] * np.ones((1, NG-1), dtype=int).T).T, \
                         [-1, 0, 1], NG-1, NG-1)
Poisson = Poisson.tocsc()
```

Main cycle

At this point we have prepared the simulation for the main cycle where the four steps of PIC are followed. First, the position is advanced, then the particles are projected to the grid to compute the density. With the density, the Poisson equation is solved. The electric field is computed to obtain the force on the particles and advance the particle velocity:

```
# Main cycle
for it in xrange(NT+1):
    # update particle position xp
    xp += vp * DT
    # Periodic boundary condition
    xp[np.where(xp < 0)] += L
    xp[np.where(xp >= L)] -= L

    # Project particles->grid
    csi = xp/dx
    g1 = np.floor(csi - 0.5) # Distance from the center of the cell
    g = np.concatenate((g1, g1+1))
    fraz1 = 1 - np.abs(xp/dx - g1 - 0.5)
    fraz = np.concatenate((fraz1, 1-fraz1))
    g[np.where(g < 0)] += NG
    g[np.where(g > NG-1)] -= NG
    mat = sparse.csc_matrix((fraz, (p, g)), shape=(N, NG))
    rho = Q / dx * mat.toarray().sum(axis=0) + rho_back

    # Compute electric field potential
    Phi = linalg.spsolve(Poisson, -dx**2 * rho[0:NG-1])
    Phi = np.concatenate((Phi,[0]))

    # Electric field on the grid

    Eg = (np.roll(Phi, 1) - np.roll(Phi, -1)) / (2*dx)

    # interpolation grid->particle and velocity update
    vp += mat * QM * Eg * DT
```

Let us comment on the particle steps in each of these phases:

1. *Position advancement.* When particles are advanced, the possibility exists for the particles to exit the system. Consistent with the periodicity used above, particles exiting form the left re-enter on the right and vice versa.

2. *Particle projection.* We introduce a *logical coordinate* defined as $\xi = x/\Delta x$. The particles and the cells have logical coordinates, where the interpolation function is more easily defined as $W_{ip} = b_\ell(\xi_p - \xi_i)$. Cell centers have coordinates $\xi_i = (i + 1/2)$, $i \in [0, NG - 1]$ and particles have $\xi_p = x/\Delta x$. We consider the case of particle shapes given by b-splines of order 0 and interpolation functions, consequently, of order 1:

$$W_{ip} = b_1(\xi_p - \xi_i) \equiv \begin{cases} 1 - |\xi_p - \xi_i|, & \text{if: } |\xi_p - \xi_i| < 1 \\ 0, & \text{otherwise.} \end{cases} \tag{4.69}$$

With this choice of interpolation, each particle can only contribute to maximum of two cells. The choice of making the particle size equal to the cell size means that it is impossible for a particle to overlap more than two cells. Figure 4.8 shows the concept. The contribution of a particle of finite length to a cell is equal to the fraction of the length of the particle that overlaps that cell. This is computed by first identifying the left-most cell using the floor command and then computing the two fractions (one being 1 minus the other since the total contribution is unity).

Once the two fractions are computed the contribution to the charge of a cell is the fraction times the charge of the particle. Periodic boundary conditions are also applied to charge projection.

The code in the example uses a matrix notation to project the particle avoiding loops but a simple loop over the particles would work as well. The matrix 'mat' is equal to the interpolation function $W_{ip} = W(x_p - x_i)$, which indeed has two indexes. The charge is then the matrix W_{ip} applied to the vector q_p:

$$q_i = W\mathbf{q} \equiv \sum_p W_{ip} \cdot q_p \tag{4.70}$$

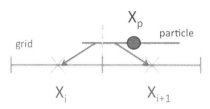

Figure 4.8. Particle interpolation in 1D: the contribution of a particle to a cell is proportional to the length of the interval of overlap with that cell. When the particle shape is a b-spline of order zero (a case called cloud-in-cell), the particle is subdivided to the cells according to its overlap.

The matrix notation can in some computer languages produce a faster execution, but this issue is largely eliminated in the latest version of most languages. The computational performance of the interpolation steps depends dramatically on the precise style of code writing, on the language, and on the hardware [67–70].

In an alternative to a matrix formulation, a loop over the particles can be used:

```
qo{alsodx = Q / dx
rho = np.zeros(NG)
for i in range(N):
        csi = xp[i]/dx
        g1 = np.floor(csi - 0.5)
        g2 = g1 + 1
        fraz1 = 1 - np.abs(csi - g1 - 0.5)
        fraz2 = 1.0 - fraz1
        rho[j1] = rhoe[j1] + qodx*fraz1
        rho[j2] = rhoe[j2] + qodx*fraz2
rho[0] += rho[NG-1]
```

In both cases the charge in the cell must be divided by Δx to obtain the density.

3. *Field solver.* In 1D electrostatics there are many alternatives. Some are very simple [5], but we use here the most general approach that would work also with more complex boundary conditions and in more dimensions. We use a linear matrix solver provided by Python. The boundary conditions are then imposed (setting the last cell to 0) and the electric field is computed from the potential:

$$E_i = -\frac{\phi_{i+1} - \phi_{i-1}}{2\Delta x}. \tag{4.71}$$

4. *Velocity update.* The electric field is used to advance the particle position. The same interpolation method used for the charge is used for the electric field.

Note that we have chosen the position update as our initial step in the cycle but the starting point is completely arbitrary, the cycle can start at any point. This also means that, logically, position and velocity advancement happen in sequence, the last operation of one cycle being followed by the first of the next.

Bibliography

[1] Frenkel D and Smit B 2002 *Understanding Molecular Simulation* (San Diego, CA: Academic)

[2] Barnes J and Hut P 1986 A hierarchical $O(N \log N)$ force-calculation algorithm *Nature* **324** 4446–9

[3] Salmon J K and Warren M S 1994 Skeletons from the treecode closet *J. Comput. Phys.* **111** 136–55

[4] Birdsall C and Langdon A 2004 *Plasma Physics via Computer Simulation* (London: Taylor and Francis)

[5] Hockney R and Eastwood J 1988 *Computer Simulation using Particles* (London: Taylor and Francis)

[6] Nicholson D 1983 *Introduction to Plasma Theory* (New York: Wiley)

[7] Lapenta G 2012 Particle simulations of space weather *J. Comput. Phys.* **231** 795–821

[8] Dawson J M 1983 Particle simulation of plasmas *Rev. Mod. Phys.* **55** 403

[9] Coppa G G M, Lapenta G, Dellapiana G, Donato F and Riccardo V 1996 Blob method for kinetic plasma simulation with variable-size particles *J. Comput. Phys.* **127** 268–84

[10] Boor C D 1978 *A Practical Guide to Splines* (Berlin: Springer)

[11] Thomée V 1984 *Galerkin Finite Element Methods for Parabolic Problems* vol 1054 (New York: Springer)

[12] Bateson W B and Hewett D W 1998 Grid and particle hydrodynamics: beyond hydrodynamics via fluid rlement particle-in-cell *J. Comput. Phys.* **144** 358–78

[13] Grigoryev Y N, Vshivkov V A and Fedoruk M P 2005 *Numerical Particle-in-Cell Methods: Theory and Applications* (Utrecht: VSP BV)

[14] Sulsky D and Brackbill J 1991 A numerical method for suspension flow *J. Comput. Phys.* **96** 339–68

[15] Taflove A and Hagness S C 2005 *Computational Electrodynamics: The Finite-Difference Time-Domain Method* (Norwood: Artech House)

[16] Vu H X and Brackbill J U 1992 CELEST1D: an implicit, fully-kinetic model for low-frequency, electromagnetic plasma simulation *Comput. Phys. Commun.* **69** 253

[17] Lapenta G 2011 DEMOCRITUS: an adaptive particle in cell (PIC) code for object-plasma interactions *J. Comput. Phys.* **230** 4679–95

[18] Delzanno G L, Camporeale E, Moulton J D, Borovsky J E, MacDonald E A and Thomsen M F 2013 CPIC: a curvilinear particle-in-cell code for plasma-material interaction studies *IEEE Trans. Plasma Sci.* **41** 3577–87

[19] Jacobs G B and Hesthaven J S 2006 High-order nodal discontinuous Galerkin particle-in-cell method on unstructured grids *J. Comput. Phys.* **214** 96–121

[20] Esirkepov T 2001 Exact charge conservation scheme for particle-in-cell simulation with an arbitrary form-factor *Comput. Phys. Commun.* **135** 144–53

[21] Langdon A B 1992 On enforcing Gauss' law in electromagnetic particle-in-cell codes *Comput. Phys. Commun.* **70** 447–50

[22] Isaacson E and Keller H B 1994 *Analysis of Numerical Methods* (New York: Dover)

[23] Dendy R O 1990 *Plasma Dynamics* (Oxford: Oxford University Press)

[24] Morton K and Mayers D 1995 *Iterative Methods for Linear and Nonlinear equations* (Philadelphia, PA: SIAM)

[25] Morse R L and Nielson C W 1971 Numerical simulation of the Weibel instability in one and two dimensions *Phys. Fluids* **14** 830–40

[26] Fujimoto K and Sydora R D 2008 Electromagnetic particle-in-cell simulations on magnetic reconnection with adaptive mesh refinement *Comput. Phys. Commun.* **178** 915–23

[27] Innocenti M E, Lapenta G, Markidis S, Beck A and Vapirev A 2013 A multi level multi domain method for particle in cell plasma simulations *J. Comput. Phys.* **238** 115–40

[28] Vay J-L, Colella P, Friedman A, Grote D P, McCorquodale P and Serafini D 2004 Implementations of mesh refinement schemes for particle-in-cell plasma simulations *Comput. Phys. Commun.* **164** 297–305

[29] Degond P and Raviart P A 1992 An analysis of the Darwin model of approximation to Maxwell's equations *Forum Math.* **4** 13–44

[30] Lipatov A S S 2002 *The Hybrid Multiscale Simulation Technology* (Berlin: Springer)

[31] Lee W 1987 Gyrokinetic particle simulation model *J. Comput. Phys.* **72** 243–69

[32] Mason R 1981 Implicit moment particle simulation of plasmas *J. Comput. Phys.* **41** 233

[33] Denavit J 1981 Time-filtering particle stimulations with $\omega_{pe}\Delta t \gg 1$ *J. Comput. Phys.* **42** 337

[34] Brackbill J and Forslund D 1982 Simulation of low frequency, electromagnetic phenomena in plasmas *J. Comput. Phys.* **46** 271

[35] Langdon A, Cohen B and Friedman A 1983 Direct implicit large time-step particle simulation of plasmas *J. Comput. Phys.* **51** 107–38

[36] Kelley C T 2003 *Solving Nonlinear equations with Newton's Method* vol 1 (Philadelphia, PA: SIAM)

[37] Knoll D A and Keyes D E 2004 Jacobian-free Newton–Krylov methods: a survey of approaches and applications *J. Comput. Phys.* **193** 357–97

[38] Markidis S and Lapenta S 2011 The energy conserving particle-in-cell method *J. Comput. Phys.* **230** 7037–52

[39] Chen G, Chacón L and Barnes D C 2011 An energy-and charge-conserving, implicit, electrostatic particle-in-cell algorithm *J. Comput. Phys.* **230** 7018–36

[40] Lapenta G and Markidis S 2011 Particle acceleration and energy conservation in particle in cell simulations *Phys. Plasmas* **18** 072101

[41] Chen G and Chacon L 2015 A multi-dimensional, energy-and charge-conserving, non-linearly implicit, electromagnetic Vlasov–Darwin particle-in-cell algorithm *Comput. Phys. Commun.* **197** 73–87

[42] Grigoryev Y N, Vshivkov V A and Fedoruk M P 2002 *Numerical Particle-in-Cell Methods: Theory and Applications* (Berlin: De Gruyter)

[43] Ueda H, Omura Y, Matsumoto H and Okuzawa T 1994 A study of the numerical heating in electrostatic particle simulations *Comput. Phys. Commun.* **79** 249–59

[44] Lewis H R 1970 Energy-conserving numerical approximations for Vlasov plasmas *J. Comput. Phys.* **6** 136–41

[45] Langdon A B 1973 Energy-conserving' plasma simulation algorithms *J. Comput. Phys.* **12** 247–68

[46] Chen G, Chacón L, Leibs C A, Knoll D A and Taitano W 2014 Fluid preconditioning for Newton-Krylov-based, fully implicit, electrostatic particle-in-cell simulations *J. Comput. Phys.* **258** 555–67

[47] Hewett D W and Langdon A B 1987 Electromagnetic direct implicit plasma simulation *J. Comput. Phys.* **72** 121–55

[48] Cohen B I, Langdon A B, Hewett D W and Procassini R J 1989 Performance and optimization of direct implicit particle simulation *J. Comput. Phys.* **81** 151

[49] Drouin M, Gremillet L, Adam J-C and Héron A 2010 Particle-in-cell modeling of relativistic laser-plasma interaction with the adjustable-damping, direct implicit method *J. Comput. Phys.* **229** 4781–812

[50] Welch D R, Rose D V, Clark R E, Genoni T C and Hughes T 2004 Implementation of an non-iterative implicit electromagnetic field solver for dense plasma simulation *Comput. Phys. Commun.* **164** 183–8

[51] Barnes D, Kamimura T, Leboeuf J-N and Tajima T 1983 Implicit particle simulation of magnetized plasmas *J. Comput. Phys.* **52** 480–502

[52] Brackbill J U and Cohen B I 1985 *Multiple Time Scales* vol 3 (New York: Academic)

[53] Lapenta G, Brackbill J U and Ricci P 2006 Kinetic approach to microscopic–macroscopic coupling in space and laboratory plasmas *Phys. Plasmas* **13** 055904

[54] Markidis S, Camporeale E, Burgess D, Rizwan-Uddin and Lapenta G 2009 Parsek2D: an implicit parallel particle-in-cell code *ASP Conf. Ser.* **406** 237

[55] Markidis S, Lapenta G and Rizwan-uddin 2010 Multi-scale simulations of plasma with iPIC3D *Math. Comput. Simul.* **80** 1509–19

[56] Yee K S and Chen J S 1997 The finite-difference time-domain (FDTD) and the finite-volume time-domain (FVTD) methods in solving Maxwell's equations *IEEE Trans. Antennas Propag.* **45** 354–63

[57] Saad Y and Schultz M H 1986 GMRES: a generalized minimal residual algorithm for solving nonsymmetric linear systems *SIAM J. Sci. Stat. Comput.* **7** 856–69

[58] Ricci P, Lapenta G and Brackbill J 2002 A simplified implicit Maxwell solver *J. Comput. Phys.* **183** 117–41

[59] Mallon D A, Eicker N, Innocenti M E, Lapenta G, Lippert T and Suarez E 2012 On the scalability of the clusters-booster concept: a critical assessment of the deep architecture *Proc. Future HPC Systems: the Challenges of Power-Constrained Performance* (New York: ACM) article 3

[60] Kumar P, Markidis S, Lapenta G, Meerbergen K and Roose D 2013 High performance solvers for implicit particle in cell simulation *Proc. Comput. Sci.* **18** 2251–8

[61] Munz C, Omnes P, Schneider R, Sonnendrücker E and Voß U 2000 Divergence correction techniques for Maxwell solvers based on a hyperbolic model *J. Comput. Phys.* **161** 484–511

[62] Peng I B, Markidis S, Vaivads A, Vencels J, Amaya J, Divin A, Laure E and Lapenta G 2015 The formation of a magnetosphere with implicit particle-in-cell simulations *Proc. Comput. Sci.* **51** 1178–87

[63] Noguchi K, Tronci C, Zuccaro G and Lapenta G 2007 Formulation of the relativistic moment implicit particle-in-cell method *Phys. Plasmas* **14** 042308

[64] Chapman S and Cowling T G 1970 *The Mathematical Theory of Non-Uniform Gases: an Account of the Kinetic Theory of Viscosity, Thermal Conduction and Diffusion in Gases* (Cambridge: Cambridge University Press)

[65] Lapenta G 2016 Exactly energy conserving semi-implicit particle in cell formulation arXiv: 1602.06326

[66] Goldston R and Rutherford P 1995 *Introduction to Plasma Physics* (London: Taylor and Francis)

[67] Markidis S, Lapenta G, VanderHeyden W and Budimlićcc Z 2005 Implementation and performance of a particle-in-cell code written in Java *Concurr. Comput.: Pract. Exper.* **17** 821–37

[68] Bowers K J, Albright B J, Yin L, Daughton W, Roytershteyn V, Bergen B and Kwan T J T 2009 Advances in petascale kinetic plasma simulation with VPIC and Roadrunner *J. Phys. Conf. Ser.* **180** 012055

[69] Vapirev A, Deca J, Lapenta G, Markidis S, Hur I and Cambier J-L 2005 Initial results on computational performance of Intel many integrated core, sandy bridge, and graphical processing unit architectures: implementation of a 1D c++/OpenMP electrostatic particle-in-cell code *Concurr. Comput.: Pract. Exper.* **27** 581–93

[70] Decyk V K 2015 Skeleton particle-in-cell codes on emerging computer architectures *Comput. Sci. Eng.* **17** 47–52

IOP Publishing

Plasma Modeling
Methods and Applications
Gianni Coppa, Luis O Silva, Antonio D'Angola, Fabio Peano and Federico Peinetti

Chapter 5

The ergodic method: plasma dynamics through a sequence of equilibrium states

5.1 Introduction to the ergodic method

In problems of plasma physics, characterized by the simultaneous presence of phenomena spanning well-separated time scales, theoretical methods can be devised in which time averages over the more rapid phenomenon are performed in order to calculate the parameters for the study of the evolution of the slower one. In some respects, the idea behind these methods is similar to the concept of ergodic theory in kinetic theory [1].

In this chapter, investigations of two different problems are described. The first (section 5.2) concerns the expansion of a spherical plasma, in which the dynamics of ions and electrons have well-separated characteristic times, due to the mass ratio of the order of $10^{-3} \div 10^{-4}$. In practice, at any time in which ion dynamics is followed, an appropriate time average replaces the real position of electrons.

The second problem (section 5.3) regards the electron dynamics in a Penning trap in which a rarefied gas is present. In this case, the slow phenomenon is represented by the collisions of the electrons with atoms and with the walls of the trap, which modify the trajectories of the electrons and their number through ionizations and the phenomena of absorption and secondary emission from the walls. The separation of multiple time scales leads to a significant computational advantage, as one can use suitable time steps to follow only the slow phenomenon of interest. The two cases presented in this chapter show the potential of the ergodic method.

5.2 Expansion of spherical nanoplasmas

The irradiation of solid targets with ultra-intense lasers can induce the prompt formation of hot, dense plasmas, which rapidly expand in a vacuum [2]. While these phenomena have been extensively studied, the expansion of spherical nanoplasmas, which are generated upon interaction of ultra-intense lasers with atomic or

doi:10.1088/978-0-7503-1200-4ch5
© IOP Publishing Ltd 2016

molecular clusters [3–7], has not been analyzed as thoroughly. Accurate knowledge of the expansion, accounting for the self-consistent dynamics of ions and electrons, is relevant in specific situations where control over the expansion is necessary [8]. In fact, exact solutions for spherical expansions exist only for ideal cases, such as the Coulomb explosion [9] of a pure ion plasma, which occurs when all the electrons are suddenly swept away from the cluster by the laser field. Under opposite conditions, when most of the electrons are not pulled away from the cluster, hydrodynamic models can be employed to estimate the basic features of the expansion [10]. In the quasi-neutral limit, a kinetic solution for the adiabatic expansion of plasma bunches into a vacuum has also been derived [11]. However, in more general situations, in which a fraction of the electrons remains bound to the cluster, the expansion process is strongly dependent on the self-consistent dynamics of ions and of trapped electrons, and can be described accurately only by the Vlasov–Poisson theory.

In this section, a kinetic analysis is presented for the collisionless expansion of spherical plasmas, based on a specific ergodic model, which accounts for the radial motion of the ions and for the three-dimensional motion of the electrons [12, 13]. Furthermore, a procedure to study the initial charging transient, during which the faster electrons leave the cluster, is presented within the framework of the ergodic model. Precise knowledge of the electron distribution after the transient is important to set the correct initial conditions for the long-term plasma expansion.

Kinetic models for the expansion
In the electrostatic, non-relativistic limit, the dynamics of a collisionless electron–ion plasma is described rigorously by the Vlasov–Poisson set of equations

$$
\left\{
\begin{aligned}
\frac{\partial f_e}{\partial t} &= -\frac{\mathbf{p}}{m} \cdot \frac{\partial f_e}{\partial \mathbf{x}} - e\frac{\partial \Phi}{\partial \mathbf{x}} \cdot \frac{\partial f_e}{\partial \mathbf{p}}, \\
\frac{\partial f_i}{\partial t} &= -\frac{\mathbf{p}}{M} \cdot \frac{\partial f_i}{\partial \mathbf{x}} + e\frac{\partial \Phi}{\partial \mathbf{x}} \cdot \frac{\partial f_i}{\partial \mathbf{p}}, \\
\nabla^2 \Phi &= 4\pi e\left(\int f_e \, \mathrm{d}^3 p - \int f_i \, \mathrm{d}^3 p \right),
\end{aligned}
\right.
\tag{5.1}
$$

where $f_e(\mathbf{x}, \mathbf{p}, t)$ and $f_i(\mathbf{x}, \mathbf{p}, t)$ are the distribution functions in phase space for electrons (having mass m and charge $-e$) and ions (having mass M and charge $+e$; for simplicity, only $Z = 1$ ions are considered), respectively, and $\Phi(\mathbf{x}, t)$ is the electrostatic potential; Φ is set to zero at infinity, so that the energy of a single electron, $\mathscr{E} = \frac{1}{2}mv^2 - e\Phi$, is negative if it is trapped. In the following, attention is focused on the expansion of a plasma sphere with initial radius R_0 composed of cold ions with initial uniform density n_{i0} and hot electrons with initial uniform density $n_{e0} = n_{i0}$ and temperature T_0. The general initial conditions for equation (5.1) can be cast in the form

$$
f_{e0}(\mathbf{x}, \mathbf{p}) = n_{e0} g(v)\, \Theta\!\left(1 - \frac{r}{R_0}\right), \qquad f_{i0}(\mathbf{x}, \mathbf{p}) = n_{i0}\delta(\mathbf{p})\, \Theta\!\left(1 - \frac{r}{R_0}\right),
\tag{5.2}
$$

where Θ is the step function and g is the Maxwell–Boltzmann distribution

$$g(v) = \left(\frac{m}{2\pi k_B T_0}\right)^{3/2} \exp\left(-\frac{mv^2}{2k_B T_0}\right). \tag{5.3}$$

In fact, the dynamics of the system is fully determined by the mass ratio m/M and by $\hat{T}_0 = 3\lambda_{D,0}^2/R_0^2$, being $\lambda_{D,0}$ the electron Debye length at $t = 0$ [12]. In practice, the electrons are supposed to be instantaneously heated by an infinitely short laser pulse, without expanding. The expansion process can be split into two stages: the first is a rapid expansion of the electrons, which leads to an equilibrium before the ions move appreciably; the second is a slow expansion of the plasma bulk, driven by the positive charge buildup formed in the first stage. Due to the large mass disparity between ions and electrons, a simplified model can be derived, in which the ions are assumed to be fixed during the former process, whereas the electrons can be considered instantaneously at equilibrium with the electrostatic potential during the latter stage. A self-consistent theoretical framework can be developed, which allows one to determine accurately both the initial equilibrium and the bulk expansion by treating the electron dynamics as a sequence of equilibrium configurations with frozen ions. The model is obtained by exploiting the functional relation existing between n_e and Φ at equilibrium, and by calculating the energy variation of the electrons under the hypothesis of slow variations of Φ.

Equilibrium solutions of the Vlasov equation for the electrons must depend on $\mathbf{x}$ and $\mathbf{p}$ only through the invariants of motion. Under the hypothesis of spherical symmetry, the only invariants of motion to be considered are the Hamiltonian $\mathcal{H}(\mathbf{x}, \mathbf{p}) = p^2/2m - e\Phi(r)$ and the angular momentum, $\mathbf{L} = \mathbf{x} \times \mathbf{p}$. Consequently, the equilibrium distribution function can be written as $f_e(\mathbf{x}, \mathbf{p}) = F(\mathcal{H}(\mathbf{x}, \mathbf{p}), \mathbf{x} \times \mathbf{p})$. If a generic space point $\mathbf{x} = r\hat{\mathbf{e}}_r$ and a generic momentum $\mathbf{p} = p_r\hat{\mathbf{e}}_r + p_\perp\hat{\mathbf{e}}_\perp$ are considered, the phase-space density is $F(\frac{1}{2m}(p_r^2 + p_\perp^2) - e\Phi(r), L, \hat{\mathbf{e}}_r \times \hat{\mathbf{e}}_\perp)$ being $L = |\mathbf{L}| = mrv_\perp$. Due to the symmetry with respect to rotations around $\hat{\mathbf{e}}_r$, the phase-space density cannot depend upon $\hat{\mathbf{e}}_\perp$, and, consequently, F depends only on $\mathcal{H}$ and L.

The energy–angular momentum distribution for the electrons, $\sigma_e(\mathscr{E}, L)$, is defined as

$$\sigma_e(\mathscr{E}, L) = \iint F(\mathcal{H}(\mathbf{x}, \mathbf{p}), L)\delta(\mathcal{H}(\mathbf{x}, \mathbf{p}) - \mathscr{E})\delta(|\mathbf{x} \times \mathbf{p}| - L)\mathrm{d}^3x\mathrm{d}^3p$$

$$= \frac{8\pi^2\sqrt{2}}{m^{3/2}}F(\mathscr{E}, L)\int_{R_1(\mathscr{E},L)}^{R_2(\mathscr{E},L)}\left[\mathscr{E} - \frac{L^2}{2mr^2} + e\Phi(r)\right]^{-\frac{1}{2}}\mathrm{d}r, \tag{5.4}$$

where $R_1(\mathscr{E}, L)$ and $R_2(\mathscr{E}, L)$ are the radial turning points such that $\mathscr{E} - \frac{L^2}{2mr^2} + e\Phi(r) = 0$. The quantity $\sigma_e(\mathscr{E}, L)\mathrm{d}\mathscr{E}\mathrm{d}L$ represents the number of electrons having energy in $[\mathscr{E}, \mathscr{E} + \mathrm{d}\mathscr{E}]$ and angular momentum in $[L, L + \mathrm{d}L]$. The electron space density, n_e, can be written as

$$n_e(r) = \frac{1}{4\pi r^2} \iint \sigma_e(\mathscr{E}, L)\mathcal{P}(r, \mathscr{E}, L; \{\Phi\})\mathrm{d}\mathscr{E}\mathrm{d}L, \tag{5.5}$$

being

$$\mathcal{P}(r, \mathscr{E}, L) = \frac{\left[\mathscr{E} - \dfrac{L^2}{2mr^2} + e\Phi(r)\right]^{-\frac{1}{2}}}{\displaystyle\int_{R_1(\mathscr{E},L)}^{R_2(\mathscr{E},L)} \left[\mathscr{E} - \dfrac{L^2}{2mr'^2} + e\Phi(r')\right]^{-\frac{1}{2}} \mathrm{d}r'}; \tag{5.6}$$

$\mathcal{P}(r; \mathscr{E}, L)\mathrm{d}r$ gives the probability for an electron with energy $\mathscr{E}$ and angular momentum L, to be found in $[r, r + \mathrm{d}r]$. If variations of Φ, due to the ion motion, are slow with respect to the period of the radial oscillation of the electrons, the mean value of $\mathrm{d}\mathscr{E}/\mathrm{d}t$ can be evaluated by using the ensemble average of $\frac{\partial\Phi}{\partial t}$, as

$$\left\langle \frac{\mathrm{d}\mathscr{E}}{\mathrm{d}t} \right\rangle = \left\langle -e\frac{\partial\Phi}{\partial t}(r(t), t) \right\rangle = -e \int_{R_1(\mathscr{E},L)}^{R_2(\mathscr{E},L)} \frac{\partial\Phi}{\partial t}(r, t)\mathcal{P}(r, \mathscr{E}, L)\mathrm{d}r; \tag{5.7}$$

this is equivalent [14] to preserving the adiabatic invariant

$$\mathcal{I}(\mathscr{E}(t), L, t) = \text{Const} \cdot \int_{R_1(\mathscr{E},L)}^{R_2(\mathscr{E},L)} \left[\mathscr{E} - \frac{L^2}{2mr^2} + e\Phi(r, t)\right]^{\frac{1}{2}} \mathrm{d}r. \tag{5.8}$$

Equations (5.4)–(5.7), coupled with Poisson's equation and the dynamic equation for the radial motion of the ions, provide a self-consistent model for the expansion of a spherical plasma.

In general, there is a precise relationship between time scales and the proper number of parameters to be used to describe a phenomenon: in the present case, the system (5.1) allows one to follow the expansion dynamics on the time scale of the fastest particles. To study the ion expansion, a quasi-equilibrium model can be used, in which the stationary solution of the Vlasov equation for the electrons is employed (equations (5.4)–(5.7)). In fact, as the kinetic model is collisionless, the equations are time-reversible and a physical mechanism leading towards the equilibrium does not exist. To justify the use of the equilibrium distribution, $f(\mathbf{x}, \mathbf{p}) = f(\mathscr{H}(\mathbf{x}, \mathbf{p}), |\mathbf{x} \times \mathbf{p}|)$, one must assume that the stationary solution of the Vlasov equation is a good representation of the real electron distribution, once high-frequency fluctuations are eliminated. Formally, this can be performed by introducing a dissipation mechanism, i.e. a suitable collision term into the Vlasov equation. In general, an approximate kinetic model can be regarded as the result of introducing a particular collision term. For example, by using a binary collision term with sufficiently high collision frequency, f_e tends towards the Maxwell–Boltzmann distribution. In this case, a hydrodynamic description is obtained, whose domain of validity is confined to situations where $\hat{T}_0 \ll 1$. For larger values of $\hat{T}_0$, the use of a proper energy spectrum is fundamental; in fact, the energy distribution presents a cutoff for $\mathscr{E} = 0$, because for $\mathscr{E} > 0$ the electrons are not confined and their stationary density must vanish. Instead, a Maxwellian distribution has a non-negligible fraction of electrons with $\mathscr{E} > 0$. Within this framework, the approach of [12] can be introduced by considering a collision term of the form

$$J(f_e) = -\nu(f_e - \bar{f}_e), \qquad \bar{f}_e = \frac{1}{4\pi} \int f_e(\mathbf{x}, p\mathbf{\Omega}, t)\mathrm{d}^2\Omega, \tag{5.9}$$

where $\mathbf{\Omega}$ is the unit vector indicating the electron direction and ν is the collision frequency; $1/\nu$ must be smaller than the characteristic time of the ion expansion. In this case, the kinetic Vlasov equation for the electrons is replaced by the collisional equation

$$\frac{\partial f_e}{\partial t} = -\mathbf{p} \cdot \frac{\partial f_e}{\partial \mathbf{x}} - \frac{e}{m}\frac{\partial \Phi}{\partial \mathbf{x}} \cdot \frac{\partial f_e}{\partial \mathbf{p}} - \nu f + \frac{\nu}{4\pi} \int f(\mathbf{x}, \nu\mathbf{\Omega}, t)\mathrm{d}^2\Omega. \tag{5.10}$$

The collisions do not alter the electron energy, but randomly change their direction $\mathbf{\Omega}$, driving f_e towards an equilibrium distribution having the form $f_e(\mathbf{x}, \mathbf{p}) = f[\mathscr{H}(\mathbf{x}, \mathbf{p})]$, a sort of ergodic density such that each electron has an equal probability to be found in every point of the surface of phase space having equation $\mathscr{H}(\mathbf{x}, \mathbf{p}) = \mathscr{E}$. This fact can be proved rigorously by defining the electron entropy, $S_e = - \iint f_e \log(f_e)\mathrm{d}^3x\mathrm{d}^3p$, and noticing that its derivative

$$\frac{\mathrm{d}S_e}{\mathrm{d}t} = \nu \iint \log\left(\frac{f_e}{\bar{f}_e}\right) \cdot \left(f_e - \bar{f}_e\right)\mathrm{d}^3x\mathrm{d}^3p \tag{5.11}$$

is always non-negative unless $f_e = \bar{f}_e$. Therefore, a necessary condition for the distribution to be stationary is that f_e must not depend on $\mathbf{\Omega}$. Finally, the equilibrium distribution must be a function of $\mathscr{H}$ and L; as it does not depend on $\mathbf{\Omega}$, it is a function of $\mathscr{H}$ only. The approach can be called the single-particle ergodic method.[1] According to this approach, the equilibrium distribution function can be written simply as $f_e(\mathbf{x}, \mathbf{p}) = f[\mathscr{H}(\mathbf{x}, \mathbf{p})]$. The dependence on L is lost, and equations (5.4)–(5.6) are replaced by

$$\begin{aligned} \rho_e(\mathscr{E}) &= \iint f(\mathscr{H}(\mathbf{x}, \mathbf{p}))\delta(\mathscr{H}(\mathbf{x}, \mathbf{p}) - \mathscr{E})\mathrm{d}^3x\mathrm{d}^3p \\ &= 16\pi^2\sqrt{2}\,m^{3/2}f(\mathscr{E}) \int_{\mathscr{D}(\mathscr{E})} [\mathscr{E} + e\Phi(r)]^{\frac{1}{2}} r^2\mathrm{d}r, \end{aligned} \tag{5.12}$$

$$n_e(r) = \frac{1}{4\pi r^2} \int \rho_e(\mathscr{E})Q(r, \mathscr{E})\mathrm{d}\mathscr{E}, \tag{5.13}$$

$$Q(r, \mathscr{E}) = \frac{r^2[\mathscr{E} + e\Phi(r)]^{\frac{1}{2}}}{\displaystyle\int_{\mathscr{D}(\mathscr{E})} r'^2[\mathscr{E} + e\Phi(r')]^{\frac{1}{2}}\mathrm{d}r'}, \tag{5.14}$$

---

[1] This is different from the usual ergodic theory of statistical mechanics, in which the state of a system of N particles can be found with equal probability on the phase-space surface $\mathscr{H}(\mathbf{x}_1, \mathbf{x}_2, ..., \mathbf{x}_N; \mathbf{p}_1, \mathbf{p}_2, ..., \mathbf{p}_N) = $ Const.

where $\mathscr{D}(\mathscr{E})$ is the integration domain, such that $r \in \mathscr{D}(\mathscr{E}) \Rightarrow \mathscr{E} + e\Phi(r) \geqslant 0$ (for monotonic potentials, $\mathscr{D}(\mathscr{E}) = [0, R(\mathscr{E})]$, where $\mathscr{E} + e\Phi(R(\mathscr{E})) = 0$), and the adiabatic invariant, equation (5.8), is replaced by the ergodic invariant [15]

$$J[\mathscr{E}(t), t] = \text{Const} \cdot \int_{\mathscr{D}(\mathscr{E})} [\mathscr{E} + e\Phi(r, t)]^{\frac{3}{2}} r^2 dr, \tag{5.15}$$

which represents the volume of the region of $\mathbb{R}^6$ enclosed by the surface of equation $\frac{1}{2m}p^2 - e\Phi(r) = \mathscr{E}$. Moreover, Q represents the density distribution associated to an electron of energy $\mathscr{E}$.

As shown in [12, 13], this approach provides excellent results for the expansion of a spherical plasma in a wide range of the parameter $\hat{T}_0$. There is a number of reasons to explain its success. First, even though hydrodynamic models can provide a qualitative agreement with the real expansion dynamics, the ergodic model is more flexible in describing the energy distribution of the electrons. In addition, the initial phase-space distribution (5.2) is assumed to be an ergodic function. Finally, it must be noticed that the angular momentum is invariant only in the case of perfect spherical symmetry. In practical situations, perturbations breaking that symmetry, such as an initial shape which is not perfectly spherical, or collisions with heavy particles, would cause a mixing in the L distribution, and their effect can be taken into account by introducing a collision term such as the one in equation (5.9).

Single-particle ergodic model

Using the hypothesis of single-particle ergodic distribution, a self-consistent model for the expansion of a spherical plasma can be formulated, starting from equations (5.12)–(5.15), as follows [12]. A Lagrangian approach can be used both for the ions and the electrons. Ions move in the radial direction, starting from the initial position r_0 with zero velocity, while the electron energy $\mathscr{E}$ evolves in time starting from the initial value $\mathscr{E}_0$. In this way, the ion trajectories $r_i(r_0, t)$, the electron energies $\mathscr{E}(\mathscr{E}_0, t)$, the ion density $n_i(r, t)$, the electron density, $n_e(r, t)$, the electron energy distributions, $\rho_e(\mathscr{E}, t)$, and the potential $\Phi(r, t)$ are determined by solving the set of equations

$$\begin{cases} \dfrac{\partial^2 r_i}{\partial t^2} = -\dfrac{e}{M}\dfrac{\partial \Phi}{\partial r}(r_i), \\[2ex] \dfrac{1}{r^2}\dfrac{\partial}{\partial r}\left(r^2\dfrac{\partial \Phi}{\partial r}\right) = 4\pi e(n_e - n_i), \\[2ex] n_i(r_i) = n_{i,0}(r_0)\dfrac{(r_0/r_i)^2}{\partial r_i/\partial r_0}, \\[2ex] n_e = \displaystyle\int \rho_e(\mathscr{E})Q(r, \mathscr{E}, \{\Phi\})d\mathscr{E}, \\[2ex] \rho_e(\mathscr{E}, t) = \dfrac{\rho_{e,0}(\mathscr{E}_0)}{\partial \mathscr{E}/\partial \mathscr{E}_0}, \\[2ex] \dfrac{\text{d}}{\text{d}t}J(\mathscr{E}(\mathscr{E}_0, t), t) = 0. \end{cases} \tag{5.16}$$

In equation (5.16), the evolution equations for the radial coordinates of the ions and the electron energies are coupled via Poisson's equation. The expansion dynamics is determined once the initial ion density $n_{i,0}$ and the electron energy distribution $\rho_{e,0}$ are given. In equation (5.16), the electron density n_e is calculated by summing the number of electrons having energy in $[\mathcal{E}, \mathcal{E} + d\mathcal{E}]$, $\rho_e(\mathcal{E})d\mathcal{E}$, multiplied by the probability for an electron to be found at the radius r. For simplicity, the ion density n_i is written under the hypothesis of no ion overtaking ($\partial r_i / \partial r_0 \neq 0$).

The set of equations (5.16) describes the expansion dynamics on the ion time scale; therefore, its numerical solution is much faster than solving the full Vlasov–Poisson model, where the electron time scale must be resolved. The model is solved by calculating the radial trajectories of a set of representative ions and the energy variations of a set of computational particles. Each computational particle represents a given number of electrons, whose radial distribution is given by equation (5.14). This description of the energy dependence corresponds to a suitable discretization of the integral in equation (5.13), which is similar to the description of the spatial dependence commonly adopted in the particle-in-cell approach [16].

Drift–diffusion approximation

As an alternative to solving equation (5.10), one can consider the drift–diffusion equation

$$\frac{\partial \Psi}{\partial t} = e \frac{\partial \Phi}{\partial t} \frac{\partial \Psi}{\partial \mathcal{E}} + \frac{2}{3m\upsilon} \frac{1}{r^2} \frac{\partial}{\partial r} \left[r^2 \left((\mathcal{E} + e\Phi) \frac{\partial \Psi}{\partial r} - \frac{e}{2} \frac{\partial \Phi}{\partial r} \Psi \right) \right], \tag{5.17}$$

obtained from equation (5.10) by approximating f_e as $f_0(r, p) + \mathbf{p} \cdot \mathbf{f}_1(r, p)$ [17]. In equation (5.17), the quantity $\Psi(r, \mathcal{E}, t)$ represents the space–energy distribution of the electrons. The self-consistent potential Φ is determined by solving Poisson's equation

$$\frac{1}{r^2} \frac{\partial}{\partial r} \left(r^2 \frac{\partial \Phi}{\partial r} \right) = 4\pi e \left(\int \Psi d\mathcal{E} - n_{i0} \right). \tag{5.18}$$

Asymptotically, for $t \to \infty$, the solution approaches a stationary solution of equation (5.17), Ψ_∞, such that

$$\frac{1}{2} e \frac{d\Phi_\infty}{dr} \Psi_\infty - (\mathcal{E} + e\Phi_\infty) \frac{\partial \Psi_\infty}{\partial r} = 0. \tag{5.19}$$

By solving equation (5.19) with respect to Ψ_∞, one finds

$$\Psi_\infty(r, \mathcal{E}) = \frac{\rho_\infty(\mathcal{E})[\mathcal{E} + e\Phi_\infty(r)]^{\frac{1}{2}}}{4\pi \int_{\mathcal{D}(\mathcal{E})} r'^2 [\mathcal{E} + e\Phi_\infty(r')]^{\frac{1}{2}} dr'}, \tag{5.20}$$

where $\rho_\infty(\mathcal{E}) = 4\pi \int \Psi_\infty(r, \mathcal{E}) r^2 dr$ is the energy distribution, which corresponds to the ergodic distribution expressed by equations (5.12)–(5.14).

Charging transient

Since the initial configuration considered here (5.2) is far from equilibrium, the proper energy spectrum $\rho_{e,0}$ to be used in equation (5.16) must be determined as the equilibrium configuration after the initial charging transient. Apparently, the ergodic method cannot be of help for this purpose, since it is valid only for sufficiently smooth variations of Φ in time, a condition which is not met in the early stage when the hot electrons are suddenly allowed to expand (as if a rigid wall, initially confining them, were instantaneously brought to infinity). However, a procedure can be envisaged which makes these equations suitable also for the analysis of the initial electron equilibrium.

To this end, the real charging transient is replaced by a virtual one, in which an external potential barrier, initially confining the electrons, is gradually moved from $R_b = R_0$ to infinity with a series of small radial displacements. Each time the barrier is moved farther by δR_b the new self-consistent potential Φ is calculated and the energy of the electrons is updated. In order to actually simulate an expansion into vacuum (which the real transient is), the electrons and the expanding barrier must not exchange energy, i.e. the electron energy must vary only because of Φ variations. This implies that the ergodic invariant (5.15) is not conserved during the initial stage. Should one conserve $\mathcal{J}$ when displacing the barrier from a given radius R_b to $R_b + \delta R_b$, the corresponding electron energy variation, $\delta\mathcal{E}$, would be

$$\delta\mathcal{E} = -e \int_0^{R_b} \delta\Phi Q(r, \mathcal{E})\mathrm{d}r - \delta W, \tag{5.21}$$

where δW, defined as

$$\delta W = \frac{2}{3}[\mathcal{E} + e\Phi(R_b)]Q(R_b; \mathcal{E})\delta R_b, \tag{5.22}$$

represents the expansion work, done by an electron having energy $\mathcal{E}$, against the expanding barrier. Thus, conserving $\mathcal{J}$ would cause the overestimation of the electron cooling as the system would lose an extra amount of energy corresponding to the expansion work. In order to obtain an energy balance equivalent to that of a vacuum expansion, the energy loss associated to the expansion work is set to zero in equation (5.21).

The physical process simulated with the barrier method can be regarded as an infinitely slow expansion during which some external energy source compensates exactly for the expansion work δW against the barrier. Alternatively, it can be considered as a series of instantaneous, small displacements of the barrier, where, after each displacement, one waits for a new equilibrium configuration to establish itself.

Reference results for the transient leading to the initial space-charge distribution of the plasma have been determined by solving equation (5.1) numerically, in the hypothesis of immobile ions, using the so-called 'shell method'.[2] The same

---

[2] According to this scheme [9, 18], computational particles representative of a given number of electrons are moved in space, under the action of the sum of the self-consistent electric field and the electric field due to the ion distribution. By resorting to the spherical symmetry of the system, the field generated by the electrons is evaluated using Gauss' law, as if each particle were actually a spherical shell, thus avoiding the use of a computational grid for solving Poisson's equation and allowing for an infinite radial domain.

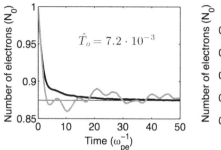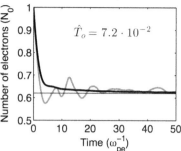

Figure 5.1. Evolution of the electronic charge contained within the ion sphere ($r < R_0$), for $\hat{T}_0 = 7.2 \times 10^{-3}$ and $\hat{T}_0 = 7.2 \times 10^{-2}$. Thick gray lines refer to the collisionless case, thick black lines refer to the collisional case ($\nu \sim \omega_{pe}$). Thin horizontal lines indicate the results obtained using the barrier method. (Reproduced with permission from [13]. Copyright 2007 The American Physical Society.)

framework has also been used to investigate numerically the effect of the presence of the collision term, equation (5.10), which forces the system towards an ergodic distribution. Such perturbations are introduced in the model by scattering randomly the computational particles, without changing their energy, according to the collision frequency ν.

Figure 5.1 shows the evolution of the electronic charge contained within the ion sphere, for low-temperature ($\hat{T}_0 = 7.2 \times 10^{-3}$) and high-temperature ($\hat{T}_0 = 7.2 \times 10^{-2}$) cases, as obtained with the Vlasov–Poisson equations, i.e. with $\nu = 0$, and with the collisional model (5.10), using $\nu = \omega_{pe}$. In the collisionless case, the charge transient exhibits small-amplitude oscillations. In the collisional model, for $\nu \gtrsim \omega_{pe}$, the oscillations are strongly damped and the system rapidly reaches an equilibrium configuration, as predicted theoretically. In figure 5.2, the electron density is plotted, along with the corresponding electric field. Figure 5.3 shows the equilibrium energy distribution $\rho_{e,0}$ to be used as the initial condition for the bulk expansion. The excellent agreement between different models confirms the validity of the barrier method. Figure 5.4 shows the asymptotic solution of equations (5.17) and (5.18), $\Psi_\infty(r, \mathscr{E})$. The corresponding electron density and energy distributions (figures 5.2 and 5.3) have been calculated as $\int \Psi_\infty(r, \mathscr{E}) \, d\mathscr{E}$ and $4\pi \int \Psi_\infty(r, \mathscr{E}) r^2 dr$, respectively.

Bulk expansion

Using the tools presented above, the expansion of ions and electrons can be investigated for a wide range of the parameter $\hat{T}_0$. The results of the study show that the expansion dynamics changes smoothly from a hydrodynamic-like regime ($\hat{T}_0 \ll 1$) to a Coulomb expansion regime ($\hat{T}_0 \sim 1$). In the former case the outer ions expand first and a rarefaction front propagates inward, in the latter all ions start expanding at the same time. The evolution of the ion phase-space profile and of the electron and ion densities, starting from the initial equilibrium of figure 5.2, are shown in figures 5.5 and 5.6, respectively. In case (a), the ion expansion starts from the periphery and a rarefaction front is clearly observed to propagate inward until it

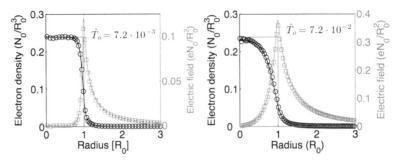

Figure 5.2. Equilibrium electron density (black) and radial electric field (gray), for $\hat{T}_0 = 7.2 \times 10^{-3}$ and $\hat{T}_0 = 7.2 \times 10^{-2}$. Solid lines refer to results from the ergodic model, markers to results from the Vlasov–Poisson model, and dotted lines to results from the drift–diffusion model of equations (5.17) and (5.18) (the curves obtained with the ergodic model and those obtained with the drift–diffusion model are practically indistinguishable). (Reproduced with permission from [13]. Copyright 2007 The American Physical Society.)

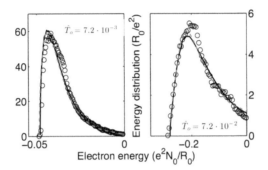

Figure 5.3. Equilibrium energy spectrum of trapped electrons, (i.e. with $\mathscr{E} = m\mathbf{p}^2/2 - e\Phi < 0$) for $\hat{T}_0 = 7.2 \times 10^{-3}$ and $\hat{T}_0 = 7.2 \times 10^{-2}$. Solid lines refer to results from the ergodic model, dots to results from the Vlasov–Poisson model, and dotted lines to results from the drift–diffusion model. (Reproduced with permission from [13]. Copyright 2007 The American Physical Society.)

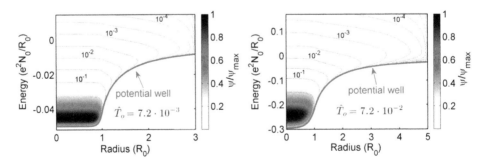

Figure 5.4. Equilibrium distribution $\Psi(r, \mathscr{E})$ (normalized to its maximum value) as obtained from the drift–diffusion model of equations (5.17) and (5.18), for $\hat{T}_0 = 7.2 \times 10^{-3}$ and $\hat{T}_0 = 7.2 \times 10^{-2}$. (Reproduced with permission from [13]. Copyright 2007 The American Physical Society.)

reaches the center of the distribution. During the expansion the plasma remains approximately neutral, apart from the thin double-layer at the ion front. These features, typical of quasi-neutral, hydrodynamic expansions, are lost in case (b), in which all the ions are promptly involved in the expansion (figure 5.5(b)) and the distribution remains non-neutral during the whole process (figure 5.6(b)). In both cases, as the ions expand and gain kinetic energy, the electrons cool down and the charge buildup within the ion sphere decreases. Asymptotically, the sphere enveloped by the expanding ion front encloses all trapped electrons, and a ballistic regime is reached for both species [19]. The self-consistent behavior of the electrons strongly

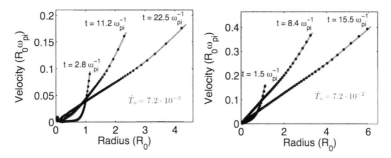

Figure 5.5. Evolution of the ion phase-space profile, for $\hat{T}_0 = 7.2 \times 10^{-3}$ and $\hat{T}_0 = 7.2 \times 10^{-2}$. Lines refer to results from the ergodic model and dots to results from the full Vlasov–Poisson model. (Reproduced with permission from [13]. Copyright 2007 The American Physical Society.)

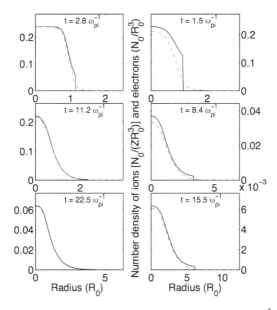

Figure 5.6. Evolution of ion (solid line) and electron (dashed curve) densities, for $\hat{T}_0 = 7.2 \times 10^{-3}$ (left) and $\hat{T}_0 = 7.2 \times 10^{-2}$ (right). (Reproduced with permission from [13]. Copyright 2007 The American Physical Society.)

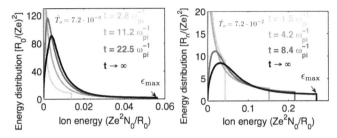

Figure 5.7. Evolution of the ion energy spectrum (from gray to black), for $\hat{T}_0 = 7.2 \times 10^{-3}$ and $\hat{T}_0 = 7.2 \times 10^{-2}$. (Reproduced with permission from [13]. Copyright 2007 The American Physical Society.)

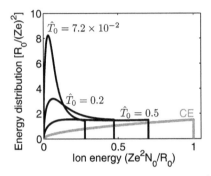

Figure 5.8. Asymptotic energy spectrum of the ions for different values of $\hat{T}_0$, compared to the asymptotic spectrum for a Coulomb explosion (gray). (Reproduced with permission from [13]. Copyright 2007 The American Physical Society.)

affects the ion dynamics and their resulting energy spectrum. Figure 5.7 illustrates the evolution of the ion energy spectrum towards its asymptotic form: in both cases, the spectrum develops a local maximum, far from the maximum value of the ion energy. This feature is absent in Coulomb explosion, where the asymptotic spectrum is always monotonic, so the maximum in the spectrum is expected to disappear when increasing $\hat{T}_0$ further. The transition from non-monotonic to monotonic ion spectra occurs about $\hat{T}_0 = 0.5$, as shown in figure 5.8.

5.3 Electron dynamics in a Penning trap for technology applications

A Penning trap is a device used to confine non-neutral plasmas, in which an axial magnetic field and a voltage applied to cylindric electrodes provide radial and axial confinement. Penning traps are used in plasma physics investigations [20–23] and in many advanced technological applications, such as in ultra-high vacuum systems [24, 25], to determine fundamental physical constants [26, 27], in high-precision mass spectrometry [28–32], in Zeeman spectroscopy [33], and in antihydrogen production and trapping [34, 35].

The electron motion inside a Penning trap is the superposition of Larmor and diocotron orbits with frequencies ω_c and ω_D, respectively, and axial bounce between the cathodes, with frequency ω_b. In ideal traps [20–23, 34, 36, 37], the three characteristic frequencies of these orbits differ by orders of magnitude [38], in particular $\omega_c \gg \omega_b \gg \omega_D$. These differences arise from the high intensity of the magnetic field and from the large value of the aspect ratio $L_z{:}R$ ($2L_z$ and R being the length and the radius of the trap). Moreover, collisions between electrons and atoms are negligible due to the ultra-high vacuum condition. In contrast, in the Penning traps used in some technological applications, such as for ion pumps, L_z and R are comparable. Moreover, the magnetic field, which is usually generated by permanent magnets, is significantly lower and electron–neutral collisions are important. In fact, the working principle of ion pumps is the electron-induced ionization of gas atoms which are then removed by the electric field [24]. For these devices, Larmor, bouncing, and diocotron frequencies are of the same order of magnitude, while they differ by several orders of magnitude with respect to the electron–gas collision frequency, ν_{el} (for a gas with a pressure of 10^{-6} torr at room temperature, the collision frequency for an electron is $\nu_{el} \sim 10^4$ Hz [39]). The large difference between the two characteristic time scales of the phenomena under examination, along with the need to analyze the system on the slow time scale, suggests the use of a suitable average of the electron dynamics on the fast time scale.

In spite of its apparent simplicity, the physics involved in the Penning traps used in ion pumps is quite rich. If electrons are introduced into the device and a discharge is set up, gas atoms or molecules can be ionized inside the cylindrical cavity. The biasing of the electrodes creates a potential well for the electrons, and hence the positive ions produced in the electron–gas collisions are accelerated toward the negatively biased planar cathodes. Consequently, ions impinge on these electrodes with a relevant kinetic energy and are removed from the system [40–42]. In contrast, electrons are confined by the electric and magnetic fields, and therefore they can induce further ionizations. In this way, an increasing negative charge is set up, which induces large modifications to the electrostatic potential. Eventually, stationary configurations are achieved due to the equilibrium between the electrons generated by ionization and secondary emission (induced by the ions impinging on the cathodes) and electron radial transport to the anode. Between two successive collisions, the electron trajectory is characterized by two constants of motion: the total energy and the canonical angular momentum. The knowledge of these constants provides useful information. In particular, for each electron a space region can be determined in which the trajectory is constrained. Therefore, each electron can be regarded as a space-charge distribution, which represents the time average of the positions spanned during the motion. The effect of collisions of the electrons with neutral atoms or molecules can be calculated in a simple way, as long as the gas density is sufficiently low. In practice, the mean time between two collisions must be higher with respect to the time required for an electron trajectory to fill the region in which it is constrained. In this case, the effect of an elastic collision is simply a variation of the constants of motions of the electron, and, consequently, of the shape of the region that is associated with them. Ionization collisions and secondary emissions can be taken into account in a similar way.

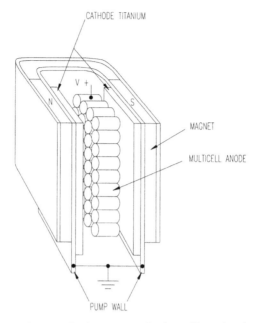

Figure 5.9. Scheme of a Penning trap for ion pump technology. (Reproduced with permission from [43]. Copyright Agilent.)

A typical Penning trap for ion pumps is made of a cylindrical anode of radius R which is inserted between two planar cathodes whose distance is L_z (R and L_z are of the order of a few centimeters), biased at a negative potential $-V_0$, typically of the order of a few kilovolts. The electric field confines the electrons in the z-direction, causing a bounce motion between the two planar cathodes. In addition, the cell is immersed in a uniform magnetic field B_0 of the order of 10^{-1} T, parallel to the axis of the cylinder, whose purpose is to confine the electrons in the radial direction (figure 5.9) [21].

Hyperbolic trap

In order to study the general properties of the electron trajectories, one can start from a simplified two-dimensional case, in which a constant magnetic field in the z-direction is present and an electrostatic potential $\Phi(x, z)$ is generated by the electrodes [44].

Between two successive collisions, the motion of a particle of charge q is governed by the equations

$$
\begin{cases}
\dfrac{dv_x}{dt} = -\dfrac{q}{m}\dfrac{\partial \Phi}{\partial x} - \omega_c v_z, \\[2mm]
\dfrac{dv_z}{dt} = -\dfrac{q}{m}\dfrac{\partial \Phi}{\partial z}, \\[2mm]
\dfrac{dv_z}{dt} = \omega_c v_x
\end{cases}
\tag{5.23}
$$

where $\omega_c = qB/(mc)$ is the Larmor frequency. From the third equation (5.23), one obtains immediately $v_z = \omega_c x + \text{Const}$, which states that the canonical momentum

$$p_z = m(v_z - \omega_c x) \tag{5.24}$$

is a constant of motion. Therefore v_z can be eliminated from the first equation (5.23) and the equations of motion in the (x, z) plane can be written as

$$\frac{\mathrm{d}p_x}{\mathrm{d}t} = -q\frac{\partial\Phi}{\partial x} - m\omega_c(\omega_c x + p_z/m), \qquad \frac{\mathrm{d}p_z}{\mathrm{d}t} = -q\frac{\partial\Phi}{\partial z}. \tag{5.25}$$

By defining the effective potential, $\mathcal{U}$, as

$$\mathcal{U}(x, z) = q\Phi(x, z) + \frac{1}{2m}(m\omega_c x + p_z)^2, \tag{5.26}$$

the equations of motion can be deduced from the Hamiltonian

$$\begin{aligned}
\mathcal{H}(x, z, p_x, p_z) &= \frac{1}{2m}\left(p_x^2 + p_z^2\right) + \mathcal{U}(x, z) \\
&= \frac{1}{2m}\left[p_x^2 + p_z^2 + (m\omega_c x + p_z)^2\right] + q\Phi(x, z).
\end{aligned} \tag{5.27}$$

As $\mathcal{H}$ does not depend explicitly on time, its value is a constant of motion

$$\mathcal{H}\big(x(t), z(t), p_x(t), p_z(t)\big) = \text{Const} = \mathcal{E}, \tag{5.28}$$

and it coincides with the total energy of an electron in the potential Φ. The level curves for $\mathcal{U}(x, z)$ and $\Phi(x, z)$ have the following property: given a generic point (x_0, z_0), if $p_z = -m\omega_c x_0$ then the two curves $\Phi(x, z) = \Phi(x_0, z_0)$ and $\mathcal{U}(x, z) = q\Phi(x_0, z_0)$ are tangent in (x_0, z_0). This can be proved by considering that the tangent line to $\Phi = \text{Const}$ is such that $\mathrm{d}\Phi = 0$ and so $\frac{\partial\Phi}{\partial x}\mathrm{d}x + \frac{\partial\Phi}{\partial z}\mathrm{d}z = 0$, while the tangent line to $\mathcal{U} = \text{Const}$ is such that $0 = \mathrm{d}\mathcal{U} = \mathrm{d}\Phi + m\omega_c(\omega_c x - \omega_c x_0)\mathrm{d}x$, so, if $x = x_0$, then $\mathrm{d}\mathcal{U} = \mathrm{d}\Phi$. Therefore the two lines have the same equation. The property can be interpreted as follows: if (x_0, z_0) is a point of the cathode and in this point an electron with zero velocity is emitted ($p_x = 0$, $p_z = -m\omega_c x_0$) then the electron moves in a part of the phase space such that $\mathcal{H} = q\Phi(x_0, z_0) = \mathcal{U}(x_0, z_0)$. Being $\mathcal{H}(x, z, p_x, p_z) \geqslant \mathcal{U}(x, z)$, the region of $\mathbb{R}^2$ in which the electron can be found must satisfy the condition

$$\mathcal{U}(x, z) < \mathcal{E} = \mathcal{U}(x_0, z_0), \tag{5.29}$$

and it is delimited by the curve $\mathcal{U}(x, z) = \mathcal{U}(x_0, z_0)$. In addition, if the curves $\Phi(x, z) = \Phi(x_0, z_0)$ and $\mathcal{U}(x, z) = \mathcal{U}(x_0, z_0)$ were not tangent in (x_0, z_0), the region of motion, equation (5.29), should have points in the region $q\Phi(x, z) > q\Phi(x_0, z_0)$, but this is impossible, as

$$\mathcal{E} = q\Phi(x, z) + \frac{1}{2}mv^2 = q\Phi(x_0, z_0) \Rightarrow q\Phi(x, z) < q\Phi(x_0, z_0). \tag{5.30}$$

In the case of potential $\Phi(x, z)$ produced by using a pair of cathodes and anodes having a hyperbolic shape, i.e.

$$q\Phi(x, z) = a^2\big(z^2 - x^2\big), \tag{5.31}$$

the effective potential $\mathcal{U}$ assumes the form

$$\mathcal{U}(x, z) = \mathcal{U}_x(x) + \mathcal{U}_z(z) + \mathscr{E}_m(p_z), \tag{5.32}$$

with

$$\mathcal{U}_x(x) = b^2(x - x_m)^2, \quad \mathcal{U}_z(z) = a^2 z^2, \quad \mathscr{E}_m(p_z) = -\frac{a^2 p_z^2}{2mb^2}, \tag{5.33}$$

with $b^2 = m\omega_c^2/2 - a^2$ and $x_m(p_z) = -\omega_c p_z/(2b^2)$. Although the potential $\Phi(x, z)$ confines electron only in the z-direction, a sufficiently strong magnetic field can also provide the confinement in the x-direction if $b^2 > 0$, i.e. if $|\omega_c| > a(2/m)^{\frac{1}{2}}$.

The electron motion is the composition of two harmonic oscillations in the x- and z-directions, according to the following equations:

$$\frac{d^2 x}{dt^2} = -\omega_x^2(x - x_m), \quad \omega_x = b\left(\frac{2}{m}\right)^{\frac{1}{2}},$$
$$\frac{d^2 z}{dt^2} = -\omega_z^2 z, \quad \omega_z = a\left(\frac{2}{m}\right)^{\frac{1}{2}}, \tag{5.34}$$

which are obtained starting from the potential equation (5.32). Therefore, the motion is confined in the rectangular region $\mathscr{R} \subset \mathbb{R}^2$ defined by

$$\mathscr{R} = \left[x_m - \frac{\mathscr{E}_x^{1/2}}{b}, x_m + \frac{\mathscr{E}_x^{1/2}}{b}\right] \times \left[-\frac{\mathscr{E}_z^{1/2}}{a}, \frac{\mathscr{E}_z^{1/2}}{a}\right], \tag{5.35}$$

where $\mathscr{E}_x$, $\mathscr{E}_z$ are the energies associated with the oscillatory motion in the x- and z-directions, respectively. The total energy, $\mathscr{E}$, is simply given by $\mathscr{E} = \mathscr{E}_x + \mathscr{E}_z + \mathscr{E}_m(p_z)$. Four constants of motion exist, i.e. the energies along x and z and the two phases, φ_x and φ_z, for the oscillations in the x- and z-directions. While the values of two energies determine the rectangular region $\mathscr{R}$ in which the motion is allowed, the difference $\varphi_x - \varphi_z$ determines the particular kind of trajectory. Only in the very specific case in which ω_x/ω_z is a rational number is the trajectory a closed Lissajous curve, otherwise, for $t \to +\infty$, the curve tends to 'cover' the entire rectangular region. Figure 5.10 shows the elliptic region defined by equation (5.29) and the corresponding region $\mathscr{R}$ for an electron emitted by the cathode. According to the theory, the ellipse is tangent to the cathode. In summary, each electron can be associated with a phase-space distribution depending on three parameters $\{p_z, \mathscr{E}_x, \mathscr{E}_z\}$. Alternatively, a simplified 'ergodic' model can be used, in which each particle is associated with the distribution $\delta(\frac{m}{2}(v_x^2 + v_z^2) + \mathcal{U}(x, z) - \mathscr{E})$, with $\mathscr{E}$ the total energy. The last assumption is rigorously correct in the presence of perturbations of the potential, otherwise it represents a simplified description of the real physical system.

A real Penning trap

In a real Penning trap, the motion of an electron can be described by the Lagrangian $\mathscr{L}$ of a particle of mass m and electric charge $-e$ moving in a

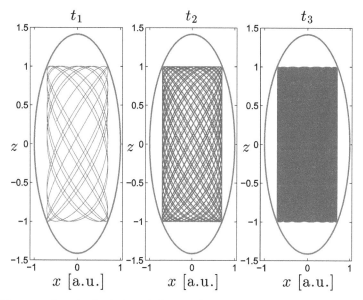

Figure 5.10. Electron trajectories for the hyperbolic potential, equation (5.31), for $t \in [0, t_i]$ with $t_1 < t_2 \ll t_3$. The ellipse $\mathscr{U}(x, z) = \mathscr{E}$ is also shown.

homogeneous magnetic field $\boldsymbol{B} = B_0 \boldsymbol{e}_z$ and in an electric field $\boldsymbol{E}(r, z) = -\nabla \Phi(r, z)$, as [21]

$$\mathscr{L}\left(r, \vartheta, z, \dot{r}, \dot{\vartheta}, \dot{z}\right) = \frac{m}{2} v^2 + e\left[\Phi(r, z) - \frac{\boldsymbol{v}}{c} \cdot \boldsymbol{A}\right]$$
$$= \frac{m}{2}\left(\dot{r}^2 + r^2 \dot{\vartheta}^2 + \dot{z}^2\right) + e\left[\Phi(r, z) - \frac{1}{2}\frac{B_0}{c} r^2 \dot{\vartheta}\right], \tag{5.36}$$

where $\boldsymbol{A} = \frac{1}{2} B_0 r \boldsymbol{e}_\vartheta$ is the vector potential . As the Lagrangian defined by (5.36) does not depend on ϑ, the corresponding canonical momentum $p_\vartheta = mr^2(\dot{\vartheta} - \frac{1}{2}\omega_c)$ is a constant of motion and therefore v_ϑ can be expressed as function of r and p_ϑ, and an effective potential, $\mathscr{U}$, can be defined as

$$\mathscr{U}\left(r, z; p_\vartheta\right) = -e\Phi(r, z) + \frac{m}{2} v_\vartheta^2\left(r, p_\vartheta\right)$$
$$= -e\Phi(r, z) + \frac{m}{2}\left(\frac{p_\vartheta}{mr} + \frac{\omega_c r}{2}\right)^2. \tag{5.37}$$

Introducing the Hamiltonian, $\mathscr{H}$:

$$\mathscr{H}\left(r, z, p_r, p_z; p_\vartheta\right) = \frac{m}{2}\left(v_r^2 + v_z^2\right) + \frac{m}{2} v_\vartheta^2\left(r, p_\vartheta\right) - e\Phi(r, z)$$
$$= \frac{1}{2m}\left[p_r^2 + p_z^2\right] + \mathscr{U}(r, z; p_\vartheta) \tag{5.38}$$

the dynamics of the electrons can be reduced to the (r, z) domain. The Hamiltonian is a constant of motion as it does not depend explicitly on time. Its value, $\mathscr{E}$, coincides with the total energy of an electron in the potential field Φ: $\mathscr{E} = \frac{1}{2}mv^2 - e\Phi$. Between two successive collisions, electrons conserve the total energy $\mathscr{E}$ and the canonical angular momentum p_ϑ and, being $\mathscr{H}(r, z, p_r, p_z; p_\vartheta) \geq \mathscr{U}(r, z; p_\vartheta)$, the particle can be found in the region $\mathscr{R}_E \subset \mathbb{R}^2$ defined by

$$\mathscr{U}(r, z; p_\vartheta) \leqslant \mathscr{E}. \tag{5.39}$$

$\mathscr{R}_E$ is delimited by the closed curve of equation $\mathscr{U}(r, z; p_\vartheta) = \mathscr{E}$. Actually, due to the existence of other constants of motion, the electron trajectories are constrained in a subdomain $\mathscr{R}_0$ of $\mathscr{R}_E$. Figure 5.11 shows the electron trajectories and the associated density distributions (density distributions are calculated by sampling the trajectories at constant time intervals). In order to approximate the accessible region, a simple model can be used in which the potential $\mathscr{U}$ is decoupled along directions r and z. In this way, the rectangular region in which the electron is confined is determined by calculating the turning points. For an electron located in (r_0, z_0) and with velocity (v_{r0}, v_{z0}), the energy associated with the motion along the z-direction has been defined as $\frac{1}{2}mv_{z0}^2 + \mathscr{U}(r_0, z_0; p_\vartheta)$. Then, the axial turning points, $\pm z_b$, have been calculated by solving the equation

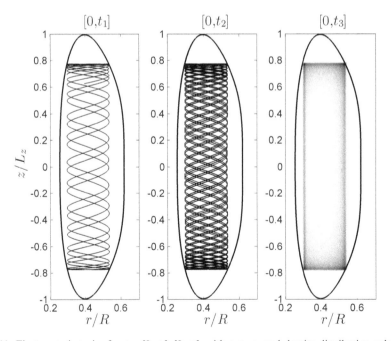

Figure 5.11. Electron trajectories for $t = [0, t_1]$, $[0, t_2]$, with $t_2 \gg t_1$, and density distribution calculated by sampling the trajectory at constant time intervals for $t = [0, t_3]$ with $t_3 \gg t_2$ ($B_0 = 0.1$ T). The curve $\mathscr{U}(r, z; p_\vartheta) = \mathscr{E}$ is also shown (reproduced from [45] with permission).

$$\mathcal{U}(r_0, z_b; p_\vartheta) = \frac{1}{2}mv_{z0}^2 + \mathcal{U}(r_0, z_0; p_\vartheta). \tag{5.40}$$

Then, the turning points r_1 and r_2 along the r-direction have been obtained as the solutions of the equation

$$\mathcal{U}(r_{1,2}, z_b; p_\vartheta) = \frac{1}{2}mv_{r0}^2 + \mathcal{U}(r_0, z_b; p_\vartheta). \tag{5.41}$$

The turning points belong to the boundary of the region $\mathcal{R}_E$. Making use of equation (5.40), equation (5.41) can be rewritten as $\mathcal{U}(\pm r_{1,2}, z_b; p_\vartheta) = \mathcal{E}$. In general, the effective potentials $\mathcal{U}$ may present a non-monotonic behavior and, depending on the value of total energy, from equations (5.40) and (5.41) multiple turning points can be found along the r- and z-directions. In this case, the domain $\hat{\mathcal{R}}$ can be univocally calculated taking into account the place where collisions happen.

The effectiveness of the approximation can be proved by comparing the rectangular domain $\hat{\mathcal{R}} = [r_1, r_2] \times [-z_b, z_b]$ with the real trajectory. In fact, a more significant test is provided by the probability distribution of finding an electron at a given spatial position. This quantity is very useful in the study of the dynamics of electrons: as collision frequencies are much smaller with respect to all the characteristic frequencies of motion, one can consider the time average of the trajectory instead of calculating the position of the electron at each time [44–46]. The calculation is simple in the case of the separable potential (i.e. $\mathcal{U}_r(r) + \mathcal{U}_z(z)$), for which one can express the time-averaged phase-space distribution associated with the electron motion as

$$f(r, z, p_r, p_z) = C \cdot \delta\left(\frac{p_r^2}{2m} + \mathcal{U}_r(r) - \mathcal{E}_r\right)\delta\left(\frac{p_z^2}{2m} + \mathcal{U}_z(z) - \mathcal{E}_z\right), \tag{5.42}$$

with $\mathcal{E}_r$, $\mathcal{E}_z$ the energies associated with the motion along r and z, respectively, (i.e. the values of $\mathcal{U}_r$ and $\mathcal{U}_z$ at turning points), and C is such that $\iint f \, d^2x d^2p = 1$. In general, $\mathcal{U}$ is not separable but, according to the above-introduced approximation, the following expressions can be used:

$$\mathcal{U}_r(r; p_\vartheta) = \mathcal{U}(r, z_b; p_\vartheta)$$
$$\mathcal{U}_z(z; p_\vartheta) = \frac{1}{2}\left[\mathcal{U}(r_1, z; p_\vartheta) + \mathcal{U}(r_2, z; p_\vartheta)\right] \tag{5.43}$$
$$\mathcal{E}_r = \mathcal{E}_z = \mathcal{E}.$$

From the density f, the probability for a particle to be in $[r - \Delta r/2, r + \Delta r/2] \times [z - \Delta z/2, z + \Delta z/2]$ is evaluated as $\mathcal{P}(r, z)\Delta r\Delta z$ where

$$\mathcal{P}(r, z) = \iint f(r, z, p_r, p_z)dp_r \, dp_z = \mathcal{P}_r(r)\mathcal{P}_z(z) \tag{5.44}$$

where

$$\mathcal{P}_r(r) = \begin{cases} C_r\left[\mathscr{E} - \mathcal{U}_r(r; p_\vartheta)\right]^{-1/2}, & r_1 < r < r_2 \\ 0 & , \quad \text{otherwise} \end{cases} \tag{5.45}$$

$$\mathcal{P}_z(z) = \begin{cases} C_z\left[\mathscr{E} - \mathcal{U}_z(z; p_\vartheta)\right]^{-1/2}, & |z| < z_b \\ 0 & , \quad \text{otherwise.} \end{cases} \tag{5.46}$$

The constants C_r, C_z are determined by requiring that $\int \mathcal{P}_r(r)\mathrm{d}r = \int \mathcal{P}_z(z)\mathrm{d}z = 1$. Figures 5.12 and 5.13 show the excellent agreement between the normalized probability $\mathcal{P}(r, z)/\overline{\mathcal{P}}$ (with $\overline{\mathcal{P}} = 1/(2RL_z)$) as given by equations (5.45) and (5.46) and the probability obtained by solving numerically equations of motion (as before, the 'exact' probability has been calculated by considering a grid in the (r, z) domain and sampling the solution at constant time intervals).

In addition to the method presented above, a simpler description of electron dynamics can be considered, in which the time average of the trajectories is calculated using an ergodic distribution for each particle. The use of a phase-space ergodic distribution is rigorously justified if a physical system is perturbed in a way that only 'strong' constants of motion, such as the energy, are conserved. As an example, in figure 5.14 the trajectory of an electron is shown for a case in which a suitable perturbation $\delta\Phi(r, z)$ is added to the electrostatic potential [45]. In this case,

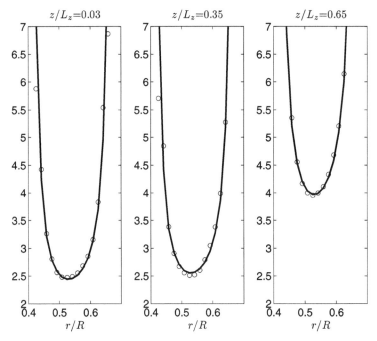

Figure 5.12. Example of comparisons between the probability density $\mathcal{P}(r, z)/\overline{\mathcal{P}}$ at different z given by equation (5.45) (solid line) with the one obtained by solving numerically the equations of motion (∘) (reproduced from [45] with permission).

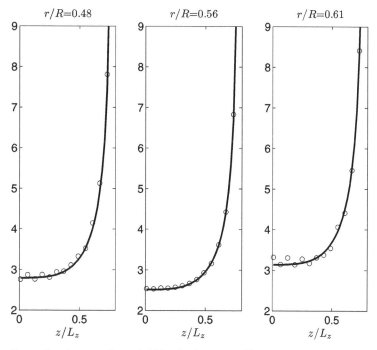

Figure 5.13. Comparison between the probability density $\mathscr{P}(r, z)/\overline{\mathscr{P}}$ at different r given by equation (5.46) (solid line) with the one obtained by solving numerically the equations of motion (∘) for the same case as reported in figure 5.12 (reproduced from [45] with permission).

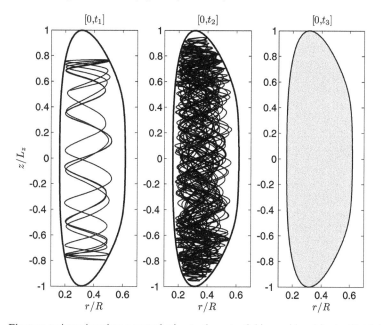

Figure 5.14. Electron trajectories when a perturbation to the potential is considered for $t = [0, t_1]$, $[0, t_2]$, with $t_2 \gg t_1$, and density distribution calculated by sampling the trajectory at constant time intervals for $t = [0, t_3]$ with $t_3 \gg t_2$ ($B_0 = 0.08$ T). The curve $\mathscr{U}(r, z; p_\vartheta) = \mathscr{E}$ is also shown (reproduced from [45] with permission).

the trajectory tends to cover uniformly the region $\mathscr{R}_E$ and not only $\hat{\mathscr{R}}$. This fact can be explained by considering that a weak perturbation does not appreciably change the total energy $\mathscr{E}$; instead, it produces a different distribution between the axial and the radial energies. In real situations, as the perturbation may involve also the ϑ coordinate, one should control the sensitivity of p_ϑ to the perturbation. In other terms, p_ϑ may not be a 'strong' constant. However, independent of the presence and the possible effects of a perturbation to the potential, the use of an ergodic distribution provides a useful tool for a simplified description of the electron density [12, 13]. In this case the time-averaged phase-space density has the form

$$f_E(\boldsymbol{x}, \boldsymbol{p}; p_\vartheta) = C \cdot \delta\left(\frac{\boldsymbol{p}^2}{2m} + \mathscr{U}(\boldsymbol{x}; p_\vartheta) - \mathscr{E}\right), \qquad (5.47)$$

which depends only on two parameters, $\mathscr{E}$ and p_ϑ, and the probability $\mathscr{P}_E$ is given by

$$\mathscr{P}_E(\boldsymbol{x}; p_\vartheta) = \begin{cases} \dfrac{1}{\mathscr{A}(\mathscr{E}, p_\vartheta)}, & \mathscr{U}(\boldsymbol{x}; p_\vartheta) \leqslant \mathscr{E}, \\ 0, & \mathscr{U}(\boldsymbol{x}; p_\vartheta) > \mathscr{E}, \end{cases} \qquad (5.48)$$

where $\mathscr{A}(\mathscr{E}, p_\vartheta) = \displaystyle\int_{\mathscr{U} \leqslant \mathscr{E}} \mathrm{d}^2 x$ is the area of the permitted region of motion. In figure 5.15, the probability densities of time-average and ergodic models, given respectively by equations (5.44) and (5.48), are shown for trajectories with the same initial conditions.

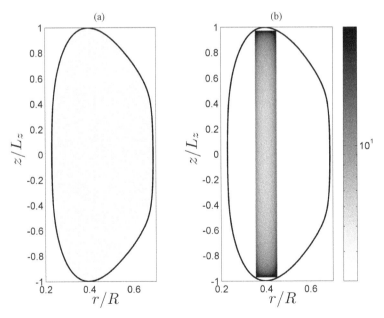

Figure 5.15. Normalized densities $\mathscr{P}_E(r, z)/\overline{\mathscr{P}}$ (a) and $\mathscr{P}(r, z)/\overline{\mathscr{P}}$ (b) for an electron with the same initial conditions (reproduced from [45] with permission).

Collisions with atoms

When phenomena having time scale much larger than the ones characterizing the electron motion are considered, each electron can be usefully regarded not as a point but as an 'object' having a spatial charge distribution, $\rho(r, z)$, given by

$$\rho(r, z) = -\frac{e\mathscr{P}(r, z)}{2\pi r}. \tag{5.49}$$

In this way, each electron is described by its constants of motion, $\mathscr{E}$, p_ϑ, and the turning point z_b. In practice, the probabilities given by equations (5.45) and (5.46) can be calculated if $\mathscr{E}$, p_ϑ, and z_b are known, considering that radial turning points $r_{1,2}$ are solutions of equation (5.41). The density distribution associated with the electron can be described by a set of parameters $\{\mathscr{E}, p_\vartheta, z_b\}$, whose values change after each collision. The instantaneous collision frequency, $\nu(t)$, is given by $\nu(t)\sigma(\nu(t))\mathscr{N}_a$, with $\mathscr{N}_a$ the atom density and $\sigma(\nu)$ the cross section of the considered process. Instead, the time-averaged collision frequency, $\bar{\nu}$, can be calculated as

$$\bar{\nu} = \iint \nu\sigma(\nu)\mathscr{N}_a f(\boldsymbol{x}, \boldsymbol{p}; p_\vartheta) \mathrm{d}^2 x \mathrm{d}^2 p, \tag{5.50}$$

with $\boldsymbol{x} = (r, z)$, $\boldsymbol{p} = (p_r, p_z)$ and $\nu(\mathscr{E}, \boldsymbol{x}) = [2(\mathscr{E} + e\Phi(\boldsymbol{x}))/m]^{\frac{1}{2}}$. Using equations (5.42) and (5.44), one obtains

$$\bar{\nu}(\mathscr{E}, p_\vartheta, z_b) = \int \nu(\mathscr{E}, \boldsymbol{x})\sigma(\nu(\mathscr{E}, \boldsymbol{x}))\mathscr{N}_a\mathscr{P}(\boldsymbol{x}; \mathscr{E}, p_\vartheta, z_b)\mathrm{d}^2 x. \tag{5.51}$$

The frequency $\bar{\nu}$ is the sum of the frequency of elastic collisions, $\bar{\nu}_{\mathrm{el}}$ and the frequency of ionizations, $\bar{\nu}_i$. The two frequencies are calculated using equation (5.51) by putting $\sigma = \sigma_{\mathrm{el}}$ or $\sigma = \sigma_i$ in it. If an elastic collision has place in $\boldsymbol{x}$ (with density probability $\nu(\mathscr{E}, \boldsymbol{x})\sigma_{\mathrm{el}}(\nu(\mathscr{E}, \boldsymbol{x}))\mathscr{N}_a/\bar{\nu}_{\mathrm{el}})$, by considering that the masses of the atoms are much larger than the electron mass, after the collision the electron has velocity $\boldsymbol{v}' = \nu(\mathscr{E}, \boldsymbol{x})\boldsymbol{n}'$, where $\boldsymbol{n}'$ is a randomly directed unit vector. After the collision, the electron trajectory is described by a set of new constants of motion, $\{\mathscr{E}', p'_\vartheta, z'_b\}$, as

$$p'_\vartheta = mr\left(v'_\vartheta - \frac{1}{2}\omega_c r\right), \quad \mathscr{E}' = \mathscr{E}, \quad z'_b \quad \text{from equation (5.40).} \tag{5.52}$$

Ionizations happen in $\boldsymbol{x}$ with density probability $\nu(\mathscr{E}, \boldsymbol{x})\sigma_i(\nu(\mathscr{E}, \boldsymbol{x}))/\bar{\nu}_i$ only if $\frac{1}{2}mv^2(\mathscr{E}, \boldsymbol{x})$ is greater than the ionization energy, ϵ_i. In this case, one must specify the constants of motion of the primary (I_1) and secondary (I_2) electrons as

$$\begin{cases} v'_{I_1} = \left(v^2 - 2\epsilon_i/m\right)^{1/2}, \quad \boldsymbol{v}'_{I_1} = v'_{I_1}\boldsymbol{n}', \quad v'_{I_2} = 0, \\[2mm] \mathscr{E}'_{I_1} = \mathscr{E}'_{r,I_1} = \mathscr{E} - \epsilon_i, \quad p'_{\vartheta,I_1} = mr\left(v'_{I_1,\vartheta} - \frac{1}{2}\omega_c r\right), \quad z'_{b,I_1} \text{ from equation (5.40),} \\[2mm] \mathscr{E}'_{I_2} = \mathscr{E}'_{r,I_2} = q\Phi(\boldsymbol{x}), \quad p'_{\vartheta,I_2} = -\frac{1}{2}m\omega_c r^2, \quad z'_{b,I_2} \quad \text{from equation (5.40).} \end{cases} \tag{5.53}$$

The positive ions produced from ionizations impact on the cathodes and, according to the mechanism of secondary emission, they emit electrons. If an ion impacts on the cathode in $x_s = (r_s, \pm L_z)$, the constants of motion of the secondary emitted electron (S) are

$$\begin{cases} v'_S = 0 \\ \mathscr{E}'_S = \mathscr{E}'_{r,S} = q\Phi(x_s), \quad p'_{\vartheta,S} = -\dfrac{1}{2}m\omega_c r_s^2, \quad z'_{b,S} \quad \text{from equation (5.40).} \end{cases} \tag{5.54}$$

If the region occupied by an electron intersects the anode, the particle is removed as it is absorbed by the electrode. The time-averaged frequency, $\bar{\nu}_E(\mathscr{E}, p_\vartheta)$, is obtained from Equation (5.51) by replacing $\mathscr{P}$ with $\mathscr{P}_E$. Using the ergodic approximation each electron is described by indicating only the two constants of motion, $\{\mathscr{E}, p_\vartheta\}$ and after elastic collision, ionization or secondary emission, the electrons change their parameters as follows:

- for elastic collision,

$$\mathscr{E}' = \mathscr{E}, \quad p'_\vartheta = mr\left(v'_\vartheta - \frac{1}{2}\omega_c r\right); \tag{5.55}$$

- for ionization,

$$\begin{cases} v'_{I_1} = \left(v^2 - 2\epsilon_i/m\right)^{1/2}, \quad \boldsymbol{v}'_{I_1} = v'_{I_1}\boldsymbol{n}', \quad \mathscr{E}'_{I_1} = \mathscr{E} - \epsilon_i, \quad p'_{\vartheta,I_1} = mr\left(v'_{I_1,\vartheta} - \frac{1}{2}\omega_c r\right), \\ v'_{I_2} = 0, \quad \mathscr{E}'_{I_2} = q\Phi(x), \quad p'_{\vartheta,I_2} = -\frac{1}{2}m\omega_c r^2; \end{cases} \tag{5.56}$$

- for secondary emission,

$$v'_S = 0, \quad \mathscr{E}'_S = q\Phi(x_s), \quad p'_{\vartheta,S} = -\frac{1}{2}m\omega_c r_s^2. \tag{5.57}$$

Monte Carlo simulation of the charge loading and saturation

Making use of the time-averaged trajectories, the initial stage of the charge loading can be investigated numerically on the time scale of electron–neutral collisions. A particle technique can be adopted, in which electron–neutral elastic collisions, ionizations, and secondary emissions are taken into account by using a Monte Carlo approach. During this initial stage, the electron density is sufficiently low and the self-consistent field can be neglected. Therefore, only the potential Φ due to the electrodes has been considered by solving the Laplace equation. When the charge density inside the trap becomes higher, the self-consistent potential must be considered by solving a self-consistent Poisson equation for Φ, which is non-linear, as the charge density for each electron depends on the accessible region $\hat{\mathscr{R}}$, which is a function of Φ itself.

Initially, a set of N computational particles is generated having zero velocity at different positions of the cathode and the relative set of $\mathscr{E}$, p_ϑ, and z_b parameters is calculated. Then, at each time step a subset of colliding particles is selected. For each electron the type of collision (elastic or ionization), the spatial position of the event, and the new direction $\boldsymbol{n}'$ are chosen by using Monte Carlo techniques. In this way, the set of characteristic parameters is updated at each time step.

The calculations are simpler, as long as the effect of the self-consistent electric field can be neglected. When the charge density increases, the electric field generated by the electron charge must be included into the calculation by solving the non-linear Poisson equation

$$\nabla^2 \Phi(r, z) = 4\pi e \sum_{p=1}^{N} \frac{\mathscr{P}\left(r, z; \mathscr{E}_p, p_{\vartheta,p}, z_{b,p}; \{\Phi\}\right)}{2\pi r}, \tag{5.58}$$

where probabilities $\mathscr{P}$ are given by equations (5.44)–(5.46). Moreover, a change $\delta\Phi$ in the potential induces a variation of the electron energy, $\delta\mathscr{E}$, given by

$$\delta\mathscr{E} = -e \int_{\hat{\mathscr{R}}} \delta\Phi(\mathbf{x})\mathscr{P}(\mathbf{x})\mathrm{d}^2x. \tag{5.59}$$

Finally, from $\delta\Phi$ and $\delta\mathscr{E}$ the new turning points are calculated. As $\mathscr{E}_p$ depends on Φ and Φ depends on $\mathscr{E}_p$, equations (5.58) and (5.59) can be solved numerically by using simple iterative procedures.

Typical results obtained with the proposed methods are shown in figure 5.16, in which the time evolution of the charge inside a trap has been reported for different values of the magnetic field ($B_0 = 0.05$, 0.08, and 0.3 T). For higher values of the

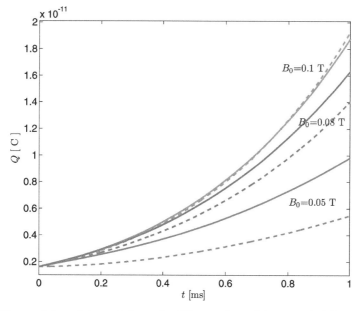

Figure 5.16. Time evolution of the total charge for different values of the magnetic field ($B_0 = 0.05$, 0.08, 0.3 T). Dashed and solid lines refer, respectively, to the ergodic and to the three-parameter model (reproduced from [45] with permission).

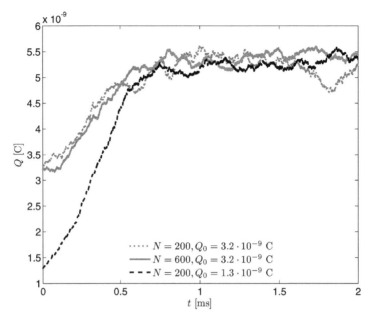

Figure 5.17. Time evolution of the total charge Q by using the three-parameter description for $B_0 = 0.06$ T, using a different number of particles and different initial charge (reproduced from [45] with permission).

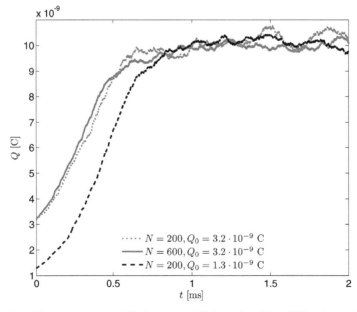

Figure 5.18. The same as figure 5.17, for $B_0 = 0.1$ T (reproduced from [45] with permission).

magnetic field, for which $\hat{\mathscr{R}} \sim \mathscr{R}_E$, the results obtained with the approximate ergodic technique are close to those of the three-parameter model. In the simulation, 10^3 computational particles have been initially emitted by the cathodes in the region $[0, 0.4R]$. Figure 5.17 shows the phenomenon of saturation of the total charge, for $B_0 = 0.06$ T. Two calculations have been performed using the three-parameter description with the same initial charge and a different number of computational particles ($N = 200$ and $N = 600$). From the two simulations, the behavior of the transient is the same (apart from the numerical noise, which is greater for the case of $N = 200$) and, in particular, the same value of the saturation charge is obtained. In the third calculation, the initial charge is a fraction (2/5) of that of the previous simulations and, although the transient is different, the asymptotic value of the total charge is the same. Similar calculations have been repeated for $B_0 = 0.1$ T; the results are presented in figure 5.18. As shown in these examples, due to its simplicity the method provides a useful tool to study the electron dynamics in Penning traps on long time scales.

References

[1] Huang K 1987 *Statistical Mechanics* (New York: Wiley)
[2] Dawson J M 1964 *Phys. Fluids* **7** 981–7
[3] Ditmire T *et al* 1997 *Nature* **386** 54
    Ditmire T *et al* 1999 *Nature* **398** 489
[4] Zweiback J *et al* 2000 *Phys. Rev. Lett.* **84** 2634
    Zweiback J *et al* 2000 *Phys. Rev. Lett.* **85** 3640
    Madison K W *et al* 2004 *Phys. Plasmas* **11** 270
    Grillon G *et al* 2002 *Phys. Rev. Lett.* **89** 065005
    Madison K W *et al* 2004 *Phys. Rev. A* **70** 053201
    Sakabe S *et al* 2004 *Phys. Rev. A* **69** 023203
    Hirokane M *et al* 2004 *Phys. Rev. A* **69** 063201
[5] Sakabe S *et al* 2006 *Phys. Rev. A* **74** 043205
[6] Last I and Jortner J 2004 *J. Chem. Phys.* **120** 1336
    Last I and Jortner J 2004 *J. Chem. Phys.* **120** 1348
[7] Last I and Jortner J 2006 *Phys. Rev. Lett.* **97** 173401
    Heidenreich A, Jortner J and Last I 2006 *Proc. Natl Acad. Sci. USA* **103** 10589
[8] Peano F, Fonseca R A and Silva L O 2005 *Phys. Rev. Lett.* **94** 033401
[9] Boella E *et al* 2016 *J. Plasma Phys.* **82** 905820110
[10] Ditmire T *et al* 1996 *Phys. Rev. A* **53** 3379
     Ditmire T *et al* 1998 *Phys. Rev. A* **57** 369
     Krainov V P and Smirnov M B 2002 *Phys. Rep.* **370** 237
     Milchberg H M, McNaught S J and Parra E 2001 *Phys. Rev. E* **64** 056402
     Lezius M *et al* 1998 *Phys. Rev. Lett.* **80** 261
     Zweiback J *et al* 2002 *Phys. Plasmas* **9** 3108
     Liu C S and Tripathi V K 2003 *Phys. Plasmas* **10** 4085
     Murakami M and Basko M M 2006 *Phys. Plasmas* **13** 012105
[11] Kovalev V F and Bychenkov V Y 2003 *Phys. Rev. Lett.* **90** 185004
[12] Peano F *et al* 2006 *Phys. Rev. Lett.* **96** 175002

[13] Peano F *et al* 2007 *Phys. Rev.* E **75** 066403

[14] Goldstein H 1980 *Classical Mechanics* 2nd edn (New York: Addison-Wesley)

[15] Ott E 1979 *Phys. Rev. Lett.* **42** 1628

[16] Birdsall C K and Langdon A B 1985 *Plasma Physics via Computer Simulations* (New York: McGraw-Hill)

[17] Raizer Yu P 1991 *Gas Discharge Physics* (Berlin: Springer)

[18] D'Angola A, Boella E and Coppa G 2014 *Phys. Plasmas* **21** 082116

[19] Manfredi G, Mola S and Feix M R 1993 *Phys. Fluids* B **90** 388

[20] Coppa G, D'Angola A, Delzanno G L and Lapenta G 2001 *Phys. Plasmas* **8** 1133–40

[21] Davidson R C 2001 *Physics of Nonneutral Plasmas* (Singapore: World Scientific)

[22] Huang X- P *et al* 1997 *Phys. Rev. Lett.* **78** 875–8

[23] Kriesel J M and Driscoll C F 2001 *Phys. Rev. Lett.* **87** 135003

[24] Audi M and de Simon D 1987 *Vacuum* **37** 629

[25] Jepsen R L 1968 The physics of sputter-ion pumps *Proc. 4th Int. Vacuum Congr. (Manchester)* pp 317–24

[26] Beier T *et al* 2001 *Phys. Rev. Lett.* **88** 011603

[27] Van Dyck R S, Schwinberg P B and Dehmelt H G 1987 *Phys. Rev. Lett.* **59** 26–9

[28] Bollen G, Moore R B, Savard G and Stolzenberg H 1990 *J. Appl. Phys.* **68** 4355–74

[29] Cornell E A *et al* 1989 *Phys. Rev. Lett.* **63** 1674–7

[30] Gerz C, Wilsdorf D and Werth G 1990 *Nucl. Instrum. Methods Phys. Res.* B **47** 453–61

[31] Gerz C, Wilsdorf D and Werth G 1990 *Zeitschr. Phys.* D **17** 119–21

[32] Moore F L, Farnham D L, Schwinberg P B and van Dyck R S Jr 1989 *Nucl. Instrum. Methods Phys. Res.* B **43** 425–30

[33] Wineland D J, Bollinger J J and Itano W M 1983 *Phys. Rev. Lett.* **50** 628–31

[34] O'Neil T M 1995 *Phys. Scri.* T **59** 341

[35] Trapp S, Tommaseo G, Revalde G, Stachowska E, Szawiola G and Werth G 2003 *Eur. Phys. J.* D **26** 237–44

[36] Andresen G *et al* 2007 *Phys. Rev. Lett.* **98** 023402

[37] Fajans J *et al* 2005 *Phys. Rev. Lett.* **95** 155001

[38] Werth G 1985 *Contemp. Phys.* **26** 241–56

[39] Lieberman M A and Lichtenberg A J 2005 *Principles of Plasma Discharges and Materials Processing* 2nd edn (Hoboken, NJ: Wiley)

[40] Helmer J C and Jepsen R L 1961 Electrical characteristic of a Penning discharge *Proc. IRE* **49** 1920–5

[41] Jepsen R L 1961 *J. Appl. Phys.* **32** 2619

[42] Redhead P A 1988 *Vacuum* **38** 901–6

[43] Agilent technologies *Agilent Ion Pumps Catalog* www.agilent.com/cs/library/catalogs/public/06_Ion_Pumps.pdf

[44] Coppa G, D'Angola A and Mulas R 2010 *Nuovo Cimento* C **33** 87–94

[45] Coppa G, D'Angola A and Mulas R 2012 *Phys. Plasmas* **19** 062507

[46] Coppa G and Ricci P 2002 *Phys. Rev.* E **66** 046409

Part II

Fluid and Hybrid Models

IOP Publishing

Plasma Modeling
Methods and Applications
Daniela Grasso

Chapter 6

Fluid equations: derivation and simulation

The kinetic approach is essential in plasmas where, due to the low collisionality, non-Maxwellian distribution functions give the most adequate description. However, for slow and large scale phenomena, such as the ones encountered both in laboratory and astrophysical plasmas, a fluid description can be very useful and is quite common. A fluid description is obtained from the kinetic one, averaging the Boltzmann (Vlasov in the absence of collisions) equation on the velocity space and introducing the moments of the distribution function. In this chapter we will introduce the moments of the distribution functions and then average the Boltzmann (Vlasov) equation to obtain a potentially n-fluid description of the plasma. We will restrict our further considerations to the two-fluid model and single-fluid approximation. The basic assumption leading to the magnetohydrodynamical (MHD) model will then be discussed. Finally the reduced version of the MHD model will be presented. The fluid approach introduces a great simplification in the tractability of the problem, although losing details that might depend on the tails of the distribution function. Despite this simplification, the system of equations that can be obtained in the fluid context is difficult to solve analytically and the numerical approach is mandatory in this research field. In the second half of this chapter we give a schematic overview of the finite difference and spectral numerical schemes commonly used in the fluid context. Finally an application of the reduced MHD model to the basic problem of magnetic reconnection will be presented.

6.1 Moments of the distribution function

The distribution function f obeys the Boltzmann equation

$$\frac{\partial f}{\partial t} + \mathbf{v} \cdot \nabla f + \frac{e}{m}\left(\mathbf{E} + \frac{1}{c}\mathbf{v} \times \mathbf{B}\right) \cdot \nabla_v f = \left(\frac{\partial f}{\partial t}\right)_{\text{coll}}, \tag{6.1}$$

where the electric and magnetic field are coupled with the Maxwell equations, which determine the charge and current density distributions that must be expressed in terms of the particles that generate the plasma and, in the end, of the distribution function itself. It is clear that the analytic and numerical solutions of this system of equations are cumbersome. Moreover we must wonder if one did obtain a solution for the distribution function, whether one would then be able to verify its validity. The key question is if and when one really needs all the information contained in the distribution function, or if one can instead rely on the information obtained from easily measurable quantities. For instance one could decide that the essential information is related to the physical geometrical space and average the distribution function over the velocity space. Let us consider the following integral:

$$\int f(\mathbf{r}, \mathbf{v}; t)d\mathbf{v} = \iiint f(\mathbf{r}, v_x, v_y, v_z; t)dv_x dv_y dv_z. \tag{6.2}$$

Since the distribution function represents the average number of particles within a infinitesimal cell in the phase space, it follows that equation (6.2) represents the average number of particles with a position between $\mathbf{r}$ and $\mathbf{r} + d\mathbf{r}$ regardless of their velocity. In other words the density in ordinary space is

$$n(\mathbf{r}; t) = \int f(\mathbf{r}, \mathbf{v}; t)d\mathbf{v}. \tag{6.3}$$

In the same way one defines the average value of any function of the velocity, $\xi(\mathbf{v})$ as

$$\langle \xi \rangle \equiv \frac{\int \xi(\mathbf{v})f(\mathbf{r}, \mathbf{v}; t)d\mathbf{v}}{\int f(\mathbf{r}, \mathbf{v}; t)d\mathbf{v}} = \frac{1}{n(\mathbf{r}; t)}\int \xi(\mathbf{v})f(\mathbf{r}, \mathbf{v}; t)d\mathbf{v}, \tag{6.4}$$

where the division by the density comes from the definition of the distribution function itself, which is not normalized to unity but to the total number of particles N.

The most interesting functions are represented by $\xi = \mathbf{vv}...\mathbf{v}$, which define the kth moment of the distribution function as

$$\mathbf{M}_k(\mathbf{r}; t) = \frac{1}{n(\mathbf{r}; t)}\int \mathbf{vv}...\mathbf{v}f(\mathbf{r}, \mathbf{v}; t)d\mathbf{v}. \tag{6.5}$$

From equation (6.5) it follows that the density is the zero-order moment. In the same way the first-order moment represents the average velocity

$$\mathbf{u}(\mathbf{r}; t) = \frac{1}{n(\mathbf{r}; t)}\int \mathbf{v}f(\mathbf{r}, \mathbf{v}; t)d\mathbf{v}. \tag{6.6}$$

If we define f_i and f_e the distribution functions of ions and electrons, respectively, from equations (6.2) and (6.6) one obtains the charge and the current density:

$$q(\mathbf{r}; t) = \int e(f_i - f_e)d\mathbf{v} \quad \text{and} \quad \mathbf{j}(\mathbf{r}; t) = \int e\mathbf{v}(f_i - f_e)d\mathbf{v}. \tag{6.7}$$

The second-order moment gives the stress tensor

$$\mathbf{P} = \frac{1}{n(\mathbf{r};\,t)} \int m\mathbf{v}\mathbf{v}f(\mathbf{r},\,\mathbf{v};\,t)\mathrm{d}\mathbf{v}, \qquad (6.8)$$

while the third-order moment gives the energy flux

$$\mathbf{Q} = \frac{1}{n(\mathbf{r};\,t)} \int \frac{mv^2}{2}\mathbf{v}f(\mathbf{r},\,\mathbf{v};\,t)\mathrm{d}\mathbf{v}. \qquad (6.9)$$

From the definitions above it follows that the moments of the distribution function have a clear physical meaning, being related to measurable quantities in the experiments. Obviously, averaging on the velocity space, the information related to particles whose velocities differ significantly from the average value are lost.

The moments of the distribution function themselves obey to an evolution equation, named general equation for the moments. To obtain this equation one has to multiply the Boltzmann equation times $\xi(\mathbf{v})$ and integrate on the velocity space. Since ξ does not depend explicitly on time and using the definition (6.4), the first term gives

$$\int \xi(\mathbf{v})\frac{\partial f(\mathbf{r},\,\mathbf{v};\,t)}{\partial t}\mathrm{d}\mathbf{v} = \frac{\partial}{\partial t}\int \xi f \mathrm{d}\mathbf{v} = \frac{\partial}{\partial t}(n\langle\xi\rangle).$$

Likewise, since ξ does not depend on $\mathbf{r}$ either, the second term gives

$$\int \xi(\mathbf{v})\mathbf{v}\cdot\nabla f \mathrm{d}\mathbf{v} = \nabla\cdot\int \mathbf{v}\xi f \mathrm{d}\mathbf{v} = \nabla\cdot(n\langle\mathbf{v}\xi\rangle).$$

For the third term we treat first the velocity free force, obtaining

$$\frac{e}{m}\int \mathbf{E}\cdot\xi(\mathbf{v})\nabla_v f \mathrm{d}\mathbf{v} = \frac{e}{m}\mathbf{E}\cdot\int \xi\nabla_v f \mathrm{d}\mathbf{v}$$

$$= \frac{e}{m}\mathbf{E}\cdot\left(\int \nabla_v(\xi f)\mathrm{d}\mathbf{v} - \int f\nabla_v(\xi)\mathrm{d}\mathbf{v}\right)$$

$$= -\frac{ne}{m}\mathbf{E}\langle\nabla_v\xi\rangle,$$

where using Gauss's theorem the first integral can be transformed in a surface integral, which goes to zero if we assume that

$$\lim_{\xi\to\infty}(\xi f) = 0.$$

This is a reasonable assumption if one thinks that a surface integral will grow with v^2 and a real distribution $f(\mathbf{v})$ (for example a Maxwellian) will likely go to zero much faster, since no particle can have infinite velocity.

Similarly, keeping in mind that the ith component of the Lorentz force is independent of the ith velocity, the second force term gives

$$\frac{e}{mc} \int \mathbf{v} \times \mathbf{B} \cdot \xi(\mathbf{v}) \nabla_\mathbf{v} f \, d\mathbf{v} = -\frac{ne}{mc} \langle (\mathbf{v} \times \mathbf{B}) \cdot \nabla_\mathbf{v} \xi \rangle.$$

For the collisional term we assume that in analogy with the temporal variation of the distribution function the following expression holds:

$$\int \xi(\mathbf{v}) \left(\frac{\partial f}{\partial t} \right)_{\text{coll}} d\mathbf{v} = \left(\frac{\partial}{\partial t} (n \langle \xi \rangle) \right)_{\text{coll}}.$$

Gathering all together the previous terms the general equation for the moments is written as

$$\frac{\partial}{\partial t} (n \langle \xi \rangle) + \nabla \cdot (n \langle \mathbf{v} \xi \rangle) - \frac{n}{m} \langle \mathbf{F} \cdot \nabla_\mathbf{v} \xi \rangle = \left(\frac{\partial}{\partial t} (n \langle \xi \rangle) \right)_{\text{coll}}, \tag{6.10}$$

where we have introduced $\mathbf{F} \equiv e \left(\mathbf{E} + \frac{1}{c} \mathbf{v} \times \mathbf{B} \right)$.

The averaging process we have introduced, although destroying a piece of information, allows us to find some general laws that the plasma obeys. To show this we now evaluate the general equation for the moments on the zeroth, first, and second orders.

6.2 Neutral gas

For the sake of simplicity, we first apply the procedure described in the previous section to a neutral gas, where all the particles have the same mass m.

From equation (6.10), posing $\xi(\mathbf{v}) = \mathbf{v}^0 m$, we obtain

$$\frac{\partial}{\partial t} (mn) + \nabla \cdot (mn\mathbf{u}) = \left(\frac{\partial}{\partial t} (n \langle \xi \rangle) \right).$$

We have not yet specified the collisional operator, but we may assume, according to Boltzmann theory, to consider only binary collisions that conserve the number of particles. This leaves us with a density continuity equation

$$\frac{\partial \rho}{\partial t} + \nabla \cdot (\rho \mathbf{u}) = 0, \tag{6.11}$$

where we have introduced the density mass $\rho = nm$.

Let us consider now the first-order moments of the general equation by choosing $\xi = mv_i$ and considering thus the ith component. Again from equation (6.10) we obtain

$$\frac{\partial}{\partial t} (mnu_i) + \frac{\partial}{\partial x_k} (mn \langle v_k v_i \rangle) - mn \langle \mathbf{F} \cdot \nabla_\mathbf{v} v_i \rangle = \left(\frac{\partial}{\partial t} (mn \langle v_i \rangle) \right)_{\text{coll}}.$$

Noting that

$$\langle \mathbf{F} \cdot \nabla_v v_i \rangle = \left\langle F_k \frac{\partial v_i}{\partial v_k} \right\rangle = \langle F_k \delta_{ik} \rangle = \langle F_i \rangle,$$

and that the momentum must be invariant under collisions, we end up with the following momentum conservation equation:

$$\frac{\partial}{\partial t}(mn\mathbf{u}) + \nabla \cdot \left(nm\langle \mathbf{vv} \rangle\right) - mn\langle \mathbf{F} \rangle = 0. \tag{6.12}$$

The first term is the rate of increase of momentum per unit volume, the second term represents the loss of momentum of the volume element due to escaping particles, i.e. the momentum flow. These two terms together represent the total variation of momentum density and they are balanced by the third term which represents the density force. We observe that the second term can be rewritten as the divergence of the stress tensor, $\mathbf{P} = n_s m_s \langle \mathbf{vv} \rangle_s$. It is important to highlight that in equation (6.12) a second-order moment, $nm\langle \mathbf{vv} \rangle$, appears.

We now proceed with the second-order moment equation taking $\xi = \frac{1}{2}mv^2$. Equation (6.10) gives

$$\frac{\partial}{\partial t}\left(n\left\langle \frac{1}{2}mv^2 \right\rangle\right) + \nabla \cdot \left(n\left\langle \frac{1}{2}mv^2 \right\rangle \mathbf{v}\right) - n\langle \mathbf{F} \cdot \mathbf{v} \rangle = \left(\frac{\partial}{\partial t} n\left\langle \frac{1}{2}mv^2 \right\rangle\right)_{\text{coll}}.$$

If we consider that the kinetic energy is unaltered by collisions, we are left with the energy conservation law

$$\frac{\partial}{\partial t}\left(n\left\langle \frac{1}{2}mv^2 \right\rangle\right) + \nabla \cdot \left(n\left\langle \frac{1}{2}mv^2 \right\rangle \mathbf{v}\right) - n\langle \mathbf{F} \cdot \mathbf{v} \rangle = 0, \tag{6.13}$$

which, since the magnetic field does not work, reduces to

$$\frac{\partial}{\partial t}\left(n\left\langle \frac{1}{2}mv^2 \right\rangle\right) + \nabla \cdot \left(n\left\langle \frac{1}{2}mv^2 \right\rangle \mathbf{v}\right) - emn\mathbf{E} \cdot \mathbf{u} = 0. \tag{6.14}$$

We note that the set of conservation equations we have derived is useful since they allow general considerations on the plasma behavior but they do not represent a closed system, for the number of unknown functions always exceeds the number of equations. So far we cannot solve the system represented by the three conservation equations without knowing the distribution function explicitly. And no help comes from considering higher-order moments, which would involve more and more unknown functions. For neutral gas dominated by collisions we can assume that the system will evolve through an equilibrium represented by a Maxwellian distribution. In both cases of global and local equilibrium it can be verified that the off-diagonal terms of the stress tensor are null and the system of equations (6.11), (6.12), and (6.14) reduces to a closed system of five equations for five unknown functions, ρ, P and the three components of the velocity $\mathbf{u}$.

6.3 The two-fluid model

We now consider a plasma composed of two species, protons and electrons. In this case we must consider two distribution functions, f_i and f_e, respectively, which obey

$$\frac{\partial f_s}{\partial t} + \mathbf{v} \cdot \nabla f_s + \frac{e}{m}\left(\mathbf{E} + \frac{1}{c}\mathbf{v} \times \mathbf{B}\right) \cdot \nabla_v f_s = \left(\frac{\partial f_s}{\partial t}\right)_{coll} \qquad s = i,\, e. \qquad (6.15)$$

The first-order moment equations give two equations similar to equation (6.11):

$$\frac{\partial \rho_s}{\partial t} + \nabla \cdot (\rho \mathbf{u_s}) = 0, \qquad (6.16)$$

where $\rho_s = n_s m_s$ and the average values $<\cdots>_s$ are defined respect to their own specific distribution function.

In the same way, for the momentum equations one obtains

$$\frac{\partial}{\partial t}(m_s n_s \mathbf{u_s}) + \nabla \cdot \left(n_s m_s \langle \mathbf{vv} \rangle_s\right) - n_s \langle \mathbf{F} \rangle_s = \mathbf{R}_s, \qquad (6.17)$$

where $\mathbf{R_s}$ remains for the collisional term, which can now be different from zero, since we are now also taking into account the collisions with the other species.

We observe that expressing the second term through the stress tensor, $\mathbf{P} = \sum_{s=i,e} n_s m_s \langle \mathbf{vv} \rangle_s$, and writing the third term as

$$n_s \langle \mathbf{F}_s \rangle_s = \int q_s\left(\mathbf{E} + \frac{1}{c}\mathbf{v} \times \mathbf{B}\right) f_s \, d\mathbf{v}$$

$$= q_s n_s \mathbf{E} + \frac{q_s}{c}\int f_s \mathbf{v} d\mathbf{v} \times \mathbf{B} = q_s n_s \mathbf{E} + \frac{q_s}{c} n_s \mathbf{u_s} \times B,$$

the momentum equations can be reformulated as

$$\frac{\partial}{\partial t}(m_s n_s \mathbf{u_s}) + \nabla \cdot \mathbf{P} - q_s n_s \mathbf{E} + \frac{q_s}{c} n_s \mathbf{u_s} \times B = \mathbf{R_s}. \qquad (6.18)$$

Finally one also obtains two equations for the energy:

$$\frac{\partial}{\partial t}\left(n_s \left\langle \frac{1}{2}m_s v^2 \right\rangle\right) + \nabla \cdot \left(n_s \left\langle \frac{1}{2}m_s v^2 \mathbf{v} \right\rangle\right) - q_s m_s n_s \mathbf{E} \cdot \mathbf{u_s}$$
$$= \frac{\partial}{\partial t}\left(n_s \left\langle \frac{1}{2}m_s v^2 \right\rangle\right)_{s\,coll}. \qquad (6.19)$$

We also have to face the closure problem for the two-fluid model. We can assume that both fluids are in thermodynamic equilibrium at different temperatures. This is a realistic hypothesis in many physical situations, that simplify the equations as for the neutral gas case. While the conservation of the number of particles is guaranteed for each species separately, the terms that represent the exchange of momentum and energy between different species must still be expressed through macroscopic variables in order to be simplified.

6.4 The single-fluid model

It must be said that even when a closure for the two-fluid model can be found, the solution is not straightforward. For this reason when the temperatures of the two fluids are not substantially different a single-fluid model is desirable. To this aim we assume, for the sake of simplicity, that the two fluids have the same particle charge, q, with opposite sign, and define the total number of particles, the mass, and the charge and current density as

$$n = n_i + n_e \tag{6.20}$$

$$\rho(\mathbf{r}, t) = n_e m_e + n_i m_i \tag{6.21}$$

$$\rho_q(\mathbf{r}, t) = q(n_i - n_e) \tag{6.22}$$

$$\mathbf{J}(\mathbf{r}, t) = q(n_i \mathbf{u}_i - n_e \mathbf{u}_e). \tag{6.23}$$

By summing the density equations (6.16) for the two species one obtains

$$\frac{\partial \rho}{\partial t} + \nabla \cdot (m_e n_e \mathbf{u}_e + m_i n_i \mathbf{u}_i) = 0. \tag{6.24}$$

Introducing the local mean velocity of particles defined as

$$\mathbf{V} = \frac{m_e n_e \mathbf{u}_e + m_i n_i \mathbf{u}_i}{n_e m_e + n_i m_i}, \tag{6.25}$$

we can rewrite this equation as a continuity equation for the total density of a fluid advected with the velocity $\mathbf{V}$,

$$\frac{\partial \rho}{\partial t} + \nabla \cdot \rho \mathbf{V} = 0. \tag{6.26}$$

Likewise, multiplying equations (6.16) for the charge and adding them, we obtain a continuity equation for the charge density, written as

$$\frac{\partial \rho_q}{\partial t} + \nabla \cdot \mathbf{J} = 0. \tag{6.27}$$

Adding the momentum equations (6.18) one obtains

$$\frac{\partial}{\partial t}(\rho \mathbf{V}) + \nabla \cdot \mathbf{P} - \rho_q \mathbf{E} + \frac{1}{c}\mathbf{J} \times \mathbf{B} = 0, \tag{6.28}$$

where $\mathbf{P} = \mathbf{P}_i + \mathbf{P}_e$ is the total stress tensor and we have used the fact that the momentum lost from one species is gained by the other.

The electromagnetic force density term in equation (6.28) can be reformulated using the Maxwell equations in such a way to highlight the contributions coming from the magnetic and electric fields separately. In this derivation we follow [1]. Using Ampere's law

$$\nabla \times \mathbf{B} = \frac{4\pi}{c}\mathbf{J} + \frac{1}{c}\frac{\partial \mathbf{E}}{\partial t}$$

and the vector identities, one can write

$$\frac{1}{c}\mathbf{J} \times \mathbf{B} = \frac{1}{4\pi}(\nabla \times \mathbf{B}) \times \mathbf{B} - \frac{1}{4\pi c}\frac{\partial \mathbf{E}}{\partial t} \times \mathbf{B}$$

$$= \nabla\left(\frac{B^2}{8\pi}\right) - \frac{1}{4\pi}(\mathbf{B} \cdot \nabla\mathbf{B}) - \frac{1}{4\pi c}\frac{\partial \mathbf{E}}{\partial t} \times \mathbf{B}.$$

We note that due to the fact that the magnetic field is divergence free, the following relation holds:

$$[(\mathbf{B} \cdot \nabla\mathbf{B})]_k = B_i\frac{\partial B_k}{\partial x_i} = \frac{\partial B_i B_k}{\partial x_i}. \tag{6.29}$$

Using the above relation eventually one obtains

$$\left[\frac{1}{c}\mathbf{J} \times \mathbf{B}\right]_k = \frac{\partial}{\partial x_i}\left(\frac{B^2}{8\pi}\delta_{ik} - \frac{B_i B_k}{4\pi}\right) - \frac{1}{4\pi c}\left(\frac{\partial \mathbf{E}}{\partial t} \times \mathbf{B}\right)_k, \tag{6.30}$$

where the first term between brackets on the right-hand side is a tensor.

We now show that the electric density term can also be rearranged, exploiting Gauss's law $\nabla \cdot \mathbf{E} = 4\pi\rho_q$, and obtaining

$$\rho_q E_k = \frac{\partial}{\partial x_i}\left(\frac{E^2}{8\pi}\delta_{ik} - \frac{E_i E_k}{4\pi}\right) + \frac{1}{4\pi c}\left(\frac{\partial \mathbf{B}}{\partial t} \times \mathbf{E}\right)_k. \tag{6.31}$$

Finally we introduce the tensor $\mathbf{T}$, whose components are defined as

$$(\mathbf{T})_{ik} = -\frac{E^2 + B^2}{8\pi}\delta_{ik} + \frac{E_i E_k + B_i B_k}{4\pi} \tag{6.32}$$

and we also define the vectors

$$\mathbf{G} = \frac{1}{4\pi c}\mathbf{E} \times \mathbf{B} \qquad \mathbf{M} = \rho\mathbf{V}. \tag{6.33}$$

All the identities and the definitions in equations (6.29)–(6.33) allow one to reformulate the momentum equation in the following conservative form:

$$\frac{\partial}{\partial t}(\mathbf{M} + \mathbf{G}) + \nabla \cdot (\mathbf{P}+\mathbf{T}) = 0, \tag{6.34}$$

in which the electromagnetic external forces are self-consistently incorporated and the interchangeability of particles and fields is highlighted.

Let us now add the second-order moment equations for the two species and obtain

$$\frac{\partial}{\partial t}\left(n_i\left\langle\frac{1}{2}m_iv^2\right\rangle_i + n_e\left\langle\frac{1}{2}m_ev^2\right\rangle_e\right) + \nabla\cdot\left(n_i\left\langle\frac{1}{2}m_iv^2\mathbf{v}\right\rangle_i + n_e\left\langle\frac{1}{2}m_ev^2\mathbf{v}\right\rangle_e\right) \quad (6.35)$$
$$- q(m_in_i\mathbf{u}_i - m_en_e\mathbf{u}_e)\cdot\mathbf{E} = 0,$$

where the right-hand side is equal to zero since the energy gained from one species is lost from the other one. Using the definition of the current density given above, the last term on the left-hand side can be rewritten as $\mathbf{J}\cdot\mathbf{E}$, and then, with the help of Maxwell's equations and some vectorial algebra, as

$$\mathbf{J}\cdot\mathbf{E} = \left(\frac{c}{4\pi}\nabla\times\mathbf{B} - \frac{1}{4\pi}\frac{\partial\mathbf{E}}{\partial t}\right)\cdot\mathbf{E} = -\frac{c}{4\pi}\nabla\cdot(\mathbf{E}\times\mathbf{B}) - \frac{1}{8\pi}\frac{\partial}{\partial t}(E^2 + B^2).$$

In the first term on the right-hand side one recognizes the Poynting vector, $\mathbf{S} = \mathbf{E}\times\mathbf{B}$, which represents the energy flux related to the electromagnetic field.

Defining the energy density of the plasma and of the electromagnetic field, respectively, as

$$W = \sum_{s=i,e}n_s\left\langle\frac{1}{2}m_sv^2\right\rangle_s \qquad W_{\text{em}} = \frac{E^2 + B^2}{8\pi}, \qquad (6.36)$$

finally equation (6.35) can be rewritten as

$$\frac{\partial}{\partial t}(W + W_{\text{em}}) + \nabla\cdot(\mathbf{Q} + \mathbf{S}) = 0, \qquad (6.37)$$

where we have introduced the energy flux $\mathbf{Q}$ defined as

$$\mathbf{Q} = \sum_{s=i,e}n_s\left\langle\frac{1}{2}m_sv^2\mathbf{v}\right\rangle_s. \qquad (6.38)$$

6.4.1 The hydrodynamic formulation

The system given by equations (6.27), (6.34), and (6.37) can also be cast in a form closer to the hydrodynamic formulation we are used to when dealing with fluids. For this purpose, it is useful to separate the velocity $\mathbf{v}$ into the mean fluid velocity $\mathbf{V}$ plus a thermal velocity $\mathbf{w}$, such that

$$\mathbf{v} = \mathbf{V} + \mathbf{w}. \qquad (6.39)$$

According to this definition it follows that $\langle \mathbf{w} \rangle = 0$, since $\langle \mathbf{v} \rangle = \mathbf{V}$. Moreover, by observing that

$$\langle v_i v_k \rangle = \left\langle (V_i + w_i)(V_k + w_k) \right\rangle = V_i V_k + \langle w_i w_k \rangle, \tag{6.40}$$

it is possible to rewrite the stress tensor in terms of the pressure tensor, $\mathbf{p} = \sum_{s=i,e} m_s n_s \langle ww \rangle$, as

$$\mathbf{P} = \rho \mathbf{VV} + \mathbf{p}. \tag{6.41}$$

All these definitions allow one to write equation (6.34) as

$$\frac{\partial}{\partial t}(\rho \mathbf{V}) + \nabla \cdot \mathbf{p} + \rho \nabla \cdot (\mathbf{VV}) - \rho_q \mathbf{E} + \frac{1}{c} \mathbf{J} \times \mathbf{B} = 0. \tag{6.42}$$

Then exploiting the continuity equations (6.26) and introducing the convective derivative

$$\frac{d}{dt} = \frac{\partial}{\partial t} + \mathbf{V} \cdot \nabla, \tag{6.43}$$

one finally obtains

$$\frac{d}{dt}(\rho \mathbf{V}) = -\nabla \cdot \mathbf{p} + \rho_q \mathbf{E} - \frac{1}{c} \mathbf{J} \times \mathbf{B}. \tag{6.44}$$

Once one associates the mean particle velocity $\mathbf{V}$ with a fluid velocity, this formulation of the momentum equation makes it easy to recognize the motion equation for a fluid. In particular, if the pressure is a scalar, i.e. $\mathbf{p} = p\mathbf{I}$, with $\mathbf{I}_{ij} = \delta_{ij}$, this equation reduces to the well known Euler equation, while in the presence of viscosity, i.e. when the off-diagonal terms of the pressure terms are different from zero, the Navier–Stokes equation is recovered.

Also for the energy equation (6.37) a different formulation can be given. It is sufficient to observe that for each species the following relations hold (we omit the s-index for the sake of clarity of the formulas):

$$\frac{1}{2} mn \langle v^2 \rangle = \frac{1}{2} mnu^2 + \frac{1}{2} mn \langle w^2 \rangle = \frac{1}{2} mnu^2 + \frac{1}{2} \sum_i P_{ii} = \frac{1}{2} \rho u^2 + \frac{3}{2} P$$

$$\frac{1}{2} mn \left\langle v_k v^2 \right\rangle = \frac{1}{2} \rho u^2 u_k + \frac{3}{2} P u_k + u_i P_{i,k} + q_k - n F_i u_i.$$

The resulting energy equation of the one fluid model is

$$\frac{\partial}{\partial t} \left(\frac{1}{2} \rho V^2 + \frac{3}{2} P \right) + \frac{\partial}{\partial x_i} \left[V_i \left(\frac{1}{2} \rho V^2 + \frac{5}{2} P \right) + \Pi_{i,k} V_k + q_i \right] - J_k E_k = 0, \tag{6.45}$$

where the total heat flux $\mathbf{q} = \mathbf{q}_i + \mathbf{q}_e$ has been introduced.

6.4.2 The closure of the system

We still need an equation to close the system. We observe that equation (6.44) has been obtained by adding the momentum equations for each species. An independent equation can be obtained by subtracting the same equations. To this aim we first multiply each momentum equation by q/m_s and then subtract them obtaining for the kth component

$$\frac{\partial}{\partial t}\left(qn_i(u_i)_k - qn_e(u_e)_k\right) + \frac{\partial}{\partial x_j}\left[qn_i\left((u_i)_k V_j + (u_i)_j V_k\right)\right.$$

$$\left. - qn_e\left((u_e)_k V_j + (u_e)_j V_k\right) + q\left(\frac{(P_i)_{jk}}{m_i} - \frac{(P_e)_{jk}}{m_e}\right)\right] - q^2\left(\frac{n_i}{m_i} + \frac{n_e}{m_e}\right)E_k \qquad (6.46)$$

$$- \frac{q^2}{c}\left(\frac{n_i}{m_i}(\mathbf{u}_i \times \mathbf{B})_k + \frac{n_e}{m_e}(\mathbf{u}_e \times \mathbf{B})_k\right) = q\left(\frac{(R_i)_k}{m_i} - \frac{(R_e)_k}{m_e}\right),$$

where we have introduced for the collisional term the definition

$$\mathbf{R}_s = \left(\frac{\partial}{\partial t}n_s\left\langle\frac{1}{2}mv_s^2\right\rangle\right)_{\text{coll}} \qquad s = i, e. \qquad (6.47)$$

At this point we use the definition of density charge and current density given in equations (6.22) and (6.23) and the fact that at each collision the motion gained from one species is lost from the other ($\mathbf{R}_i = -\mathbf{R}_e$). Then, since $m_e \ll m_i$ and $n_e \simeq n_i$, we rewrite equation (6.46) as

$$\frac{\partial J_k}{\partial t} + \frac{\partial}{\partial x_j}\left(J_k V_j + J_j V_k\right) - \frac{q}{m_e}\frac{\partial(P_e)_{kj}}{\partial x_j} - \frac{q^2 n_e}{m_e}E_k - \frac{q^2 n_e}{m_e m_i c}[(m_e\mathbf{u}_i + m_i\mathbf{u}_e) \times \mathbf{B}]_k$$

$$= \frac{q(R_i)_k}{m_e}. \qquad (6.48)$$

Then we use the following relation:

$$m_e\mathbf{u}_i + m_i\mathbf{u}_e = (m_e + m_i)\mathbf{V} + (m_e - m_i)\frac{\mathbf{J}}{qn_e} \simeq m_i\left(\mathbf{V} - \frac{\mathbf{J}}{qn_e}\right), \qquad (6.49)$$

obtaining

$$\frac{\partial J_k}{\partial t} + \frac{\partial}{\partial x_j}\left(J_k V_j + J_j V_k\right) - \frac{q}{m_e}\frac{\partial(P_e)_{kj}}{\partial x_j} - \frac{q^2 n_e}{m_e}E_k$$

$$- \frac{q^2 n_e}{m_e c}\left[\left(\mathbf{V} - \frac{\mathbf{J}}{qn_e}\right) \times \mathbf{B}\right]_k = \frac{q(R_i)_k}{m_e}. \qquad (6.50)$$

We now need to make some hypothesis on the collisional term in order to relate it to the fluid quantities we have introduced so far and to gain some physical insight on this expression. It seems reasonable to assume that this term is related to the relative motion of electrons and ions and to write it as

$$\mathbf{R}_s = -n_s m_s \nu_{s,s'}(\mathbf{u}_s - \mathbf{u}_{s'}), \tag{6.51}$$

where the parameter $\nu_{s,s'}$ represents the average collision frequency between particles of one species with the particles of the other species. If we multiply equation (6.50) by $(m_e/q^2 n_e)$, introduce the parameter $\eta = \frac{m_e \nu_{ei}}{q^2 n_e}$, that represents the electrical resistivity, and recall that $\mathbf{R}_i = -\mathbf{R}_e$, we obtain the following equation:

$$E_k + \frac{1}{c}(V \times B)_k - \eta J_k = \frac{m_e}{q^2 n_e}\left[\frac{\partial J_k}{\partial t} + \frac{\partial}{\partial x_j}\left(J_k V_j + J_j V_k\right)\right]$$
$$+ \frac{1}{q n_e c}(J \times B)_k - \frac{q}{n_e}\frac{\partial (P_e)_{kj}}{\partial x_j}. \tag{6.52}$$

Rewriting the electron stress tensor in terms of the pressure tensor and using the quasi-neutrality condition, $\rho_q \approx 0$, equation (6.52) assumes the following vectorial form:

$$\mathbf{E} + \frac{1}{c}\mathbf{V} \times \mathbf{B} - \eta \mathbf{J} = \frac{m_e}{q^2 n_e}\left[\frac{\partial \mathbf{J}}{\partial t} + \nabla \cdot (\mathbf{V}\mathbf{J} + \mathbf{J}\mathbf{V})\right]$$
$$+ \frac{1}{q n_e c}(\mathbf{J} \times \mathbf{B}) - \frac{q}{n_e}\nabla \mathbf{p}_e, \tag{6.53}$$

which is called the *generalized Ohm's law*, since it relates the electric field to the current density. When neglecting the right-hand side, equation (6.53) reduces to Ohm's law for a fluid conductor, where $\mathbf{E}' = \mathbf{E} = \frac{1}{c}(\mathbf{V} \times \mathbf{B})$ is the electric field in the reference frame moving with the fluid. It is worth remarking that for Coulomb collisions $\nu_{ei} \propto T^{-3/2}$, hence the electric resistivity η decreases with the temperature, opposite to what happens for ordinary conductors, where by increasing the temperature the ions lattice through which the electrons move oscillates, leading to an increase of the resistivity. If we now compare the inertial terms, i.e. the terms proportional to the electron mass in equation (6.53), with the resistive one we find that the former can be neglected when the current density varies slowly on the collisional time scale $\tau_{ie} = 1/\nu_{ei}$. We must compare the term proportional to $\mathbf{J} \times \mathbf{B}$, usually called the *Hall term*, with the $\mathbf{V} \times \mathbf{B}$ term. The unique component of the current density that contributes to the Hall term is the one perpendicular to the magnetic field. We may estimate this current with the drift current due to the presence of gradients of the magnetic field, meaning that this term can be neglected in the approximation of small Larmor radius. Finally another assumption usually made is that of isotropic pressure, in such a way that the tensor pressure reduces to a scalar and viscosity effects are neglected.

6.5 The MHD equations

Often, it is found that the scale length of many instabilities and waves able to grow and propagate in a system is comparable to the plasma size. This circumstance holds both for interstellar plasmas and thermonuclear plasmas. For this reason it is possible to further simplify the single-fluid description by considering only phenomena occurring on a large scale ($L \to \infty$), that must be much larger than the Debye length, allowing the quasi-neutrality condition to hold ($\rho_q \approx 0$). We also consider only low-frequency ($\omega \to 0$) phenomena that allow the displacement current to be neglected in the Maxwell's equations. If in addition to these restrictions we consider only collisions dominated phenomena, i.e. that keep the system isotropic at all the times ($\nabla \mathbf{p} = \nabla p$), and consider $m_e/m_i \to 0$, a further simplified set of equations, the MHD equations, is obtained. The MHD equations are summarized as follows:

$$\frac{\partial \rho}{\partial t} + \nabla \cdot \rho \mathbf{V} = 0, \tag{6.54}$$

$$\frac{d}{dt}(\rho \mathbf{V}) = -\nabla p + -\frac{1}{c}\mathbf{J} \times \mathbf{B}, \tag{6.55}$$

$$\mathbf{E} + \frac{\mathbf{V} \times \mathbf{B}}{c} = \eta \mathbf{J}, \tag{6.56}$$

$$\nabla \times \mathbf{E} = -\frac{1}{c}\frac{\partial \mathbf{B}}{\partial t}$$

$$\nabla \times \mathbf{B} = \frac{4\pi \mathbf{J}}{c}. \tag{6.57}$$

This set of equations is closed by an equation of state or by some other assumption relating the pressure to the density. Usually MHD uses the approximation of an incompressible fluid, which reduces the continuity equation (6.54) to the condition

$$\nabla \cdot \mathbf{V} = 0, \tag{6.58}$$

or the approximation of adiabatic fluid

$$\frac{dp\rho^{-\gamma}}{dt} = 0 \tag{6.59}$$

or the approximation of isothermal fluid

$$\frac{d}{dt}\frac{p}{\rho} = 0. \tag{6.60}$$

6.6 The reduced MHD model

One often encounters situations in which the magnetic field is strong and almost unidirectional. Examples are the magnetic fields in loops in the solar corona and tokamaks. These circumstances allow one to consider approximations of the full MHD equations based on physical and/or geometrical considerations. Since the magnetic field is almost uniform and unidirectional, the field has one almost uniform component that is much larger than the other components and is called the *guide field*. It is customary to denote this field as $\mathbf{B}_0$ and to write the total magnetic field as $\mathbf{B} = \mathbf{B}_0 + \mathbf{B}_\perp$, where $\mathbf{B}_\perp$ represents the components of $\mathbf{B}$ perpendicular to the strong, nearly uniform component. In the following we make use of a Cartesian reference frame and assume that the strong uniform component is along the z-direction and the other components are in the (x, y)-plane. The magnetic field intensity, denoted by $B \approx |\mathbf{B}_z| = B_0$, is thus characterized by the condition

$$\frac{B_\perp}{B_0} = \epsilon \ll 1. \tag{6.61}$$

Under this assumption the MHD equations can be further simplified. Formally, the variables appearing in the MHD equations are ordered as some power of the small parameter ϵ. This ansatz is introduced into the MHD equations, and only the lowest powers of ϵ are retained. The resulting equations have been extensively exploited for both analytic and numerical calculations. The model is called reduced MHD (RMHD). It describes the dynamics of the system in the plane perpendicular to the mean field.

The assumption of a strong guide field leads us to assume that the thermal and kinetic energy are much smaller than the magnetic energy:

$$p \sim \rho v^2 \ll \frac{B_0^2}{8\pi} \Rightarrow \beta \equiv \frac{8\pi p}{B_0^2} \ll 1. \tag{6.62}$$

This assumption leads to the following orderings:

$$v \sim \epsilon, \quad p \sim \epsilon^2. \tag{6.63}$$

Moreover the relation (6.62) together with the magnetic field being divergence free imply we are considering only phenomena with a slow variation along the direction of the strong guide field, since

$$\frac{B_\perp}{B_0} \sim \frac{\nabla_\parallel}{\nabla_\perp} \sim \epsilon,$$

which in our coordinate system can be read as

$$\frac{\partial}{\partial z} \sim \epsilon \quad \nabla_\perp \sim 1. \tag{6.64}$$

We also assume that the dynamics parallel to the mean magnetic field occurs on a much shorter time scale than the dynamics in the perpendicular plane. (For example, sound waves will propagate rapidly along the field and smooth out significant variations in that direction.) Hence, we expect approximate force balance to be maintained in the parallel direction on the time scale of the perpendicular dynamics. From equation (6.55) it then follows that

$$\frac{\mathbf{B}}{B_0} \cdot \left(\nabla p + \frac{B^2}{8\pi} \right) \approx 0 \tag{6.65}$$

in such a way that $dV_z/dt \approx 0$, which means $V_z = \text{const}$, and without any loss of generality we can assume $V_z = 0$. Since

$$\frac{\mathbf{B}}{B_0} = \frac{B_0 \mathbf{e}_z + \mathbf{B}_\perp}{\sqrt{B_0^2 + B_\perp^2}} = \mathbf{e}_z + \epsilon \frac{\mathbf{B}_\perp}{B_0} + O\left(\epsilon^2\right) \approx \mathbf{e}_z, \tag{6.66}$$

from equation (6.65) it follows that

$$\frac{\partial p}{\partial z} + \frac{B_0}{4\pi} \frac{\partial B_z}{\partial z} = 0. \tag{6.67}$$

Then since B_z is almost uniform it can be rewritten as

$$B_z = B_0 + \tilde{B}_z$$

and eventually one finds that $\tilde{B}_z \sim \epsilon^2$.

In conclusion the RMHD orderings can be summarized in the following formulas:

$$\begin{aligned}
V_z &= 0, \quad \nabla_\perp \sim 1 \\
\frac{\partial}{\partial z} &\sim \epsilon, \quad V_\perp \sim \epsilon \\
p &\sim \epsilon^2, \quad B_0 \sim \epsilon^2.
\end{aligned} \tag{6.68}$$

Then we assume the incompressible approximation closure which, due to the relations (6.64), reduces to

$$\nabla_\perp \cdot V_\perp = 0. \tag{6.69}$$

According to this assumption, from the continuity equation (6.54), it follows that if the density is constant at the initial time, then it stays constant at all times. It is customary to assume $\rho = \rho_0 = 1$.

To proceed with the derivation we exploit equation (6.69) and the $\nabla \cdot \mathbf{B} = 0$ property, coupled with the constancy of B_z, to introduce the magnetic flux and the stream functions, ψ and ϕ, respectively, and to write

$$\mathbf{B} = B_z \mathbf{e}_z + \nabla \psi \times \mathbf{e}_z \quad \text{and} \quad \mathbf{V} = \mathbf{e}_z \times \nabla \phi. \tag{6.70}$$

Note that ψ coincides with the component along the z-direction of the vector potential $\mathbf{A}$, such that $\mathbf{B} = \nabla \times \mathbf{A}$. As a consequence of this choice for the magnetic field the current density is directed along z and defined as

$$\mathbf{J} = \frac{c}{4\pi}\nabla \times \mathbf{B} = J\mathbf{e}_z = -\frac{c}{4\pi}\nabla^2\psi\,\mathbf{e}_z. \tag{6.71}$$

Considering now the parallel component of the generalized Ohm's law (6.56) we obtain

$$\frac{\partial\psi}{\partial t} = \mathbf{V}_\perp \cdot \nabla\psi = \frac{\eta c^2}{4\pi}\nabla^2\psi, \tag{6.72}$$

where we have used the fact that Faraday's equation reduces to

$$E_z = -\frac{1}{c}\frac{\partial\psi}{\partial t}. \tag{6.73}$$

We now need to find an equation for the flow. To this aim we take the z-component of the curl of the motion (equation (6.55)) and obtain

$$\mathbf{e}_z \cdot \nabla \times \left[\frac{\mathrm{d}}{\mathrm{d}t}(\mathbf{V}) = -\nabla p - \frac{1}{c}\mathbf{J} \times \mathbf{B}\right]. \tag{6.74}$$

We now introduce the vorticity ω, defined as

$$\boldsymbol{\omega} \equiv \nabla \times \mathbf{V} = \nabla \times \mathbf{V}_\perp = \omega\mathbf{e}_z = \nabla^2\phi\mathbf{e}_z. \tag{6.75}$$

Using the vector identity $(\mathbf{V} \cdot \nabla)\mathbf{V} = -\mathbf{V} \times (\nabla \times \mathbf{V}) + \frac{1}{2}\nabla V^2$ with some straightforward algebra one obtains

$$\frac{\partial\omega}{\partial t} + \mathbf{V}_\perp \cdot \nabla\omega = \frac{1}{c}\mathbf{e}_z \cdot \nabla J \times \nabla\psi. \tag{6.76}$$

Finally we normalize equations (6.72) and (6.76) to the typical time scale of MHD phenomena, the Alfvèn time, $\tau_A = L/v_A$, where $v_A = B_{y0}/\sqrt{4\pi}$, and to the equilibrium magnetic field variation scale, L. The RMHD model equations are then

$$\begin{aligned}\frac{\mathrm{d}\psi}{\mathrm{d}t} &= \epsilon_\eta\nabla^2\psi \\ \frac{\mathrm{d}\omega}{\mathrm{d}t} &= \frac{1}{c}\mathbf{e}_z \cdot \nabla J \times \nabla\psi,\end{aligned} \tag{6.77}$$

where $\epsilon_\eta = \eta c^2\tau_A/(4\pi L)$. These equations are closed by the subsidiary equations

$$J = -\nabla^2\psi, \quad \omega = \nabla^2\phi. \tag{6.78}$$

An alternative form of the RMHD equations can be given introducing the so called *Poisson bracket*, defined for two generic fields f and g as

$$[f, g] \equiv \mathbf{e}_Z \cdot \nabla f \times \nabla g = \frac{\partial f}{\partial x}\frac{\partial g}{\partial y} - \frac{\partial g}{\partial x}\frac{\partial f}{\partial y}. \quad (6.79)$$

The terms of the type $\mathbf{V}_\perp \cdot \nabla \xi$ can be manipulated with some algebra and rewritten as

$$\mathbf{V}_\perp \cdot \nabla \xi = (\mathbf{e}_z \times \nabla \phi) \cdot \nabla \xi = [\phi, \xi], \quad (6.80)$$

allowing the following formulation for the magnetic flux function and the vorticity equations:

$$\begin{aligned}
\frac{\partial \psi}{\partial t} + [\phi, \psi] &= \epsilon_\eta \nabla^2 \psi \\
\frac{\partial \omega}{\partial t} + [\phi, \omega] &= [J, \psi].
\end{aligned} \quad (6.81)$$

6.7 Numerical simulations

It is clear that although there is a great simplification introduced from the fluid approach with respect to the kinetic description, the system of equations obtained in the previous sections still remain cumbersome and difficult to solve. The high non-linearity of the equations makes them difficult to treat analytically and research in this field can proceed only if accompanied by numerical simulations. These are necessary to validate theories and to perform numerical experiments when data are inaccessible, as for instance in the astrophysics context.

6.7.1 Spatial discretization schemes

When computers come into play the continuum space must be discretized and replaced by a finite set of values, while the differential equations are approximated by algebraic equations. Basically we can distinguish between three kinds of discretization methods: finite difference, finite element, and spectral schemes. In the following we take a brief look at the essential features of these different computational approaches without pretending to be exhaustive on the subject, and refer the reader to the copious literature for deeper knowledge and insight into the subject (see, e.g. [2] and references therein).

The finite difference method
Under this category are gathered all the schemes in which derivatives, such as $\partial f/\partial x$, are replaced by ratios of differences, $\Delta f/\Delta x$. How these differences are computed depends on the approximation scheme. For example, a Taylor expansion or a polynomial fitting can be used. The main advantages of this method are that the equations are simply to derive, there are few operations for grid points to be

performed, and the coding is easy for a regular mesh. The main limitations are related to the stability of this scheme, which depends strongly on the grid size and its diffusion. This may alter the solution of the equations, introducing spurious effects. Moreover these schemes are not easily adaptable to irregular domains.

The finite element method
This method consists of replacing the functions by piecewise polynomials with nodal values and the derivatives by derivatives of these approximating polynomials. This kind of method originated in engineering applications, when the integration domain has complex structures. The main advantage of this method is its efficiency on irregular shaped regions. The limitations come from the fact that the resulting equations are usually implicit and require great computational effort.

The spectral method
In the spectral method the continuum is replaced by an expansion using a certain basis function. Often the Fourier expansion is adopted and the use of the fast Fourier transform is involved. The derivatives are replaced by derivatives of the approximated functions and the equations are replaced by projections onto the basis functions or onto meshes. In the latter case the approach is called *pseudo-spectral*. The main advantages are represented by high accuracy in the solution and high speed in the computation. The drawback of this method is that, in the most common Fourier version, periodic boundary conditions are required.

We conclude by noting that usually the finite difference and spectral method schemes mentioned above are coupled into the codes, which adopt one scheme for some spatial direction and the other for the remaining directions, depending on the applicable boundary conditions.

It is worth mentioning that to improve the accuracy of the finite difference schemes, it is possible to make use of *compact* [3] finite differences, which are the equivalent of an implicit time scheme (see next section) for spatial derivatives. Compact schemes make it possible to describe a wide range of spatial scales, likewise the spectral methods, while retaining the computational advantages of the finite difference schemes for non-periodic situations.

6.8 Time discretization schemes

Time discretization is usually based on a finite difference approach. When speaking about time discretization one encounters two distinct approaches: the explicit and the implicit schemes. When the advancing in time of a quantity is based only on the mean of the variable values at previous time steps, then the scheme is known as explicit. The main advantage of this approach is its simplicity, but it suffers from very restrictive constraints on time steps. On the other hand implicit schemes, which have better numerical properties, involve solving a matrix equation to evolve the quantities in time, and can be computationally heavy (see, e.g. [4]).

6.9 An application of RMHD equations: magnetic reconnection

In this section we provide an example of an application of fluid models to the study of a basic fundamental process in plasma physics, *magnetic reconnection*. We start by analyzing a characteristic feature of the MHD equations that lead us to the definition of a reconnection process. Then we provide the main ingredients of the analytical treatment of this phenomena through the reduced MHD equations, together with the results of numerical simulations.

An important property of the MHD equations when the resistivity is very low, which is a common feature in many astrophysical and laboratory plasmas, consists in the conservation of the magnetic flux. This characteristic follows in a straightforward manner from Ohm's law, when we let the resistivity vanish. Taking the curl of equation (6.56), where we have set $\eta \to 0$, and using the induction equation, one obtains an evolution equation for the magnetic field,

$$\frac{\partial \mathbf{B}}{\partial t} = \nabla(\mathbf{V} \times \mathbf{B}), \tag{6.82}$$

that expresses the coupling of the magnetic field with the fluid motion and is called the *frozen in condition* (see, e.g. [1]). Equation (6.82) implies that two points connected by magnetic field line at a certain time will stay connected at any later time. In the two-dimensional setting defined by the reduced model introduced previously, this topological constraint is recovered simply by letting the right-hand side of the equation for the magnetic flux, ψ, vanish in the system equation (6.81) and obtaining

$$\frac{\partial \psi}{\partial t} + [\phi, \psi] = 0. \tag{6.83}$$

This condition implies that the magnetic flux is advected with the fluid and, since fluid elements cannot compenetrate each other, the magnetic field lines cannot reconnect, i.e. they cannot tear and connect according to a different topological configuration.

Relaxing this constraint, by permitting the right-hand side of equation (6.83) to be different from 0, opens new scenarios for the magnetic field that can decouple its motion from the fluid one. Under this assumption the evolution equation for the magnetic field becomes

$$\frac{\partial \mathbf{B}}{\partial t} = \nabla(\mathbf{V} \times \mathbf{B}) + \epsilon_\eta \nabla^2 \mathbf{B}, \tag{6.84}$$

or in terms of the magnetic flux for the reduced model as

$$\frac{\partial \psi}{\partial t} + [\phi, \psi] = \epsilon\eta \nabla^2 \psi. \tag{6.85}$$

This phenomenon is called *magnetic reconnection*. Since resistivity is very low one may expect that over a long time scale magnetic field lines will always diffuse through the plasma and reconnection will eventually take place. From equation (6.85) it is clear that there are exceptional regions where reconnection can take place even for low resistivity values. These are the regions where the magnetic or the velocity fields vanish. In these regions an exceptional release of magnetic energy into heat and ordered kinetic energy of accelerated particles occurs on a very fast time scale. Solar flares in the solar corona and sawtooth oscillations in tokamak devices are typical phenomena related to such a reconnection event. A complete discussion of this phenomenon is outside the scope of this chapter and we refer the reader to the extensive literature for further information (see, e.g. [5] and references therein). We have introduced this phenomenon here since it is one of the phenomena that can be studied both analytically and numerically by the reduced MHD equations. We use this process to illustrate the power of the fluid approach in describing a complex phenomenon. In the following we will sketch the main results that can be obtained by analytically linearizing the set of equations (6.81) and by solving them numerically.

6.9.1 Linear analysis

As pointed out previously, an essential feature of this instability is the distinction between the local scale on which it takes place and the global scale on which its effects are felt. For high-temperature plasmas ϵ_η is usually very small and it is clear from equation (6.84) that we can neglect it except in those regions close to a null of the magnetic or velocity fields. Although the process takes place on a local scale its effects are felt on a global scale, since the rearrangement of the magnetic topology extends up to the equilibrium magnetic field scale length. When treated analytically such phenomena involve the use of boundary layer theory and asymptotic matching. The prototype of an instability involving reconnection is the tearing mode (suggested in the ground-breaking paper by Furth *et al* [6]). The authors consider there a plane current sheet located in an infinite domain that undergoes spontaneous reconnection. The classical reference configuration is the Harris sheet [7], where the equilibrium magnetic field, **B**, the flux function, ψ, and the stream function, ϕ, are given by

$$\mathbf{B} = -\tanh(x)\mathbf{e}_y, \quad \psi = \log(\cosh(x)), \quad \phi = 0. \tag{6.86}$$

This magnetic field has field lines in opposite directions with respect to the null line located at $x = 0$, called the *rational surface*. In the two-dimensional setting defined by the reduced MHD equations the change of topology caused by the reconnection process appears via formation of a magnetic island around the rational surface.

Linearizing the equations of the system (6.81) and solving them for the perturbed quantities

$$\tilde{\psi} = \delta\psi(x)\exp(iky + \gamma t), \quad \tilde{\phi} = \delta\phi(x)\exp(iky + \gamma t),$$

where $k = 2\pi m/L_y$ and m is the mode number, gives two ordinary differential equations for $\delta\psi$ and $\delta\phi$. This set of equations is then solved distinguishing between an inner layer, close to the magnetic flux surface where the magnetic field vanishes and the reconnection takes place, and an outer region far from the rational surface, where it is possible to solve the system of equations neglecting ϵ_η. We do not enter into the details of all the calculations. We simply return to the dispersion relation [6]

$$\frac{\pi}{\epsilon_\eta^{1/3}\Delta'} = \frac{8}{Q^{5/6}}\frac{\Gamma\left(\dfrac{Q+5}{4}\right)}{\Gamma\left(\dfrac{Q-1}{4}\right)}, \tag{6.87}$$

where Δ' is a parameter which depends on the solution for the magnetic flux in the outer region and expresses its jump across the rational surface, located at $x = 0$, and

$$Q = \frac{\tilde{\gamma}}{\epsilon_\eta^{1/3}}, \quad \tilde{\gamma} = \frac{\gamma}{k^{2/3}\epsilon_\eta^{1/3}}.$$

The Δ' parameter, first introduced in [6], is a measure of the instability, which depends only on the wave vector k along the y-direction. The larger the parameter the more unstable is the mode we are considering. The solution of the dispersion relation gives the growth rate γ of the reconnection instability in terms of the wave vector k. In particular we highlight two limits of relation (6.87):

- **Small Δ' regime**
  This regime, known as the *tearing regime*, is characterized by $\Delta'\epsilon_\eta^{1/3} \ll 1$ and has a growth rate and an inner layer width given by

$$\gamma = 1.37\Delta'^{4/5}\epsilon_\eta^{3/5} \quad \delta_{\text{in}} = \epsilon_\eta^{2/5}. \tag{6.88}$$

  The flow cells are localized around the rational surface and the magnetic island grows slowly. The non-linear regime is characterized by an algebraic growth [8] before reaching saturation.

- **Large Δ' regime**
  This regime is characterized by $\Delta'\epsilon_\eta^{1/3} \gg 1$ and has a growth rate and an inner layer width given by

$$\gamma \sim \epsilon_\eta^{1/2} \quad \delta_{\text{in}} \sim \epsilon_\eta^{1/3}. \tag{6.89}$$

  The flow cells here have a macroscopic size and the magnetic island grows up to a width of the order of the equilibrium scale length.

6.9.2 Numerical simulation

In the following we are going to show the numerical solution for this reconnection problem. The boundary conditions have been chosen in such a way that the solution goes to zero at the edge of the integration domain along the x-direction and periodic in the y-direction. The code we use [9] is based on a combination of numerical techniques: an implicit finite difference discretization scheme has been implemented on a variable size grid along the x-direction, while a pseudo-spectral discretization has been chosen for the periodic y-direction. The time discretization is explicit. The resistivity value is fixed to $\epsilon_\eta = 0.00028$. The integration domain is $L_x \times L_y$.

In particular, $x \in [-11.32, 11.32]$ and its range extends up to values where the perturbed fields vanish, in order to simulate an infinite domain. Along the periodic y-direction two values of L_y have been chosen in such a way that the two simulations presented fall into the two different regimes described above. In particular for the small Δ' regime we have $L_y = 4\pi$ and $\Delta'\epsilon_\eta^{1/3} = 0.24$, while for the large Δ' regime we have $L_y = 15.32\pi$ and $\Delta'\epsilon_\eta^{1/3} = 2$. In figure 6.1 we show the growth rate for the two regimes depicted above. After a linear phase, characterized by a constant γ value, the

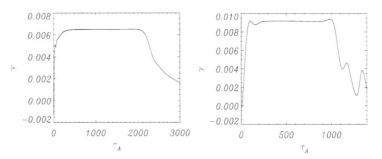

Figure 6.1. Growth rate, γ, as a function of time. In the left frame the behavior for the small Δ' regime is shown, while in the right frame the evolution in the large Δ' regime is plotted.

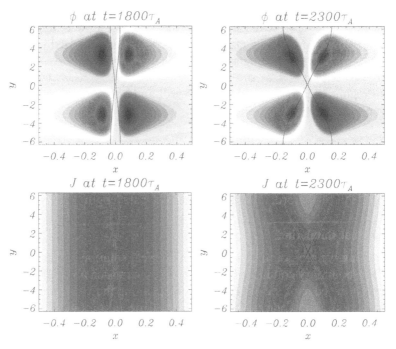

Figure 6.2. Contour plots of the stream function, ϕ (upper row), and the current density, J (lower row), for two different time steps for the case $\Delta' = 3$.

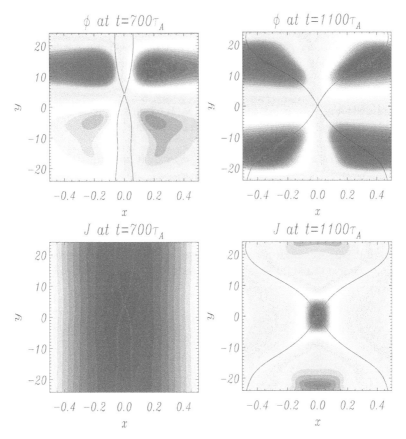

Figure 6.3. Contour plots of the stream function, ϕ (upper row), and the current density, J (lower row), for two different time steps for the case $\Delta' = 15$.

system enters into the non-linear phase in two different ways. For the small Δ' regime the growth rate slows down according to an algebraic law [8], while for the large Δ' regime there is an increase of the reconnection rate before the final slowing down that eventually leads to saturation.

In figures 6.2 and 6.3 a comparison between the two regimes is shown through the contour plots of the stream function and of the current density. Specifically, figure 6.2 (figure 6.3) shows the small (large) Δ' regime. The corresponding magnetic island is superimposed on all the contour plots. For both regimes two different time steps corresponding to the end of the linear phase and well into the non-linear phase are shown. From these figures we can see that while in the small Δ' regime the magnetic island remains small also in the non-linear phase, in the large Δ' regime the island grows up to a size comparable to the integration domain. This difference in the island size is explained in terms of the flow cells that remain microscopic in the first regime, while they have a macroscopic size in the latter regime. Note that along the x-direction only a small portion, close to the rational surface, of the integration domain is shown.

Bibliography

[1] Schmidt G 1979 *Physics of High Temperature Plasmas* (New York: Academic)

[2] Jardin S 2010 *Computational Methods in Plasma Physics* (Boca Raton, FL: CRC Press)

[3] Lele S 1992 Compact finite difference schemes with spectral-like resolution *J. Comput. Phys.* **103** 16–41

[4] Chacón L 2004 A non-staggered, conservative, $\nabla \cdot \mathbf{B} = 0$, finite-volume scheme for 3D implicit extended magnetohydrodynamics in curvilinear geometries *Comput. Phys. Commun.* **163** 143–71

[5] Biskamp D 2000 *Magnetic Reconnection in Plasmas* (Cambridge: Cambridge University Press)

[6] Furth H, Killeen J and Rosenbluth M 1963 Finite-resistivity instabilities of a sheet pinch *Phys. Fluids* **6** 459–84

[7] Harris 1962 On a plasma sheaths separating regions of oppositely directed magnetic field *Nuovo Cimento* **23** 115–21

[8] Rutherford P 1903 Nonlinear growth of the tearing mode *Phys. Fluids* **16** 1903

[9] Militello F, Grasso D and Borgogno D 2014 The deceiving Δ': on the equilibrium dependent dynamics of nonlinear magnetic islands *Phys. Plasmas* **21** 102514

IOP Publishing

Plasma Modeling
Methods and Applications
Andrea Cristofolini

Chapter 7

Magnetohydrodynamics equations

Magnetohydrodynamics (MHD) is a branch of physics and engineering that investigates the effects of a magnetic field on the motion of an electrically conducting fluid. Although the term 'magnetohydrodynamics' and its acronym MHD may be somewhat misleading, it is often used to designate physical phenomena in which the conducting fluid is not a liquid, but rather a compressible ionized gas or a plasma. Some authors prefer more specific terms like magneto-fluid-dynamics or magneto-gas-dynamics.

Since the pioneering experiment of Michael Faraday in 1832, MHD has appeared as an appealing way to manipulate a fluid without the intervention of mechanical means. Faraday failed in his attempt to measure the electric current produced by the MHD interaction taking place in the water of the river Thames flowing through the Earth's magnetic field. However, his experimental activities opened the way to the study of a mechanism which has fundamental implications in plasma physics, astrophysics, and engineering.

MHD interaction principles have been proposed for a number of technological applications. A first example can be found in MHD generators, in which the MHD interaction is used for the direct conversion of the working fluid energy into electrical energy [1, 2]. MHD generators have been intensively investigated as a means to increase the efficiency of electrical power plants and as a viable alternative to gas turbines. MHD electric generators has also been proposed in nuclear-powered spacecraft for deep space missions.

Reversing the power flow, the MHD interaction can be exploited to produce a body force on a conducting fluid. Electromagnetic pumps based on the MHD principle have been used in fast nuclear reactors, in which a liquid such as sodium is used to remove the heat produced by fission from the core. The application of electromagnetic forces has also been successfully implemented in materials processing, in particular in the fields of steel-manufacture and advanced metallurgy [3, 4]. In these fields many processes involve the handling of liquid metal in a technologically difficult environment. Electromagnetic stirring and flow control in continuous casting are examples of the application of MHD processes in metallurgical processing.

There are also examples of MHD propulsion in naval and aerospace applications. In 1992, Mitsubishi Industries presented the Yamato 1 [5], the first prototype of an MHD propelled ship. The Yamato 1 included a large number of innovative technological solutions, including the adoption of superconducting magnets to generate the magnetic field necessary to produce the MHD interaction. The ship took its maiden voyage in the waters of Kobe Harbour, and during the tests, the vessel reached a velocity of 7.5 knots, demonstrating the feasibility of MHD ship propulsion [6]. In space propulsion, the magnetoplasmadynamic (MPD) thruster has been actively investigated since the the the late 1960s in ground-based test facilities [7]. The MPD thruster is a plasma accelerator, in basic terms composed of two coaxial electrodes between which a high current discharge ionizes a gas and accelerates it to high exhaust velocities by means of the MHD interaction. The magnetic field responsible for the interaction can be self-induced by the current. Although the magnetic field induced by the electric current is sufficient to support an MHD interaction, many thrusters have been proposed and studied in which the field is reinforced via external coils, primarily for stabilization purpose.

In addition to electric propulsion, during the last few decades the MHD mechanism has received significant interest in the fields of aerothermodynamics and aerospace, in applications concerning the control of hypersonic flows of an ionized gas [8, 9]. The utilization of the MHD interaction in hypersonic flight has been investigated since the late 1950s [8], and a wide range of possible applications has been proposed and investigated. Many of the applications are directed at the solution of specific problems faced by a spacecraft during the re-entry phase in the planetary atmosphere, such as thermal flux mitigation [10, 11], trajectory control through the parachute effect [12, 13], and blackout time reduction [14]. Other investigations have focused on shock control at the inlet of an air-breathing thruster. In the AJAX project, MHD techniques are utilized to bypass the kinetic energy of the working fluid from the supersonic diffuser to the nozzle. By doing this, the flow velocity in the combustion chamber is reduced to acceptable values, even for high Mach numbers.

In this chapter, a general formulation of the MHD model will be derived and discussed. The approximations underlying the model will be examined, and two formulations of the model will be obtained from the general one, in order to adequately describe the limit cases of convective and diffusive magnetic regimes. Finally, an MHD model based on a low magnetic Reynolds number approximation will be described, and some numerical results will be presented and discussed.

7.1 MHD models

7.1.1 Model foundation

A magneto-fluid-dynamics model is generally based on the assumption that the plasma can be regarded as a continuum, and thus may be characterized by relatively few macroscopic quantities. The requirement for the applicability of this approach is that the plasma is collision dominated. Under this assumption, the basis for an MHD model is constituted by the continuity equations for mass, momentum, and energy, and by the Maxwell equations. In order to evaluate the plasma properties in thermo-chemical non-equilibrium, microscopic scale phenomena are taken into account by means of continuity equations for relevant species. For each continuity equation, the proper production rates have to be known.

In order to consider non-equilibrium plasmas, particularly plasmas where electron temperature can deviate from bulk temperature, a supplementary conservation equation for energy has to be added. A variety of approaches, based on different assumptions and addressing the needs of different scientific and technological fields, are available in the literature. The assumption of a purely convective magnetic regime (ideal MHD with infinite electrical conductivity) is widely utilized in astrophysics and plasma physics. In this case, the equations of the fluid dynamics and the electrodynamics are integrated into a single set of equations which yields a propagation system of seven waves. The magnetic field and the flow field are strongly coupled and jointly react to perturbations. The complementary limit case is the low magnetic Reynolds number regime. This approach, very popular in the MHD power generation field, yields a separate set of equations for fluid dynamics and electrodynamics. The intermediate case is the diffusive–convective magnetic regime which is characterized by a disperse propagation system.

Regardless of the specific assumptions, an MHD model is usually obtained by treating the plasma as a continuum medium, and then coupling fluid dynamics with electrodynamics. In the framework of the continuum assumption, the fluid dynamics is described by the Navier–Stokes equations. Treating the plasma as a viscous fluid subject to a force per unit volume $\mathbf{f}$ and supplied with a rate of heat per unit volume $\dot{Q}$, the balance equations for mass, momentum, and internal energy can be written as

$$\frac{\partial \rho}{\partial t} + \nabla \cdot (\rho \mathbf{v}) = 0 \tag{7.1}$$

$$\frac{\partial \rho \mathbf{v}}{\partial t} + \nabla \cdot (\rho \mathbf{v}\mathbf{v} + \mathbf{I}p - \boldsymbol{\tau}) = \mathbf{f} \tag{7.2}$$

$$\frac{\partial \rho e_i}{\partial t} + \nabla \cdot (\rho e_i \mathbf{v}) + p \, \nabla \cdot \mathbf{v} - \boldsymbol{\tau} : \mathbf{v} + \nabla \cdot \boldsymbol{\Gamma}_T = \dot{Q}. \tag{7.3}$$

An equation for the energy per unit volume $\rho(e_i + \rho v^2/2)$ can be derived from equations (7.1), (7.2), and (7.3) after some manipulation:

$$\frac{\partial}{\partial t}\left[\rho\left(e_i + \frac{v^2}{2}\right)\right] + \nabla \cdot \left[\rho\left(e_i + \frac{v^2}{2}\right)\mathbf{v} + (\mathbf{I}p - \boldsymbol{\tau}) \cdot \mathbf{v} + \boldsymbol{\Gamma}_T\right] = \dot{Q} + \mathbf{f} \cdot \mathbf{v}. \tag{7.4}$$

The viscosity tensor $\boldsymbol{\tau}$ describes the momentum fluxes due to diffusion and depends on the mobility of the species constituting the fluid. Assuming that the fluid behaves as a Newtonian fluid, the viscous stress tensor can be expressed as

$$\boldsymbol{\tau} = \mu[\nabla \mathbf{v} + (\nabla \mathbf{v})^T] + (\lambda \nabla \cdot \mathbf{v})\mathbf{I}, \tag{7.5}$$

where μ is the dynamic viscosity and λ the bulk viscosity. According to the Stokes hypothesis, $\lambda = -2\mu/3$.

A description is also needed to model the thermal diffusion mechanisms. The diffusive thermal flux $\boldsymbol{\Gamma}_T$ can be expressed by Fourier's law

$$\boldsymbol{\Gamma}_T = -k\nabla T. \tag{7.6}$$

The fluid-dynamics model is closed by adding the equation of state, which specifies the thermodynamic behavior of the fluid. Assuming a calorically perfect gas, we can relate the pressure p and the mass density ρ to the temperature T,

$$p = \rho R T, \tag{7.7}$$

and we can express the internal energy as

$$e_i = c_v T, \tag{7.8}$$

where the ideal gas constant can be expressed as $R = c_p - c_v$ and $c_p/c_v = \gamma$. In the framework of classical theory, the formulation of electromagnetics in a continuum medium is described by the Maxwell equations

$$\nabla \times \mathbf{E} = -\frac{\partial \mathbf{B}}{\partial t} \tag{7.9}$$

$$\nabla \times \mathbf{H} = \mathbf{J} + \frac{\partial \mathbf{D}}{\partial t} \tag{7.10}$$

$$\nabla \cdot \mathbf{J} = -\frac{\partial \rho_c}{\partial t}. \tag{7.11}$$

From equations (7.9)–(7.11), three further equations may then be derived:

$$\nabla \cdot \mathbf{D} = \rho_c \tag{7.12}$$

$$\nabla \cdot \mathbf{B} = 0 \tag{7.13}$$

$$\nabla \cdot \left(\mathbf{J} + \frac{\partial \mathbf{D}}{\partial t}\right) = 0. \tag{7.14}$$

The electromagnetic model is completed by three constitutive relations. Neglecting magnetization and electric polarization, one may write

$$\mathbf{D} = \epsilon_0 \mathbf{E} \tag{7.15}$$

$$\mathbf{B} = \mu_0 \mathbf{H}. \tag{7.16}$$

The third relevant constitutive relation is obtained considering the momentum balance for the electric charge carriers, i.e. electrons and ions. In a weakly ionized gas, assuming that the characteristic time of the macroscopic phenomenon is much greater than the average electron collision time, one can obtain the generalized Ohm's law [15]

$$\mathbf{J} = \sigma\left(\mathbf{E} + \mathbf{v} \times \mathbf{B} + \frac{\nabla T_e}{e n_e}\right) - \beta_e \frac{\mathbf{J} \times \mathbf{B}}{B} - s\frac{\mathbf{B} \times (\mathbf{J} \times \mathbf{B})}{B^2}. \tag{7.17}$$

The electrical conductivity σ can be expressed as

$$\sigma = \frac{n_e^2 e}{m_e \nu_{e\mathrm{H}}}. \tag{7.18}$$

Since the electron mobility is $\mu_e = e/(m_e \nu_{e\mathrm{H}})$, we can also write

$$\sigma = n_e e \mu_e. \tag{7.19}$$

The electron Hall parameter β_e is defined as the radians around the gyration center an electron travels between two collisions in its cyclotron motion induced by the presence of the magnetic field. The electron Hall parameter can formally be expressed as the ratio between the cyclotron angular frequency ω_c and the electron-heavy particle collision frequency $\nu_{e\mathrm{H}}$:

$$\beta_e = \frac{\omega_c}{\nu_{e\mathrm{H}}} = \frac{eB}{m_e \nu_{e\mathrm{H}}}, \tag{7.20}$$

and, recalling the definition of μ_e,

$$\beta_e = \mu_e B. \tag{7.21}$$

We will assume then that the contribution of the gradient of electron pressure $p_e = n_e k_\mathrm{B} T_e$ to the conduction electric current density is negligible. We will neglect as well the ion slip factor term ($s = \beta_e \beta_i \ll 1$), which is equivalent to considering the electrons as the only species responsible for electric charge transport. Under these assumptions, equation (7.17) can be rewritten in the form

$$\mathbf{J} = \sigma(\mathbf{E} + \mathbf{v} \times \mathbf{B}) - \beta_e \frac{\mathbf{J} \times \mathbf{B}}{B}, \tag{7.22}$$

or, introducing the reduced electric field $\mathbf{E}' = \mathbf{E} + \mathbf{v} \times \mathbf{B}$, i.e. the electric field as seen in a reference system moving with the fluid, we obtain

$$\mathbf{E}' = \frac{1}{\sigma}\left(\mathbf{J} - \beta_e \frac{\mathbf{J} \times \mathbf{B}}{B}\right). \tag{7.23}$$

In a Cartesian coordinate system, defining the resistivity tensor $\boldsymbol{\eta}$ as

$$\boldsymbol{\eta} = \frac{1}{\sigma}\begin{bmatrix} 1 & \beta_{e,z} & -\beta_{e,y} \\ -\beta_{e,z} & 1 & -\beta_{e,x} \\ \beta_{e,y} & -\beta_{e,x} & 1 \end{bmatrix}, \tag{7.24}$$

where

$$\beta_{e,x} = \frac{\beta_e}{B}B_x = \mu_e B_x, \quad \beta_{e,y} = \frac{\beta_e}{B}B_y = \mu_e B_y, \quad \beta_{e,x} = \frac{\beta_e}{B}B_z = \mu_e B_z,$$

we can rewrite equation (7.23) in a more compact form:

$$\mathbf{E}' = \boldsymbol{\eta}\mathbf{J}, \tag{7.25}$$

or

$$\mathbf{J} = \boldsymbol{\Sigma}\mathbf{E}', \tag{7.26}$$

where $\boldsymbol{\Sigma} = \boldsymbol{\eta}^{-1}$ is the conductivity tensor

$$\boldsymbol{\Sigma} = \frac{\sigma}{\left(1 + \beta_e^2\right)}\begin{bmatrix} 1 + \beta_{e,x}^2 & \beta_{e,x}\beta_{e,y} - \beta_{e,z} & \beta_{e,x}\beta_{e,z} + \beta_{e,y} \\ \beta_{e,y}\beta_{e,x} + \beta_{e,z} & 1 + \beta_{e,y}^2 & \beta_{e,y}\beta_{e,z} - \beta_{e,x} \\ \beta_{e,z}\beta_{e,x} - \beta_{e,y} & \beta_{e,z}\beta_{e,y} + \beta_{e,x} & 1 + \beta_{e,z}^2 \end{bmatrix}. \tag{7.27}$$

In order to define the terms appearing on the right-hand sides of equations (7.2), (7.3), and (7.4), a convenient expression for $\mathbf{f}$ and $\dot{Q}$ due to the action of the electromagnetic field has to be developed. Indeed, the body force per unit volume in a fluid subject to an electric field $\mathbf{E}$ and moving with a velocity $\mathbf{v}$ across a magnetic field $\mathbf{B}$ can be expressed as [15]

$$\mathbf{f} = \rho_c \mathbf{E}' + \mathbf{J} \times \mathbf{B}. \tag{7.28}$$

The total work per time and volume unit p_{EM} done by the electromagnetic field can then be expressed as

$$p_{EM} = \rho_c \mathbf{v} \cdot \mathbf{E} + \mathbf{J} \cdot \mathbf{E}, \tag{7.29}$$

while the work per time and volume unit delivered to the fluid by the force $\mathbf{f}$ is

$$\mathbf{v} \cdot \mathbf{f} = \rho_c \mathbf{v} \cdot \mathbf{E}' + \mathbf{u} \cdot (\mathbf{J} \times \mathbf{B}). \tag{7.30}$$

The difference between the total power density delivered to the fluid and the mechanical work per time and volume unit on the fluid represents the thermal power per unit volume $\dot{Q}$ that the charge carriers transmits to the fluid through collisions:

$$\dot{Q} = p_{EM} - \mathbf{v} \cdot \mathbf{f} = \rho_c \mathbf{v} \cdot \mathbf{E} + \mathbf{J} \cdot \mathbf{E} - \left[\rho_c \mathbf{v} \cdot \mathbf{E}' + \mathbf{v} \cdot (\mathbf{J} \times \mathbf{B}) \right].$$

Since

$$\mathbf{v} \cdot \mathbf{E}' = \mathbf{v} \cdot (\mathbf{E} + \mathbf{v} \times \mathbf{B}) = \mathbf{v} \cdot \mathbf{E}$$

and

$$\mathbf{v} \cdot (\mathbf{J} \times \mathbf{B}) = -\mathbf{J} \cdot (\mathbf{v} \times \mathbf{B}),$$

we obtain that

$$\dot{Q} = \mathbf{J} \cdot (\mathbf{E} + \mathbf{v} \times \mathbf{B}) = \mathbf{J} \cdot \mathbf{E}'. \tag{7.31}$$

From equation (7.23), noting that $\mathbf{J} \times \mathbf{B} \perp \mathbf{J}$, we obtain

$$\dot{Q} = \frac{J^2}{\sigma}. \tag{7.32}$$

Thus, equations (7.31) and (7.32) can be used to express the right-hand side term in equation (7.3). The right-hand side term of equation (7.4) $\dot{Q} + \mathbf{f} \cdot \mathbf{v}$ is given by the total electromagnetic work per time and volume unit p_{EM} expressed by equation (7.29).

7.1.2 The MHD approximation

We will now briefly describe treatment of the MHD approximation [15]. Since equations (7.22) and (7.23) have been derived assuming that electrons are the only charge carriers, the relevant collision time is given by the inverse of the electron collision frequency ν_{eH}, and the condition that has to be verified is thus

$$t_c \gg \nu_{eH}. \tag{7.33}$$

By comparing the ratio between the displacement current density, given by the time derivative of the displacement field $\mathbf{D}$, and the conduction current density, we obtain

$$\frac{\frac{\partial D}{\partial t}}{J} = \frac{\frac{\partial D}{\partial t}}{\sigma E} \approx \frac{\epsilon_0}{t_c \sigma}.$$

Thus, the displacement current is neglected assuming that

$$t_c \gg \frac{\epsilon_0}{\sigma}. \tag{7.34}$$

From equations (7.18), (7.33), and (7.34), remembering also that the plasma angular frequency as $\omega_p = \sqrt{(n_e e^2)/(\epsilon_0 m_e)}$, the following condition can be derived:

$$t_c \gg \frac{1}{\omega_p}. \tag{7.35}$$

When condition (7.34) holds, the ionized gas may be assumed as globally neutral, and the effects of the electric charge density on the charge transport and on the electromagnetic body force can be neglected. Indeed,

$$\frac{\rho_c v}{J} \approx \frac{\epsilon_0}{t_c \sigma}$$

and

$$\frac{\rho_c E}{|\mathbf{J} \times \mathbf{B}|} \approx \frac{\epsilon_0}{t_c \sigma}.$$

Consequently, the force per unit volume becomes

$$\mathbf{f} = \mathbf{J} \times \mathbf{B}, \tag{7.36}$$

and the total work per unit time and volume delivered to the fluid becomes

$$p_{\mathrm{EM}} = \mathbf{v} \cdot \mathbf{f} + \dot{Q} = \mathbf{J} \cdot \mathbf{E}. \tag{7.37}$$

As a result, the equations describing the behavior of the system can be modified in accordance with the assumptions of the MHD approximation discussed thus far. The general formulations of the governing equations under the MHD approximation are

$$\frac{\partial \rho}{\partial t} + \nabla \cdot (\rho \mathbf{v}) = 0 \tag{7.38}$$

$$\frac{\partial \rho \mathbf{v}}{\partial t} + \nabla \cdot (\rho \mathbf{v} \mathbf{v} + \mathbf{I} p - \boldsymbol{\tau}) = \mathbf{J} \times \mathbf{B} \tag{7.39}$$

$$\frac{\partial}{\partial t} \left[\rho \left(e_i + \frac{v^2}{2} \right) \right] + \nabla \cdot \left[\rho \left(e_i + \frac{v^2}{2} \right) \mathbf{v} + (\mathbf{I} p - \boldsymbol{\tau}) \cdot \mathbf{v} + \boldsymbol{\Gamma}_r \right] = \mathbf{E} \cdot \mathbf{J} \tag{7.40}$$

$$\nabla \times \mathbf{E} = -\frac{\partial \mathbf{B}}{\partial t} \tag{7.41}$$

$$\nabla \times \mathbf{H} = \mathbf{J} \tag{7.42}$$

$$\nabla \cdot \mathbf{J} = 0 \tag{7.43}$$

$$\nabla \cdot \mathbf{B} = 0. \tag{7.44}$$

7.1.3 Non-equilibrium conditions

As previously stated, a set of closure equations is necessary to close the system of the governing equations of fluid dynamics. The equations of state for ideal gases, equations (7.7) and (7.8), are well-suited to describe a wide range of physical situations. However, in ionized reactive gas flows the thermodynamic description of the system increases its complexity, since the equations relating the thermodynamic quantities depend on the local composition of the fluid, which is not uniform and varies over time following the evolution of the system. Diffusive transport parameters (i.e. viscosity, thermal and electrical conductivity, the Hall parameter, etc) also depend on the temperature and on the number density of the species in the mixture. When a condition of thermochemical equilibrium can be assumed, the composition of the gas can be derived as a function of the thermodynamic macroscopic quantities. However, the characteristic times of the reactions involved in the investigated phenomena are very often comparable with the diffusion and convection times. In this case, the relevant species have to be tracked down by writing a balance equation for their number densities. For the generic species s with number density n_s and a the rate of production per unit volume $\dot{\Omega}_s$ we have

$$\frac{\partial n_s}{\partial t} + \nabla \cdot \left(n_s \mathbf{v} + \mathbf{\Gamma}_s \right) = \dot{\Omega}_s. \tag{7.45}$$

The diffusive fluxes can be expressed as $\mathbf{\Gamma}_s = -D_s \nabla n_s$, where D_s is the diffusivity of the species.

In ionized flows where the elastic collision rate of the electrons with the heavy particles is not sufficient, a condition of partial thermal equilibrium occurs. When this happens, the electrons temperature T_e can significantly deviate from bulk temperature T. Since the electronic transport parameters and the pressure gradient in equation (7.17) depend on T_e, an energy equation for electrons has to be added to the model:

$$\frac{\partial}{\partial t}\left(\frac{3}{2}kT_e\right) + \nabla \cdot \left(\frac{5}{2}kT_e\mathbf{v}\right) + \nabla \cdot \mathbf{\Gamma}_e$$
$$= \mathbf{E}' \cdot \mathbf{J} - \sum_s \nu_{es} \frac{2\,m_e}{m_s} n_e \frac{3}{2}k(T_e - T) - \dot{N}. \tag{7.46}$$

The sum in the right-hand side of equation (7.46) represents the rate of energy per unit volume which electrons lose to heavy species in the plasma due to elastic collisions. The rate of energy loss per unit volume due to inelastic collisions is taken into account by the term $\dot{N}$.

7.1.4 Magneto-quasi-statics

The mathematical description of electromagnetics constituted by equations (7.9), (7.10), (7.12), and (7.13) leads to the well-known wave equations for the scalar and vector potentials. The retarded potentials solution of wave equations shows how the electromagnetic field propagates through space with a velocity $c = (\mu_0 \epsilon_0)^{-1/2}$. Thus, a traveling time L/c exists between the instant in which the field is produced and the instant in which the same field is observed at a distance L from its source. The electromagnetic wave mechanism is related to the interaction between the Faraday–Neumann–Lenz law (7.9) and the Ampère–Maxwell law (7.10): in a plane wave, a wave of electric field $\mathbf{E}$ traveling at a velocity c is seen by a fixed observer as a time-varying field, thus associated with a displacement current density $\frac{\partial \mathbf{D}}{\partial t}$. The displacement current density, due to equation (7.10), produces a space-varying magnetic field $\mathbf{H}$, which travels with $\mathbf{E}$ at the same velocity c. The traveling wave of the magnetic field is seen by a fixed observer as a time-varying magnetic field, which closes the loop producing an electric field due to equation (7.9). In many physical and technological applications, however, the propagative behavior of the electromagnetic field can be neglected. This is the case when the time L_c/c it takes an electromagnetic wave to travel a distance L_c equal to the characteristic length of the problem considered is much less then the characteristic time t_c. In these cases, it can be safely assumed that the electromagnetic waves instantaneously propagates through the entire space constituting the domain of the considered problem, and that the time behavior is governed by other mechanisms. When the condition $L_c/c \ll t_c$ is verified, it is convenient to refer to a quasi-static formulation of the electromagnetism, which is obtained when one of the two time derivatives appearing in equations (7.9) and (7.10) is neglected. In doing so, the previously described mechanism which sustains the electromagnetic wave is interrupted, and the propagation time no longer exists. The quasi-static formulation of electromagnetism obtained neglecting the displacement current density is called magneto-quasi-statics, as opposed to electro-quasi-statics, which is obtained by neglecting the time derivative of the magnetic induction field $\mathbf{B}$ in equation (7.9).

According to the assumptions constituting the MHD approximation, electromagnetism is described by a magneto-quasi-static formulation. We are interested in finding an equation for the magnetic induction field $\mathbf{B}$, which will uniquely describe the electromagnetic behavior oh the problem. Introducing equation (7.25) in equation (7.42), we obtain

$$\mathbf{E}' = \left(\mathbf{E} + \mathbf{v} \times \mathbf{B} \right) = \frac{1}{\mu} \boldsymbol{\eta} \cdot \nabla \times \mathbf{B}.$$

Applying the curl operator to both sides, expressing $\nabla \times \mathbf{E}$ by means of equation (7.9) and remembering that $\boldsymbol{\Sigma}^{-1} = \boldsymbol{\eta}$, we derive

$$-\frac{\partial \mathbf{B}}{\partial t} + \nabla \times (\mathbf{v} \times \mathbf{B}) = \nabla \times \left[\frac{1}{\mu}\boldsymbol{\eta} \cdot (\nabla \times \mathbf{B}) \right].$$

Reordering and applying the vector identity,

$$\nabla \times (\mathbf{a} \times \mathbf{b}) = \nabla \cdot (\mathbf{ba} - \mathbf{ab}),$$

we can write the convection–diffusion equation for the $\mathbf{B}$ field as

$$\frac{\partial \mathbf{B}}{\partial t} + \nabla \times \left[\frac{1}{\mu}\boldsymbol{\eta} \cdot (\nabla \times \mathbf{B}) \right] - \nabla \cdot (\mathbf{Bv} - \mathbf{vB}) = 0. \tag{7.47}$$

It is also useful to develop a balance equation for the magnetic energy density $B^2/(2\mu_0)$, which will be conveniently utilized to derive an equation for the total energy. Taking the dot product of equation (7.41) with $\mathbf{H}$ and subtracting from the equation obtained by dot multiplying equation (7.42) by $\mathbf{E}$ we obtain

$$\mathbf{E} \cdot (\nabla \times \mathbf{H}) - \mathbf{H} \cdot (\nabla \times \mathbf{E}) = \mathbf{E} \cdot \mathbf{J} + \mathbf{H} \cdot \frac{\partial \mathbf{B}}{\partial t}.$$

Introducing the vector identity

$$\nabla \cdot (\mathbf{a} \times \mathbf{b}) = \mathbf{b} \cdot (\nabla \times \mathbf{a}) - \mathbf{a} \cdot (\nabla \times \mathbf{b})$$

and reordering, one obtains

$$-\frac{\partial}{\partial t}\left(\frac{B^2}{2\,\mu_0} \right) - \nabla \cdot (\mathbf{E} \times \mathbf{H}) = \mathbf{E} \cdot \mathbf{J}. \tag{7.48}$$

Equation (7.48) is the formulation of the Poynting theorem in the magneto-quasi-static assumption and constitutes a balance of electromagnetic energy for an infinitesimal volume $\mathrm{d}V$. The Poynting vector $\mathbf{S} = \mathbf{E} \cdot \mathbf{H}$ appearing in equation (7.48) represents the rate of electromagnetic energy per unit area and its divergence takes into account the net balance energy per unit time leaving the volume. The theorem states that the total work per unit time $(\mathbf{E} \cdot \mathbf{J})\mathrm{d}V$ delivered to the volume by the electromagnetic field is equal to the sum of the rate of decrease of the magnetic energy $B^2/(2\mu)\mathrm{d}V$ and the electromagnetic energy per unit time $-\nabla \cdot (\mathbf{E} \times \mathbf{H})\mathrm{d}V$ entering in the infinitesimal volume through its bounding surface. Since, from equation (7.23),

$$\mathbf{E} = \boldsymbol{\eta} \cdot \mathbf{J} - \mathbf{v} \times \mathbf{B},$$

the Poynting vector $\mathbf{S}$ can be expressed as

$$\mathbf{S} = \boldsymbol{\eta} \cdot \mathbf{J} \times \mathbf{H} - (\mathbf{v} \times \mathbf{B}) \times \mathbf{H}$$

and, using equations (7.16) and (7.10),

$$\mathbf{S} = \frac{1}{\mu_0}\left[\frac{1}{\mu_0}\boldsymbol{\eta} \cdot (\nabla \times \mathbf{B}) \times \mathbf{B} - (\mathbf{v} \times \mathbf{B}) \times \mathbf{B}\right]. \tag{7.49}$$

The term $(\nabla \times \mathbf{B}) \times \mathbf{B}$ appearing in equation (7.49) can be treated using the vector identity

$$\nabla(\mathbf{a} \cdot \mathbf{b}) = \mathbf{a} \cdot (\nabla \mathbf{b}) + \mathbf{b} \cdot (\nabla \mathbf{a}) + \mathbf{a} \times (\nabla \times \mathbf{b}) + \mathbf{b} \times (\nabla \times \mathbf{a}).$$

We can then write

$$(\nabla \times \mathbf{B}) \times \mathbf{B} = -\mathbf{B} \times (\nabla \times \mathbf{B}) = \mathbf{B} \cdot (\nabla \mathbf{B}) - \frac{1}{2}\nabla(B^2).$$

Noting that for equation (7.13) we have

$$\nabla \cdot (\mathbf{BB}) = \mathbf{B} \cdot (\nabla \mathbf{B}) + \mathbf{B}(\nabla \cdot \mathbf{B}) = \mathbf{B} \cdot (\nabla \mathbf{B}),$$

we can finally write

$$(\nabla \times \mathbf{B}) \times \mathbf{B} = \nabla \cdot \left[\mathbf{BB} - \frac{1}{2}\nabla(B^2)\right].$$

Introducing now the Maxwell stress tensor $\mathbf{T}$ which, under the current approximation, is constituted only by its magnetic component

$$\mathbf{T} = \frac{1}{\mu_0}\left(\mathbf{BB} - \frac{1}{2}\mathbf{I}B^2\right), \tag{7.50}$$

we can conclude that

$$\frac{1}{\mu}(\nabla \times \mathbf{B}) \times \mathbf{B} = \nabla \cdot \mathbf{T}. \tag{7.51}$$

Using the vector identity

$$(\mathbf{a} \times \mathbf{b}) \times \mathbf{c} = \mathbf{a} \cdot (\mathbf{cb}) - \mathbf{b} \cdot (\mathbf{ca})$$

and recalling that $\mathbf{b} \cdot (\mathbf{ca}) = (\mathbf{b} \cdot \mathbf{c})\mathbf{a}$, we can manipulate the term $(\mathbf{v} \times \mathbf{B}) \times \mathbf{B}$ in equation (7.49),

$$(\mathbf{v} \times \mathbf{B}) \times \mathbf{B} = \mathbf{v} \cdot (\mathbf{BB}) - \mathbf{B} \cdot (\mathbf{Bv}) = \mathbf{v} \cdot (\mathbf{BB}) - B^2\mathbf{v},$$

and, since $\mathbf{v} \cdot (\mathbf{BB}) = (\mathbf{BB}) \cdot \mathbf{v}$,

$$(\mathbf{v} \times \mathbf{B}) \times \mathbf{B} = (\mathbf{BB} - \mathbf{I}B^2) \cdot \mathbf{v}.$$

We can now introduce the definition of the Maxwell stress tensor given in equation (7.50) to conclude that

$$\frac{1}{\mu_0}(\mathbf{v} \times \mathbf{B}) \times \mathbf{B} = \left(\mathbf{T} - \frac{1}{2\mu_0}\mathbf{I}B^2\right) \cdot \mathbf{v}. \tag{7.52}$$

Using equations (7.51) and (7.52), the expression for the Poynting vector given in equation (7.49) can be rewritten as

$$\mathbf{S} = \left[\frac{1}{\mu_0}\boldsymbol{\eta} \cdot (\nabla \cdot \mathbf{T}) - \left(\mathbf{T} - \frac{1}{2\mu_0}\mathbf{I}B^2\right) \cdot \mathbf{v}\right], \tag{7.53}$$

while equation (7.48) becomes

$$-\frac{\partial}{\partial t}\left(\frac{B^2}{2\,\mu_0}\right) - \nabla \cdot \left[\frac{1}{\mu_0}\boldsymbol{\eta} \cdot (\nabla \cdot \mathbf{T}) - \left(\mathbf{T} - \frac{1}{2\mu_0}\mathbf{I}B^2\right) \cdot \mathbf{v}\right] = \mathbf{E} \cdot \mathbf{J}. \tag{7.54}$$

7.1.5 The general model

The general model for a resistive MHD flow is obtained by coupling the fluid-dynamics equations (7.38), (7.39), and (7.40) with the magnetic convection–diffusion equation (7.47) and the magnetic energy balance equation (7.54). The right-hand side term in equation (7.39) is expressed according to equation (7.51),

$$\mathbf{J} \times \mathbf{B} = \nabla \cdot \mathbf{T},$$

and an equation for the total energy per unit volume E_T defined as

$$E_T = \rho\left(e_i + \frac{v^2}{2}\right) + \frac{B^2}{2\mu_0} \tag{7.55}$$

is obtained introducing equation (7.54) in equation (7.40). The complete set of equations reads as follows:

$$\frac{\partial \rho}{\partial t} + \nabla \cdot (\rho\mathbf{v}) = 0 \tag{7.56}$$

$$\frac{\partial \rho\mathbf{v}}{\partial t} + \nabla \cdot (\rho\mathbf{vv} + \mathbf{I}p - \boldsymbol{\tau} - \mathbf{T}) = 0 \tag{7.57}$$

$$\frac{\partial E_T}{\partial t} + \nabla \cdot \left[(E_T + p)\mathbf{v} - (\boldsymbol{\tau} + \mathbf{T}) \cdot \mathbf{v} + \frac{1}{\mu_0}\boldsymbol{\eta} \cdot (\nabla \cdot \mathbf{T}) + \boldsymbol{\Gamma}_T\right] = 0 \tag{7.58}$$

$$\frac{\partial \mathbf{B}}{\partial t} + \nabla \times \left[\frac{1}{\mu_0} \boldsymbol{\eta} \cdot (\nabla \times \mathbf{B}) \right] - \nabla \cdot (\mathbf{B}\mathbf{v} - \mathbf{v}\mathbf{B}) = 0. \tag{7.59}$$

One can recognize that the only electromagnetic quantity in the above set of equations is the magnetic induction field $\mathbf{B}$. The current density and the electric field can be derived from equations (7.42) and (7.25), respectively.

The transport of the magnetic induction and of the magnetic energy density occurs due to two mechanisms, convection and diffusion, that can be observed in equations (7.58) and (7.59). The physical mechanism of magnetic diffusion is more easily understood as a diffusion of the electric current density. Indeed, a non-uniform distribution of current density in a conductive medium tends to spread over all the available paths. However, a redistribution in the current density lines causes a time variation in the magnetic induction field produced by the current, which in turn generates through an induced electric field. The induced electric field generates a current density countering the initial variation, hindering the redistribution of the current density. Assuming a scalar conductivity, one can recognize that the parameter governing the magnetic diffusion is the magnetic diffusivity $1/(\sigma\mu_0)$.

An important parameter characterizing the MHD regime of the considered problem is the magnetic Reynolds number Re_m which, analogously to its fluid dynamic counterpart, is defined as the ratio of the magnetic convection and diffusion effects. Considering the equation for the magnetic induction, the order of magnitude of the convective term is

$$\nabla \cdot (\mathbf{B}\mathbf{v} - \mathbf{v}\mathbf{B}) \sim \frac{Bv}{L_c},$$

while, for the convective term,

$$\nabla \times \left[\frac{1}{\mu_0} \boldsymbol{\eta} \cdot (\nabla \times \mathbf{B}) \right] \sim \frac{B}{\mu_0 \sigma L_c^2}.$$

We have then

$$\mathrm{Re}_m = \frac{(Bv)/L_c}{B/(\mu_0 \sigma L_c^2)} = \mu_0 \sigma v L_c. \tag{7.60}$$

In high Re_m MHD regimes, the magnetic convection is the dominant transport mechanism, and the magnetic behavior of the system is strongly coupled with the flow field. Conversely, for low Re_m, the magnetic induction field equation is mainly governed by diffusion, and is rather insensitive to the velocity distribution. Two representative cases will now be briefly presented and discussed.

7.1.6 Ideal MHD

The ideal MHD model is obtained from the viscosity and the conductive thermal fluxes in the fluid, and assuming an infinite electrical conductivity. The only transport mechanisms represented in the model are then related to convection.

The governing equation are derived by dropping all the diffusive terms in the MHD governing equations:

$$\frac{\partial \rho}{\partial t} + \nabla \cdot (\rho \mathbf{v}) = 0 \tag{7.61}$$

$$\frac{\partial \rho \mathbf{v}}{\partial t} + \nabla \cdot (\rho \mathbf{v}\mathbf{v} + \mathbf{I}p - \mathbf{T}) = 0 \tag{7.62}$$

$$\frac{\partial E_T}{\partial t} + \nabla \cdot \left[(E_T + p)\mathbf{v} - \mathbf{T} \cdot \mathbf{v} \right] = 0 \tag{7.63}$$

$$\frac{\partial \mathbf{B}}{\partial t} - \nabla \cdot (\mathbf{B}\mathbf{v} - \mathbf{v}\mathbf{B}) = 0. \tag{7.64}$$

The basic assumptions of the ideal MHD model do not allow one to adequately treat the presence of viscous and thermal boundary layers. Additionally, ideal MHD equations are applicable for describing physical problems in which the magnetic Reynolds number can be considered infinite. High-density astrophysical plasmas such those found in a star are good examples of physical systems that can be treated by ideal MHD. In fact, the model was first formulated by Alfvén for purposes related to astrophysics [16]. It is worth noting that the infinite electrical conductivity assumption involves the cancellation of the $\dot{Q} = \mathbf{J} \cdot \mathbf{E}'$ term at the right-hand side in the equation of internal energy (7.4). We can then manipulate equation (7.4) assuming a calorically perfect gas to obtain an equation for pressure p:

$$\frac{\partial p}{\partial t} + \mathbf{v} \cdot \nabla p + \gamma p \nabla \cdot \mathbf{v} = 0, \tag{7.65}$$

which can substitute equation (7.63). Introducing the convective derivative $\frac{\mathrm{d}\{\cdot\}}{\mathrm{d}t} = \frac{\partial\{\cdot\}}{\partial t} + \mathbf{v} \cdot \nabla\{\cdot\}$, equation (7.65) can be rewritten in the form

$$\frac{\mathrm{d}}{\mathrm{d}t}\left(\frac{p}{\rho^\gamma} \right) = 0. \tag{7.66}$$

Equation (7.66) expresses the fact that an element $\mathrm{d}V$ moving with the fluid does not receives any energy contributions and therefore undergoes only adiabatic processes.

A distinctive feature of MHD regimes described by the ideal model is that the induction flux linked to a curve that moves with the fluid cannot undergo variations in time. This condition, known as the 'frozen flow' condition, is a direct consequence of the electromagnetic induction law (7.9), and gives insight into the mechanism underlying the magnetic convection phenomenon: a closed curve Γ moving within an electrically conducting fluid through a region where a non-uniform magnetic field is present will experience a time variation in the linked magnetic flux. The time variation of the magnetic flux will produce, due to equation (7.9), an electric field and a system of electric current densities that will in turn produce a reaction magnetic induction field in order to counterbalance the initial variation. In a perfectly conductive medium, the reaction field perfectly matches the initial time variation; the linked magnetic flux is not allowed to vary in time and is convected along with the Γ curve.

The ideal MHD model also constitutes the model for developing the theory of wave propagation in a plasma. While the conservative formulation given by equations (7.61), (7.62), (7.63), and (7.64) can be attractive from the numerical point of view, a more convenient form of the governing equation can be obtained by reformulating the problem in the primitive variables λ, $\mathbf{v}$, p, and $\mathbf{B}$. The equations for λ and $\mathbf{B}$ are given by equations (7.61) and (7.64), respectively, and we already derived an equation for pressure (7.65). The equation for velocity can be derived after some manipulation from equation (7.62):

$$\frac{\partial \mathbf{v}}{\partial t} + \mathbf{v} \cdot \nabla \mathbf{v} + \frac{1}{\rho} \nabla \cdot (\mathbf{I}p - \mathbf{T}) = 0. \tag{7.67}$$

We now assume a one-dimensional problem, where the primitive variables depend only on the time t and on the space variable x. Noting that $\mathbf{B}$ is a solenoidal field, in the chosen geometry we must have

$$\frac{\partial B_x}{\partial x} = 0,$$

and we can rewrite equations (7.61), (7.67), (7.65), and (7.64) in the quasilinear form:

$$\frac{\partial \mathbf{u}}{\partial t} + A \frac{\partial \mathbf{u}}{\partial x} = 0, \tag{7.68}$$

where

$$\mathbf{u} = \begin{Bmatrix} \rho \\ v_x \\ v_y \\ v_z \\ p \\ B_y \\ B_z \end{Bmatrix} \quad A = \begin{bmatrix} v_x & \rho & 0 & 0 & 0 & 0 & 0 \\ 0 & v_x & 0 & 0 & 1/\rho & B_y/(\rho\mu_0) & B_z/(\rho\mu_0) \\ 0 & 0 & v_x & 0 & 0 & -B_x/(\rho\mu_0) & 0 \\ 0 & 0 & 0 & v_x & 0 & 0 & -B_x/(\rho\mu_0) \\ 0 & \gamma p & 0 & 0 & v_x & 0 & 0 \\ 0 & B_y & -B_x & 0 & 0 & v_x & 0 \\ 0 & B_z & 0 & -B_x & 0 & 0 & v_x \end{bmatrix}.$$

The eigenvalues of matrix A can be expressed defining the speeds c_a, c_a, x, and c_s as

$$c_a^2 = \frac{B^2}{\rho\mu_0}, \quad c_{a,x}^2 = \frac{B_x^2}{\rho\mu_0}, \quad c_s^2 = \frac{\gamma p}{\rho}.$$

The seven eigenvalues are then

$$\lambda_{1,7} = v_x \pm c_{m,f}; \quad \lambda_{2,6} = v_x \pm c_{ax}; \quad \lambda_{3,5} = v_x \pm c_{f,s}; \quad \lambda_4 = v_x.$$

The speed $c_{a,x}$ is called the Alfvén speed, while c_m, f and c_m, s are the speeds of the fast and slow magnetoacoustic (or magnetosonic) waves, respectively:

$$c_{m,f}^2 = \frac{1}{2}\left[\left(c_a^2 + c_s\right) + \sqrt{\left(c_a^2 + c_s\right)^2 - 4c_a^2 c_s^2}\right]$$

$$c_{m,s}^2 = \frac{1}{2}\left[\left(c_a^2 + c_s\right) - \sqrt{\left(c_a^2 + c_s\right)^2 - 4c_a^2 c_s^2}\right].$$

In the given geometry, Alfvén waves [17] propagate along the x-direction with speed $c_{a,x}$, introducing a transverse perturbation to the B_x field. Alfvén waves do not produce pressure or density variations in the fluid. On the other hand, magneto-acoustic waves traveling in the x-direction produce compressions and rarefactions in pressure and in the magnetic field components perpendicular to x. A more detailed discussion of MHD waves can be found in [18] and [19].

The numerical solution of ideal MHD has been intensively investigated, and various schemes are available in the literature. Many of them have been obtained as an extension of well known fluid-dynamics solvers (Roe [20–22], HLL [23], HLLC [24], and HLLD [25]).

7.1.7 The low magnetic Reynolds number model

Under the assumption of $\mathrm{Re}_m \ll 1$, some useful approximations can be applied. Most notably, when an externally generated magnetic flux density $\mathbf{B}_0$ is applied to the plasma flow (i.e. a field produced by a magnet), the magnetic flux density $\mathbf{B}_0$ produced by the current density flowing in the plasma can be neglected. Indeed, as $J \sim \sigma v B$ is the order of magnitude of the current density, then from equation (7.42) we obtain that B_i is of the order of

$$B_i \sim \mu_0 \sigma v B L_c = \mathrm{Re}_m B.$$

Thus, when $\mathrm{Re}_m \ll 1$ the total magnetic induction $\mathbf{B}$ can be considered equal to the externally applied field $\mathbf{B}_0$. The electromagnetic behavior of the system is described by equation (7.59), dropping the convecting term $\nabla \cdot (\mathbf{B}\mathbf{v} - \mathbf{v}\mathbf{B})$. Comparing the order of magnitude of the time derivative with the diffusive term at the right-hand side of equation (7.59) and taking $t_c = v/L_c$ as the characteristic time, we obtain

$$\frac{\dfrac{\partial \mathbf{B}}{\partial t}}{\nabla \times \left[\dfrac{1}{\mu_0}\boldsymbol{\eta} \cdot (\nabla \times \mathbf{B})\right]} \sim \frac{B/t_c}{B/(\mu_0 \sigma L_c^2)} \sim \mu_0 \sigma v L_c = \mathrm{Re}_m,$$

which shows that in a low Re_m regime the rate of change of the magnetic field is negligible, and that electromagnetics is basically described by a steady-state formulation. In other words, the electromagnetic characteristic time is negligible when compared with the fluid dynamics, and electromagnetic quantities are assumed to adapt themselves instantaneously to the time variations of the flow field. As a result, equation (7.59) is no longer representative of the problem, as it relates to the **B** field which is essentially a known quantity, and the description of the electro-magnetic behavior has to be rewritten in an electrodynamic formulation. Under the assumption of $Re_m \ll 1$, the electrodynamic formulation is derived assuming that the applied magnetic induction does not vary in time. The electric field **E** then becomes a curl free field due to equation (7.41) and can be expressed as the gradient of an electric scalar potential φ:

$$\mathbf{E} = -\nabla\varphi. \tag{7.69}$$

Introducing equation (7.69) in equation (7.26), we have

$$\mathbf{J} = -\mathbf{\Sigma} \cdot (\nabla\varphi + \mathbf{v} \times \mathbf{B}). \tag{7.70}$$

and taking into account equation (7.43), a steady-state electrodynamics described by means of an elliptical partial differential equation is derived:

$$\nabla \cdot \left[\mathbf{\Sigma} \cdot (-\nabla\varphi + \mathbf{v} \times \mathbf{B}) \right] = 0. \tag{7.71}$$

Once the velocity, the applied magnetic induction field, and the conductivity tensor are known at a generic instant t, equation (7.71) allows one to evaluate the distribution of the electric potential and, consequently, the electric field and the current density through equations (7.69) and (7.70). The MHD model in the low Re_m approximation is then constituted by the fluid-dynamics equations (7.38), (7.39), and (7.40) leading the evolution in time and driving a steady-state electro-dynamics described by equation (7.71). These quantities can then be utilized to evaluate the right-hand side terms $\mathbf{J} \times \mathbf{B}$ and $\mathbf{J} \cdot \mathbf{E}$ appearing in the momentum and energy equations.

7.2 Numerical model

In this section a brief description is given of a numerical model of the MHD interaction developed around a body immersed in a hypersonic flow. The model is oriented toward the design and the analysis of experiments aimed at investigating the MHD phenomena involved in the re-entry phase of a space vehicle.

Since the expected value of electrical conductivity ranges between a few to hundreds of siemens per meter, and the typical flow velocity is in the range of some thousands of meters per second, a low Re_m approach has been chosen.

As discussed before, the physical coupling between between the fluid-dynamic and the electrodynamic regimes in a low Re_m condition is rather weak. As a result, the two problems can be solved by means of two distinct solvers which reciprocally interact by exchanging data according to the scheme shown in figure 7.1. The fluid dynamics can be solved by means of an appropriate time-dependent component which, at each time step, receives from the electrodynamic solver the distribution of the electric field and of the current density. These quantities are utilized to compute the right-hand side terms in the momentum and energy equations. On the other side,

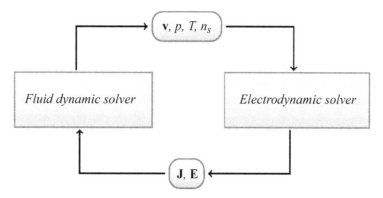

Figure 7.1. Coupling of the solvers in a low Re_m MHD numerical scheme.

the electrodynamic solver obtains from the fluid dynamic solver the velocity field needed to compute the forcing term in equation (7.71) and the data that allow the evaluation of the conductivity tensor.

The fluid-dynamics solver utilized to obtain the results presented in the following section is based on a cell-centered finite volume formulation. The control cells are defined as a dual tessellation of Delaunay triangulation of the domain. Convective fluxes are evaluated by means of an Osher's approximate Riemann solver [26] with a second-order accurate MUSCL approach. Diffusive flux computation is performed using a finite element approach on the domain triangulation. Time integration is performed utilizing an explicit scheme, based on a fourth-order Runge–Kutta method.

The electrodynamics is solved by means of a finite element approach. The details of the formulation are given in [27, 28] and [29]. Defining N as the set of shape functions on the given domain triangulation, we can approximate the unknown φ with a piecewise polynomial representation

$$\tilde{\varphi} = N^T \boldsymbol{\varphi}. \tag{7.72}$$

Introducing equation (7.72) in equation (7.71) and applying a weighted residual approach we obtain the weak formulation of the electrodynamic problem:

$$\int_\Omega \nabla W \cdot (\boldsymbol{\Sigma} \cdot \nabla \tilde{\varphi}) \mathrm{d}\Omega = \int_\Omega \nabla W \cdot \left[\boldsymbol{\Sigma} \cdot (\mathbf{v} \times \mathbf{B}) \right] \mathrm{d}\Omega - \oint_{\delta\Omega} W J_n \mathrm{d}(\delta\Omega). \tag{7.73}$$

A linear algebraic system is obtained using the shape functions in N as a weighting function W:

$$\left[\int_\Omega [\nabla N]^T [\boldsymbol{\Sigma}][\nabla N] \mathrm{d}\Omega \right] \boldsymbol{\varphi} = \int_\Omega [\nabla N]^T [\boldsymbol{\Sigma}](\mathbf{v} \times \mathbf{B}) \mathrm{d}\Omega - \oint_{\delta\Omega} N J_n \mathrm{d}(\delta\Omega). \tag{7.74}$$

The system (7.74) has to be defined and solved at each time step. Great attention therefore needs to be paid to optimizing the solution of this system, as poor performance of the electrodynamic solver would dramatically impair the overall efficiency of the whole code. We can observe in equation (7.27) that the Hall parameter determines the anisotropy of the conductivity tensor Σ. Thus the anisotropy of the electrodynamics is more pronounced for high values of the electron mobility. These conditions may be expected in plasmas at relatively low pressure and density, as in MHD aerospace applications in the hypersonic regimes of interest. An ILU-type preconditioned GMRES solver has been adopted for the solution of equation (7.74). The convergence of the solver can become very critical for high values of the Hall parameters in the plasmas, and high-quality preconditioning is needed in order to enhance the performance. A relatively straightforward way to obtain a higher accuracy in preconditioning is to perform a convenient numbering of the mesh nodes. In [29] the authors report a reverse Cuthill–McKee ordering weighted on the local direction of the magnetic induction field with respect to the topology of the triangular mesh utilized.

7.3 Applications

The low Re_m numerical model described in the previous section will now be applied to the analysis of the MHD interaction around a body immersed in a hypersonic flow. The discussion is aimed at aerospace applications related to re-entry in a planetary atmosphere. As already mentioned, the fundamentals of the MHD interaction have been investigated by many authors in the aerospace community, both experimentally [30–33] and numerically [34–37], and the efforts to develop reliable numerical models have received great momentum.

The low Re_m approximation can be verified computing the MHD regime around a body immersed in a hypersonic flow and comparing the applied magnetic field with the field produced by the current density due to the MHD interaction. We consider then an axisymmetric body immersed in a Mach 15 argon flow (see figure 7.2(a)). The applied magnetic induction field B_e, shown in figure 7.2(b), is assumed to be produced by a 56.5 kA electric current in a circular loop embedded in the body. The value of the electrical current is chosen so as to generate a 1 T induction field at

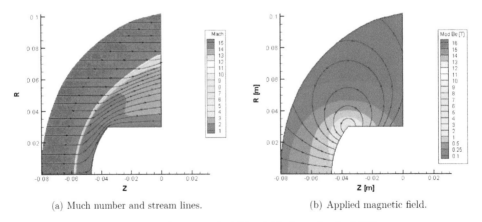

(a) Much number and stream lines.　　　　(b) Applied magnetic field.

Figure 7.2. Applied magnetic field and Mach number for MHD.

the stagnation point. For our current purposes, a scalar conductivity is adequate, and thus the Hall parameter β_e is forcibly set to zero. In this case the only component of the current density is the azimuthal one, and can be promptly evaluated as the cross product between the velocity and the B field times the conductivity. In figure 7.3 are shown some results of the calculation using the low Re_m model described in the previous section with a constant conductivity $\sigma = 100$ S m^{-1} in the shock layer. The azimuthal current distribution is shown in figure 7.3(a). It is worth noting that in the case of scalar conductivity, the azimuthal component of the electric field is equal to zero due to the axial symmetry of the problem. If not, there would exist a circular path centered on the symmetry axis along which the circulation of the electric field would not be zero, violating condition (7.69) which states that the electric field is conservative. As a result the term $\mathbf{E} \cdot \mathbf{J}$ describing the contribution of the MHD interaction to the energy equation (7.58) is zero. This means that MHD interaction directly and locally converts kinetic energy into internal energy and vice versa, leaving the total energy $\rho(e_i + v^2/2)$ unaffected. The magnetic induction field B_p generated by the current density in the shock layer can be evaluated using the Biot–Savart law. The percentage ratio of the induced field magnitude B_p to the applied field magnitude B_e is plotted in figure 7.3(b). As we can easily observe, in the given conditions the induced field B_p is well below 1% of the applied field B_e.

In order to appreciate the impact that the tensorial characteristic has on the electrodynamics and, as a consequence, on the whole MHD regime, we now consider the same problem taking a constant scalar conductivity $\sigma = 100$ S m^{-1} and and a 100 T^{-1} electron mobility. In figure 7.4 some results of the calculations are shown. The modifications induced in the flow by the MHD interaction can be observed comparing the Mach number distribution and the streamlines in figure 7.4(a) with the same quantities in the unmodified case in figure 7.2(a). The increase in the shock stand-off distance caused by the MHD interaction is evident. One of the main consequences of the tensorial characteristic of the conductivity is the appearance of a Hall electric field, having a direction approximately opposite to that of the velocity in

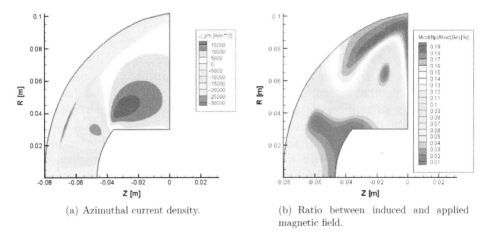

(a) Azimuthal current density.

(b) Ratio between induced and applied magnetic field.

Figure 7.3. Calculation results for a scalar conductivity $\sigma = 100$ S m^{-1}.

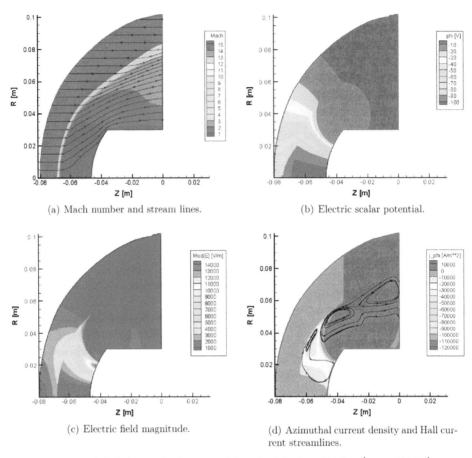

(a) Mach number and stream lines.

(b) Electric scalar potential.

(c) Electric field magnitude.

(d) Azimuthal current density and Hall current streamlines.

Figure 7.4. Calculation results for a tensorial conductivity ($\sigma = 100$ S m^{-1}, $\mu_e = 100$ T^{-1}).

the gas. This phenomenon is responsible for the electric potential distribution depicted in figure 7.4(b). A plot of the electric field magnitude is reported in 7.4(c). Since in high β_e regimes (i.e. in plasmas with low density and high electron mobility) the electric current density tends to be perpendicular to both the magnetic induction field **B** and the reduced electric field **E**′, the Hall electric field is actually the main field responsible for the azimuthal current density, which in turn produces the body force **J** × **B**. The azimuthal current density contour plot is shown in figure 7.4(d). In the same figure, the streamlines of the electric current density developing in the r–z-plane due to the Hall effect are reported.

The formulation of models and the consequent implementation of numerical codes goes hand-in-hand with the experimental study of a given phenomenon. Numerical codes are useful in the process of designing an experiment to assess the extent of the expected results. They are of course also essential tools for interpreting experimental data and for achieving insight into the phenomenon under investigation. On the other hand, the data produced by experiments are needed to carry out validation of the numerical models and increase confidence in the results that models provide. The investigations carried out in the field of MHD applied to hypersonic re-entry are no exception to this general rule, and many experiments in the last few decades have been accompanied by activities directed at the development of appropriate numerical tools.

The results in figure 7.6 report the numerical rebuilding of an experiment [38] that aimed at evaluating the MHD effects in a Mach 6 ionized argon flow around a conical test body. The test article incorporates an arrangement of permanent magnets producing the magnetic induction field depicted in figure 7.5(a). The magnetic configuration of the test article is designed so as to have regions in which a high MHD interaction is expected. Indeed, in these regions, corresponding to the bright rings in figure 7.5(a), the magnetic induction field is roughly perpendicular to the the the flow velocity, as can be observed comparing the **B** field streamlines in figure 7.5(a) with the flow streamlines in figure 7.5(b). The electric transport properties for the calculations has been evaluated by utilizing the results of a model based on a two-term approximation of the Boltzmann equation, in the assumption of a Maxwellian distribution of the electrons [39]. Due to the high electron mobility (μ_e ranges in the order of some tens of T^{-1}) the general behavior of the phenomenon is similar to that seen in the previous illustrations. Also in this case a Hall electric field roughly opposed to the direction of the flow is observed (see the electric potential plot in figure 7.6(a)). The azimuthal currents are concentrated in the high MHD interactions regions, as shown in figure 7.6(b). The azimuthal current density, interacting with the magnetic induction field, produces the force per unit volume responsible for the modification of the flow field. Current densities, as well, are in the main responsible for the heating of the ionized gas flowing past the test article. These effects are easily recognized in figures 7.6(c) and 7.6(d), where the computed pressure field around the cone and the temperature are shown, respectively. In the interaction region, experimental measurements have shown a change in pressure induced by MHD above 10%. Good agreement has been observed between experimental data and numerical rebuilding [40, 41].

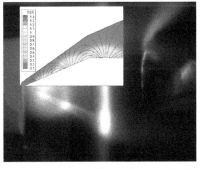

(a) Magnetic induction field distribution around the cone.

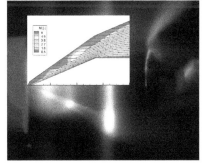

(b) Mach number and flow streamlines around the cone.

Figure 7.5. MHD experiment in an argon flow: the oblique shock is clearly visible. The bright rings around the cone correspond to the region where the MHD interaction is stronger.

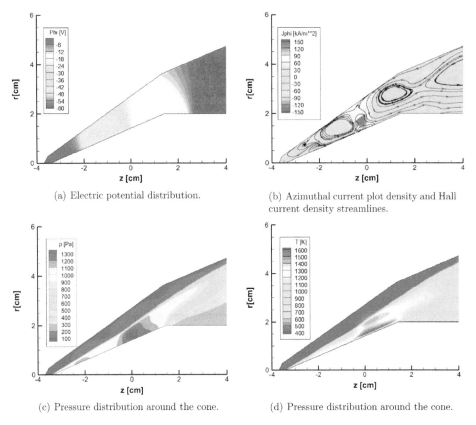

(a) Electric potential distribution.

(b) Azimuthal current plot density and Hall current density streamlines.

(c) Pressure distribution around the cone.

(d) Pressure distribution around the cone.

Figure 7.6. Numerical rebuilding of the MHD experiment.

Bibliography

[1] Rosa R J, Krueger C H and Shioda S 1991 Plasmas in MHD power generation *IEEE Trans. Plasma Sci.* **19** 1180–90

[2] Rosa R J 1987 *Magnetohydrodynamic Energy Conversion* (New York: Hemisphere)

[3] Birat J and Chone J 1983 Electromagnetic stirring on billet, bloom, and slab continuous casters: state of the art in 1982 *Ironmaking Steelmaking* **10** 269–81

[4] Vives C and Ricou R 1985 Experimental study of continuous electromagnetic casting of aluminum alloys *Metallurg. Mater. Trans.* B **16** 377–84

[5] Motora S, Takezawa S and Tamama H 1991 Development of the MHD ship YAMATO-1 *Oceans Conf. Record* vol 3 (NJ: Piscataway IEEE) pp 1636–41

[6] Yohei S, Takezawa S, Sugawara Y and Kyotani Y 1995 The superconducting MHD-propelled ship YAMATO-1 *Proc. 4th Int. Conf. and Exhibition: World Congress on Superconductivity* vol 1 pp 167–76

[7] Jahn R G 1968 *Physics of Electric Propulsion* (New York: McGraw-Hill)

[8] Ziemer R 1959 Experimental investigation in magneto-aerodynamics *Am. Rocket Soc. J.* **29** 642–7

[9] Bityurin V, Lineberry J, Potebnia V, Alferov V, Kuranov A and Kuranov A 1997 Assessment of hypersonic MHD concepts *Meeting Papers of 28th Plasmadynamics and Lasers Conf. AIAA Paper* 1997–2393

[10] Bityurin V, Vatazhin A, Gus'kov O and Kopchenov V 2004 Hypersonic flow past the spherical nose of a body in the presence of a magnetic field *Fluid Dyn.* **39** 657–66

[11] Shang J 2001 Recent research in magneto-aerodynamics *Prog. Aerospace Sci.* **37** 1–20

[12] Bityurin V, Bocharov A, Baranov D, Popov N, Alferov V and Tikhonov V 2011 Progress in MHD parachute concept study *49th AIAA Aerospace Sciences Meeting (Orlando, FL) AIAA Paper* 2011–902

[13] Fomichev V and Yadrenkin M 2013 Experimental investigation of the magnetohydrodynamic parachute effect in a hypersonic air flow *Tech. Phys.* **58** 144–7

[14] Rybak J and Churchill R 1971 Progress in reentry communications *IEEE Trans. Aerospace and Electronic Systems* **7** 879–94

[15] Mitchner M and Kruger C H J 1973 *Partially Ionized Gases* (New York: Wiley)

[16] Alfvén H and Fälthammar C G 1963 *Cosmical Electrodynamics* (Oxford: Oxford University Press)

[17] Alfvén H 1942 Existence of electromagnetic-hydrodynamic waves *Nature* **150** 405–6

[18] Goedbloed H P and Poedts 2004 *Principles of Magnetohydrodynamics* (Cambridge: Cambridge University Press)

[19] Hosking R J and Dewar R L 2016 *Fundamental Fluid Mechanics and Magnetohydrodynamics* (Singapore: Springer)

[20] Brio M and Wu C 1988 An upwind differencing scheme for the equations of ideal magnetohydrodynamics *J. Comput. Phys.* **75** 400–22

[21] Cargo P and Gallice C 1997 Roe matrices for ideal MHD and systematic construction of Roe matrices for systems of conservation laws *J. Comput. Phys.* **136** 446–66

[22] Powell K, Roe P, Linde T, Gombosi T and De Zeeuw D 1999 A solution-adaptive upwind scheme for ideal magnetohydrodynamics *J. Comput. Phys.* **154** 284–309

[23] Janhunen P 2000 A positive conservative method for magnetohydrodynamics based on HLL and Roe methods *J. Comput. Phys.* **160** 649–61

[24] Gurski K 2004 An HLLC-type approximate Riemann solver for ideal magnetohydrodynamics *SIAM J. Sci. Comput.* **25** 2165–87

[25] Miyoshi T and Kusano K 2005 A multi-state HLL approximate Riemann solver for ideal magnetohydrodynamics *J. Comput. Phys.* **208** 315–44

[26] Harten A, Engquist B, Osher S and Chakravarthy S 1987 Uniformly high order accurate essentially non-oscillatory schemes, III *J. Comput. Phys.* **71** 231–303

[27] Borghi C, Carraro M and Cristofolini A 2003 Numerical modeling of MHD interaction in the boundary layer of hypersonic flows *IEEE Trans. Magn.* **39** 1507–10

[28] Borghi C, Carraro M and Cristofolini A 2005 Analysis of magnetoplasmadynamic interaction in the boundary layer of a hypersonic wedge *J. Spacecraft Rockets* **42** 45–50

[29] Cristofolini A, Latini C and Borghi C 2011 A weighted reverse Cuthill–McKee procedure for finite element method algorithms to solve strongly anisotropic electrodynamic problems *J. Appl. Phys.* **109** 033301

[30] Matsuda A, Wakatsuki K, Takizawa Y, Kawamura M, Otsu H, Konigorski D, Sato S and Abe T 2006 Shock layer enhancement by electro-magnetic effect for spherically blunt body *Meeting Papers of 37th Plasmadynamics and Lasers Conf. AIAA Paper* 2006–3573

[31] Gülhan A, Esser B, Koch U, Siebe F, Riehmer J, Giordano D and Konigorski D 2009 Experimental verification of heat-flux mitigation by electromagnetic fields in partially-ionized-argon flows *J. Spacecraft Rockets* **46** 274–83

[32] Cristofolini A, Borghi C, Neretti G, Passaro A, Fantoni G and Paganucci F 2008 Magnetohydrodynamics interaction over an axisymmetric body in a hypersonic flow *J. Spacecraft Rockets* **45** 438–44

[33] Cristofolini A, Borghi C, Neretti G, Schettino A, Trifoni E, Battista F, Passaro A and Baccarella D 2012 Experimental investigations on the magneto-hydro-dynamic interaction around a blunt body in a hypersonic unseeded air flow *J. Appl. Phys.* **112** 093304

[34] Poggie J and Gaitonde D 2000 Magnetic control of hypersonic blunt body flow *AIAA Paper* 2000–0452

[35] Gaitonde D and Poggie J 2002 Elements of a numerical procedure for 3-D MGD flow control analysis *40th Aerospace Sciences Meeting and Exhibit, AIAA Paper* 2002-0198

[36] MacCormack R 2002 Three dimensional magneto-fluid dynamics algorithm development *38th Aerospace Sciences Meeting and Exhibit, AIAA Paper* 2002–0197

[37] D'Ambrosio D and Giordano D 2004 Electromagnetic fluid dynamics for aerospace applications. Part I: Classification and critical review of physical models *40th Aerospace Sciences Meeting and Exhibit, AIAA Paper* 2002-0198

[38] Cristofolini A, Borghi C, Carraro M, Neretti G, Passaro A, Fantoni G and Biagioni L 2008 Hypersonic MHD interaction on a conical test body with a Hall electrical connection *IEEE Trans. Plasma Sci.* **36** 530–41

[39] Colonna G and Capitelli M 2008 Boltzmann and master equations for MHD in weakly ionized gases *J. Thermophys. Heat Transfer* **22** 414–23

[40] Cristofolini A, Borghi C and Colonna G 2007 Numerical analysis of the experimental results of the MHD interaction around a sharp cone *Meeting Papers of the 38th AIAA Plasmadynamics and Lasers Conf. AIAA Paper* 2007–4252

[41] Cristofolini A, Neretti G and Borghi C 2012 Plasma parameters and electromagnetic forces induced by the magneto hydro dynamic interaction in a hypersonic argon flow experiment *J. Appl. Phys.* **112** 033302

IOP Publishing

Plasma Modeling
Methods and Applications
Gianpiero Colonna, Lucia Daniela Pietanza and Giuliano D'Ammando

Chapter 8

Self-consistent kinetics

An important aspect of modeling weakly ionized gases is to describe the chemical transformations occurring in the system. Thermal and non-thermal plasmas are commonly used in material processing [1], etching [2], deposition [3], waste treatment [4–6], combustion [7–9], pollutant reduction [10], and so on, applications where chemistry plays a fundamental role [11].

In non-thermal plasmas, the understanding of chemical kinetics is not a trivial issue, because of non-equilibrium conditions: commonly, the gas temperature is low, usually $T_g < 1000$ K, while the electron temperature, T_e, is above 10 000 K. The population of internal states is determined, in a first approximation, by the balance between excitation induced by free electron collisions, quenching by heavy particle interaction, and radiative decay. Another contribution comes from chemical reactions which can have a preferential path through excited states [12, 13]. On the other hand, the electron energy distribution is strongly influenced by the population of excited levels, in particular vibrational and long-living metastable states, not decaying by radiative emission [14, 15].

This scenario demonstrates the complex interaction among the different plasma components, in particular, the interplay between heavy-particle internal levels and free electrons. To describe these features, the *self-consistent* approach must be used, determining, at the same time, the chemical composition, the internal level, and the free electron energy distributions [16], the latter calculated by solving the Boltzmann equation, as described in chapter 2. The main difficulty of this approach is the size of the chemical problem. For diatomic molecules, such as N_2, the number of vibrational levels in the ground electronic state is of the order of one hundred, while for triatomic molecules, such as CO_2, this number reaches the order of ten thousand [17].

Level kinetics should also be coupled with the radiative transport equation. Part of the emitted photons are reabsorbed, exciting atoms and molecules, or inducing stimulated emission, and the speed of these processes depends linearly on the density of photons. Usually, collisional–radiative (CR) models avoid the solution of the radiative transport equation by introducing the *escape factor* [18–20], the fraction of emitted photons reabsorbed locally. This approach cannot account for some interesting phenomena observed in non-homogeneous plasmas, such as the modification of the line shape,[1] a consequence of the reabsorption of photons far from the emission zone, introducing *non-local effects* that can be described only by solving the radiative transport equation. This aspect is relevant not only to optical emission spectroscopy, but also to vehicles entering planetary atmospheres, where part of the global heat flux to the vehicle surface is due to radiation, especially in regions where the conductive and convective fluxes are small [22, 23].

The *line-by-line* solution of the photon transport equation is a highly demanding computational problem due to the large number of wavelength sampling points to be considered. To reduce the computational load for calculating radiative properties, a *multi-group* approach can be used [24], considering a limited number of spectral intervals where emissivity and absorbance are estimated.

This chapter introduces the reader to the state-to-state (StS) approach, providing at the same time the fundamentals of the self-consistent coupling and showing its application in modeling discharges and high-enthalpy flows.

8.1 The state-to-state approach

The StS kinetics was born together with computer modeling of cold plasmas [25–29], when it was understood that some behaviors could be explained only by considering non-Boltzmann internal level distributions. The StS approach was also applied in modeling high-enthalpy flows [30–33]. The peculiarity of the StS approach is the inclusion in the kinetic model of equations describing the evolution of internal levels, considered as independent chemical species, under the action of internal transitions[2] and chemical reactions. To extract global information on an internal degree of freedom from its level distribution χ_v, it is sufficient to calculate the weighted mean of a given state-specific quantity Q_v as

$$\langle Q \rangle = \sum_{v=0}^{v_M} Q_v \chi_v,$$ (8.1)

with the normalization condition

$$\sum_{v=0}^{v_M} \chi_v = 1,$$ (8.2)

where $v = 0, v_M$ are the ground and the last level of the considered internal degree of freedom. As an example, equation (8.1) can be used to calculate the internal energy by putting $Q_v = \varepsilon_v$, the energy of the vth level.

---

[1] As extreme condition of line-shape modification is the double peak line [21], due to emission from hot plasma and absorption in a cold region.
[2] In the framework of StS kinetics, internal transitions are equalized to chemical reactions.

For non-Boltzmann distributions, the internal temperature loses its meaning, even if it remains one of the measured quantities in plasma experiments. In non-equilibrium conditions, different definitions of temperature are possible. The most commonly used is the so called *first-level temperature* $T_{1,0}$, calculated as the temperature of the Boltzmann function connecting the ground and the first excited states

$$T_{1,0} = \frac{\varepsilon_1 - \varepsilon_0}{\ln\left(\dfrac{\chi_0/g_0}{\chi_1/g_1}\right)}. \tag{8.3}$$

This definition is based on the fact that low energy levels usually follow a Boltzmann distribution. Equation (8.3) can be generalized for any couple of levels by defining $T_{v,\omega}$ as the temperature of the Boltzmann distribution connecting the vth and ωth levels. An improvement of the previous definition can be obtained by modeling a group of levels with a Boltzmann distribution and considering the internal temperature as a fitting parameter. The definitions of temperature given above are adequate to compare numerical calculations with values measured by spectroscopy. In some experiments, temperature is obtained from the measurement of thermodynamic quantities,[3] in this case, a suitable value can be estimated by solving the following algebraic non-linear equation:

$$E_{\text{int}} = \sum_{v=0}^{v_M} \varepsilon_v \chi_v = \sum_{v=0}^{v_M} \varepsilon_v g_v e^{-\frac{\varepsilon_v - \varepsilon_0}{T_{\text{int}}}} \bigg/ \sum_{v=0}^{v_M} g_v e^{-\frac{\varepsilon_v - \varepsilon_0}{T_{\text{int}}}}, \tag{8.4}$$

where E_{int} is the mean energy of the non-equilibrium distribution and T_{int} is the temperature corresponding to a Boltzmann distribution having the same mean energy as the actual distribution. This definition gives a different value for the internal temperature than equation (8.3). If the energy of excited states is very high, as in noble gases or atomic hydrogen, the temperature calculated by equation (8.4) is weakly sensitive to the upper level population, because the contribution of these states to the mean energy is very small except at very high temperature. On the other hand, equation (8.3) can give negative temperatures in the case of population inversion, and therefore without the possibility of relating internal energy, always positive, to the corresponding temperature.

Another useful quantity connected to data measured in experiments is the global rate of a given process, calculated by extending equation (8.1) to a generic reaction

$$R_1(v_1) + \cdots + R_n(v_n) \xrightarrow{K_{v_1,\ldots,v_n}^{\omega_1,\ldots,\omega_m}} P_1(\omega_1) + \cdots + P_m(\omega_m). \tag{8.5}$$

It should be noted that the above reaction in fact represents a whole family of processes obtained by varying the internal levels $v_1, \ldots, v_n$ and $\omega_1, \ldots, \omega_m$ of the reactants and products. All these elementary processes contribute to the same global chemical reaction

---

[3] Shadowgraphy, based on the variation of density, can be considered a thermodynamic measurement.

$$R_1 + \cdots + R_n \xrightarrow{K_p} P_1 + \cdots + P_m, \tag{8.6}$$

whose global rate is obtained by weighting over the reactant distributions and summing over the states of the products:

$$K_p = \sum_{v_1} \cdots \sum_{v_n} \chi_{R_1,v_1} \cdots \chi_{R_n,v_n} \sum_{\omega_1} \cdots \sum_{\omega_m} K^{\omega_1, \ldots, \omega_m}_{v_1, \ldots, v_n}. \tag{8.7}$$

The same expression can be adapted to calculate the rate of internal energy variation of the Sth chemical species involved in the set of elementary processes described by equation (8.5):

$$K^{en}_{p,S} = z_S \sum_{v_1} \cdots \sum_{v_n} \chi_{R_1,v_1} \cdots \chi_{R_n,v_n} \sum_{\omega_1} \cdots \sum_{\omega_m} K^{\omega_1, \ldots, \omega_m}_{v_1, \ldots, v_n} \varepsilon_{S,v_S}, \tag{8.8}$$

where v_s is the index for the internal state, $z_s = -1$ for reactants, and $z_s = 1$ for products.

8.2 The collisional–radiative model

The first example of StS kinetics is the CR model developed by Bates in the 1960s [34, 35] to describe the spectra of atomic hydrogen and helium in conditions where the Saha equation [36] failed. The simplest version of the CR model considers a plasma containing an atomic species $\mathcal{A}$, its parent ion $\mathcal{A}^+$, and electrons e^-. Only electron impact collisional processes are considered, in particular excitation/de-excitation

$$e^- + \mathcal{A}(i) \underset{K_{j,i}}{\overset{K_{i,j}}{\rightleftharpoons}} e^- + \mathcal{A}(j), \tag{8.9}$$

and ionization/three-body recombination

$$e^- + \mathcal{A}(i) \underset{K^{rec}_i}{\overset{K^{ion}_i}{\rightleftharpoons}} e^- + \mathcal{A}^+ + e^-. \tag{8.10}$$

The collisional processes must fulfill the detailed balance principle, therefore the stationary solution for internal level population is the Boltzmann distribution. However, by including radiative decay

$$\mathcal{A}(i) \xrightarrow{A'_{i,j}} \mathcal{A}(j) + h\nu \tag{8.11}$$

a source of non-equilibrium is introduced, since it is not balanced by the reverse process of radiation reabsorption, resulting in an irreversible energy loss outside the boundaries of the system. The usual approximation to model the reabsorption of radiation is to introduce an *escape factor* $\vartheta_{i,j}$ for each transition, with $0 \leqslant \theta_{ij} \leqslant 1$:

$$A'_{i,j} = A_{i,j}\vartheta_{i,j}, \tag{8.12}$$

which reduces the spontaneous emission Einstein coefficients $A_{i,j}$ by the fraction of photons reabsorbed very close to the emission position. For each transition, the $\vartheta_{i,j}$

coefficient is a function of the emission line shape, the source geometry, the optical depth and the space distribution of emitting atoms [37, 38]. Another important process to be included in the CR model is the radiative recombination

$$A^+ + e^- \xrightarrow{\beta_i} A(i) + h\nu, \tag{8.13}$$

where β_i is corrected by its own escape factor, as in equation (8.12). Radiative recombination is responsible for the continuum spectrum, together with Bremsstrahlung radiation. In the traditional approach to the CR model, two assumptions are considered:

1. The electron energy distribution is Maxwellian.
2. The level distribution relaxes faster than the electron density, therefore the quasi-steady-state (QSS) distribution can be considered, and the time evolution is calculated only for the ground state and electron (and ion) densities.

As a consequence of these hypotheses, the electron impact rate coefficients are a function of T_e, while the electron density can be considered constant for the level kinetics.

In a two-level system, i.e. the ground and first excited level, undergoing only electron-impact induced and radiative processes in equations (8.9) and (8.11), the kinetic equations for the atomic level population are given in matrix form

$$\frac{d}{dt}\begin{pmatrix} n_1 \\ n_2 \end{pmatrix} = \begin{vmatrix} -K_{1,2}n_e & K_{2,1}n_e + A'_{2,1} \\ K_{1,2}n_e & -K_{2,1}n_e - A'_{2,1} \end{vmatrix} \begin{pmatrix} n_1 \\ n_2 \end{pmatrix}, \tag{8.14}$$

where n_1, n_2, n_e are particle densities of ground state (1), excited level (2), and electrons (e), respectively, while $K_{1,2}$ and $K_{2,1}$ are the rate coefficients for the $1 \to 2$ and $2 \to 1$ electron impact induced transitions, which are a function of T_e. Assuming constant n_e and T_e, it becomes a linear problem that can be solved by calculating the eigenvalues (λ) and eigenvectors ($\mathbf{v}$) of the matrix. With simple math, we obtain the following sets:

$$\lambda_1 = 0$$
$$\mathbf{v}_1 = \frac{1}{K_{1,2}n_e + K_{2,1}n_e + A'_{2,1}} \begin{pmatrix} K_{2,1}n_e + A'_{2,1} \\ K_{1,2}n_e \end{pmatrix} \tag{8.15}$$

$$\lambda_2 = -\left(K_{1,2}n_e + K_{2,1}n_e + A'_{2,1}\right)$$
$$\mathbf{v}_2 = \begin{pmatrix} 1 \\ -1 \end{pmatrix}, \tag{8.16}$$

which give the general solution of the differential equation

$$\begin{pmatrix} n_1 \\ n_2 \end{pmatrix} = C_1\mathbf{v}_1 e^{\lambda_1 t} + C_2\mathbf{v}_2 e^{\lambda_2 t} = C_1\mathbf{v}_1 + C_2\mathbf{v}_2 e^{-\left(K_{1,2}n_e + K_{2,1}n_e + A'_{2,1}\right)t}, \tag{8.17}$$

where C_1 and C_2 are constants that should be determined from the initial composition. From this simple problem some general properties of the solution can be deduced, which are extendable to multi-level atoms:

1. There is always the null eigenvalue and its corresponding eigenvector is the stationary solution. This property assures the existence of a stationary value.
2. All the other eigenvalues are negative. As a consequence, a system will always evolve to the stationary solution.
3. The sum of the elements of the eigenvectors corresponding to the negative eigenvalue is null. This means that the sum of levels is constant along the whole evolution.

These properties also have consequences on the numerical solution of the general problem. However, for an atom with n_{lev} levels, there are $n_{lev} - 1$ negative eigenvalues, usually different by orders of magnitude. Therefore, the numerical solution should be determined using an algorithm adequate for problems with large stiffness.

In the case of the two-level system, we can understand if the stationary solution is an equilibrium Boltzmann distribution or not. This can be done by calculating the ratio between the two elements of the first eigenvector, i.e.

$$\frac{n_1}{n_2} = \frac{K_{2,1}n_e + A'_{2,1}}{K_{1,2}n_e} = \frac{K_{2,1}}{K_{1,2}} + \frac{A'_{2,1}}{K_{1,2}n_e}. \tag{8.18}$$

The first fraction on the right-hand side is the ratio between the rates of direct and reverse electron-induced processes. Due to the detailed balance principle,

$$\frac{K_{2,1}}{K_{1,2}} = \frac{g_1}{g_2}e^{(\varepsilon_2-\varepsilon_1)/T_e} = \frac{n_1^B(T_e)}{n_2^B(T_e)}, \tag{8.19}$$

where ε_i is the energy of the ith level, g_i the corresponding statistical weight, and n_i^B the Boltzmann distribution. The second fraction on the right-hand side of equation (8.18) represents the deviation from the equilibrium distribution due to the radiation emission, which reduces the population of the excited state, i.e. $n_2 < n_2^B$. If the plasma is thick ($A'_{2,1} \approx 0$) or if the electron density is high enough ($n_e \gg A'_{2,1}/K_{1,2}$), the radiative loss is negligible with respect to the up-pumping of the excited level and therefore the equilibrium condition is reached.

In the previous example, we have neglected ionization/recombination (equation (8.10)) processes, as well as radiative recombination (equation (8.13)). In the QSS approximation, we do not need the equation for the ground state, but only that for the excited state, i.e.

$$\frac{dn_2}{dt} = -\left(K_{2,1}n_e + A'_{2,1} + K_2^{ion}n_e\right)n_2 + K_{1,2}n_1n_e + \left(K_2^{rec}n_e + \beta_2\right)n_e n_{A^+}, \tag{8.20}$$

where, by considering n_1, n_e, and n_{A^+} constant, the following general solution is obtained:

$$n_2(t) = Ce^{-\left(K_{2,1}n_e + A'_{21} + K_2^{\text{ion}}n_e\right)t} + \frac{K_{1,2}n_1n_e + \left(K_2^{\text{rec}}n_e + \beta_2\right)n_e n_{A^+}}{K_{2,1}n_e + A'_{2,1} + K_2^{\text{ion}}n_e}, \tag{8.21}$$

with C a constant determined by the initial condition. The stationary solution is given by

$$n_2 = \frac{K_{1,2}n_1n_e + \left(K_2^{\text{rec}}n_e + \beta_2\right)n_e n_{A^+}}{K_{2,1}n_e + A'_{2,1} + K_2^{\text{ion}}n_e}. \tag{8.22}$$

Let us analyze the solution in the case where n_e is much lower than the equilibrium value, so that the recombination terms can be neglected. We obtain

$$\frac{n_1}{n_2} = \frac{K_{2,1}n_e + A'_{2,1} + K_2^{\text{ion}}n_e}{K_{1,2}n_e} = \frac{n_1^B}{n_2^B} + \frac{A'_{2,1}}{K_{1,2}n_e} + \frac{K_2^{\text{ion}}}{K_{1,2}}, \tag{8.23}$$

showing that the excited state population is further reduced with respect to equation (8.18) by a term proportional to the ionization rate from the excited state.

When the electron density is high enough to neglect the radiative terms, considering the detailed balance for the ionization equilibrium

$$K_i^{\text{rec}} = \frac{K_i^{\text{ion}}}{K_{\text{eq}}'} \frac{n_2^B}{n_1^B}, \tag{8.24}$$

where K_{eq}' is the equilibrium constant for the two level system, from equation (8.22) we have that the population of the excited state is given by

$$\frac{n_2}{n_1} = \frac{n_2^B}{n_1^B}\left[1 + \frac{K_2^{\text{ion}}n_2^B}{n_1^B K_{1,2} + K_2^{\text{ion}}n_2^B}\left(\frac{n_e n_{A^+}}{n_1 K_{\text{eq}}'} - 1\right)\right]. \tag{8.25}$$

According to this equation, the internal distribution shows a deviation from the Boltzmann one, whose sign depends on the departure from equilibrium of the ionization. If the electron density is higher than the equilibrium value ($n_e n_{A^+} > n_1 K_{\text{eq}}'$), the excited state is overpopulated with respect to the Boltzmann distribution. On the other hand, if electron density is lower than the equilibrium value ($n_e n_{A^+} < n_1 K_{\text{eq}}'$), the excited state is underpopulated. These two conditions are defined as ionization and recombination regimes, respectively. Equation (8.25) can easily be extended by adding the radiative terms

$$\frac{n_2}{n_1} = \frac{n_2^B}{n_1^B}\left[1 + \frac{K_2^{\text{ion}}n_2^B\left(\frac{n_e n_{A^+}}{n_1 K_{\text{eq}}'} - 1\right) + \beta_2 n_{A^+} - \frac{A'_{2,1}n_2^B}{n_e n_1^B}}{n_1^B K_{1,2} + K_2^{\text{ion}}n_2^B + \frac{A'_{2,1}n_2^B}{n_e n_1^B}}\right]. \tag{8.26}$$

Therefore, the radiative recombination gives a positive contribution, while that of spontaneous emission is negative, confirming the result obtained in equation (8.23).

In a multi-level system, the kinetic equations are more complex, because all the transitions between the levels give a contribution. However, the general behavior discussed in the two-level system is still valid, because the main contribution usually comes from the direct interaction with the ground state and the ionization. Therefore, this picture can be considered as a first-order approximation of the CR problem.

It should be pointed out that the deviation from the Boltzmann distribution is not the same for all the levels, depending on the state-specific rates or cross sections. Fixing T_e, n_e, and n_0, we have a linear set of equations [39].

As an example, figure 8.1 shows the time evolution of hydrogen atom level distribution in (a) ionization and (b) recombination regimes, calculated in the conditions reported in table 8.1 at atmospheric pressure. In the ionization regime, the distribution tails are depleted with respect to a Boltzmann distribution, while in the recombination regime they are overpopulated and plateaux appear.

Table 8.1. Conditions for the calculations reported in figure 8.1. χ_e is the electron molar fraction, T_e the electron temperature and $T_H(0)$ the initial temperature for the H internal distribution.

	Case (a)	Case (b)
χ_e	10^{-3}	10^{-3}
T_e	20 000 K	1000 K
$T_H(0)$	1000 K	20 000 K

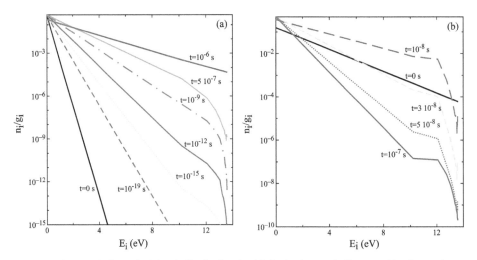

Figure 8.1. Time evolution of H level distribution in (a) ionization and (b) recombination regimes at atmospheric pressure. The conditions for temeratures and electron molar fraction are listed in table 8.1.

8.3 Vibrational kinetics

The StS approach has also been used to investigate the evolution of the vibrational distribution of diatomic molecules. For homonuclear molecules, such as N_2, O_2, and H_2, vibrational kinetics allows one to theoretically determine the vibrational distribution of the ground electronic state, experimentally observed only by sophisticate techniques [40], as the radiative vibrational transitions are optically forbidden. StS models were used to predict the behavior of high-enthalpy flows in hypersonic wind tunnels [41, 42], shock tubes [43], and during the entry of space vehicles in planetary atmospheres [44–46]. In these conditions, the relaxation of vibrational levels to equilibrium passes through states characterized by non-Boltzmann distributions. If an N_2/N mixture is considered, all internal transitions

$$N_2(v) + N_2(\omega) \rightleftharpoons N_2(v') + N_2(\omega') \tag{8.27}$$

are allowed and any quadruplet (v, ω, v', ω') is possible. However, many of these processes have a very small probability and only monoquantum transitions are usually considered[4]:

$$N_2(v) + N_2(\omega - 1) \rightleftharpoons N_2(v - 1) + N_2(\omega) \quad VV \tag{8.28}$$

$$N_2(v) + N_2(\omega) \rightleftharpoons N_2(v - 1) + N_2(\omega) \quad VTm. \tag{8.29}$$

VV rate coefficients are written as $K_{v,v-1}^{\omega-1,\omega}(T)$ as a function of the temperature, with $v < \omega$, while for $v > \omega$ the detailed balance principle is applied to obtain the rate $K_{\omega-1,\omega}^{v,v-1}(T)$. The case with $v = \omega$, known as resonant transfer, has no global effect, with the final and initial states indistinguishable. In the presence of atoms, the vibrational de-activation/excitation in atom–diatom collisions is also considered:

$$N_2(v) + N \rightleftharpoons N_2(v - \Delta v) + N \quad VTa. \tag{8.30}$$

Differently to VTm, the VTa process also includes multi-quantum transitions. The kinetic scheme is completed with the dissociation–recombination processes[4]

$$\begin{aligned} N_2(v) + N_2(\omega) &\rightleftharpoons 2N + N_2(\omega) \quad DRm \\ N_2(v) + N &\rightleftharpoons 2N + N \quad DRa. \end{aligned} \tag{8.31}$$

It should be noted that, varying v, ω over the vibrational ladder, the above reactions represent a few thousands elementary processes!

There are some aspects that should be considered in developing an StS kinetic model:

- The vibrational levels follow an anharmonic law, with the energy quanta decreasing as the levels approach the dissociation limit.
- The construction of a reliable, accurate, and consistent database of rate coefficients and cross sections for the huge number of reactive channels

---

[4] In VTm and DRm (equation (8.31)) processes, the rate coefficient is usually considered independent of ω and therefore the state of the second molecule is omitted.

usually included in the StS vibrational kinetics is, even nowadays, a challenging problem. Semiclassical approaches are often used, with quantum calculations performed in few cases as an accuracy check. In some cases, heuristic analytical expressions are used, with a limited validity.

- Another problem is the matching of data coming from different sources. The rate coefficients are reported as a function of the vibrational quantum number. Researchers often neglect the fact that the energy of the level, corresponding to a given vibrational quantum number, is not the same for the different sources, using the data without any adaption. Scaling relations should be used to adapt the different energy level sets. The most common procedure is to interpolate the data from one set to the other.

A sufficient condition to have a Boltzmann distribution at any time of the evolution is the following:

1. the vibrational ladder is given by the truncated harmonic oscillator;
2. only mono-quantum transitions are included;
3. VT rates grow linearly with the vibrational quantum number, i.e. $K_{v \to v-1}^{VT} = vK_{1 \to 0}^{VT}(T)$.

These conditions are not fulfilled by real molecules.

VV transitions, negligible for harmonic oscillators, are an important source of non-equilibrium for the anharmonic ladder. This process conserves the number of quanta but not the total energy. For a cooling gas, in the presence of VV only, the final solution is a distribution with the same number of quanta as the initial one, but with the minimum total vibrational energy. This behavior can be explained easily by considering the detailed balance principle for VV in equation (8.28),

$$K_{v,v-1}^{\omega-1,\omega} = K_{\omega,\omega-1}^{v-1,v} e^{-\frac{\varepsilon_\omega + \varepsilon_{v-1} - \varepsilon_{\omega-1} + \varepsilon_v}{T}} = K_{\omega,\omega-1}^{v-1,v} e^{-\frac{\Delta_\omega - \Delta_v}{T}}, \tag{8.32}$$

where $\Delta_v = \varepsilon_v - \varepsilon_{v-1}$. For a harmonic oscillator, $\Delta_v = \Delta_\omega$ with the consequence that direct and reverse rates are always equal, independent of the temperature, giving the Boltzmann distribution as stationary solution. On the other hand, for real molecules, with Δ_v a decreasing function of v, when $\Delta_v > \Delta_\omega$ the process $v \to v-1, \omega-1 \to \omega$ is preferred with respect to its reverse, resulting in a parabolic-like distribution, first proposed by Treanor [47]. When both the VV and VT processes are present, during the time evolution the distribution is characterized by an S-shape, generated by the interplay between the two processes.

In figure 8.2, the cooling of the N_2 vibrational distribution under the action of (a) VV and (b) $VV + VTm$ is shown. From inspection of the figure, inclusion of VTm processes drives the final distribution toward the Boltzmann one at the gas temperature, in a time scale of seconds.

Dissociation and recombination are another cause of non-equilibrium, as these processes favor highly excited states, which are underpopulated in the dissociation regime and overpopulated in the recombination one [48].

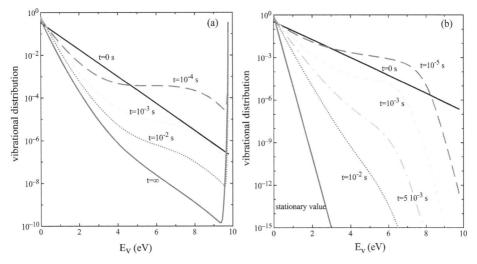

Figure 8.2. Time evolution of N_2 vibrational distribution, with initial vibrational temperature $T_v(0) = 8000$ K, for $P = 1$ atm and $T_{gas} = 1000$ K, calculated by considering only (a) VV and also including (b) VTm processes.

As an example, let us estimate the behavior of the tail of the vibrational distribution in the recombination regime. Let us consider that the gas in equilibrium at temperature T_0 instantaneously changed to $T < T_0$. For each vibrational level, the following evolution equation can be written:

$$\frac{dN_v}{dt} = G_v - L_v, \tag{8.33}$$

where N_v is the concentration of molecules in the v state and G and L are, respectively, total gains and losses. As a general assumption, the VV processes (equation (8.28)) can be neglected [48]. Moreover, we will consider the evolution of the vibrational distribution at fixed dissociation degree, supposing that the distribution tail relaxes faster than the dissociation degree. The system is cooling and the pumping from lower states can be neglected. The pumping from higher states can be neglected as well, because the level population is usually lower. As a consequence, the *gain* term is only due to recombination

$$G_v = N_N^2 \left(K_v^{Rm} N_{N_2} + K_v^{Ra} N_N \right) = K_v^r N_{tot} N_N^2, \tag{8.34}$$

where N_S is the global concentration of the Sth species. Losses come from all the processes, apart from VV, starting from the v level, specifically vT and dissociation

$$L_v = N_v \left[\left(K_v^{Dm} + \sum_{\omega \neq v} K_{v,\omega}^{VTm} \right) N_{N_2} + \left(K_v^{Da} + \sum_{\omega \neq v} K_{v,\omega}^{VTa} \right) N_N \right] = K_v^d N_{tot} N_v. \tag{8.35}$$

K_v^r and K_v^d are global constants that depend on the temperature and on the composition, not on the distribution, and N_{tot} is the total gas density. Substituting in equation (8.33) the vibrational distribution and the dissociation degree

$$\chi_v = \frac{N_v}{N_{N_2}}; \quad \alpha_d = \frac{N_N}{N_N + N_{N_2}} = \frac{N_N}{N_{tot}}; \quad \frac{N_{N_2}}{N_{tot}} = 1 - \alpha_d, \tag{8.36}$$

we obtain

$$\frac{d\chi_v}{dt} = N_{tot}^2 \frac{\alpha_d^2}{1 - \alpha_d} K_v^r - N_{tot} K_v^d \chi_v = a - b\chi_v, \tag{8.37}$$

with a, b constants, i.e. independent of the distribution. The solution of this equation, neglecting the dependence on time of composition and temperature, is

$$\chi_v(t) = \left(\chi_v^0 - \frac{a}{b}\right)e^{-bt} + \frac{a}{b} = \left(\chi_v^0 - \chi_v^\infty\right)e^{-bt} + \chi_v^\infty, \tag{8.38}$$

where χ_v^0 is the initial distribution and χ_v^∞ is the QSS solution

$$\chi_v^\infty = \frac{a}{b} = N_{tot} \frac{\alpha_d^2}{1 - \alpha_d} \frac{K_v^r}{K_v^d}. \tag{8.39}$$

It should be noted that K_v^r and K_v^d are linear functions of α_d and therefore the QSS value for a given level is a rational function of the type $\chi_v^\infty \propto P_3(\alpha_d)/P_2(\alpha_d)$ where P_n is a generic polynomial of degree n. Differently to the collisional radiative model, the QSS value of equation (8.39) could not be considered as a global solution because the approximations considered are not valid for all the levels, but only for those with high v. In particular, the equation is very accurate for the last vibrational level v_M. Global dissociation rates, as well as those of endothermic processes, are favored from highly excited states, and therefore they are not related to the vibrational temperature, but to the population of the last vibrational level. When QSS condition is reached, considering that the population of the last vibrational level is a function of the dissociation degree (see equation (8.39)), the global dissociation rate coefficient is directly related to α_d. From these considerations, a simplified model, called the *two-level distribution* has been applied to pure nitrogen [49] and air [50]. An alternative, the *multi-internal temperature* model, was proposed [51, 52], which divides the vibrational distribution in a given number of Boltzmann distributions with their own temperature, each one describing the population in a given energy interval. These models can be regarded as the evolution of the traditional multi-temperature approach [53, 54], which considers a single temperature for each internal degree of freedom, and the chemical rate coefficients as analytical functions of the gas and relevant internal temperatures.

One difficulty of the StS model is the lack of state-specific rates, in particular for molecule–molecule interaction. To overcome this problem, some heuristic approaches are used, such as SSH theory [55] or the forced harmonic oscillator approach [56]. An interesting example is the dissociation (and the relative recombination) by molecule–molecule collisions, whose state-selective rates are unknown. Commonly, these rates are estimated by the so-called *ladder-climbing* approximation, considering the dissociation as a virtual level ($v_d = v_M + 1$) instantaneously dissociating:

$$N_2(v) + N_2(\omega) \xrightarrow{\text{slow}} N_2(v') + N_2(v_d) \xrightarrow{\text{fast}} N_2(v') + 2N. \qquad (8.40)$$

The rate is estimated extending the dependence of the vibrational excitations to the virtual level. This approach underestimates the dissociation rates by orders of magnitude [57]. Another approach assumes the same dependence on T and v of the dissociation induced by atomic collisions, multiplied by a factor depending on the temperature to reproduce the experimental thermal rate [49].

8.4 The self-consistent approach

The previous sections offer only a partial view of the possible applications of the StS models. To characterize realistic systems, vibrational kinetics for molecules and level kinetics for atoms must be included in the same chemical model [58–60], together with electronically excited states of molecular species [14, 16], single and multiple ionized atoms and molecules [61], and polyatomic species [62, 63], interacting through relevant processes. Among these, electron-induced processes play an important role, not only during discharges, but also in post-discharge and in high-enthalpy flows. Electron collisions with atoms and molecules in excited states can strongly affect the kinetics, introducing different mechanisms, such as:

- the ladder-climbing mechanism for dissociation and ionization, with successive excitation from one level to the next up to the highest level which can be easily dissociated or ionized [62–64];
- the excited level gives its energy to the impinging electron that acquires enough energy to induce high-threshold chemical processes [65–67].

This last class of processes (called super-elastic or second-kind collisions) produces non-equilibrium electron energy distributions, with bumps and plateaux. These examples illustrate how complex the link between electrons and excited states can be in affecting the chemistry of a plasma and its relaxation during post-discharge [68]. These mechanisms are not accounted for when the E/N approach[5] is used. In this case, the electron-impact rate coefficients are pre-calculated for different values of the reduced electric field from the electron energy distribution obtained by solving the stationary Boltzmann equation and neglecting the contribution of excited states [69–71]. This approach is justified under the assumption that in the presence of strong electric fields, electrons reach a steady state in a very short time and the contribution of super-elastic collisions is negligible. This approximation is sometimes also extended to post-discharge conditions, though it has been well assessed that super-elastic collisions significantly warm up the distribution of electrons when the electric field is switched off [39].

---

[5] This approach is also known as the *cold gas approximation*.

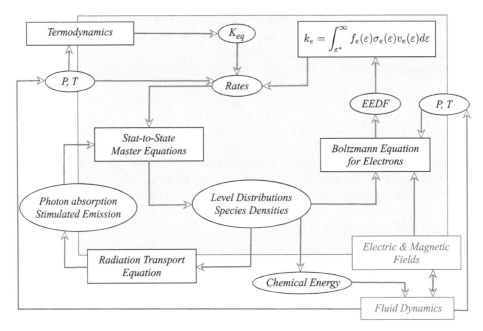

Figure 8.3. A scheme of the self-consistent approach. The self-consistent cycle is in red. Blue lines indicate information coming from external modules. Purple lines show information passed to external modules.

To properly account for the synergy between electrons and excited states, the self-consistent approach must be used. The general scheme is sketched in figure 8.3, where the self-consistent kinetics is represented by the red-arrow's path. The electron energy distribution function (EEDF) is calculated solving the Boltzmann equation (see chapter 2), which needs the gas composition and the internal distributions. From the EEDF, electron impact rate coefficients are calculated, input data for the system of master equations, whose solution updates composition and internal distributions for the next cycle. This procedure should be followed at any time step, and, depending on the solution algorithm, also more than one time.

Moreover, the self-consistent kinetics could be coupled to the fluid dynamic and/or field equations. At the same time, the radiation transport equation, which in principle is a Boltzmann equation for photons, is solved to determine the rates of processes involving photons, such as stimulated emission, absorption, photo-ionization, and photo-dissociation. To determine the bremsstrahlung radiation and radiative recombination, the radiation transport equation also needs the EEDF, being processes involving free electrons.

Shock tubes have been used since the beginning of space exploration to simulate the extreme conditions space vehicles experience when entering the atmosphere of a planet [72–74]. Some calculations have been performed for hydrogen [75] and helium–hydrogen mixtures [58] in order to simulate Jupiter entry conditions, coupling the StS self-consistent model with the fluid dynamic equations for the shock tube and with the equation of radiation transport. The chemical model [75] includes the vibrational kinetics for molecules, as discussed in section 8.3, the CR model, section 8.2, completed by processes involving electronically excited states of molecules, dissociation, and ionization induced by electron collisions. Let us analyze the evolution of species and the distributions relative to the shock front at Mach 20. A fundamental role in initiating the ionization is played by atom–atom collisions in the region just after the shock, where the temperature is very high and the electron and ion densities are still very low.

Figure 8.4 shows the spatial profiles of the plasma composition and temperatures as a function of the distance from the shock front. The vibrational temperature of molecules and the electronic temperature of atoms are calculated from internal distribution using equation (8.3), while the electron temperature is obtained from the mean energy (2.11) by $T_e = \frac{2}{3}\varepsilon_e$. Just after the front shock, the pressure suddenly increases from about 100 Pa to 0.5 atm, while the gas temperature T reaches about 24 000 K, mainly heating the translational degree of freedom, with the internal distribution and the chemical composition remaining those of the free stream condition.

As the distance from the shock front increases, collisions between molecules first promote vibrational excitation, and, after dissociation, reduce the gas temperature, as can be deduced from the evolution of the vibrational temperature of H_2 and from the atomic molar fraction. At the same time, the free electron molar fraction also increases, activating processes induced by electron collisions. The electron molar fraction presents a maximum, a behavior that can be reproduced only by using an StS approach. The energy stored in the internal degrees of freedom is released slowly and therefore the ionization process can proceed, even if the gas temperature is not sufficient to sustain it. When the inertia due to internal states is exhausted, electrons and ions recombine. Another anomalous behavior can be observed in the internal temperature of atoms, which presents two maxima and a minimum. The first maximum and minimum are produced by internal collisions trying to equilibrate with the translational temperature. The internal temperature remains a little lower than T because of the radiative losses. However, at about 3 cm from the shock front the gas temperature is still decreasing, while T_H starts increasing and then decreases

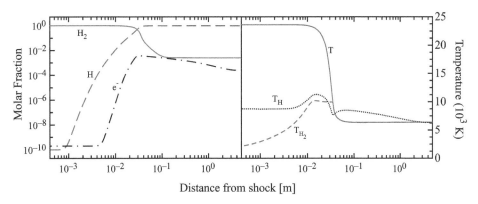

Figure 8.4. Spatial profiles of molar fractions and temperatures as a function of the distance from the shock for free stream velocity Mach = 20. T_H is the vibrational temperature of molecules, T_H the electronic temperature of atoms, and T the gas temperature.

again to equilibrate with T. This behavior, demonstrated in figure 8.5, can be explained by considering that the recombination $H^+ + e^-$ is favored on excited states, increasing the internal temperature.

The evolution of the gas temperature and compositions is reflected in the EEDF and internal distributions (figure 8.6). During the relaxation, the EEDF presents a non-Maxwellian shape, due to super-elastic collisions with the first excited state of atomic hydrogen, showing a bump ($x = 2$ cm) and a plateau ($x = 10$ cm) at an energy of about 10 eV. The level distribution presents the characteristic shape observed in the ionization regime up to $x = 2$ cm and the recombination regime for larger distances. Comparing these results with those shown in figure 8.1 for the CR model, a similar trend can be observed.

Focusing again on figures 8.4 and 8.5, we should note that the temperature of atomic hydrogen just after the shock front is quite high, about 9000 K, a result that cannot be explained by the CR models, but only as non-local effects due to radiation emitted in regions where the atomic hydrogen has a large internal temperature and absorbed in the region just after the shock. Therefore the plasma behaves as if it

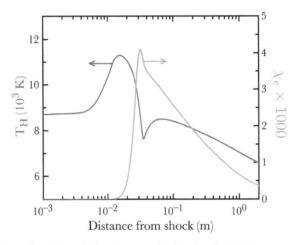

Figure 8.5. Spatial profiles of T_H and the electron molar fraction for the free stream speed Mach = 20.

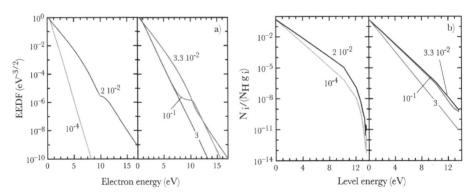

Figure 8.6. (a) EEDF and (b) atomic level distribution $N_i/(g_i N_H)$ at different distances from the shock front.

were thick, i.e. the radiation emitted is reabsorbed. Figure 8.7 shows the spatial distribution of T_H calculated with the self-consistent approach (including radiation transport) calculated with radiation transport compared with results in the thin and thick cases, as described in section 8.2. The temperature in the self-consistent case behaves as in the thick case but for short distances from the shock, where the temperature is much higher than in the thin and thick cases. In this system, the optical depth is long enough to make the escape factor approach inapplicable, and this behavior can be observed only through the solution of the radiation transport.

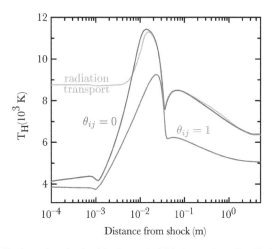

Figure 8.7. Spatial distribution of T_H in the thin ($\theta_{ij} = 1$), thick ($\theta_{ij} = 0$), and radiation-coupled calculations for free stream velocity of Mach 20.

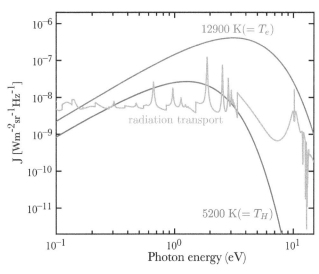

Figure 8.8. Monochromatic average J_ν (green line) and the local Planck distributions at local T_e (blue line) and at T_H (red line) in the post-shock region just behind the shock wave (0.1 nm from the shock).

These non-local effects are clear also in the spectra close to the shock front (see figure 8.8), where absorbing lines are observed for photons above 10 eV (Lyman lines). Another important feature observed in the spectra is the departure from the Plank distributions calculated at the local electron and atomic hydrogen temperatures.

Bibliography

[1] Lieberman M and Lichtenberg A J 2005 *Principles of Plasma Discharges and Materials Processing* (New York: Wiley)

[2] Manos D and Flamm D (ed) 1989 *Plasma Etching: an Introduction* (New York: Academic)

[3] Bruno G, Capezzuto P and Madan A (ed) 1995 *Plasma Deposition of Amorphous Silicon-Based Materials* (New York: Academic)

[4] Murphy A B and McAllister T 2001 Modeling of the physics and chemistry of thermal plasma waste destruction *Phys. Plasmas* **8** 2565–71

[5] Heberlein J and Murphy A B 2008 Thermal plasma waste treatment *J. Phys. D: Appl. Phys.* **41** 053001

[6] Durme J V, Dewulf J, Leys C D and Van Langenhove H 2007 Combining non-thermal plasma with heterogeneous catalysis in waste gas treatment: A review *Appl. Catal.* B **78** 324–33

[7] Starikovskaia S M 2006 Plasma assisted ignition and combustion *J. Phys. D: Appl. Phys.* **39** R265

[8] Starikovskiy A and Aleksandrov N 2013 Plasma-assisted ignition and combustion *Prog. Energy. Combust. Sci.* **39** 61–110

[9] Adamovich I V, Lempert W R, Rich J W, Utkin Y G and Nishihara M 2008 Repetitively pulsed nonequilibrium plasmas for magnetohydrodynamic flow control and plasma-assisted combustion *J. Propul. Power* **24** 1198–215

[10] Kim H H 2004 Nonthermal plasma processing for air-pollution control: a historical review, current issues, and future prospects *Plasma Process. Polym.* **1** 91–110

[11] Fridman A 2008 *Plasma Chemistry* (Cambridge: Cambridge University Press)

[12] Esposito F and Capitelli M 1999 Quasiclassical molecular dynamic calculations of vibrationally and rotationally state selected dissociation cross-sections: $N + N_2$ (v, j) $\rightarrow$ 3N *Chem. Phys. Lett.* **302** 49–54

[13] Bose D and Candler G V 1996 Thermal rate constants of the $N_2 + O \rightarrow NO + N$ reaction using *ab initio* $3A$ and $3A$ potential energy surfaces *J. Chem. Phys.* **104** 2825–33

[14] Colonna G and Capitelli M 2001 The influence of atomic and molecular metastable states in high-enthalpy nozzle expansion nitrogen *J. Phys. D: Appl. Phys.* **34** 1812–8

[15] D'Ammando G, Colonna G, Capitelli M and Laricchiuta A 2015 Superelastic collisions under low temperature plasma and afterglow conditions: a golden rule to estimate their quantitative effects *Phys. Plasmas* **22** 034501

[16] Colonna G and Capitelli M 2001 Self-consistent model of chemical, vibrational, electron kinetics in nozzle expansion *J. Thermophys. Heat Transfer* **15** 308–16

[17] Armenise I and Kustova E 2013 State-to-state models for CO_2 molecules: from the theory to an application to hypersonic boundary layers *Chem. Phys.* **415** 269–81

[18] Bogaerts A, Gijbels R and Vlcek J 1998 Collisional-radiative model for an argon glow discharge *J. Appl. Phys.* **84** 121–36

[19] Drawin H W 1971 *Reaction Under Plasma Condition* vol 1 (New York: Wiley) pp 53–238 ch 3

[20] Griem H R 1997 *Principles of Plasma Spectroscopy* (Cambridge: Cambridge University Press)

[21] Ozaki N 1971 Temperature distribution of the high-pressure sodium vapour discharge plasma *J. Quant. Spectrosc. Radiat. Transfer* **11** 1111–23

[22] Shang J and Surzhikov S 2011 Simulating stardust Earth reentry with radiation heat transfer *J. Spacecraft Rockets* **48** 385–96

[23] Shang J and Surzhikov S 2012 Nonequilibrium radiative hypersonic flow simulation *Prog. Aerospace Sci.* **53** 46–65

[24] Surzhikov S, Capitelli M, Colonna G and Defilippis F 2002 Comparative investigation of the multigroup optical models for the radiation characteristics prediction in the arc and laser plasma generators *Fourth Symp. on Aerothermodynamics for Space Vehicles* **487** 153

[25] Rich J W 1971 Kinetic modeling of the high-power carbon monoxide laser *J. Appl. Phys.* **42** 2719–30

[26] Capitelli M and Dilonardo M 1977 Nonequilibrium vibrational populations of diatomic species in electrical discharges: effects on the dissociation rates *Chem. Phys.* **24** 417–27

[27] Cartwright D C 1978 Vibrational populations of the excited states of N_2 under auroral conditions *J. Geophys. Res.: Space Phys.* **83** 517–31

[28] Gordiets B F, Osipov A I and Shelepin L A 1980 *Kinetic Processes in Gases and Molecular Lasers* (Moscow: Izdatel Nauka)

[29] Aleksandrov N, Konchakov A and Son E 1979 Electron energy distribution and kinetic coefficients of a CO plasma. II. Vibrationally excited molecules *Sov. Phys.–Tech. Phys.* **24** 6

[30] Rich J W and Treanor C E 1970 Vibrational relaxation in gasdynamic flows *Annu. Rev. Fluid Mech.* **2** 355–96

[31] Kewley D 1975 Numerical study of anharmonic diatomic relaxation rates in shock waves and nozzles *J. Phys. B: Mol. Phys.* **8** 2564–79

[32] Josyula E and Bailey W F 2001 Vibration-dissociation coupling using master equations in nonequilibrium hypersonic blunt-body flow *J. Thermophys. Heat Transfer* **15** 157–67

[33] Colonna G 2014 Problems and perspectives of state-to-state kinetics for high enthalpy plasma flows *Proc. 29th Int. Symp. Rarefied Gas Dynamics* vol 1628 (College Park, MD: AIP) pp 1176–85 doi:10.1063/1.4902726

[34] Bates D R and Kingston A E 1962 Recombination between electrons and atomic ions. I. Optically thin plasmas *Proc. R. Soc. Lond.* A **267** 685–9

[35] Bates D R and Kingston A E 1964 Recombination and energy balance in a decaying plasma. 2. He-He$^+$-e plasma *Proc. R. Soc. Lond.* A **279** 32–8

[36] Capitelli M, Colonna G and D'Angola A 2011 *Fundamental Aspects of Plasma Chemical Physics: Thermodynamics* vol 66 (Berlin: Springer)

[37] Irons F E 1979 The escape factor in plasma spectroscopy–I. The escape factor defined and evaluated *J. Quant. Spectrosc. Radiat. Transfer* **22** 1–20

[38] Pestehe S J and Tallents G J 2002 Escape factors for laser-plasmas *J. Quant. Spectrosc. Radiat. Transfer* **72** 853–78

[39] Capitelli M, Celiberto R, Colonna G, Esposito F, Gorse C, Hassouni K, Laricchiuta A and Longo S 2015 *Fundamental Aspects of Plasma Chemical Physics: Kinetics* vol 85 (Berlin: Springer)

[40] Sharma S P, Ruffin S M, Gillespie W D and Meyer S A 1993 Vibrational relaxation measurements in an expanding flow using spontaneous Raman scattering *J. Thermophys. Heat Transfer* **7** 697–703

[41] Colonna G, Tuttafesta M, Capitelli M and Giordano D 1999 Non-Arrhenius NO formation rate in one-dimensional nozzle airflow *J. Thermophys. Heat Transfer* **13** 372–5

[42] Cutrone L, Tuttafesta M, Capitelli M, Schettino A, Pascazio G and Colonna G 2014 3D nozzle flow simulations including state-to-state kinetics calculation *AIP Conf. Proc.* **1628** 1154

[43] Lordet F, Meolans J, Chauvin A and Brun R 1995 Nonequilibrium vibration-dissociation phenomena behind a propagating shock wave: vibrational population calculation *Shock Waves* **4** 299–312

[44] Armenise I, Capitelli M, Colonna G, Kudriavtsev N and Smetanin V 1995 Nonequilibrium vibrational kinetics during hypersonic flow of a solid body in nitrogen and its influence on the surface heat flux *Plasma Chem. Plasma Proc.* **15** 501–28

[45] Armenise I, Capitelli M, Colonna G and Gorse C 1996 Nonequilibrium vibrational kinetics in the boundary layer of re-entering bodies *J. Thermophys. Heat Transfer* **10** 397–405

[46] Giordano D, Bellucci V, Colonna G, Capitelli M, Armenise I and Bruno C 1997 Vibrational relaxing flow of N_2 past an infinite cylinder *J. Thermophys. Heat Transfer* **11** 27–35

[47] Treanor C E, Rich J and Rehm R 1968 Vibrational relaxation of anharmonic oscillators with exchange-dominated collisions *J. Chem. Phys.* **48** 1798–807

[48] Colonna G, Armenise I, Bruno D and Capitelli M 2006 Reduction of state-to-state kinetics to macroscopic models in hypersonic flows *J. Thermophys. Heat Transfer* **20** 477–86

[49] Colonna G, Pietanza L D and Capitelli M 2008 Recombination assisted nitrogen dissociation rates under nonequilibrium conditions *J. Thermophys. Heat Transfer* **22** 399–406

[50] Colonna G, Pietanza L D, Capitelli M, Levin D A, Wysong I J and Garcia A L 2011 Reduced two-level approach for air kinetics in recombination regime *AIP Conf. Proc.* **1333** 1365

[51] Bourdon A, Annaloro J, Bultel A, Capitelli M, Colonna G, Guy A, Magin T, Munafóaaa A, Perrin M and Pietanza L 2014 Reduction of state-to-state to macroscopic models for hypersonics *Open Plasma Phys. J.* **7** M4

[52] Guy A, Bourdon A and Perrin M-Y 2015 Consistent multi-internal-temperature models for vibrational and electronic nonequilibrium in hypersonic nitrogen plasma flows *Phys. Plasmas* **22** 043507

[53] Park C 1993 Review of chemical-kinetic problems of future NASA missions. I: Earth entries *J. Thermophys. Heat Transfer* **7** 385–98

[54] Chernyi G G, Losev S A, Macheret S O and Potapkin B V 2002 *Physical and Chemical Processes in Gas Dynamics: Cross sections and Rate Constants for Physical and Chemical Processes* vol 1 (Reston, VA: AIAA)

[55] Schwartz R, Slawsky Z and Herzfeld K 1952 Calculation of vibrational relaxation times in gases *J. Chem. Phys.* **20** 1591–9

[56] Adamovich I V, MacHeret S O, Rich J W and Treanor C E 1998 Vibrational energy transfer rates using a forced harmonic oscillator model *J. Thermophys. Heat Transfer* **12** 57–65

[57] Capitelli M, Colonna G and Esposito F 2004 On the coupling of vibrational relaxation with the dissociation-recombination kinetics: from dynamics to aerospace applications *J. Phys. Chem.* A **108** 8930–4

[58] Colonna G, D'Ammando G, Pietanza L and Capitelli M 2014 Excited-state kinetics and radiation transport in low-temperature plasmas *Plasma Phys. Control. Fusion* **57** 014009

[59] D'Ammando G, Capitelli M, Esposito F, Laricchiuta A, Pietanza L D and Colonna G 2014 The role of radiative reabsorption on the electron energy distribution functions in H_2/He plasma expansion through a tapered nozzle *Phys. Plasmas* **21** 093508

[60] Perrin M, Colonna G, D'Ammando G, Pietanza L, Riviere P, Soufani A and Surzhikov S 2014 Radiation models and radiation transfer in hypersonics *Open Plasma Phys. J.* **7** 114–26

[61] Pietanza L, Colonna G, De Giacomo A and Capitelli M 2010 Kinetic processes for laser induced plasma diagnostic: a collisional-radiative model approach *Spectrochim Acta* B **65** 616–26

[62] Pietanza L, Colonna G, D'Ammando G, Laricchiuta A and Capitelli M 2015 Vibrational excitation and dissociation mechanisms of CO_2 under non-equilibrium discharge and post-discharge conditions *Plasma Sources Sci. Technol.* **24** 042002

[63] Pietanza L, Colonna G, D'Ammando G, Laricchiuta A and Capitelli M 2016 Electron energy distribution functions and fractional power transfer in 'cold' and excited CO_2 discharge and post discharge conditions *Phys. Plasmas* **23** 013515

[64] Capitelli M, Colonna G, D'Ammando G, Laporta V and Laricchiuta A 2014 Nonequilibrium dissociation mechanisms in low temperature nitrogen and carbon monoxide plasmas *Chem. Phys.* **438** 31–6

[65] Colonna G, Capitelli M, DeBenedictis S, Gorse C and Paniccia F 1991 Electron energy distribution functions in CO_2 laser mixture: the effects of second kind collisions from metastable electronic states *Contrib. Plasma Phys.* **29** 575–9 991

[66] Capriati G, Colonna G, Gorse C and Capitelli M 1992 A parametric study of electron energy distribution functions and rate and transport coefficients in nonequilibrium helium plasmas *Plasma Chem. Plasma Process.* **12** 237–60

[67] D'Ammando G, Colonna G, Capitelli M and Laricchiuta A 2015 Superelastic collisions under low temperature plasma and afterglow conditions: a golden rule to estimate their quantitative effects *Phys. Plasmas* **22** 034501

[68] Colonna G, D'Ammando G and Pietanza L 2016 The role of molecular vibration in nanosecond repetitively pulsed discharges and in DBD's in hydrogen plasmas *Plasma Sources Sci. Technol.* **25** 054001

[69] Aerts R, Martens T and Bogaerts A 2012 Influence of vibrational states on CO_2 splitting by dielectric barrier discharges *J. Phys. Chem.* C **116** 23257–73

[70] Nighan W L 1970 Electron energy distributions and collision rates in electrically excited N_2, CO, and CO_2 *Phys. Rev.* A **2** 1989–2000

[71] Silva T, Britun N, Godfroid T and Snyders R 2014 Optical characterization of a microwave pulsed discharge used for dissociation of CO_2 *Plasma Sources Sci. Technol.* **23** 025009

[72] Nerem R M and Stickford G H 1965 Shock-tube studies of equilibrium air radiation *AIAA J.* **3** 1011–8

[73] Livingston F R and Poon P T Y 1976 Relaxation distance and equilibrium electron density measurements in hydrogen-helium plasmas *AIAA J.* **14** 1335–7

[74] Johnson C O, Brandis A M and Bose D 2013 Radiative heating uncertainty for hyperbolic Earth entry. Part 3: Comparisons with electric arc shock-tube measurements *J. Spacecraft Rockets* **50** 48–55

[75] Colonna G, Pietanza L D and D'Ammando G 2012 Self-consistent collisional-radiative model for hydrogen atoms: atom–atom interaction and radiation transport *Chem. Phys.* **398** 37–45

Chapter 9

Hybrid models

A computational approach extensively used in plasma computational physics is the hybrid method [1, 2]. It consists of solving the fluid laws (chapters 6 and 7) for the electrons using a kinetic approach (usually discrete-particle, particle-in-cell with Monte Carlo collisions (PIC-MCC) methods [3], described in chapter 4) to represent the evolution of the heavy species. Thus, the hybrid code combines a Lagrangian and an Eulerian approach. It is suitable for studying the low-frequency/high-wave-number phenomena that are not captured by full fluid theory because ion kinetic effects (ion beam instability, shock by back-streaming ions, ion–neutral charge exchange (CX), etc) are at the origin of the phenomenon itself. A full kinetic particle code must resolve the electron behavior at scales much smaller than for ions, studying smaller regions for shorter times. Moreover, full particle techniques require a much larger computer memory. Therefore, hybrid approach allows one to save computational time and, as a result, it has gained considerable popularity. The first one-dimensional hybrid codes appeared in the 1970s thanks to the pioneering works of Mason [4], Chodura [5], and Sgro-Nielsen [6] on fusion plasmas. The extension to two-dimensional models dates back to the early 1980s to the work by Hewett [7] and Harned [8]. Finally, the first full three-dimensional representations can be considered the work of Brecht and Thomas [9] and Moore *et al* [10] presented in the late 1980s/early 1990s.

This chapter is organized as follows. The next section is devoted to the description of the physics laws and hypothesis behind the hybrid description. Fluid electron and ion kinetic equations will be discussed. In section 9.2 the general numerical procedure necessary for the implementation of a hybrid code will be presented. Finally, in section 9.3 some numerical variants of the general scheme will be presented for selected space and laboratory applications of low-temperature plasmas. The corresponding results will be discussed.

9.1 Basic assumptions and governing equations

Hybrid models describe phenomena occurring on space and time scales much larger than those typical of electrons; they are used to simulate and resolve processes that occur on ion gyro-radius or ion inertial spatial scales, and inverse ion gyro-frequency or ion transit times. As a consequence, the space scales are much larger than the Debye length and the electron Larmor radius, and the time scales are much larger than the plasma and electron gyration periods.

The hybrid approximation treats the electrons as a fluid while the ions are modeled kinetically. The most used kinetic formulation is based on particle representation (the PIC-MCC scheme [3], see chapter 4). It represents the ion sub-system as a collection of macro-ions (a cloud with a fixed number of real ions) whose position $\mathbf{x}_p$ and velocity $\mathbf{v}_p$ are updated, solving the usual equations of motion

$$m_i \frac{d\mathbf{v}_p}{dt} = q_i\left(\mathbf{E} + \mathbf{v}_p \times \mathbf{B}\right) \qquad (9.1)$$

$$\frac{d\mathbf{x}_p}{dt} = \mathbf{v}_p, \qquad (9.2)$$

where $\mathbf{E}$ and $\mathbf{B}$ (defined on a spatial grid through the coupling with the electron sub-system) have been interpolated to the macro-particle location. The ion distribution function f_i and the corresponding generic momenta $\mathcal{M}_i$ (charge density ρ_i, bulk flow velocity $\mathbf{u}_i$, current $\mathbf{j}_i$, temperature T_i, etc) are reconstructed on the grid points through an inverse interpolation process using the same weighting function used for the gathering procedure, in order to avoid self-forces on the particles [3]:

$$\mathcal{M}_i \leftarrow (\mathbf{x}_p; \mathbf{v}_p). \qquad (9.3)$$

Concerning the electron fluid description under hybrid approximation it can be shown that:
  1. quasi-neutrality is assumed

$$\rho_e = \rho_i = \rho, \qquad (9.4)$$

   so that given the ion density from the ion PIC module, the electron density is specified (here no negative ion species are assumed for simplicity);
  2. neglecting the displacement currents in Ampère's law (the Darwin approximation) $\mathbf{j} = \nabla \times \mathbf{B}/\mu_0$ (μ_0 is the vacuum permeability) and the electron mass in the electron momentum equation, this results in an equation of state (generalized Ohm's law) for the electric field

$$\mathbf{E} = -\mathbf{u}_i \times \mathbf{B} + \frac{(\nabla \times \mathbf{B}) \times \mathbf{B}}{\mu_0 \rho} - \frac{\nabla p_e}{\rho} \qquad (9.5)$$

once the electron pressure (considered isotropic) is known. Through the ideal gas law $p_e = n_e \kappa T_e$, this corresponds to handling the electron temperature and the following different options are available:

(a) isothermal (electrons distribute their energy fast in the whole domain because of high thermal conductivity),

$$T_e = \text{const};\qquad\qquad(9.6)$$

(b) adiabatic (small collision frequency),

$$T_e \propto n_e^{\gamma-1},\qquad\qquad(9.7)$$

where γ is the polytropic index; and

(c) solve for the massless energy equation

$$\frac{\partial T_e}{\partial t} + \mathbf{u}_e \cdot \nabla T_e + \frac{3}{2}\nabla \cdot \mathbf{u}_e = \frac{2}{3}\frac{\eta |J|^2}{n_e},\qquad\qquad(9.8)$$

where $\mathbf{u}_e$ is the electron velocity.

In the collisional case, the right-hand side of equation (9.5) also contains a force term $C_{\text{coll}} = \eta \nabla \times \mathbf{B}/\mu_0$ due to electron–ion, electron–neutral, and anomalous (i.e. wave–particle interaction) collisions (η is the plasma resistivity, considered for simplicity as scalar). Similarly, the electric field in the ion equation of motion (9.1) becomes $\mathbf{E}' = \mathbf{E} - \eta \nabla \times \mathbf{B}$ unless an explicit Monte Carlo module is used to take into account collisions.

It should be noted that under the Darwin approximation there is no equation for the time evolution of the electric field. It is instead determined by the requirement that the electric force balances the other forces acting on the massless electron fluid: the electric field is a state function of the ion moments ρ and $\mathbf{u}_i$, the magnetic field $\mathbf{B}$, and the electron temperature T_e.

Finally, in the electromagnetic case, the electric field is used to advance the magnetic field in time through Faraday's law $\frac{\partial \mathbf{B}}{\partial t} = -\nabla \times \mathbf{E}$ which, substituting equation (9.5), becomes

$$\frac{\partial \mathbf{B}}{\partial t} = \nabla \times (\mathbf{u}_i \times \mathbf{B}) - \nabla \times \frac{(\nabla \times \mathbf{B}) \times \mathbf{B}}{\mu_0 \rho}.\qquad\qquad(9.9)$$

The first term on the right-hand side describes the induction, while the second represents the dispersion. Note that the electron pressure term of equation (9.5) does not influence the magnetic field evolution. Under previous assumptions the other Maxwell equations

$$\nabla \cdot \mathbf{E} = -\frac{\rho}{\epsilon_0}\qquad\qquad(9.10)$$

$$\nabla \cdot \mathbf{B} = 0,\qquad\qquad(9.11)$$

are automatically satisfied.

9.2 Numerical implementation

In this section the implementation of the general hybrid approach will be presented. The different variations of the general scheme will be presented in section 9.3 for a number of applications ranging from plasma-based technological devices to astrophysical problems.

The governing laws presented in the previous section are solved on a computational grid discretizing the time. This requires defining a time step Δt and a cell size Δx characteristic of the typical case studied. Usually the time step Δt is defined by ion plasma or cyclotron frequencies, while the cell size Δx is chosen such as to resolve ion gradients or to verify the Courant–Friedrichs–Lewy (CFL) condition: $v_{i,\max} \Delta t < \Delta x$, where $v_{i,\max}$ represents the maximum speed of the ions. The computational grid can be uniform (Cartesian, spherical, or even in general orthogonal curvilinear coordinates) or non-uniform (usually in Cartesian coordinates). In the latter case, hierarchical mesh or adaptive mesh refinement [11, 12] is used to generate a non-uniform grid.

Plasma in the hybrid scheme has two time-dependent components: the positions/velocities of ion macroparticles and electromagnetic fields specified at the nodes of the computational grid. The bridge between the two kinds of quantities is carried by ion charge density ρ and ion flow velocity $\mathbf{u}_i$ interpolated from the particle location to the grid points. The magnetic field $\mathbf{B}$ is located on the grid (for example the cell centers). The electric fields $\mathbf{E}$, the plasma density ρ, the ion current densities $\mathbf{u}_i$, and the electron temperature T_e are all staggered from the locations of $\mathbf{B}$ (such as at the vertices of the cell).

9.2.1 The time-advance algorithm

Here we follow the implementation discussed in [11, 13, 14] and show it to be of second-order accuracy. It is an explicit time-centered leap-frog algorithm for both particles and fields, consisting of the following computational cycle (repeated until stationarity is reached):

1. The electromagnetic fields and particle positions are defined at full time steps $\mathbf{E}^{(n)}$, $\mathbf{B}^{(n)}$, and $\mathbf{x}_p^{(n)}$, as well as the density $\rho^{(n)}$ gathered from the particle positions

$$\rho^{(n)} \leftarrow \left(\mathbf{x}_p^{(n)}\right). \tag{9.12}$$

Usually linear weighting is used for particle-to-grid location interpolation, but more complex functions are used for cylindrical or spherical metrics [15–17].

2. The particle velocities defined at half time steps are advanced by solving equation (9.1):

$$\mathbf{v}_p^{(n+1/2)} = \mathbf{v}_p^{(n-1/2)}\left(1 - \frac{h^2}{2}\mathbf{B}^{(n)} \cdot \mathbf{B}^{(n)}\right) + h\left(\mathbf{E}^{(n)} + \mathbf{v}_p^{(n-1/2)} \times \mathbf{B}^{(n)}\right) \\ + \frac{h^2}{2}(\mathbf{E}^{(n)} \times \mathbf{B}^{(n)}) + \frac{h^2}{2}(\mathbf{v}^{(n-1/2)} \cdot \mathbf{B}^{(n)})\mathbf{B}^{(n)}, \tag{9.13}$$

where $h = q_i \Delta t / m_i$. For unmagnetized ions, this equation is greatly simplified.

3. Introduce the ion velocity $\mathbf{u}_i^{(-)}$ gathered from full time step position $\mathbf{x}_p^{(n)}$ and half time step velocity $\mathbf{v}_p^{(n+1/2)}$,

$$\mathbf{u}_i^{(-)} \leftarrow \left(\mathbf{x}_p^{(n)}; \mathbf{v}_p^{(n+1/2)}\right). \tag{9.14}$$

4. Advance $\mathbf{x}_p$ by solving equation (9.2)

$$\mathbf{x}_p^{(n+1)} = \mathbf{x}_p^{(n)} + \mathbf{v}^{(n+1/2)}\Delta t. \tag{9.15}$$

5. Gather $\rho^{(n+1)}$ and $\mathbf{u}_i^{(+)}$ from $\mathbf{x}_p^{(n+1)}$ and $\mathbf{v}_p^{(n+1/2)}$,

$$\rho^{(n+1)} \leftarrow \left(\mathbf{x}_p^{(n+1)}; \mathbf{v}_p^{(n+1/2)}\right) \tag{9.16}$$

$$\mathbf{u}_i^{(+)} \leftarrow \left(\mathbf{x}_p^{(n+1)}; \mathbf{v}_p^{(n+1/2)}\right), \tag{9.17}$$

in order to calculate the following averaged density and velocity centered at half time steps:

$$\rho^{(n+1/2)} = \frac{\rho^{(n+1)} + \rho^{(n)}}{2} \tag{9.18}$$

$$\mathbf{u}_i^{(n+1/2)} = \frac{\mathbf{u}_i^{(+)} + \mathbf{u}_i^{(-)}}{2}. \tag{9.19}$$

6. Advance $\mathbf{B}^{(n+1)}$ by solving equation (9.9) using the previous averaged quantities $\rho^{(n+1/2)}$ and $\mathbf{u}_i^{(n+1/2)}$. This task is done by using the cyclic leap-frog technique: it consists in dividing the PIC cycle in m sub-time-steps $\Delta t_B = \Delta t_{\text{PIC}}/m$. With the notation $\mathbf{B}^{(p,n)}$ as $\mathbf{B}^{(n)}$ calculated at the pth sub-time-step and $\mathbf{E}^{(p,n)}$ as the corresponding electric field obtained from equation (9.5) using the current value $\mathbf{B}^{(p,n)}$, we have the following iterative rule:

$$\mathbf{B}^{(1,n)} = \mathbf{B}^{(0,n)} - \Delta t_B \nabla \times \mathbf{E}^{(0,n)} \tag{9.20}$$

$$\mathbf{B}^{(2,n)} = \mathbf{B}^{(0,n)} - 2\Delta t_B \nabla \times \mathbf{E}^{(1,n)} \tag{9.21}$$

$$\vdots \tag{9.22}$$

$$\mathbf{B}^{(p+1,n)} = \mathbf{B}^{(p-1,n)} - 2\Delta t_B \nabla \times \mathbf{E}^{(p,n)} \tag{9.23}$$

$$\vdots \tag{9.24}$$

$$\mathbf{B}^{(m,n)} = \mathbf{B}^{(m-2,n)} - 2\Delta t_B \nabla \times \mathbf{E}^{(m-1,n)} \tag{9.25}$$

$$\mathbf{B}^{*(m,n)} = \mathbf{B}^{(m-1,n)} - \Delta t_B \nabla \times \mathbf{E}^{(m,n)} \tag{9.26}$$

$$\mathbf{B}^{(n+1)} = \frac{1}{2}(\mathbf{B}^{(m,n)} + B^{*(m,n)}). \tag{9.27}$$

In addition, to stabilize the magnetic field integration against the influence of perturbations at short wavelengths, different smoothing procedures can be applied.

7. Extrapolate from $\mathbf{u}_i^{(n+1/2)}$ to $\mathbf{u}_i^{(n+1)}$. This can be done in several ways. First, keeping $\mathbf{u}_i^{(n-1/2)}$ as

$$\mathbf{u}_i^{(n+1)} = \frac{3\mathbf{u}_i^{(n+1/2)} - \mathbf{u}_i^{(n-1/2)}}{2} \tag{9.28}$$

or keeping $\mathbf{u}_i^{(n-3/2)}$ and improving the accuracy, as in the fourth-order Bashford–Adams extrapolation,

$$\mathbf{u}_i^{(n+1)} = \frac{4\mathbf{u}_i^{(n+1/2)} - 3\mathbf{u}_i^{(n-1/2)} + \mathbf{u}_i^{(n-3/2)}}{2}. \tag{9.29}$$

Alternatively, the velocity is calculated following the current advancement method described by Matthews [13]. It consists of performing an extra half time step push using a mixed level evaluation of the electric field

$$\mathbf{u}_i^{(n+1)} = \mathbf{u}_i^{(n+1)} + \frac{q_i \Delta t}{2m_i}\left(\mathbf{E}^* + \mathbf{u}_i^{(n+1)} \times \mathbf{B}^{(n+1)}\right), \tag{9.30}$$

where $\mathbf{E}^{(n+1)}$ is calculated from equation (9.5) using $\mathbf{B}^{(n+1)}$, $\mathbf{u}_i^{(n+1/2)}$, and $\rho^{(n+1)}$.

8. Calculate $\mathbf{E}^{(n+1)}$ from equation (9.5) using $\mathbf{B}^{(n+1)}$, $\mathbf{u}_i^{(n+1)}$, and $\rho^{(n+1)}$. In order to avoid numerical instability due to the term $(\nabla \times \mathbf{B}) \times \mathbf{B}/\mu_0\rho$ (the electric field becomes very large at low density) a minimum threshold density is defined below which the field is set equal to zero [18].

Alternative time-advance algorithms are the predictor–corrector [9] and Horowitz [19] algorithms, and a combination of velocity extrapolation and predictor–corrector [20].

9.2.2 Initialization and boundary conditions

The starting point of a simulation is always a sensitive issue. Initial conditions are specified by giving the magnetic field $\mathbf{B}$, plasma density ρ, and ion flow velocity $\mathbf{u}_i$ on the grid points. These latter two quantities are directly calculated from the initial ion distribution or loading/injection particles (only in few cases can you start from scratch), and the number of particles per cell plays a relevant role in defining the resolution of phase space.

Regarding the boundaries two different kinds of elements can be distinguished: the external surface domain and the internal obstacle. The external boundaries can represent inflow, outflow, periodic, axis of symmetry, or wall conditions, while the obstacle can be an insulator or metallic material with its own intrinsic magnetic field. For all the cases, the conditions have to be set for the particles and electromagnetic fields.

General rules have to be fulfilled. External boundaries should be as far as possible from the obstacle in order to reduce all possible perturbations coming from the obstacle itself. In addition, for periodic conditions, keep in mind that the size of the imposed periodicity can induce an artificial scale or dump high-wave-number fluctuations.

9.3 Applications

9.3.1 The electrostatic case: plasma plume expansion and Langmuir probes

In the absence of an external magnetic field and for negligible induced magnetic field, the problem is purely electrostatic. Ohm's law (equation (9.5)) reduces to the well-known Boltzmann relation

$$n_e = n_0 \exp(e\phi/\kappa T_e), \tag{9.31}$$

where n_0 is the equilibrium density and ϕ is the electric potential $\nabla\phi = -\mathbf{E}$. Equation (9.31) can be used directly to derive the electric field $\mathbf{E}$ from the ion density through the quasi-neutrality condition equation (9.4). Alternatively, without assuming quasi-neutrality, the Poisson equation

$$\nabla^2\phi = -\frac{e}{\epsilon_0}(n_i - n_0 \exp(e\phi/\kappa T_e)) \tag{9.32}$$

has to be solved when space-charge effects are important (small simulated regions, a large object immersed in plasma, etc). In this case, equation (9.32) leads to a non-linear elliptic differential equation which can be quickly solved by the iterative Newton–Raphson method [21]. The electric potential ϕ at the iteration $(n + 1)$ is the solution of a linear equation whose coefficients are functions of the solution at the previous iteration (n)

$$\left[\nabla^2 - \frac{e^2 n_0}{\epsilon_0 \kappa T_e} \exp(e\phi^{(n)}/\kappa T_e)\right]\phi^{(n+1)} = \left[1 - \frac{e}{\kappa T_e}\phi^{(n)}\right]\frac{e n_0}{\epsilon_0} \exp(e\phi^{(n)}/\kappa T_e) - \frac{e n_i}{\epsilon_0}. \tag{9.33}$$

The disadvantage of the Newton–Raphson method is that the equations have variable coefficients and direct use cannot be made of rapid elliptic solvers. Electron temperature T_e is considered to be fixed (isothermal) or coupled thought the adiabatic or energy conserving equation.

The electrostatic hybrid technique has been extensively used to study problems such as plume expansion from electric thrusters [22–24] and from laser-induced ablation [25], or in the calculation of ion current collection by an electrostatic probe [26–29].

Figure 9.1 shows an ion swarm in an (a) ordinary and (b) velocity space from an electrostatic three-dimensional hybrid simulation of the plume emitted by a Hall-effect thruster (HET) [22–24]. The model solves the non-linear Poisson's equation (9.33) coupled with an adiabatic expansion equation (9.7) where the value $\gamma = 1.3$ has been chosen for the polytropic exponent to fit experimental measurements. The important effects of the production of slow ions (the black dots in figure 9.1, while the red dots are beam ions) by resonant CX collisions with a neutral unionized propellant (where an electron is exchanged),

$$A_{\text{fast}}^{+} + A_{\text{slow}}^{0} \rightarrow A_{\text{fast}}^{0} + A_{\text{slow}}^{+}, \tag{9.34}$$

is highlighted by the plume beam divergence (see figure 9.2): the fraction of CX ions is larger for higher angles, while in the 40 degree (half-angle) cone the ion population is dominated by beam ions. In this way the production of slow CX ions reduces the thruster efficiency and is dangerous for the sensitive parts of the satellite.

The second example of a hybrid electrostatic model represents the dynamics of ion collection by a Langmuir probe. Electrostatic probes are widely used to determine the local values of plasma parameters, such as electron and ion number densities, electron temperature, electron energy distribution function, and plasma potential, from the experimentally measured current–voltage ($I–V$) probe characteristic. For a negatively biased probe (the ion collection regime) and in a middle pressure discharge, the commonly used orbit motion limited theory [30] does not explain experimental results and, in particular, the fact that the ion current reaches a maximum as a function of neutral gas pressure due to the destruction of the ion orbital motion around the probe (see figure 9.3(a) for the collisionless case). The problem can be reduced to a one-dimensional radial with ions moving in a central field of force under the effective potential, defined as the sum of electrostatic attracting and centrifugal repelling contributions

$$U_{\text{eff}}(r) = e\phi(r) + \frac{L^2}{2M_i r^2}, \tag{9.35}$$

where L is the angular momentum and ϕ is calculated by solving the non-linear Poisson's equation (9.33) and assuming electrons as isothermal. Figure 9.3(b) shows the typical ion trajectory for different Knudsen numbers k under the effect of the self-consistent electric potential reported in figure 9.4(a). The effect of ion–neutral collisions is to create a strong potential and consequently to reduce the sheath thickness. This effect is well understood and is caused by the increase of the viscous drag force that reduces the ion velocity in the collisional sheath. To satisfy the conservation of ion flux, the reduced ion velocity requires an increase in the ion density. Via the Poisson equation, this increase leads to a stronger gradient in the electric field. A larger gradient implies a smaller scale length, i.e. a smaller sheath thickness. This is confirmed in figure 9.4(b), where ion (green curves) and electron (red curves) number densities versus radial position at various k are shown. Note

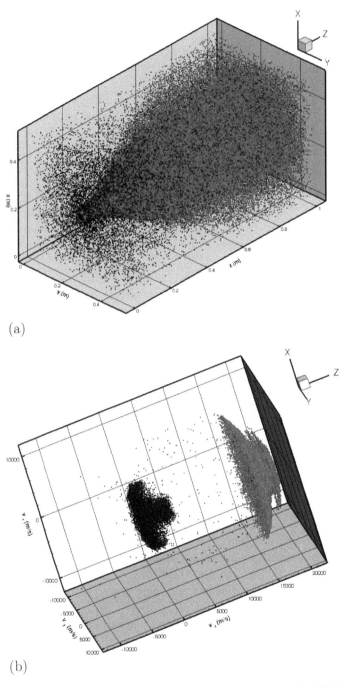

Figure 9.1. Ion snapshot in (a) ordinary and (b) velocity space at steady state from a hybrid electrostatic three-dimensional simulation of the plume emitted from an HET.

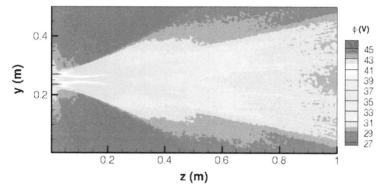

Figure 9.2. Two-dimensional (y, z) maps of electric potential ϕ (V) at the symmetric plane. The coaxial channel and plume divergency are evident.

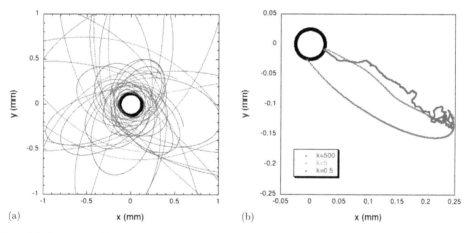

Figure 9.3. Ion trajectories around an electrostatic probe in (a) collisionless and (b) different collisional cases (k is the Knudsen number).

how the space-charged region around the probe has been resolved without assuming quasi-neutrality $n_e \neq n_i$.

9.3.2 Magnetostatic case: $E \times B$ field devices

Electron confinement with a magnetic field is often used in the $E \times B$ configuration for different reasons. It reduces plasma diffusion to the wall, thus increasing plasma uniformity and density at the same discharge power. A non-uniform magnetic field confines fast electrons better than cold ones, thus performing the role of a filter between hot and cold electrons. Finally, electron magnetic confinement is often used in the $E \times B$ configuration (HETs, magnetrons, electron cyclotron resonance (ECR), and helicon discharges) in order to increase ionization efficiency and create a strong impedance with a virtual cathode to accelerate the generated ions.

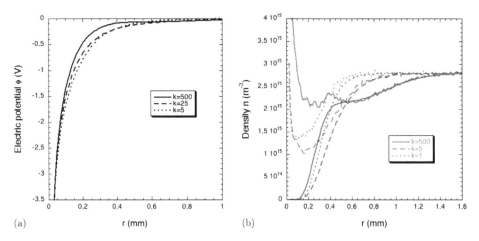

Figure 9.4. Radial profiles of (a) electric potential and (b) electron (red curves) and ion (green curves) densities for different Knudsen numbers k.

Hybrid models have often been developed for the simulation of $E \times B$ field devices. One of the first models to follow this approach was developed by Fife [31] for the simulation of the acceleration channel of an HET. Based on this model different variants have been developed [32–35].

Here, Fife's original numerical procedure is followed. The simulation is reduced to two-dimensional (r, z) cylindrical geometry using the axial-symmetry of the HET, shown schematically in figure 9.5(a). It is characterized by a DC glow discharge with a dielectric annular chamber: a magnetic circuit generates an axial-symmetric and quasi-radial magnetic field between the inner and outer poles (see figure 9.5(b)), while an electrical discharge is established between the anode (deep inside the channel), acting also as a gas propellant distributor, and the external cathode, used also as an electron emitter. In this configuration, cathode electrons are drawn to the positively charged anode, but the radial magnetic field creates a strong impedance, trapping the electrons in cyclotron motion, which follows a closed drift path inside the annular chamber along the azimuthal θ-direction. The trapped and energized (by Joule heating) electrons act as a volumetric zone of ionization for atoms and as a virtual cathode to accelerate the generated ions, which are not affected by the magnetic field due to their larger Larmor radii. The thruster is relieved of space-charge limitations by the neutralizing effect of the trapped electrons, and HETs are therefore capable of providing higher thrust densities than gridded ion engines. The resulting external jet composed of the high-speed ion beam is subsequently neutralized by part of electrons coming from the external cathode-compensator (see section 9.3.1).

The system can be considered magnetostatic: the magnetic field variation due to plasma currents and changing electric fields are assumed to be small compared to the externally generated magnetic field. Ampère's law reduces to $\nabla \times \mathbf{B} = 0$ which allows one to define a magnetic potential function $\mathbf{B} = \nabla \sigma$ satisfying the Laplace equation

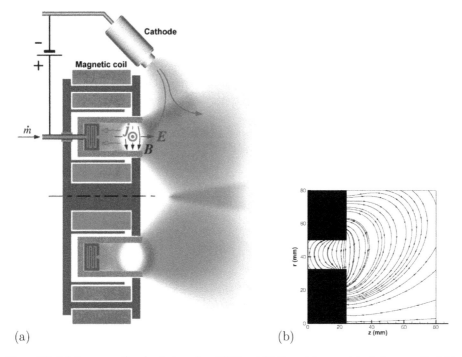

Figure 9.5. (a) A cross-sectional cut view of an HET and (b) the axial-symmetric static magnetic field.

$$\nabla^2 \sigma = 0. \tag{9.36}$$

Also, since $\nabla \cdot \mathbf{B} = 0$, it is possible to define a magnetic stream function λ whose gradient is everywhere orthogonal to $\mathbf{B}$:

$$\frac{\partial \lambda}{\partial z} = r\frac{\partial \sigma}{\partial r} = rB_r \tag{9.37}$$

$$\frac{\partial \lambda}{\partial r} = -r\frac{\partial \sigma}{\partial z} = -rB_z. \tag{9.38}$$

The function λ defined by these equations is constant along magnetic field lines $\mathbf{B} \cdot \nabla \lambda = 0$ and increases monotonically from anode to cathode. The cross-field gradient (the gradient normal to the magnetic field lines) of any quantity Q can be expressed in terms of λ as

$$\nabla_\perp Q = -rB\frac{\partial Q}{\partial \lambda}. \tag{9.39}$$

This equation allows us to write spatial derivatives across magnetic field lines as derivatives with respect to values invariant along the lines. This is in order to overcome the numerical errors associated with the extreme anisotropy of electron transport: there is a large disparity (more than two orders of magnitude in regions with high values of the magnetic field) between perpendicular $\perp$ and parallel $\parallel$

electron transport. Therefore, the fluid equations are solved on a computational mesh that is aligned with the applied magnetic field: magnetic field-aligned meshes (MFAMs).

Along the magnetic field lines the problem reduces to an electrostatic one. As shown above equation (9.31), it is possible to define a quantity $\phi^*(\lambda)$ (called the thermalized potential) constant along the magnetic field line

$$\phi^*(\lambda) = \phi - \frac{\kappa T_e}{e}\ln(n_e). \tag{9.40}$$

In contrast, electron transport across the magnetic field lines obeys the generalized form of Ohm's law equation (9.5) in the drift–diffusion approximation (the electron collisional term is included). Replacing the expression of the cross-field gradient equation (9.39) and using the definition of thermalized potential equation (9.40) gives

$$u_{e,\perp} = \mu_{e,\perp}\left[-rB\frac{\partial \phi^*}{\partial \lambda} - rB\frac{\kappa}{e}(\ln(n_e) - 1)\frac{\partial T_e}{\partial \lambda}\right], \tag{9.41}$$

where $\mu_{e,\perp} = \frac{\mu_{e,\|}}{1+\beta^2}$ represents the cross-field mobility ($\mu_{e,\|} = e/\nu_{en}m_e$ is the electron mobility and $\beta = \omega_e/\nu_{en}$ the Hall parameter). In this equation, the gradient of thermalized potential and the electron temperature along the magnetic stream function are obtained from the current and the energy conservation equation (9.8), respectively. It should be noted that an artificial increased electron collision frequency $\nu_{en}^* = \nu_{en} + \nu_{ew} + \nu_{ano}$ is used to fit experimental measurements. The electron current values collected at the anode are larger than those obtained by classical collisional transport. Electron–wall collision ν_{ew} and azimuthal fluctuation ν_{ano} contribute to increasing the electron cross field mobility. Figure 9.6 shows (a) the electric potential and (b) plasma density maps in the (r, z)-plane as resulting from the magnetostatic hybrid model in [32]. The electric potential lines deviate significantly from the magnetic field lines (figure 9.5(b)) due to the second term of equation (9.40). Most of the electric field is concentrated near the exhaust, where the neutral density is so low that the electron mobility is strongly reduced. Outside the channel the

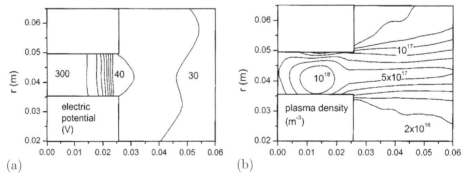

(a)   (b)

Figure 9.6. Contour plot of (a) electric potential and (b) plasma density in the (r, z)-plane of an HET. (Reproduced with permission from [32]. Copyright 2002 AIP Publishing).

larger fluctuation-induced mobility ν_{ano} imposed prevents a strong electric field. The plasma density is highest at the center of the channel and decreases at the exhaust due to the increasing ion velocity from the effect of the axial electric field.

The MFAM method is strongly sensitive to small changes in the magnetic field strength or geometry, and for this reason it has to be avoided in cases of strong magnetic field gradients or curvatures (as near ring cusps). In these cases non-MFAM approaches, such as a simple orthogonal mesh or transverse-flux method [36], are suggested.

9.3.3 The electromagnetic case: fusion and space plasmas

Hybrid models have been developed to study plasma–wall transition regions in thermonuclear fusion reactors (the scrape off layer and divertor) [37] and instabilities driven by fusion-born ions [38]. Hybrid models are commonly used to simulate the interaction of the solar wind with planets, moons, comets, asteroids, and the local interstellar medium, showing the different plasma structures that form: bow shock (BS), magnetopause, magnetotail, etc. Here, as an example, the three-dimensional global simulation of the Martian plasma environment is described. Mars represents a very challenging case for numerical simulation: the presence of multiple ion species in the ionosphere, the presence of a strong non-homogeneous crustal magnetic field at the surface, and the fact that Mars' radius and shock stands-off are smaller than the ion gyroradius. The motivation to study Mars is increased by the question whether the escape of planetary oxygen ions is the phenomenon at the base of the loss of water from the planet [39].

Following [40, 41], the hybrid model treats six ion species: solar wind H^+ and He^{++}, and planetary H^+, O^+, O_2^+, and CO_2^+. Although the density of alpha particles is smaller than the proton density, they carry a significant fraction of the solar wind mass and momentum (20%), and thus can influence the overall dynamics. The electron fluid is assumed to be adiabatic with a polytropic index $\gamma = 2$ (equation (9.7)), with two specified temperatures for the undisturbed solar wind plasma and the ionospheric plasma. Planetary ions are produced by three processes: ionization by solar extreme ultraviolet photons, resonant and non-resonant CX, and electron impact ionization. The production rate of each ion species is computed from prescribed neutral reservoirs of atomic oxygen, hydrogen, and carbon dioxide. The model is implemented on a uniform Cartesian grid with cubic cells, extending from -2.4 to $3R_M$ in x and from -5 to $5R_M$ in y and z, where R_M is the Martian radius. The coordinate system is defined as follows: the x-direction is aligned with the solar wind bulk flow velocity $\mathbf{u}_{sw}$ and points towards the Sun, z is the direction of the motional electric field of the solar wind $\mathbf{E}_{sw} = -\mathbf{u}_{sw} \times \mathbf{B}_{IMF}$, where $\mathbf{B}_{IMF}$ is the interplanetary magnetic field conveyed by the solar wind and is imposed to be directed along the z-direction.

Figure 9.7 shows two-dimensional cuts of solar wind H^+ (left) and planetary O^+ (right) density (cm^{-3}) in the noon–midnight plane $y = 0$ containing $\mathbf{u}_{sw}$ and the interplanetary magnetic field $\mathbf{B}_{IMF}$. These figures show a first steep variation (at $x = 1.7R_M$ along the Sun–Mars line $z = 0$), a clear signature of the BS. The second

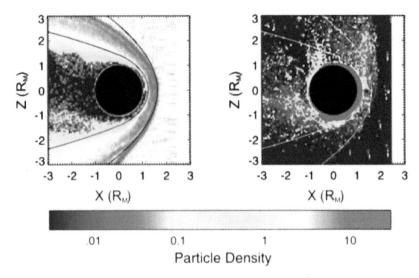

Figure 9.7. Two-dimensional cuts of H^+ (left) and O^+ (right) density (cm^{-3}) in the noon–midnight ($y = 0$) plane. The nominal locations of the BS and MPB are indicated in black in each panel. The curves indicate the location of the BS, first from the right, and MPB, second from the right. (Reproduced with permission from from [40, 41]. Copyright 2005 and 2010 Elsevier.)

steep increase (around $x = 1.3R_M$ along the Sun–Mars line $z = 0$), corresponding to the maximal value reached by the magnetic field intensity in the simulation domain, represents the magnetic pile-up boundary (MPB). It can also be considered as the boundary of the induced Martian magnetosphere. Three regions can be easily distinguished: the upstream region where the solar wind plasma density fluctuates around the undisturbed value $n_{H^+} = 2.3$ cm^{-3} (see figure 9.7(a)), the downstream region between BS and MPB, and, finally, the region below MPB where planetary ions dominate (see figure 9.7(b)). The maps show z-asymmetry, a consequence of the action of the motional electric field of the incoming plasma on the planetary ions which are accelerated mainly along the direction of the convection electric field, i.e. in the positive z-direction, during a fraction of their first gyroperiod. Since their gyroradii are of the order of the size of the Martian obstacle, this initial acceleration breaks the symmetry between the two sides of the planet. Finally, the plot of O^+ density (figure 9.7(b)) indicates the dominant pathways for escaping planetary ions: due to the absence of a global intrinsic magnetic field, the Martian atmosphere is exposed to a continuous erosion process because the ionized component can be efficiently removed by ion pickup mechanisms. The escaping proton flux comes from CX while photo-ionization is the main process which contributes to the escape of O^+ ions.

9.3.4 Spatially hybrid simulation: streamers and laser–plasma interaction

A variant of the hybrid approach is to resolve the electron subsystem as PIC-MCC in low-density/high-field regions while maintaining the fluid approach in the high-density/low-field regions. This spatially hybrid simulation is well suited for cases characterized by high gradients, where plasma density spans several orders of magnitude, such as streamer discharges [42] or ignition laser fusion [43]. A spatially hybrid approach allows an improvement in computational speed in excess of 40× for the one-dimensional case and >500× in two dimensions.

The spatially hybrid model contains two important distinct features: the calculation of transport and reaction coefficients for the fluid model taken from the particle model to keep the two models consistent; and the construction of the interface between the particle and fluid regions. The model interface can be set at the position where the electron density reaches $n_e = \eta n_{e,\text{max}}$ or where the electric field is $E = \xi E_{\text{max}}$; here $n_{e,\text{max}}$ and E_{max} stand for the maximal electron density and electric field, and η and ξ are real numbers between $(0, 1)$, and are chosen in such a manner that the calculation is efficient and the error is small. Different options for building the particle–fluid interface are reported in [42].

Usually spatially hybrid simulations begin with a few electron–ion pairs followed by the full particle model. As new electrons are generated, the number of particles eventually reaches a given threshold, after which the simulation switches from the full particle to hybrid approach. A this point the typical spatially hybrid cycle is repeated, consisting of the following steps:

1. the particle–fluid boundary is determined and, in order to ensure a reasonable interaction between the particle and fluid approaches, buffer regions are built by extending the particle regions into the fluid region;
2. the movement of electrons is first followed in the particle and then in the buffer regions, the number of electrons crossing the model interface is recorded during this time step;
3. the electron fluxes on the interface are used as the boundary condition to update the densities in the fluid region; and
4. the electron and ion densities are now known in both the particle and fluid regions, and the new electric field is then calculated. With the updated electric field and the particle densities, the new model interface is determined.

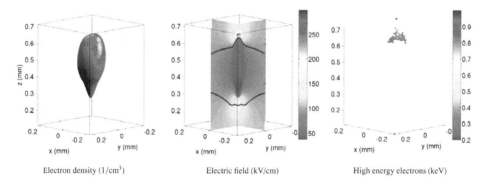

Electron density (1/cm³)   Electric field (kV/cm)   High energy electrons (keV)

Figure 9.8. Three-dimensional simulation results of the three-dimensional hybrid model for a negative streamer in air without photo-ionization developing in a background field of 100 kV cm⁻¹. The columns show from left to right: the bulk electron density isosurface at $n_e = 10^{13}$ cm⁻³, the electric field on two orthogonal planes, and electrons with energy larger than 200 eV. The position of the particle–fluid interface is indicated in red in the middle column.

Here, the example of a streamer discharge is reported as spatial hybrid model application. Streamers are ionized fingers that penetrate into a previously non-ionized region under the action of a strong electric field (larger than the breakdown threshold). The resulting electron avalanche leaves behind positive ions, so in time more and more space charge builds up. If the electric field from the space charge becomes comparable to the background electric field, new avalanches predominantly grow in the high-field regions, so a self-propagating structure, known as a streamer, can emerge. Figure 9.8 shows the three-dimensional structure of a negative streamer in air as result of a spatially hybrid simulation [42]. The columns show, from left to right: the electron density isosurface at $n_e = 10^{13}$ cm^{-3}, the electric field on two orthogonal planes, and electron macroparticles with energy larger than 200 eV. The position of the particle–fluid interface used in the simulation is indicated in red in the middle column. It should be noted how the hybrid simulation is able to trace the electrons with high energies that have the potential to run away.

Bibliography

[1] Brackbill J and Cohen B 1985 *Multiple Time Scales* (San Diego, CA: Academic)

[2] Lipatov A S 2002 *The Hybrid Multiscale Simulation Technology* (Berlin: Springer)

[3] Tskhakaya D, Matyash K, Schneider R and Taccogna F 2007 The particle-in-cell method, *Contrib. Plasma Phys.* **47** 563–94

[4] Mason R J 1971 Computer simulation of ion-acoustic shocks: the diaphragm problem *Phys. Fluids* **14** 1943–58

[5] Chodura R 1975 A hybrid fluid-particle model of ion heating in high-Mach-number shock waves *Nucl. Fusion* **15** 55

[6] Sgro A G and Nielson C W 1976 Hybrid model studies of ion dynamics and magnetic field diffusion during pinch implosions *Phys. Fluids* **19** 126–33

[7] Hewett D 1980 A global method of solving the electron-field equations in a zero-inertia-electron-hybrid plasma simulation code *J. Comput. Phys.* **38** 378–95

[8] Harned D S 1982 Quasineutral hybrid simulation of macroscopic plasma phenomena *J. Comput. Phys.* **47** 452–62

[9] Brecht S H and Thomas V A 1988 Multidimensional simulations using hybrid particles codes *Comput. Phys. Commun.* **48** 135–43

[10] Moore K R, Thomas V A and McComas D J 1991 Global hybrid simulation of the solar wind interaction with the dayside of Venus *J. Geophys. Res.: Space Phys.* **96** 7779–91

[11] Müller J, Simon S, Motschmann U, Schüle J, Glassmeier K A and Pringle G J 2011 AIKEF: adaptive hybrid model for space plasma simulations *Comput. Phys. Commun.* **182** 946–66

[12] Leclercq L, Modolo R, Leblanc F, Hess S and Mancini M 2016 3D magnetospheric parallel hybrid multi-grid method applied to planet–plasma interactions *J. Comput. Phys.* **309** 295–313

[13] Matthews A P 1994 Current advance method and cyclic leapfrog for 2D multispecies hybrid plasma simulations *J. Comput. Phys.* **112** 102–16

[14] Büchner J, Dum C and Scholer M 2003 *Space Plasma Simulation* vol 615 (Berlin: Springer)

[15] Ruyten W M 1993 Density-conserving shape factors for particle simulations in cylindrical and spherical coordinates *J. Comput. Phys.* **105** 224–32

[16] Verboncoeur J 2001 Symmetric spline weighting for charge and current density in particle simulation *J. Comput. Phys.* **174** 421–7

[17] Cornet C and Kwok D T 2007 A new algorithm for charge deposition for multiple-grid method for PIC simulations in *r–z* cylindrical coordinates *J. Comput. Phys.* **225** 808–28

[18] Amano T, Higashimori K and Shirakawa K 2014 A robust method for handling low density regions in hybrid simulations for collisionless plasmas *J. Comput. Phys.* **275** 197–212

[19] Horowitz E J, Shumaker D E and Anderson D V 1989 QN3D: a three-dimensional quasi-neutral hybrid particle-in-cell code with applications to the tilt mode instability in field reversed configurations *J. Comput. Phys.* **84** 279–310

[20] Terasawa T, Hoshino M, Sakai J-I and Hada T 1986 Decay instability of finite-amplitude circularly polarized Alfven waves: a numerical simulation of stimulated brillouin scattering *J. Geophys. Res.: Space Phys.* **91** 4171–87

[21] Hockney R W and Eastwood J W 1989 *Computer Simulation Using Particles* (New York: IOP Publishing)

[22] Taccogna F, Longo S and Capitelli M 2002 Particle-in-cell with Monte Carlo simulation of SPT-100 exhaust plumes *J. Spacecraft Rockets* **39** 409–19

[23] Taccogna F, Longo S and Capitelli M 2004 Very-near-field plume simulation of a stationary plasma thruster *Eur. Phys. J. Appl. Phys.* **28** 113–22

[24] Taccogna F, Pagano D, Scortecci F and Garulli A 2014 Three-dimensional plume simulation of multi-channel thruster configuration *Plasma Sources Sci. Technol.* **23** 065034

[25] Ellegaard O, Nedelea T, Schou J and Urbassek H 2002 Plume expansion of a laser-induced plasma studied with the particle-in-cell method *Appl. Surface Sci.* **197–198** 229–38

[26] Taccogna F, Longo S and Capitelli M 2003 A particle-in-cell/Monte Carlo model of the Ar$^+$ ion collection in He gas by a cylindrical Langmuir probe in the transition regime *Eur. Phys. J.* **22** 29–39

[27] Taccogna F, Longo S and Capitelli M 2004 PIC model of the ion collection by a Langmuir probe *Contrib. Plasma Phys.* **44** 594–600

[28] Taccogna F, Longo S and Capitelli M 2006 Ion orbits in a cylindrical Langmuir probe *Phys. Plasmas* **13** 043501

[29] Taccogna F, Longo S and Capitelli M 2008 Ion–neutral collision effects in Langmuir probe theory *Contrib. Plasma Phys.* **48** 509–14

[30] Lampe M 2001 Limits of validity for orbital-motion-limited theory for a small floating collector *J. Plasma Phys.* **65** 171–80

[31] Fife J and Martinez-Sanchez M 1995 Two-dimensional hybrid particle-in-cell (PIC) modeling of Hall thrusters *Proc. 24th Int. Electric Propulsion Conf. IEPC95-240* (Fairview Park, OH: Electric Rocket Propulsion Society) pp 1213–24

[32] Hagelaar G J M, Bareilles J, Garrigues L and Boeuf J P 2002 Two-dimensional model of a stationary plasma thruster *J. Appl. Phys.* **91** 5592–8

[33] Parra F, Ahedo E, Fife J and Martinez-Sanchez M 2006 A two-dimensional hybrid model of the Hall thruster discharge *J. Appl. Phys.* **100** 023304

[34] Sommier E, Scharfe M K, Gascon N, Cappelli M A and Fernandez E 2007 Simulating plasma-induced Hall thruster wall erosion with a two-dimensional hybrid model *IEEE Trans. Plasma Sci.* **35** 1379–87

[35] Mikellides I G and Katz I 2012 Numerical simulations of Hall-effect plasma accelerators on a magnetic-field-aligned mesh *Phys. Rev. E* **86** 046703

[36] Hagelaar G J M 2007 Modelling electron transport in magnetized low-temperature discharge plasmas *Plasma Sources Sci. Technol.* **16** S57

[37] Furubayashi M, Hoshino K, Toma M, Hatayama A, Coster D, Schneider R, Bonnin X, Kawashima H, Asakura N and Suzuki Y 2009 Comparison of kinetic and fluid neutral models for attached and detached state *J. Nucl. Mater.* **390–391** 295–8

[38] Dendy R O and McClements K G 2015 Ion cyclotron emission from fusion-born ions in large tokamak plasmas: a brief review from JET and TFTR to ITER *Plasma Phys. Control. Fusion* **57** 044002

[39] Brecht S H and Ledvina S A 2010 The loss of water from Mars: numerical results and challenges *Icarus* **206** 164–73

[40] Modolo R, Chanteur G M, Dubinin E and Matthews A P 2005 Influence of the solar EUV flux on the Martian plasma environment *Ann. Geophys.* **23** 433–44

[41] Brain D *et al* 2010 A comparison of global models for the solar wind interaction with Mars *Icarus* **206** 139–51

[42] Li C, Ebert U and Hundsdorfer W 2012 Spatially hybrid computations for streamer discharges: II. Fully 3D simulations *J. Comput. Phys.* **231** 1020–50

[43] Cohen B I, Kemp A and Divol L 2010 Simulation of laser–plasma interactions and fast-electron transport in inhomogeneous plasma *J. Comput. Phys.* **229** 4591–612

Part III

Applications

IOP Publishing

Plasma Modeling
Methods and Applications
Gianni Coppa, Antonio D'Angola, Ivo Furno, Davide Bernardi, Philippe Guittienne, Alan Howling,
Remy Jacquier and Renato Zaffina

Chapter 10

Radio frequency inductively coupled discharges in thermal plasmas

10.1 Introduction

The term 'thermal plasma' is usually used to mean a partially ionized gas in local thermodynamic equilibrium (LTE), or close to that condition [1]. Thermal plasmas are characterized by a large range of industrial applications, such as surface treatment, cutting, welding, spraying, and waste destruction [2, 3]. Under LTE condition, the plasma behaves as a conductive fluid and it can be modeled using the magnetohydrodynamics equations.

In radio frequency (RF) induction plasmas, the energy transferred into the discharge is governed by electromagnetic coupling. The energy transfer mechanism is basically the same as that used in the induction heating of metals [4]. However, the plasma gas has a much lower electrical conductivity and the frequencies required are typically in the megahertz range, while for metals they are in the kilohertz range. In the case of cylindrical torches, plasma coupling with the electromagnetic field is obtained by applying an RF voltage to coils surrounding the torch. The AC current flowing in the coil induces an axial oscillating magnetic field which generates an azimuthal electric field and, consequently, an induction current that sustains the plasma through Joule heating. The induction current flows in the opposite direction with respect to the coil current, thus inducing a magnetic field opposite to that generated by the coils. As a result, the plasma current flow is localized in a small region whose thickness is of the order of the skin depth, which is a function of the electrical conductivity of the plasma and the frequency [5]. This current shields the external magnetic field, which cannot penetrate the plasma, particularly at very high frequencies, and a typical off-axis peak of the current density is observed. Currently, innovative RF plasma planar discharges powered by resonant antennae are being developed in order to increase the processing area and to improve the power

doi:10.1088/978-0-7503-1200-4ch10
© IOP Publishing Ltd 2016

performance [6]. Coupling between the resonant antenna and gas is attained in a planar discharge following the same principle as a conventional RF plasma torch.

The power delivered to the plasma, which is proportional to the electric conductivity of the gas, must balance conductive, convective, and radiative losses to maintain the discharge at a sufficiently high temperature. Moreover, molecular gases or high-pressure regimes would require a much greater energy density than monatomic gases, or a lower pressure, to obtain a plasma. In fact, the minimum sustaining power increases with gas pressure and this explains why the ignition of inductively coupled plasma discharges is usually performed at low pressure. Once the discharge is stable, it is possible to increase the pressure with a corresponding increase in the plasma power. Since the energy density is inversely proportional to the skin depth [3], a direct relationship between the energy density and the frequency is established and, at a given pressure, the minimum power required to sustain the discharge drops with the increase of the oscillator frequency [7].

The pressure of the discharge plays a crucial role in the choice of the proper theoretical model for studying inductively coupled plasmas. Two different approaches can be used: a one-temperature model; and a two- or multi-temperature model. In the one-temperature model, both the heavy particles and electrons have the same temperature. When significant deviations from kinetic equilibrium are present, the temperature of the heavy particles and electrons are different and the plasma is defined as non-thermal. In this case, a two- or multi-temperature model must be considered. The parameter that plays a crucial role in attaining kinetic equilibrium is the ratio between the electric field E and the pressure p. Kinetic equilibrium is reached for small values of E/p, i.e. when the pressure is high or the electric field is small. Through Dalton's law, LTE can be reached for high values of gas density, when the mean free path for ions and electrons is reasonably small and, consequently, the collision frequency is high enough to thermalize plasma particles. Typically, in atmospheric or sub-atmospheric discharges plasma can be considered to be in LTE conditions and a one-temperature model can be adopted.

This chapter is organized as follows. In section 10.2, the physical model and the equations to be solved are presented. In section 10.3, the commonly employed numerical techniques are introduced. Section 10.4 is devoted to the modeling of conventional (cylindrical) torches, while in the final section an innovative type of plasma torch is presented.

10.2 Physical model for an LTE inductively coupled plasma torch

The physical model of the torch is based upon the coupling of fluid dynamics and electromagnetism equations.

10.2.1 Fluid dynamics model

Since the pressure variations within the torch are negligible, the gas properties, the density in particular, are assumed to be dependent on the temperature alone. Therefore, the continuity equation,

$$\frac{\partial \rho}{\partial t} = -\mathrm{div}\,(\rho \mathbf{u}), \tag{10.1}$$

cannot be considered as the equation that determines the evolution of ρ: $\rho = \rho(T)$ is the evolution of the temperature that determines the value of ρ. In other terms, writing equation (10.1) as

$$\frac{\mathrm{d}\rho}{\mathrm{d}t} = \frac{\partial \rho}{\partial t} + \mathbf{u} \cdot \nabla \rho = -\rho \, \mathrm{div}\,\mathbf{u}, \tag{10.2}$$

and being $\rho = \rho(T)$, one has

$$\mathrm{div}\,\mathbf{u} = -\frac{1}{\rho}\left(\frac{\partial \rho}{\partial T}\right)_p \frac{\mathrm{d}T}{\mathrm{d}t} = -\frac{1}{\rho}\left(\frac{\partial \rho}{\partial T}\right)_p \left(\frac{\partial T}{\partial t} + \mathbf{u} \cdot \nabla T\right), \tag{10.3}$$

or, equivalently,

$$\mathrm{div}(\rho \mathbf{u}) = -\left(\frac{\partial \rho}{\partial T}\right)_p \frac{\partial T}{\partial t}. \tag{10.4}$$

Therefore, equation (10.3) determines the value of div $\mathbf{u}$ that gives a fluid motion compatible with the thermal dilation of the gas.

The evolution of the velocity field, $\mathbf{u}(\mathbf{x}, t)$, is governed by the Navier–Stokes equation

$$\rho\frac{\mathrm{d}u_i}{\mathrm{d}t} = \rho\left(\frac{\partial u_i}{\partial t} + \mathbf{u} \cdot \nabla u_i\right) = \mathrm{div}\left[\mu\left(\nabla u_i + \frac{\partial \mathbf{u}}{\partial x_i}\right)\right] - \nabla p + \mathbf{j} \times \mathbf{B}, \tag{10.5}$$

where the negligible terms have been eliminated. The term $\mathbf{j} \times \mathbf{B}$ represents the Lorentz force density and it can be written as $\sigma(T)\mathbf{E} \times \mathbf{B}$, with σ the conductivity of the ionized gas. Since the fields $\mathbf{E}$ and $\mathbf{B}$ are oscillating with frequencies of the order of 10^7 Hz, the instantaneous value of the Lorentz force can be replaced by the average over the RF period: $\mathbf{j} \times \mathbf{B} \to \sigma(T)\langle\mathbf{E} \times \mathbf{B}\rangle$. In general, the pressure p is a function of ρ and T according to the state equation, but, under the hypothesis of an incompressible fluid, the strategy to determine p must be different. If one assumes to know the velocity field at instant $t - \Delta t$ and uses equation (10.5) to calculate the value at time t, it follows that

$$\rho\mathbf{u}(\mathbf{x}, t) = \rho\mathbf{u}(\mathbf{x}, t - \Delta t) + \{\mathbf{W}(\mathbf{x}, t) - \nabla p(\mathbf{x}, t)\}\Delta t, \tag{10.6}$$

where $\mathbf{W}$ contains all the terms of the Navier–Stokes equation apart from the pressure. Substituting equation (10.6) into equation (10.4), one obtains

$$\nabla^2 p = \frac{1}{\Delta t}\left\{\frac{1}{\rho}\left(\frac{\partial \rho}{\partial T}\right)_p \frac{\partial T}{\partial t}(\mathbf{x}, t) - \frac{1}{\rho}\left(\frac{\partial \rho}{\partial T}\right)_p \frac{\partial T}{\partial t}(\mathbf{x}, t - \Delta t)\right\} + \mathrm{div}\,\mathbf{W}(\mathbf{x}, t). \tag{10.7}$$

Equation (10.7) determines the pressure field that gives a velocity field compatible with the continuity equation.

Finally, the evolution of the temperature, $T(\mathbf{x}, t)$ is determined by

$$\rho c_p \frac{\mathrm{d}T}{\mathrm{d}t} = \rho c_p \left(\frac{\partial T}{\partial t} + \mathbf{u} \cdot \nabla T \right) = \mathrm{div}\,(k \nabla T) + \mathbf{j} \cdot \mathbf{E} - q_R, \qquad (10.8)$$

where $\mathbf{j} \cdot \mathbf{E} = \sigma E^2$ is the density of the power dissipated by the electromagnetic field (the Joule effect), while q_R is the power density related to the irradiation.

Instead of the temperature, one can use the specific enthalpy, $h(t)\,(\mathrm{d}h/\mathrm{d}T = c_p(T))$. In this case, equation (10.8) becomes

$$\rho \frac{\mathrm{d}h}{\mathrm{d}t} = \rho \left(\frac{\partial h}{\partial t} + \mathbf{u} \cdot \nabla h \right) = \mathrm{div}\left(\frac{k}{c_p} \nabla h \right) + \sigma \langle E^2 \rangle - q_R, \qquad (10.9)$$

having used again the averaged value of E^2. To discretize equations (10.5) and (10.9), it turns out useful to write them as

$$\begin{cases} \dfrac{\partial}{\partial t}(\rho u_i) = \mathrm{div}\left(-\rho u_i \mathbf{u} + \mu \left(\nabla u_i + \dfrac{\partial \mathbf{u}}{\partial x_i} \right) \right) - \dfrac{\partial p}{\partial x_i} + \sigma \langle \mathbf{E} \times \mathbf{B} \rangle, \\[4mm] \dfrac{\partial}{\partial t}(\rho h) = \mathrm{div}\left(-\rho h \mathbf{u} + \dfrac{k}{c_p} \nabla h \right) + \sigma \langle E^2 \rangle - q_R. \end{cases} \qquad (10.10)$$

10.2.2 Equations for the electromagnetic field

Assuming

$$\begin{aligned} \mathbf{E}(\mathbf{x}, t) &= \mathrm{Re}\!\left(\mathscr{E}(\mathbf{x}, t) \mathrm{e}^{\mathrm{i}\omega t} \right), \\ \mathbf{B}(\mathbf{x}, t) &= \mathrm{Re}\!\left(\mathscr{B}(\mathbf{x}, t) \mathrm{e}^{\mathrm{i}\omega t} \right), \\ \mathbf{j}(\mathbf{x}, t) &= \mathrm{Re}\!\left(\mathscr{J}(\mathbf{x}, t) \mathrm{e}^{\mathrm{i}\omega t} \right), \end{aligned} \qquad (10.11)$$

and considering that the time variation of the field amplitudes $\mathscr{E}$ and $\mathscr{B}$ is much slower with respect to the RF period, i.e.

$$\frac{|\partial \mathscr{E}/\partial t|}{|\mathscr{E}|}, \quad \frac{|\partial \mathscr{B}/\partial t|}{|\mathscr{B}|} \ll \omega, \qquad (10.12)$$

Maxwell's equations reduce to

$$\begin{cases} \nabla \times \mathscr{B} = \mu_0 \left(\sigma \mathscr{E} + \mathscr{J}_{\mathrm{coil}} \right) + \dfrac{\mathrm{i}\omega}{c^2} \mathscr{E}, \\[3mm] \nabla \times \mathscr{E} = -\mathrm{i}\,\omega \mathscr{B}. \end{cases} \qquad (10.13)$$

The coil current could be written as $\sigma_{coil}\mathscr{E}$, however, if one does not want to include the RF generator in the model, $\mathscr{J}_{coil}$ can be regarded as a known quantity. In particular, it will be assumed that $\mathrm{div}\,\mathscr{J}_{coil} = 0$. By taking the divergence of the first equation (10.13), one obtains

$$\left(\mu_0\sigma + \frac{i\omega}{c^2}\right)\mathrm{div}\,\mathscr{E} + \mu_0\nabla\sigma \cdot \mathscr{E} = 0. \tag{10.14}$$

In the cases considered here, $\nabla\sigma$ and $\mathscr{E}$ are always perpendicular (in the cylindrical torch (section 10.4) σ is a function of r and z, while $\mathscr{E}$ is parallel to $\mathbf{e}_\theta$; in the planar device (section 10.5) σ depends on x and y, and $\mathscr{E}$ is along the z-direction). Therefore, the condition $\mathrm{div}\mathscr{E} = 0$ is assumed as valid in the following. By taking the curl of the second equation (10.13), one has

$$-\nabla^2\mathscr{E} + \nabla\mathrm{div}\,\mathscr{E} = -i\omega\mu_0\left(\sigma\mathscr{E} + \mathscr{J}_{coil}\right) + \frac{1}{\lambda^2}\mathscr{E}. \tag{10.15}$$

with $\lambda = \omega/c$ the RF wavelength. In equation (10.15), the second term on the left-hand side is zero, while the last term on the right-hand side is negligible. In fact, $|\nabla^2\mathscr{E}|$ can be roughly evaluated as $|\mathscr{E}|/L^2$, with L a characteristic dimension of the device (of the order of $10^{-2} \div 10^{-1}$ m), while λ is of the order of 10 m. Thus, the correct equation for $\mathscr{E}$ is

$$\nabla^2\mathscr{E} = i\omega\mu_0\left(\sigma\mathscr{E} + \mathscr{J}_{coil}\right). \tag{10.16}$$

In summary, equations (10.1), (10.10), and (10.16) constitute the mathematical model for an inductively coupled plasma torch. The temperature field changes the electric conductivity and, consequently, it modifies the $\mathscr{E}$ and $\mathscr{B}$ fields. Furthermore, $\sigma\langle E^2\rangle = \frac{\sigma}{2}|\mathscr{E}^2|$ and $\langle \mathbf{E} \times \mathbf{B}\rangle = \frac{1}{2}\mathrm{Re}(\mathscr{E} \times \mathscr{B}^*)$ affect the temperature and velocity fields.

10.3 The numerical model for a plasma torch

The physical behavior of inductively coupled plasma torches at atmospheric pressure is studied by solving the mass, momentum, and energy conservation equations coupled with Maxwell's equations. Steady-state or time-dependent solutions can be obtained by using the SIMPLER algorithm (Semi-Implicit Method for Pressure Linked Equations Revisited) [8], which is implemented in commercial software tools such as ANSYS/FLUENT. These commercial codes offer important advantages, such as the possibility of studying complicated geometries and of generating both structured and non-structured meshes. The coupling between fluid dynamics and Maxwell equations can be realized by resorting to a user-defined function (UDF) or to user-defined scalars, which obey general partial differential equations that are already implemented in the solver [9].

Time-dependent codes can be used to gradually achieve a steady-state working configuration, which is difficult to obtain directly using static codes. Moreover, the time-dependent simulation of a discharge represents a tool to predict the final plasma temperature and velocity fields for some types of non-conventional torch configurations, where the presence of multiple self-sustaining equilibrium configurations may cause the failure of the numerical procedure when using static analysis [10].

By suitably setting Γ, Φ, and Λ, each of the fluid dynamics equations of the previous section can be written in the form

$$\frac{\partial(\rho\Phi)}{\partial t} + \nabla \cdot (\rho u\Phi) = \nabla \cdot (\Gamma\nabla\Phi) + \Lambda, \tag{10.17}$$

which represents a generalized conservation equation composed of the time derivative of the physical quantity, $\frac{\partial(\rho\Phi)}{\partial t}$, the convective term, $\nabla \cdot (\rho u\Phi)$, the diffusive term, $\nabla \cdot (\Gamma\nabla\Phi)$, and the source term, Λ. The whole numerical procedure will be described briefly for equation (10.17), thus obtaining a formulation that is valid for all the equations of interest. Equation (10.17) can be rewritten in a different form, as

$$\frac{\partial(\rho\Phi)}{\partial t} = -\text{div}\,\boldsymbol{J} + \Lambda, \tag{10.18}$$

where $\boldsymbol{J}$ is the total (convective and diffusive) current density for the physical quantity Φ:

$$\boldsymbol{J} = \rho u\Phi - \Gamma\nabla\Phi. \tag{10.19}$$

Equations (10.17) and (10.18) can be transformed into a set of algebraic equations by using suitable space and time discretization techniques. For simplicity, here a two-dimensional case is considered: all quantities are defined in a space domain $\mathscr{D} \subset \mathbb{R}^2$, which is decomposed in rectangular cells. With reference to figure 10.1, the central point of one of these cells is indicated by P, while the nodes of the contiguous cells are N, S, E, and W. If equation (10.17) is integrated over the cell, one obtains

$$\frac{\mathrm{d}}{\mathrm{d}t}\left(\Delta x\Delta y\overline{(\rho\phi)}_P\right) = -\Delta x\left(\bar{J}_n + \bar{J}_s\right) - \Delta y\left(\bar{J}_e + \bar{J}_w\right) + \Delta x\Delta y\bar{\Lambda}_P, \tag{10.20}$$

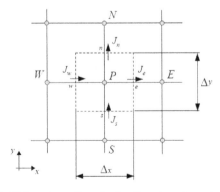

Figure 10.1. Finite-volume discretization for a two-dimensional domain.

where $\bar{f}_P$ indicates the mean value of f over the cell, while $\bar{J}_i$ is the mean value of $\mathbf{J} \cdot \mathbf{n}$ on the i-face ($\mathbf{n}$ is the outgoing normal unit vector). Now the finite volume methodology can be used. According to this technique, $\bar{f}_P$ is approximated simply as $f_P \equiv f(\mathbf{x}_P)$, while $\bar{J}_i$ can be evaluated using the finite difference method. For example, for $\bar{J}_e$ one has

$$\bar{J}_e \simeq \frac{\rho_E u_{x,E} \Phi_E + \rho_P u_{x,P} \Phi_P}{2} - \frac{\Gamma_E + \Gamma_P}{2} \frac{\Phi_E - \Phi_P}{\Delta x}. \tag{10.21}$$

In a similar way, $\bar{J}_w$, $\bar{J}_n$, and $\bar{J}_s$ can be calculated. More sophisticated approximations exist; for a complete discussion, the reader is referred to [11]. Finally, the space-discretized equation (10.20) in the cell with center $\mathbf{x}_P$ has the form

$$\frac{\mathrm{d}}{\mathrm{d}t}(\rho_P \phi_P) = a_E \Phi_E + a_W \Phi_W + a_N \Phi_N + a_S \Phi_S + a_P \Phi_P + \Lambda_P, \tag{10.22}$$

where the coefficients a_i are readily calculated using equations (10.20) and (10.22), or different approximations. In general, they must satisfy the following property: if ρ is constant, their sum must be zero. In fact, in this case

$$\Phi = \text{Const} \Rightarrow \frac{\partial}{\partial t}(\rho \Phi) = 0. \tag{10.23}$$

The next step is time discretization. The simplest way (the simplest among stable methods) to do this is the Euler implicit method:

$$\begin{aligned} \rho_P(t_m)\Phi_P(t_m) = {} & \rho_P(t_{m-1})\Phi_P(t_{m-1}) + \{a_E(t_m)\Phi_E(t_m) + a_W(t_m)\Phi_W(t_m) \\ & + a_N(t_m)\Phi_N(t_m) + a_S(t_m)\Phi_S(t_m) + \Lambda_P(t_m)\}\Delta t. \end{aligned} \tag{10.24}$$

Equation (10.24), associated with the proper boundary conditions, forms a system of algebraic equations to be solved at any time step t_m once the solution at t_{m-1} is known. In general, it is a non-linear system, as coefficients depend on the unknowns. The SIMPLER algorithm is a useful tool to solve continuity and momentum equations [8]. This algorithm, introduced in 1980 by S V Patankar, is efficient and stable; it represents an improved version of the SIMPLE algorithm, developed by Patankar himself and Spalding [12].

The SIMPLER algorithm uses the continuity equation as a constraint on the pressure field. In fact, the unknown pressure appears with its gradient in the source term of the momentum equation. With the correct pressure field in the momentum equations, the resulting velocity field satisfies the continuity equation, as described in section 10.2. For this reason the pressure field is iteratively refined until the 'correct' field is obtained, such that the velocity field that satisfies the continuity equation.

In the momentum equation, the pressure gradient appears in the source term Λ and must be integrated over the reference control volume (see figure 10.1). In the

Navier–Stokes equation, the term ∇p must be evaluated. For example, according to the finite-volume technique, $\left(\frac{\overline{\partial p}}{\partial x}\right)_p$ can be written as

$$\left(\frac{\overline{\partial p}}{\partial x}\right)_p = \frac{\bar{p}_w - \bar{p}_e}{\Delta x} \simeq \frac{1}{\Delta x}\left(\frac{p_W + p_P}{2} - \frac{p_P + p_E}{2}\right) = \frac{p_W - p_E}{2\Delta x}. \qquad (10.25)$$

Equation (10.25) shows that the momentum equation will contain the pressure difference between two alternate grid points and not between adjacent ones. This implies that the pressure is taken from a coarser grid than the one actually employed with a consequent reduction of the accuracy of the solution. Moreover, a 'checkerboard' pressure field would produce a vanishing contribution. In order to avoid these drawbacks, velocity components are calculated on a staggered grid [12]. Details of the iterative procedure can be found in [8, 11].

10.3.1 Thermodynamic and transport properties

In order to simulate inductively coupled thermal plasmas, the thermodynamic and transport properties are necessary data [13]. Several sets of thermodynamic and transport properties have been obtained in LTE conditions by means of the Chapman–Enskog approximation of the Boltzmann equation. The main difficulty is finding accurate data for a plasma mixture with a variable composition of species. Computational fluid dynamics solvers usually do not provide these data due to the huge number of possible plasma mixtures under investigation. An example of software that provides these data is EquilTheTA (Equilibrium Thermodynamics and Transport Applications) [1, 14]. EquilTheTA is an efficient computational tool for the calculation of thermodynamic properties in the framework of statistical thermodynamics, and transport properties, using the high-order Chapman–Enskog method, for gas mixtures and plasmas.

10.4 A conventional inductively coupled plasma torch at atmospheric pressure

The typical geometry of an RF induction torch is shown in figure 10.2. This system can be studied by using an axial-symmetric two-dimensional model [15].

The unknowns of the model are:
- the velocity, $\mathbf{u}(r, z) = u_r(r, z)\mathbf{e}_r + u_z(r, z)\mathbf{e}_z$;
- the temperature, $T(r, z)$;
- the pressure, $p(r, z)$;
- the electric field, $\mathscr{E}(r, z) = \mathscr{E}_\theta(r, z)\mathbf{e}_\theta$;
- the magnetic field, $\mathscr{B}(r, z) = \mathscr{B}_r(r, z)\mathbf{e}_r + \mathscr{B}_z(r, z)\mathbf{e}_z$.

As discussed in section 10.2, the density is not a real unknown, as it is a function of T, and the continuity equation provides a physical constraint to the pressure field. Moreover, as the density current of the coils is parallel to $\mathbf{e}_\theta$, it produces a magnetic field with a zero θ component. As $\mathscr{B}_r$ and $\mathscr{B}_z$ do not depend on θ, the r and z

Figure 10.2. Scheme of a cylindrical induction plasma torch (left) and plasma torches manufactured by Tekna (right; models PN35, PN50, PN70) (from [16]).

components of $\nabla \times \mathscr{B}$ vanish, and, consequently, the only non-zero component of $\mathscr{E}$ is $\mathscr{E}_\theta$. In (r, z) geometry, the fluid dynamics equations presented in section 10.2 assume the form

$$\frac{\partial \rho}{\partial t} + \frac{1}{r}\frac{\partial}{\partial r}(r\rho u_r) + \frac{\partial}{\partial z}(\rho u_z) = 0, \qquad (10.26)$$

$$\rho\left(\frac{\partial u_r}{\partial t} + u_r\frac{\partial u_r}{\partial r} + u_z\frac{\partial u_r}{\partial z}\right) = -\frac{\partial p}{\partial r} - \frac{2\mu u_r}{r^2} + \frac{2}{r}\frac{\partial}{\partial r}\left(\mu r\frac{\partial u_r}{\partial r}\right)$$
$$+ \frac{\partial}{\partial z}\left[\mu\left(\frac{\partial u_r}{\partial z} + \frac{\partial u_z}{\partial r}\right)\right] + \frac{\sigma}{2}\mathrm{Re}\left(\mathscr{E}_\theta\mathscr{B}_z^*\right) \qquad (10.27)$$

$$\rho\left(\frac{\partial u_z}{\partial t} + u_r\frac{\partial u_z}{\partial r} + u_z\frac{\partial u_z}{\partial z}\right) = -\frac{\partial p}{\partial z} + 2\frac{\partial}{\partial z}\left(\mu\frac{\partial u_z}{\partial z}\right)$$
$$+ \frac{1}{r}\frac{\partial}{\partial r}\left[r\mu\left(\frac{\partial u_z}{\partial r} + \frac{\partial u_r}{\partial z}\right)\right] - \frac{\sigma}{2}\mathrm{Re}\left(\mathscr{E}_\theta\mathscr{B}_r^*\right) \qquad (10.28)$$

$$\rho c_p\left(\frac{\partial T}{\partial t} + u_r\frac{\partial T}{\partial r} + u_z\frac{\partial T}{\partial z}\right) = \frac{1}{r}\frac{\partial}{\partial r}\left(rk\frac{\partial T}{\partial r}\right) + \frac{\partial}{\partial z}\left(k\frac{\partial T}{\partial z}\right) + \frac{\sigma}{2}|\mathscr{E}_\theta|^2 - q_R. \qquad (10.29)$$

The boundary conditions for the velocity and the temperature fields are summarized below:

- *Inlet* ($z = -L_T/2$):

$$u_r(r, -L_T/2) = 0$$

$$u_z(r, -L_T/2) = Q \cdot f(r), \qquad \int_0^{R_0} 2\pi r\mathrm{dr}\, f(r) = 1 \qquad (10.30)$$

$$T(r, -L_T/2) = T_{\mathrm{in}},$$

where $f(r)$ is a function which takes into account the velocity profiles.

- *Axis* ($r = 0$):

$$u_r(0, z) = 0, \quad \frac{\partial u_z}{\partial r}(0, z) = 0, \quad \frac{\partial T}{\partial r}(0, z) = 0 \tag{10.31}$$

- *Wall* ($r = R_0$):

$$u_r(R_0, z) = u_z(R_0, z) = 0, \quad -k\frac{\partial T}{\partial r}(R_0, z) = \frac{k_w}{dw}(T - T_{\text{ext}}), \tag{10.32}$$

where k_w and dw are the thermal conductivity and the thickness of the confinement tube; T_{ext} is the external temperature.
- *Exit* ($z = L_T/2$).

The (approximated) one-way conditions [8] can be applied:

$$\frac{\partial u_r}{\partial z}(r, L_T/2) = 0, \quad \frac{\partial (u_z)}{\partial z}(r, L_T/2) = 0, \quad \frac{\partial T}{\partial z}(r, L_T/2) = 0, \tag{10.33}$$

assuming that the post-coil region is long enough and heat convection usually prevails over conduction at the exit of the torch (i.e. the Peclet number is large).
In cylindrical coordinates, one obtains

$$\begin{cases} \dfrac{1}{r}\dfrac{\partial}{\partial r}\left(r\dfrac{\partial \mathscr{E}_\theta}{\partial r}\right) - \dfrac{\mathscr{E}_\theta}{r^2} + \dfrac{\partial^2 \mathscr{E}_\theta}{\partial z^2} = i\omega\mu_0(\sigma\mathscr{E}_\theta + \mathscr{J}_\theta), \\ \mathscr{B}_r = \dfrac{1}{i\omega}\dfrac{\partial \mathscr{E}_\theta}{\partial z}, \quad \mathscr{B}_z = -\dfrac{1}{i\omega r}\dfrac{\partial}{\partial r}(r\mathscr{E}_\theta). \end{cases} \tag{10.34}$$

To solve equations (10.34) particular attention must be paid to defining the computational domain and the proper boundary conditions. In principle, the physical domain for equation (10.34) is infinite, while it is finite for equations (10.26)–(10.29). Different strategies can be adopted. One way consists of writing $\mathscr{E}$ as $\mathscr{E}_0 + \mathscr{E}_p$, where $\mathscr{E}_0$ is generated by the coils when $\sigma \equiv 0$, i.e.

$$\nabla^2 \mathscr{E}_0 = i\omega\mu_0 \mathscr{J}_{\text{coil}}, \tag{10.35}$$

while $\mathscr{E}_p$ is due to the plasma and it satisfies the equation

$$\nabla^2 \mathscr{E}_p = i\omega\mu_0\sigma(\mathscr{E}_0 + \mathscr{E}_p). \tag{10.36}$$

As N is the number of turns of the coil, $\mathscr{E}_0$ can be evaluated analytically by summing the fields generated by N rings (each of them can be written in terms of elliptic functions [17]). The physical domain for $\mathscr{E}_p$ is still infinite, but it can be reduced to the domain of the torch using a suitable iterative procedure (as an example, one can set initially $\mathscr{E}_p = 0$ on the border and calculate the field, then the new value of $\mathscr{E}_p$ is used on the border, and so on). As an alternative, the plasma currents inside the torch can be approximated as a simple magnetic dipole (located at the center of the device and parallel to the z-direction). In this case, the electric field can be written as

$$\mathscr{E}_\theta(r, z) = \frac{\mathscr{A} \cdot r}{\left(r^2 + z^2\right)^{3/2}}. \tag{10.37}$$

By taking the radial derivative of equation (10.37) and eliminating the unknown constant, $\mathscr{A}$, one obtains the boundary condition to be applied on the lateral border of the domain ($r = R_0$):

$$\frac{\partial \mathscr{E}_\theta}{\partial r}(R_0, z) = \xi(z)\mathscr{E}_\theta(R_0, z), \quad \xi(z) = \frac{1}{R_0} - \frac{3R_0}{R_0^2 + z^2}. \tag{10.38}$$

In the same manner, by taking the z-derivative of equation (10.37), the following boundary condition is obtained for the upper and the lower borders ($z = \pm L_T/2$):

$$\frac{\partial \mathscr{E}_\theta}{\partial z}(r, \pm L_T/2) = \mp\eta(r)\mathscr{E}_\theta(r, \pm L_T/2), \quad \eta(r) = \frac{3L_T/2}{r^2 + (L_T/2)^2}. \tag{10.39}$$

On the torch axis ($r = 0$), the symmetry condition $E_\theta = 0$ is imposed.

Steady-state numerical results obtained by using ANSYS/FLUENT within the UDF technique are presented in figures 10.3 and 10.4. A structured grid with quadrilateral cells has been employed for the region inside the torch, while a non-structured mesh with triangular elements has been adopted outside (figure 10.3). The UDF approach consists of developing an external C-function to be linked to the FLUENT solver [9], which fully treats the electric field equation on the structured grid, by means of a finite difference technique. So, the coil region is accurately modeled (figure 10.3) and the fluid-dynamics equations are solved by FLUENT only

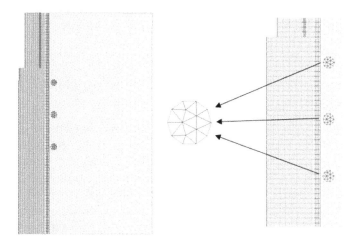

Figure 10.3. Computational meshes and details of the coil region discretization.

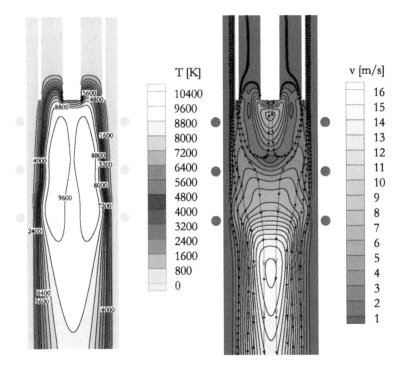

Figure 10.4. Plasma temperature [10^3 K] and velocity fields as obtained with the UDF approach.

inside the torch region on a different grid; moreover, all solid zones are treated in this way. The C-function receives the electric conductivity distribution coming from the FLUENT solver as input data and returns to FLUENT the values of the electric field after suitable interpolation on the fluid-dynamics grid. The results obtained for the temperature and velocity fields are presented in figure 10.4 where all the calculations have been performed for a two-dimensional axisymmetric torch under the assumptions of laminar flow for LTE and an optically thin argon plasma.

10.5 A resonant planar antenna as an inductive plasma source at atmospheric pressure

An RF plasma source employing a planar resonant antenna has recently been proposed [6, 18]. In industrial plasma applications (solar cell production, food packaging, waste treatment, space propulsion, plasma welding, material deposition, etc), the energy consumption of the plasma source and the possibility to increase the processed area and process rates are relevant issues. The device operates at 13.56 MHz and an RF power of 2 kW. It is located inside a low pressure vacuum vessel [6]. Argon plasmas are easily ignited and inductively coupled above a threshold power of about 60 W. This system is able to generate a large-area plasma with an excellent power efficiency. When excited at one of its resonance frequencies, the antenna develops very high currents within its structure, which can be used for inductively coupled plasma sources. A brief description of the principles of the antenna [6] and of a theoretical and numerical model of the plasma discharge at atmospheric pressure are presented in the following.

10.5.1 Basic principles of the system: the Helyssen ideal planar antenna

A resonant antenna can be planar or cylindrical. The structure and the working principles are similar, but a planar antenna is suited for large-area surface treatment, whereas the cylindrical configuration is adapted for plasma processes requiring a volume plasma source. In the ideal case, as shown in figure 10.5, the antenna is composed of N elements constituting copper tubes, called 'legs', each characterized by inductance L. Capacitors C link the legs together: they are called 'stringers'. Capacitors also present a small inductance M due to metal leads which connect them to the legs. In the ideal case, the impedances of both legs and stringers are purely imaginary.

From this parallel assembly of L and C components with open boundary conditions of the planar structure, $N - 1$ resonant modes m arise. Each of these modes is characterized by a resonance frequency ω_m and by a specific current distribution in each segment I_n, M_n, and K_n, and voltage distributions at nodes A_n and B_n. The resonance frequencies are given by [18]

$$\nu_m = \frac{1}{2\pi}\left[CM + 2LC \sin^2\left(\frac{m\pi}{2\pi}\right)\right]^{-1/2}, \quad m = 1, 2, ..., N - 1. \quad (10.40)$$

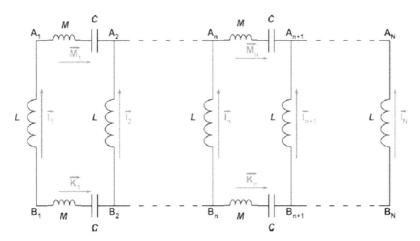

Figure 10.5. Electric circuit of the Helyssen ideal planar resonant antenna (from [6]).

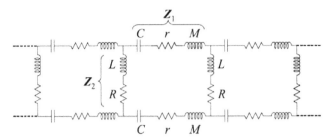

Figure 10.6. Dissipative planar antenna (from [18]).

Starting from equation (10.40) the following equations describing the current distribution in the legs and stringers can be derived:

$$M_n \propto \sin\left(n\frac{m\pi}{N}\right),\tag{10.41}$$

$$I_n \propto \sin\left(\frac{m\pi}{2N}\right)\cos\left[\left(n-\frac{1}{2}\right)\frac{m\pi}{N}\right].\tag{10.42}$$

In other terms, an N-leg ideal planar antenna presents $N - 1$ normal modes whose current amplitudes have a sinusoidal spatial distribution and oscillate in phase [6, 18]. It must be noted that the equivalent circuit of an ideal planar antenna does not include a generator; in fact, without dissipation, no power is needed to sustain voltage and currents.

In a real planar antenna, resistive elements must be considered on legs and stringers. With reference to figure 10.6, R and r represent the resistance of the copper tubes and of the metal leads which connect capacitors to copper tubes, respectively. Resonant modes are the same as calculated in equation (10.40) in the non-dissipative case, but an RF generator is needed to drive current in the antenna. More details about calculations are in [6, 18].

10.5.2 A real working system: geometrical and technical features

Working prototypes of resonant planar antennas have been designed and built at the Swiss Plasma Center of the École Polytechnique Fédérale de Lausanne [6] (figure 10.7).

A schematic view is shown in figures 10.8 and 10.9. The antenna consists of 23 copper tubes, each of length $L_b = 0.2$ m, which are assembled in parallel over a total length $L_a = 0.55$ m. The antenna is embedded in a silicone elastomer dielectric with a protective glass cover and a grounded metal baseplate. This assembly constitutes the plasma source module which is entirely placed in a vacuum vessel.

The antenna can easily be adapted to larger area processes by increasing the number of copper tubes. This does not affect the distribution of currents and voltages, which an important advantage of this system in comparison to other existing sources. The antenna operates at 13.56 MHz, which is the standard frequency for industrial applications. Consequently, the antenna has been designed

Figure 10.7. A general view of the device, in particular of the capacitors and the glass.

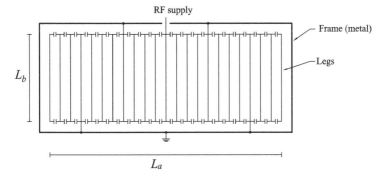

Figure 10.8. Schematic of the torch: top view (adapted from [6]).

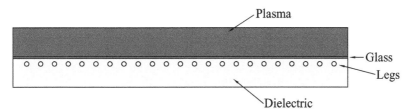

Figure 10.9. Schematic of the torch: section view (adaptetd from [6]).

to have one of its resonance frequencies at that value. Moreover, the mode $m = 6$ has been chosen because it generates higher currents in the legs with respect to lower modes and it provides better plasma uniformity.

Plasma is both a resistive and inductive load for the plasma source system and normal modes. At atmospheric pressure, higher values of plasma impedance (both resistive and inductive terms) are expected compared to low pressure conditions. Consequently, reducing the internal pressure by vacuum pumping is not necessary in order to run the system. This is an important advantage in terms of time, as the system can run directly, and in terms of cost. On the other hand, an atmospheric pressure system has two disadvantages: (i) stronger plasma–antenna coupling effects, resulting in a significant alteration of the antenna resonance modes; (ii) heavier heat loads on the the antenna, on the glass separator, and on the metallic covering of the system. The evaluation of the heat load is important to properly design a cooling system for the copper tubes of the antenna and, possibly, for the metallic walls and the glass.

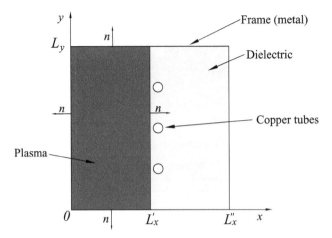

Figure 10.10. Computational domain (for simplicity, only three copper tubes are indicated).

10.5.3 Computational model

The system consists of a plasma confined in a rectangular box, excited by an electromagnetic field produced by the resonant planar antenna, and it can be modeled by a two-dimensional code in Cartesian geometry. In fact, all variables can be considered independent in the direction parallel to the copper tubes. With reference to the cross section of the system (figure 10.10), the torch can be considered a two-dimensional domain in which the plasma zone and dielectric zone can be distinguished. The boundary of the domain is represented by metallic walls surrounding the torch. They can be considered as perfect electric conductors.

The mathematical model is described again by equations (10.1), (10.10), and (10.16), which, in this case, assumes the form

$$
\begin{cases}
\dfrac{\partial \rho}{\partial t} = -\mathrm{div}(\rho \boldsymbol{u}) \\[2ex]
\dfrac{\partial (\rho u_x)}{\partial t} = \mathrm{div}\left(-\rho u_x \boldsymbol{u} + \mu \nabla u_x + \mu \dfrac{\partial \boldsymbol{u}}{\partial x}\right) - \dfrac{\partial p}{\partial x} - \dfrac{\sigma}{2}\mathrm{Re}\left(\mathscr{E}_z \mathscr{B}_y^*\right), \\[2ex]
\dfrac{\partial (\rho u_y)}{\partial t} = \mathrm{div}\left(-\rho u_y \boldsymbol{u} + \mu \nabla u_y + \mu \dfrac{\partial \boldsymbol{u}}{\partial y}\right) - \dfrac{\partial p}{\partial y} + \dfrac{\sigma}{2}\mathrm{Re}\left(\mathscr{E}_z \mathscr{B}_x^*\right), \\[2ex]
\dfrac{\partial (\rho h)}{\partial t} = \mathrm{div}\left(-\rho h \boldsymbol{u} + \dfrac{k}{c_p}\nabla h\right) + \dfrac{1}{2}\sigma |\mathscr{E}_z|^2 - q_R, \\[2ex]
\nabla^2 \mathscr{E}_z = i\omega\mu_0\left(\sigma\mathscr{E} + \mathscr{J}_{\mathrm{leg}}\right), \\[2ex]
\mathscr{B}_x = -\dfrac{1}{i\omega}\dfrac{\partial \mathscr{E}_z}{\partial y}, \qquad \mathscr{B}_y = \dfrac{1}{i\omega}\dfrac{\partial \mathscr{E}_z}{\partial x}.
\end{cases}
\tag{10.43}
$$

Again, the finite volume method and the SIMPLER algorithm can be used. In fact, two different domains must be considered:

- for fluid dynamics variables the domain is $\mathcal{D}' = [0, L_x'] \times [0, L_y]$;
- for the electric field the domain is $\mathcal{D}'' = [0, L_x''] \times [0, L_y]$;

and the boundary conditions are:

- on $\partial\mathcal{D}'$: $u_x = u_y = 0$, $-k\frac{\mathrm{d}T}{\mathrm{d}n} = q$, with $q = h_c(T - T_{\text{ext}})$ for the three sides facing the metal walls, and $q = h_c'(T - T_D)$ for the side facing the dielectric (h_c and h_c' are the convective heat flux coefficients, and T_{ext} and T_D are external and dielectric temperatures);
- on $\partial\mathcal{D}''$: $\mathcal{E}_z = 0$, as the walls are regarded as perfect conductors.

10.5.4 Some results

To calculate the threshold current to ignite the plasma at atmospheric pressure, two different approaches can be used; both are based on the simulation of the ignition transient. In the first, the velocity field is considered negligible and only the Maxwell equations coupled with the energy equation are considered. In the second approach, a more accurate solution is obtained, as convective terms are included. In the first case, results are obtained by means of a two-dimensional, time-dependent code in which the finite volume method is used for spatial discretization of the equations. In the second case, the SIMPLER algorithm has been considered using ANSYS/FLUENT and the UDF approach has been used for Maxwell's equations.

As for the conventional torches (section 10.4), the ignition transient is simulated by introducing an artificial electrical conductivity in a small area. The extra conductivity is maintained as long as the temperature inside the torch is high enough to sustain the discharge. The results can be used to acquire information on the feasibility of the system at atmospheric pressure, which must be analyzed through the complete model of the resonant antenna.

Figure 10.11 shows the minimum required current to ignite the plasma at 13.56 MHz in the framework of the pure diffusive model, when the position of the igniting zone

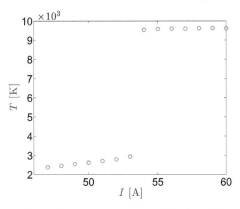

Figure 10.11. Maximum plasma temperature as a function of the supply current.

is located at 1 cm from the glass and when for simplicity one single leg is considered. In particular, the plasma temperature increases abruptly for a current value of $I = 54$ A. The steady-state temperature field is shown in figure 10.12 when the leg current is 90 A.

In general, a frequency higher than the usual value of 13.56 MHz could reduce the current and consequently the RF power. This is confirmed in figure 10.13, showing that the threshold current decreases with the frequency both at atmospheric and sub-atmospheric pressure: at atmospheric pressure, the current is reduced of about 40% when the frequency is increased from 13.56 to 30 MHz (if $p = 200$ mbar, a similar reduction has been calculated).

In the case of the complete (diffusive and convective) model, results have been obtained using ANSYS/FLUENT. Simulations have been performed in the case of the system reported in figure 10.9, where 23 legs have been considered, the current of

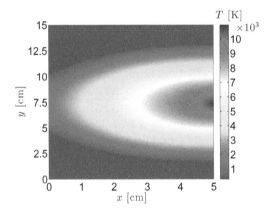

Figure 10.12. Plasma temperature distribution at steady-state condition for $I = 90$ A.

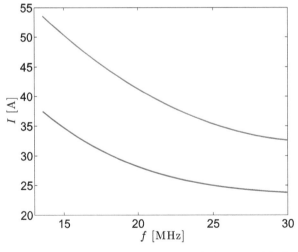

Figure 10.13. Threshold current as a function of frequency at atmospheric pressure (blue line) and at $p = 200$ mbar (red line).

the generator is 3 A, and the frequency is 13.56 MHz. The current distribution in the legs is shown in figure 10.14. Figures 10.15, 10.16, and 10.17 show the temperature and velocity fields obtained when 23 legs are considered.

Figure 10.18 shows the threshold current as a function of the frequency in the case of the simplified model with one leg and $I = 70$ A. When the convection is included, higher, more realistic values of the threshold current are obtained with respect to the pure diffusive model. The model could be improved by considering a more realistic configuration of the resonant antenna and including the mutual coupling between plasma and antenna.

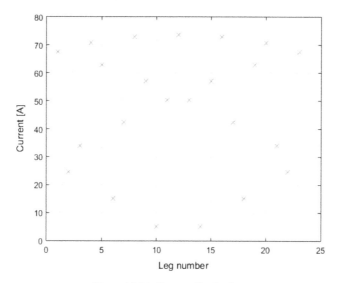

Figure 10.14. Current distribution.

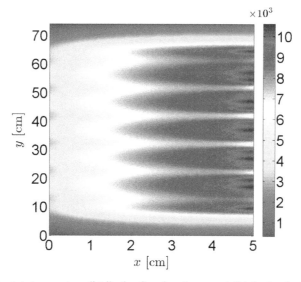

Figure 10.15. Steady-state temperature distribution for when the current distribution is given by figure 10.14.

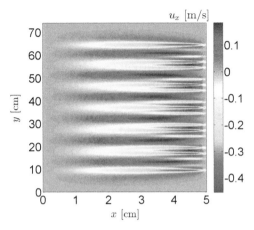

Figure 10.16. Steady-state velocity field along the *x*-direction when the current distribution is given by figure 10.14.

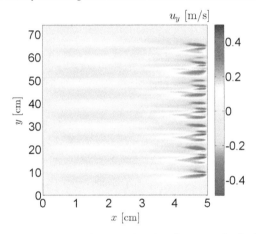

Figure 10.17. Steady-state velocity field along the *y*-direction when the current distribution is given by figure 10.14.

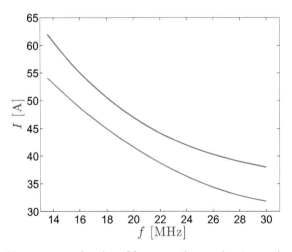

Figure 10.18. Threshold current as an function of frequency: the complete (convective and diffusive) model (red line) and the diffusive model (blue line) at atmospheric pressure.

Bibliography

[1] Capitelli M, Colonna G and D'Angola A 2011 *Fundamental Aspects of Plasma Chemical Physics* (Berlin: Springer)

[2] Lieberman M A and Lichtenberg A J 2005 *Principles of Plasma Discharges and Materials Processing* (New York: Wiley)

[3] Boulos M I, Fauchais P and Pfender E 1994 *Thermal Plasmas, Fundamentals and Applications* vol 1 (Berlin: Springer)

[4] Freeman M P and Chase J D 1968 *J. Appl. Phys.* **39** 180–90

[5] Chase J D 1969 *J. Appl. Phys.* **40** 318–25

[6] Guittienne Ph *et al* 2012 *J. Appl. Phys.* **111** 083305

[7] Thorpe M L and Scammon L W 1968 *Induction Plasma Heating: High Power, Low Frequency Operation and Pure Hydrogen Heating NAS 3-9375* (Cleveland, OH: NASA)

[8] Patankar S V 1980 *Numerical Heat Transfer and Fluid Flow* (New York: McGraw Hill)

[9] Bernardi D *et al* Comparison of different techniques for the treatment of the electromagnetic field for studying inductively-coupled plasma torches by means of the Fluent code *E-MRS European Materials Research Society Meeting, TPP7, Thermal Plasma Processes (Strasbourg, France)*

[10] Bernardi D *et al* 2003 *High Temp. Mater. Process.* **7** 71

[11] Ferziger J H and Perić M 1996 *Computational Methods for Fluid Dynamics* (Berlin: Springer)

[12] Patankar S V and Spalding D B 1972 A calculation procedure for heat, mass and momentum transfer in three-dimensional parabolic flows *Int. J. Heat Mass Transfer* **15** 1787

[13] Colonna G *et al* 2013 *Plasma Chem. Plasma Process.* **33** 401–31

[14] EquilTheTa http://phys4entrydb.ba.imip.cnr.it/EquilTHETA/index.php

[15] Bernardi D *et al* 2001 *Eur. Phys. J.* D **14** 337–48

[16] TEKNA www.tekna.com

[17] Mostaghimi J *et al* 1985 *Numer. Heat Transfer* **8** 187

[18] Guittienne Ph *et al* 2014 *Plasma Sources Sci. Technol.* **23** 015006

IOP Publishing

Plasma Modeling
Methods and Applications
Marco A Ridenti, Vasco Guerra and Jayr Amorim

Chapter 11

Atmospheric pressure plasmas operating in high-frequency fields

The study of plasmas fits well into the modern field of complex systems, however, the first scientific accomplishments in plasma physics, both theoretical and experimental, date to works from the 19th century, which then developed into quite an elaborate science by the end of the first half of the 20th century. From the first laboratory production of electrical discharges by Sir William Crookes in 1879 [1, 2] and the first attempts at modeling by Sir J J Thomson [3], plasma science drove physicists to the exercise of formulating comprehensive models which required ingredients from many areas of physics, namely, statistical physics, kinetic theory, electromagnetic theory, fluid mechanics, atomic, molecular and nuclear physics, chemistry, and, in some cases, relativity theory. It has not only required input from many fields of physics, plasma science has also required the treatment of systems containing many kinds of interacting elements, such as electrons, ions, atoms, molecules, waves, and radiation, which frequently must be treated simultaneously in order to guarantee the predictive power of the model. Another difficulty are the non-linearities often found in the resultant system of equations, which may result in chaos, emergent phenomena, self organization, feedback loops, solitons, pattern formation, and collective behavior. In addition, plasmas are usually observed and modeled at the macro-scales, but micro-scale phenomena must be studied and modeled to some degree, depending on the model purpose, in order to give obtain predictions.

As stated by Weisskopf, two different trends in science research may be identified, one of which could be called 'intensive' and the other 'extensive'. He states that 'intensive research goes for the fundamental laws and the extensive research goes for the explanation of phenomena in terms of known fundamental laws' [4, page 54]. Plasma science is almost totally devoted to extensive research, since all the phenomena in this realm result from many interacting and non-interacting micro-scale physical entities which are governed by fundamental laws. Plasma research, as were other fields of 'extensive' research such as solid-state physics, was greatly

doi:10.1088/978-0-7503-1200-4ch11

© IOP Publishing Ltd 2016

inspired by the approach of starting from fundamental laws to derive the observed collective behavior.

In this chapter, we try to follow the 'extensive' approach mentioned above where possible to give the best foundations to the model. As usually stated, the modeling process is strongly influenced by its purpose in the sense that some elements of the presumed reality of the system may be disregarded depending on the experimental conditions and measured quantities that one wishes to reproduce. For instance, if a measurement is designed to probe the electron density only, other elements of the reality, which only slightly influence electrons, such as radiation, may be completely disregarded. On the other hand, if one wishes to know the total radiative power delivered by a plasma discharge in a decontamination device, then radiation must be modeled, although its influence on the overall plasma behavior could be minimal.

Herein we address a specific kind of discharge, known as surface-wave discharge, which may be found in many technological applications [5–7]. Surface-wave discharges all have in common the property of being sustained by high-frequency electrical fields which also propagate in the medium. They may be operated at intermediate or low pressures, but here we focus on the atmospheric pressure case. Even in this specific case many differences can be found, depending on other elements such as the feeding gas, openness, flow regime, geometry, types of electromagnetic propagation, etc, that it would be very difficult to offer all the necessary tools to build any model. Nonetheless, we try to raise as many issues as we can in order to cover the most common problems, providing references to specialized literature for further details. We also try to build general equations which can be applied to as many cases as possible, leaving the most drastic simplifications to the worked example.

Many different and newly observed collective behaviors in plasmas produced by microwaves in atmospheric pressure have never been studied using extensive modeling, such as column bifurcation, which is a remarkable chaotic feature [8]. The phenomenon of plasma contraction, which is treated in the worked example, is still a matter of vigorous debate. Although some general agreement exists over the fundamental equations, its complexity almost always forces modelers to adopt some kind of approximation. Sometimes different approximations may explain the same collective phenomena, but in different ways, giving rise to such debates. Therefore this is still an open field, both experimentally and theoretically, even though the underlying fundamental laws may be well known.

Atmospheric pressure plasmas have some special features which should be pointed out, as some approximations may be applicable to shape their description into an adequate model. On the other hand, certain types of reactions which introduce non-linearities make them quite different from discharges at lower pressures. For instance, the hydrodynamic approximation, meaning that fluid equations may be used to describe the system macroscopic behavior, is valid in most cases of interest. Furthermore, the pressure is not so high as to break the *molecular chaos* hypothesis underlying the kinetic theory, which provides the transport parameters for the fluid model. However, three-body reactions may occur due to the elevated pressure, introducing additional non-linear terms in the system of differential equations describing the creation and destruction of particles through chemical reactions.

Microwave generated plasmas are particularly interesting for technological applications for many reasons. Plasmas may be kept at a *quasi*-steady-state with power delivered continuously by microwaves, because their frequency is usually higher than the plasma relaxation frequency. This means that the plasma system can be modeled using the usual steady-state fluid equations as in the DC case with some modifications in the transport parameters, as pointed out in the text. From the technological point of view, microwave plasmas have much better power efficiency than radio-frequency or kilohertz discharges, as the capacitive losses are lower. In addition, they may be produced using an electrodeless configuration, which can completely eliminate the undesired residuals from plasma–electrode interactions. Such discharges are usually warmer than their low-frequency counterparts, but this can be overcome using pulsed microwave sources or gas flows with large flow rates.

The structure of this chapter is as follows. Section 12.2 deals with the general aspects of plasma modeling at atmospheric pressure based on the fluid equations and electron kinetic theory. We provide a brief review of the classical electron kinetic theory with special emphasis on its connection with fluid modeling. The continuity equation for the reactive neutral species and ions is written explicitly and the main creation and destruction terms are described in detail. We also provide a brief description of the moment and energy equations of the gas mixture in a partially ionized gas, which would be needed in the most rigorous and comprehensive description of a non-magnetized laboratory plasma at atmospheric pressure. We also mention the usual method to account for space-charge fields using the ambipolar diffusion coefficient. Finally, we conclude the section describing the Maxwell equations for a plasma sustained by a high-frequency field in the transversal mode, TM_{00}. In section 12.3 we apply the theory described in the previous section to treat the problem of discharge contraction in a high-frequency argon discharge at atmospheric pressure.

11.1 Atmospheric pressure plasmas modeling in high-frequency fields

11.1.1 Transport properties of electrons in non-magnetized and partially ionized gases

Our goal is to derive the macroscopic equations that describe the density distributions of relevant species in the plasma, as well as some macroscopic physical quantities such as gas temperature and electric field. In this section, we discuss the connection between the kinetic theory of electrons and the fluid equations and show how electron transport parameters from two-term Boltzmann codes should be inserted into macroscopic equations.

To start, let us consider a system of electrons subject to a constant electric field, which can be non-uniform, and may collide with other electrons, neutral species, or ions. No strong restrictions over the type of collisions are imposed, as inelastic collisions, ionization, attachment and super-elastic collision can occur, apart from elastic collisions with neutrals. As the molecular chaos hypothesis applies and correlation between particles is rapidly destroyed, the system may be fully described by means of a simple distribution function $f(\mathbf{r}, \mathbf{v}, t)$ subject to the condition $\int f(\mathbf{r}, \mathbf{v}, t) \, v^2 \, d^2\Omega \, dv = N_e(\mathbf{r}, t)$ where the integration is taken over all velocities and $N_e(\mathbf{r}, t)$ is just a function that gives the electron density at instant t and a small volume centered at position $\mathbf{r}$. Here we make use of the two-term distribution and the quasi-stationary approximation developed in chapter 2. Equations (2.30) and (2.35) are particularly relevant in this context, because they are taken as the starting point in the following sections. Here we take $\hat{\Omega}$ in equation (2.30(b)) as the sum of the collision frequencies due to elastic and inelastic processes, i.e. $\hat{\Omega} = \nu_{el} + \nu_{ei} + \sum_j^{N_0} \delta_j \sum_i^{M_j} \nu_{ji}$, ν_{el} being the elastic moment transfer collision frequency between electrons and neutrals, ν_{ei} the electron–ion collision frequency, N_0 the number of species, M_j the number inelastic processes involving species j with fraction δ_j, and ν_{ji} the inelastic collision frequency of transition i. In addition, we shall consider only the non-magnetized case.

The expanded components of the distribution function $(f_0, \mathbf{f}_1)$ depend on the position coordinate in a non-trivial manner unless some reasonable assumption may be made which greatly simplify the problem. In the atmospheric pressure case, for instance, the electron mean free path (≈ 1 μm) is usually much shorter than the scale at which strong density variations occur, so that the space dependence of $f(\mathbf{r}, \mathbf{v}, t)$ is approximately equal to the space dependence of $N_e(\mathbf{r}, t)$. For instance, the distribution function f_0 may be written as $f_0 = F_0(v, t) N_e(\mathbf{r}, t)$, where $4\pi \int F_0 v^2 dv = 1$. This approximation may be corrected in terms of a gradient expansion of the electron density, and we shall just take the first-order correction, $f_0 = F_0(v, t) N_e + \boldsymbol{\xi}_0(v) \cdot \nabla N_e$, where $\int \xi_0 v^2 dv = 0$. The assumption underlying this approximation is usually referred as hydrodynamic regime in the literature. In the hydrodynamic regime the fluid equations may be derived from equations (2.30) and (2.35), as indicated in the following section.

Electron fluid equations

In the quasi-stationary approximation $(\partial \mathbf{f}_1/\partial t \approx 0)$, the integration of equation (2.30(b)) in the velocity space $(\int \{...\} v^2 dv)$ after multiplication by $(4\pi/3)v$ gives the well-known drift–diffusion equation

$$\mathbf{\Gamma}_e = -\mathbf{D}_e \nabla (N_e(\mathbf{r}, t)) + \mathbf{v}_e N_e(\mathbf{r}, t), \tag{11.1}$$

where $\mathbf{D}_e$ is the diffusion tensor and $\mathbf{v}_e$ is the electron drift velocity, which may also be written as $\mathbf{v}_e = \mu_e \mathbf{E}$, μ_e being defined as the electron mobility. Their explicit expressions, given below, are particularly important in the application of the fluid equations to model real problems.

The assumption $\partial \mathbf{f}_1/\partial t = 0$ is exact in the steady-state case and may be taken as a good approximation if $(\partial \mathbf{f}_1/\partial t)\hat{\Omega}^{-1} \ll \mathbf{f}_1$ after a small time comparable to $\hat{\Omega}^{-1}$.

It was shown in chapter 2 that when the quasi-stationary approximation is assumed one may decouple the system of equations (2.30) and write an equation for the isotropic component of the distribution function only (see equation (2.35)).

From this equation one may obtain the electron number conservation equation by integration in the velocity space ($\int \{\ldots\} 4\pi v^2 \mathrm{d}v$)

$$\frac{\partial N_e}{\partial t} + \mathbf{v}_e \cdot \nabla N_e - \mathbf{D}_e \nabla^2 N_e = \left(\tilde{\nu}_{\mathrm{ion}} - \tilde{\nu}_{\mathrm{att}} \right) N_e, \tag{11.2}$$

where the electron drift velocity, diffusion tensor, and mean ionization and attachment rates are given by the following formulas when the z-axis is taken as the acceleration direction:

$$\mathbf{v}_e = -\frac{4\pi a_z}{3} \int_0^\infty \frac{v^3}{\hat{\Omega}} \frac{\mathrm{d}F_0}{\mathrm{d}v} \, \mathrm{d}v \, \hat{\mathbf{z}}, \tag{11.3}$$

$$D_{e,ij} = 0, \ i \neq j; \ \ D_{e,xx} = D_{e,yy} = D_{e,T} = \frac{4\pi}{3} \int_0^\infty \frac{v^4}{\hat{\Omega}} F_0 \, \mathrm{d}v, \tag{11.4}$$

$$D_{e,zz} = D_{e,L} = \frac{4\pi}{3} \int_0^\infty \frac{v^4}{\nu_i} F_0 \, \mathrm{d}v + \frac{4\pi a_z}{3} \int_0^\infty \frac{v^3}{\hat{\Omega}} \frac{\mathrm{d}\xi_{0,z}}{\mathrm{d}v} \, \mathrm{d}v, \tag{11.5}$$

$$\tilde{\nu}_{\mathrm{ion}} = N \sum_j^{N_0} \delta_j \int_0^\infty 4\pi \sigma_{j,\mathrm{ion}}(v) F_0(v) v^3 \, \mathrm{d}v, \tag{11.6}$$

$$\tilde{\nu}_{\mathrm{att}} = N \sum_j^{N_0} \delta_j \int_0^\infty 4\pi \sigma_{j,\mathrm{att}}(v) F_0(v) v^3 \, \mathrm{d}v. \tag{11.7}$$

In the expressions above we immediately see that the electron diffusion tensor is diagonal, with equal diagonal components in the x- and y-directions, which are usually called the transversal diffusion coefficient, and the component in the z-direction, which is usually called the longitudinal diffusion coefficient. In the expressions of the mean ionization and attachment rates we introduced the total particle density, N, the species fraction $\delta_j = N_j/N$, and the integral ionization or attachment cross section $\sigma_{j,\mathrm{ion}}$ and $\sigma_{j,\mathrm{att}}$. Note that collision operator terms associated with conserving processes vanish after integration. In addition, the first-order correction of the hydrodynamic expansion was not taken into account on the ionization and attachment components of the collision operator, which would generate a correction on the electron drift velocity due to non-conserving processes. This correction is very important in some cases, where strongly attaching species are present, but it is usually very small in typical cases. If the second-order term was kept in the gradient expansion, corrections in the diffusion tensor would appear due to non-conserving transport. The explicit expression for these correction may be found in [9], in the context of the two-term expansion.

The longitudinal and transverse diffusion coefficient may differ in magnitude depending on the gas species and the electron mean energy. For instance, a three-dimensional fluid description of electrons in a weakly ionized gas in a uniform electric field may be based on

$$\frac{\partial N_e}{\partial t} + v_{e,z}\frac{\partial N_e}{\partial z} - D_{e,L}\frac{\partial^2 N_e}{\partial z^2} - D_{e,T}\left(\frac{\partial^2 N_e}{\partial x^2} + \frac{\partial^2 N_e}{\partial y^2}\right) = \left(\tilde{\nu}_{\text{ion}} - \tilde{\nu}_{\text{att}}\right)N_e. \qquad (11.8)$$

Finally, the electron energy equation may be obtained by multiplying equation (2.35) by $(m_e/2)v^2$, where m_e is the electron mass, and then integrating in the velocity space. The resulting equation may be written as

$$\frac{\partial(N_e\varepsilon_e)}{\partial t} + \nabla_r \cdot \mathbf{\Gamma}_{e,\text{en}} + e\mathbf{E} \cdot \mathbf{\Gamma}_e = P_{\text{coll}}, \qquad (11.9)$$

where ε_e is the mean electron energy, defined as $\varepsilon_e = (2\pi m_e/N_e)\int_0^\infty f_0 v^4 dv$, P_{coll} is the net power loss per unit volume due to collisions, and $\mathbf{\Gamma}_{e,\text{en}}$ is the energy flux, which may be written as

$$\mathbf{\Gamma}_{e,\text{en}} = -\mathbf{G}\nabla n + \boldsymbol{\beta}n, \qquad (11.10)$$

where $\mathbf{G}$ is the heat diffusion tensor and $\boldsymbol{\beta}$ is the thermoelectricity, defined as

$$G_{ij} = 0, \; i \neq j; \quad G_{xx} = G_{yy} = G_T = 2\pi m_e \int_0^\infty \frac{F_0 v^6}{3\hat{\Omega}}dv, \qquad (11.11)$$

$$G_{zz} = G_L = G_T + 2\pi m_e a_z \int_0^\infty \frac{v^5}{3\hat{\Omega}}\frac{d\xi_{0,z}}{dv}dv, \qquad (11.12)$$

$$\boldsymbol{\beta} = -2\pi m_e a_z \int_0^\infty \frac{v^5}{3\hat{\Omega}}\frac{dF_0}{dv}dv\,\hat{\mathbf{z}}. \qquad (11.13)$$

In the literature, a different form of the electron energy flux is often found, which may be written as $\mathbf{\Gamma}_{e,\text{en}} = (5/3)N_e\varepsilon_e\mathbf{v}_d - (5/3)N_eD_T\nabla\varepsilon_e$. Although this is a simpler expression, other assumptions must be made, namely, a Maxwellian electron energy distribution function (EEDF), a constant momentum transfer frequency ($\hat{\Omega} = $ cte), and constant kinetic pressure [10].

Collision operator and local field approximation
The collision operator gives the mathematical expression which determines how the collisional processes shape the EEDF. The EEDF ($n(\varepsilon)$) is simply the isotropic velocity distribution function expressed in terms of the classical kinetic energy, i.e. $n(\varepsilon) = (4\pi/m_e)(\sqrt{2/m_e})f_0(v)/N_e$, with the normalization convention $1 = \int_0^\infty n(\varepsilon)\varepsilon^{1/2}\,d\varepsilon$. This is particularly relevant when the modeler's efforts are directed towards the plasma chemistry, in applications where a good estimate of the reaction rates of electronic reactions is crucial. Here, we provide a brief description of the formulas and their

physical meaning, without going into the details of their derivation, which may be found in chapter 2 and/or in the specialized literature [11, 12].

In the previous sections, the collision operator was expanded in spherical harmonics. The isotropic term, which was designated by the symbol S_0 in chapter 2, accounts for different collisional processes, and contain corrections to the assumption of infinitely heavy species. It can describe the effect of several kinds of collisional processes, i.e.

$$S_0 = S_{0,\text{el}} + S_{0,\text{in}} + S_{0,\text{ee}} + S_{0,\text{t}}, \tag{11.14}$$

where $S_{0,\text{el}}$ is the elastic term, which accounts for elastic collisions with heavy neutrals or ions with a Maxwellian distribution function, $S_{0,\text{in}}$ is the inelastic term, $S_{0,\text{ee}}$ is the electron–electron (e–e) collision term and $S_{0,t}$ is the transformation term, which accounts for collisional processes that can create or destroy electrons, such as ionization, attachment, or recombination ($S_{0,t} = S_{0,\text{ion}} + S_{0,\text{rec}} + S_{0,\text{t}}$).

Elastic collisions
The elastic collision term may be written as [11, 13] (see also section 2.6.1)

$$S_{0,\text{el}} = N \sum_j^{N_0} \delta_j \frac{m}{m_j} \frac{1}{v^2} \frac{\partial}{\partial v}\left[\nu_{g,j} v^3 \left(f_0 + \frac{2k_B T_g}{m} \frac{\partial f_0}{\partial v^2} \right) \right], \tag{11.15}$$

where k_B is the Boltzmann constant, T_g is the equilibrium temperature of the heavy species, and m_j is the mass of species j. The collision frequency ν_g is equal to the elastic moment transfer collision frequency if species j is neutral

$$\nu_g^{(n)} = n_j v \iint (1 - \cos\theta)\,\sigma_{\text{el}}(\theta, v)\,\mathrm{d}^2\Omega. \tag{11.16}$$

If the scatterer is an ion one should write the electron–ion collision frequency instead

$$\nu_{g,j}^{(i)} = \frac{n_j}{v^3} 4\pi \left(\frac{Ze^2}{4\pi\epsilon_0 m_e} \right)^2 \ln\Lambda, \tag{11.17}$$

where ϵ_0 is the vacuum permittivity and $\lambda = 12\pi(\epsilon_0 k_B T_e)^{3/2}/(N_e e^3)$. The electron temperature, in the expression of Λ, is defined in terms of the mean energy, $T_e = (2/(3k_B))\varepsilon_e$. The electron–ion collision frequency is usually small in the case of chemical plasmas in which the degree of ionization is typically very small.

In the simple case of a discharge in a partially ionized gas, where there is only one dominant neutral species, it can be readily seen that the constant $\tau_e = (m_j/(m\nu_{g,j}))$ gives the relaxation time-scale of the isotropic component of f_0. The anisotropic component $\mathbf{f}_1$, on the other hand, relaxes with $\tau_p = 1/\hat{\Omega}$, therefore much faster than f_0.

Inelastic collisions
The general form for the inelastic term is given by the following formula [14] (see also section 2.6.2):

$$S_{0,\text{in}} = N \sum_{j=1}^{N_0} \sum_{i=1}^{M_j} \left\{ \delta_j \left[(v + v_{ij}) \sigma_{ij} (v + v_{ij}) f_0 (v + v_{ij}) - v \sigma_{ij}(v) f_0(v) \right] \right.$$
$$\left. + \delta_{ji} \left[(v - v_{ij}) \sigma'_{ij} (v - v_{ij}) f_0 (v - v_{ij}) - v \sigma'_{ij}(v) f_0(v) \right] \right\}, \tag{11.18}$$

where the 'velocity' v_{ij} is equivalent to the energy threshold ($v_{ij} = \sqrt{(2/m)\varepsilon_{ij}}$) of the transition between the lower state of species j and its upper state i and σ_{ij} (σ'_{ij}) is the inelastic (or super-elastic) cross section. The newly introduced term δ_{ji} is the fractional population of level i for a given lower level species j. Note that the lower state j does not need to be on the ground state so that metastable or resonant species may be included as distinct species as long as only levels with higher energies are considered.

Rotational, vibrational, and electronic inelastic collisions may be accounted for using equation (11.18), but the large difference between rotational thresholds and vibrational or electronic thresholds can make the numerical solution difficult or impossible. In the case of homonuclear diatomic molecules it has been proven in [15] that rotational collisions may be expressed by a continuous approximation

$$S_{0,\text{rot}} = N \sum_{j=1}^{N_0^{(r)}} \delta_j \frac{4N\sigma_0 B}{mv^2} \frac{\partial}{\partial v} \left[v^2 \left(f_0 + \frac{2k_{\text{B}}T}{m} \frac{df_0}{dv^2} \right) \right], \tag{11.19}$$

where B is the rotational constant in units of energy and $\sigma_0 \equiv 8\pi Q^2 a_0^2/15$, with Q the quadrupole moment constant in ea_0^2 units and a_0 the Bohr radius.

Non-conserving collisions
The ionization can be seen as an inelastic processes which adds one electron per collision. It is usual to assume that this extra electron is created with zero energy, so that the collision operator may be written as [16]

$$S_{0,\text{ion}} = N \sum_{j=1}^{N_0} \delta_j \left\{ \left[(v + v_{\text{ion},j}) \sigma_j^{(\text{ion})}(v + v_{\text{ion},j}) f_0 (v + v_{\text{ion},j}) - v \sigma_j^{(\text{ion})}(v) f_0(v) \right] \right.$$
$$\left. + \frac{\delta(v)}{v^2} \int_0^\infty \sigma_j^{(\text{ion})}(v') f_0(v') v'^3 dv' \right\}. \tag{11.20}$$

Other forms of this collision operator are possible, depending on the assumption over the energy partitioning between the two electrons after the collision. Unless the ionization collision rate is very high in comparison with other inelastic processes, which is unusual in most applications of plasma chemistry, then the electron energy partitioning choice makes little difference in the final shape of the EEDF [17].

Attachment processes may be simply written as [16]

$$S_{0,\text{att}} = -\sum_{j}^{N_0} v \delta_j \sigma_j^{(\text{att})}(v) f_0(v). \tag{11.21}$$

Both the ionization and recombination collision terms do not conserve the electron number density, as expected. In the cases in which ionization collisions occur at very low rates and attachment is negligible, the electron creation term (the third term on the right-hand side of equation (11.20)) may be neglected in the numerical calculation of $F_0(v)$. Electron recombination by the processes $(e + A^+ \longrightarrow A + h\nu)$ or $(e + e + A^+ \longrightarrow A + e)$ are usually not taken into account in the collision term, although the literature lacks studies showing whether they could be relevant to the EEDF calculation in plasmas at atmospheric pressure.

Electron–electron collisions
e–e collisions are important in the description of atmospheric pressure plasma as electron densities may be large enough to produce probable encounters between electrons. Collisions between elementary species of the same type do not produce power gain or loss, but e–e collisions force the EEDF towards a Maxwellian distribution, producing drastic changes in some reaction rates that strongly depend on the EEDF tail. The derivation of the $C_{0,\mathrm{ee}}$ is somewhat cumbersome and usually takes as its starting point the Fokker–Planck equation [11], given by

$$C_{0,\mathrm{ee}} = \frac{Y}{v^2} \frac{\partial}{\partial v} \left[f_0 I_0^0 + \frac{mv^2}{3} (I_2^0 + J_{-1}^0) \frac{\partial f_0}{\partial \varepsilon} \right], \quad \text{where}$$

$$I_2^0 = \frac{4\pi}{v^2} \int_0^v f_0 v'^4 \mathrm{d}v'; \quad J_{-1}^0 = 4\pi v \int_v^\infty f_0 v' \mathrm{d}v'; \quad Y = 4\pi \left(\frac{Ze^2}{4\pi\epsilon_0 m_e} \right)^2 \ln \Lambda. \tag{11.22}$$

The above equation has some integral terms that need f_0 to be evaluated, which in turn is the desired solution. Therefore, if e–e collisions are included, then an iterative scheme is required for the numerical solution of the electron Boltzmann equations (2.35).

Local field approximation
Once the collision operator is defined and chosen to describe a particular electron discharge, the electron Boltzmann (2.35) may be solved numerically and P_{coll} in equation (11.9) may be evaluated. It is generally difficult to simultaneously obtain the velocity and space dependence of the electron distribution function, even in the hydrodynamic regime. In some cases, where the electron mean free-path (l_c) is small when compared with the length in which strong densities variations occur, i.e. $l_c |\nabla n|/n \ll 1$, it is common to neglect all the terms with space derivatives in equation (2.35). In this approximation, the plasma can be viewed as a composition of small 'plasma volumes', each of them with uniform electron density. In addition, the EEDF is only determined by the average uniform field and collision processes in this small volume. The electron Boltzmann equation (2.35) and energy balance equation (11.9) could be simplified to

$$\frac{\partial f_0}{\partial t} - \frac{a_z^2}{v^2} \frac{\partial}{\partial v} \left(\frac{v^2}{3\hat{\Omega}} \frac{\partial f_0}{\partial v} \right) = S_0 \tag{11.23}$$

and

$$\frac{\partial (N_e u_e)}{\partial t} + e\mathbf{E} \cdot (n\mathbf{v}_e) = P_{\text{coll}}. \tag{11.24}$$

In the steady case, we could go even further, writing

$$-\frac{a_z^2}{v^2} \frac{\partial}{\partial v} \left(\frac{v^2}{3\hat{\Omega}} \frac{\partial F_0}{\partial v} \right) = S_0 \tag{11.25}$$

and

$$e\mathbf{E} \cdot (N_e \mathbf{v}_e) = P_{\text{coll}}. \tag{11.26}$$

In a practical application of this approximation, F_0 must be evaluated for each finite space domain of the numerical scheme and then applied to obtain the estimate of the transport parameters that are needed to evaluate the electron continuity equation (11.2). The local energy distribution is also needed to obtain the local electron reaction rates required to compute the continuity equations for neutrals and ions (see section 11.1.2).

Note that equation (11.26) has a very simple physical interpretation. The left-hand side of the equation represents the power density due to the electric field energy gained by the electrons. It equates with the power loss and gain by collisions, which means that the field energy is totally dissipated by collisions. In this situation, it is usually said that electrons are in equilibrium with the electric field.

High-frequency electric field
Let us consider now the problem of a discharge where the electric energy is provided by a high-frequency field, e.g. microwaves. In the following analysis, we consider a linearly polarized field in the z-direction, given by the expression $E_z = E_0 \cos \omega t$ (the phase is arbitrarily chosen to be zero). In general, even in the atmospheric pressure case, the energy relaxation frequency is much shorter than the frequency of a high-frequency field $\omega \gg \nu_e$, so the isotropic component of the EEDF may be considered almost constant in time. This is not so in the case of $\mathbf{f}_1$, as the varying field causes the electron mean drift velocity to alternate. The first anisotropic component $\mathbf{f}_1$ could be expanded in time harmonics, but since we are interested in the steady-state rather than the transient solution, it is enough to write $\mathbf{f}_1^1 = \mathbf{f}_{0,1}^1 \exp(j\omega t)$, where $\mathbf{f}_{0,1}^1$ is just a complex amplitude. The isotropic component, on the other hand, could be expanded as $f_0(t) = f_{0,cte} + f_{0,t}(t)$, where $f_{0,t}(t)$ is a small time-dependent perturbation. Writing the acceleration $\mathbf{a}$ as a complex number $\mathbf{a}^1 = \mathbf{a}_0 \exp j\omega t$, $\mathbf{a}_0$ being real since the phase is zero, then we may derive $\mathbf{f}_1^1$ from equation (2.30(b))

$$\mathbf{f}_1^1 = \left(-\frac{1}{\hat{\Omega} + j\omega} \right) \left[v\nabla_r f_0 + \mathbf{a}^1(t)\frac{\partial f_0}{\partial v} \right]. \tag{11.27}$$

Substitution of the above equation in equation (2.30(a)) yields

$$
\frac{\partial f_0}{\partial t} + \frac{v}{3} \nabla_r \cdot \left[\frac{v \nabla_r f_0 + \left(\cos(\omega t) - \frac{\omega}{\hat{\Omega}} \sin(\omega t) \right) \frac{\partial f_0}{\partial v}}{\hat{\Omega} \left(1 + \omega^2/\hat{\Omega}^2 \right)} \right] - \mathbf{a}_0 \cdot \cos(\omega t)
$$

$$
\times \frac{1}{3v^2} \frac{\partial}{\partial v} \left\{ v^2 \left[\frac{v \nabla_r f_0 + \mathbf{a}_0 \left(\cos(\omega t) - \frac{\omega}{\hat{\Omega}} \sin(\omega t) \right) \frac{\partial f_0}{\partial v}}{\hat{\Omega} \left(1 + \omega^2/\hat{\Omega}^2 \right)} \right] \right\} = S_0
$$

(11.28)

Note that only the real part of the solution was retained. Taking the time average over a period and neglecting the small time perturbation $f_{0,t}$, one arrives at the following equation for f_0:

$$
\frac{v^2 \nabla^2 f_0}{3 \left[\hat{\Omega} \left(1 + \omega^2/\hat{\Omega}^2 \right) \right]} - \frac{a_0^2}{6v^2} \frac{\partial}{\partial v} \left[\frac{1}{\hat{\Omega} \left(1 + \omega^2/\hat{\Omega}^2 \right)} \frac{\partial f_0}{\partial v} \right] = S_0.
$$

(11.29)

A simple continuity equation may be obtained now, which is written

$$
-D_{hf} \nabla^2 N_e = \left(\tilde{\nu}_{\text{ion}} - \tilde{\nu}_{\text{att}} \right) N_e,
$$

(11.30)

where D_{hf} is the 'high-frequency' diffusion coefficient defined as

$$
D_{hf} = \frac{4\pi}{3} \int_0^\infty \frac{v^4}{\hat{\Omega} \left(1 + \omega^2/\hat{\Omega}^2 \right)} F_0 \, dv.
$$

(11.31)

This equation shows that the higher the frequency ω is, the lower will be the effective diffusion coefficient.

In the time averaging procedure we assumed a steady-state discharge, in the sense that the electron density is constant in time, although not uniform. This is possible if during one period of the field only few non-conserving collisions occur, $\tilde{\nu}_{\text{ion,att}} \ll \omega$. If this condition is not satisfied, then significant charge creation or destruction can occur during one period of the field and the full time differential equation has to be solved instead of the time averaged one, as f_0 could appreciably change over the period.

Let us look now at the power equation. We will take the averaged equation (11.28) and proceed with the energy integration, as we did before to derive the energy conservation equation in the DC case. If we limit ourselves to the local approximation, then the power balance equation may be written as

$$
\frac{1}{2} e E_0^2 N_e \mu_{hf} = P_{\text{coll}},
$$

(11.32)

where μ_{hf} is the high-frequency electron mobility, which may be written as

$$\mu_{hf} = -\frac{4\pi}{3}\frac{e}{m}\int_0^\infty \frac{v^2}{\hat{\Omega}\left(1 + \omega^2/\hat{\Omega}^2\right)}\frac{\mathrm{d}F_0}{\mathrm{d}v}\,\mathrm{d}v. \tag{11.33}$$

Equations (11.32) and (11.33) show that no power can be transferred to the electrons if collisions do not occur. In this case, the electrons simply move in quadrature with the field, so that the net work done by the field over a period on the moving particles is zero. In the other extreme, in which both the collision and the electron energy relaxation frequency are larger than the field frequency, the energy distribution function instantaneously adapts to the varying electron field and electrons 'swing' in phase with it. The intermediate case, where $\nu_e \approx \omega$ is more complicated, as f_0 will vary with time in a non-trivial manner.

11.1.2 Treatment of ions and neutral species

Ion and neutral species are assumed to be in equilibrium so that their energy distribution may be described by a Maxwellian distribution with a temperature T. This approximation is valid as long as the plasma is strongly collisional and the energy gained or lost in exothermic or endothermic reactions rapidly relaxes to the equilibrium temperature.

Both ions and neutrals can be described by the following continuity equation:

$$\begin{aligned}
\frac{\partial\left[N_j\right]}{\partial t} &+ \nabla \cdot \left(\left[N_j\right]\mathbf{u}\right) - D_j\nabla^2\left[N_j\right] \\
&= -\sum_\alpha k_\alpha(m-p)\left[N_j\right]^m[N_i]^n - \sum_\beta k_\beta(m-p)\left[N_j\right]^m[N_i]^n[N_l]^s - \sum_{l<j} A_{jl}\left[N_j\right] \\
&\quad + \sum_\zeta k_\zeta(p'-m')\left[N_j\right]^{m'}[N_i]^{n'}[N_l]^{s'} + \sum_\eta k_\eta\,p'[N_k]^{m'}[N_i]^{n'}[N_l]^{s'} \\
&\quad + \sum_{l>j} A_{lj}[N_l] + \sum_\kappa \gamma_\kappa\,p_\kappa\left(\nabla \cdot \mathbf{\Gamma}_\kappa\right)_S - \gamma_j\left(\nabla \cdot \mathbf{\Gamma}_j\right)_S,
\end{aligned} \tag{11.34}$$

where $p < m$ and $p' > m'$, and where the square bracket is used to represent the particle density, in units of particles per cm^3. The velocity $\mathbf{u}$ is the fluid velocity of the mixture, which is mainly determined by the neutral species in a weakly ionized plasma and defined as $\mathbf{u} = \sum_j^{N_0}([N_j]m_j/\rho)\langle\mathbf{v}_j\rangle$, where ρ is the mass density, m_j the molecular mass and $\langle\mathbf{v}_j\rangle$ the mean velocity of species j. The parameter γ_κ is the probability of species conversion due to wall interaction and $\mathbf{\Gamma}_\kappa$ is the flux of species κ, given by $\mathbf{\Gamma}_\kappa = -D_\kappa\nabla n_\kappa + n_\kappa\mathbf{u}$. The divergence of the flux in equation (11.34) is taken over the vessel boundaries. All the terms on the right-hand side of the equation represent particle creation or loss due to chemical reactions, including electronic reactions and radiative processes. These terms do not exhaust all the possibilities, but include the most common types of reactions usually considered in global models.

The loss terms, represented in the first two sums in equation (11.34) with negative signal account for the destruction of species N_j by two-body and three-body reactions, respectively. These reactions may be represented explicitly as

$$mN_j + nN_i \xrightarrow{k_\alpha} pN_j + \sum_k q_k N_k,$$
(11.35)

$$mN_j + nN_i + sN_l \xrightarrow{k_\beta} pN_j + \sum_k q_k N_k,$$
(11.36)

where $p < m$, $j \neq i$, $j \neq l$, $j \neq k$, $\forall k$.

The stoichiometric coefficients are represented by the letters m, n, p e q_k, and the reaction rate by k_α or k_β. Analogously, the gain reactions may be written explicitly as

$$m'N_j + n'N_i + s'N_l \xrightarrow{k_\zeta} p'N_j + \sum_k q'_k N_k,$$
(11.37)

$$m'N_k + n'N_i + s'N_l \xrightarrow{k_\eta} p'N_j + \sum_k q'_k N_k,$$
(11.38)

where $p' > m'$, $j \neq i$, $j \neq l$, $j \neq k$, $\forall k$.

Although both equations have three species on the left-hand side, the particular case of two-body reactions is included as long as one of the stoichiometric coefficients is zero.

The third sum in the right-hand side of equation (11.34) represents the loss of species N_j by radiative decaying, which may be expressed as

$$N_j \xrightarrow{A_{jl}} N_l + h\nu,$$
(11.39)

where A_{lj} is the Einstein coefficient of the transition of state j to l. On the other hand, the spontaneous emission from an upper energy state contributes to populate the state j, i.e.

$$N_l \xrightarrow{A_{lj}} N_j + h\nu.$$
(11.40)

It is evident that the term corresponding to the radiative emission only occurs if species N_j is radiative. In addition, if the species is metastable, the radiative loss is usually negligible. Another effect that must be considered is the radiative imprisonment, which occurs due to radiative absorption by the species in the medium. The relevance of this effect depends on the plasma dimensions and the concentration of absorbing species. Usually, in weakly ionized plasmas, the ground state species from the background gas or mixture are predominant and the main absorbers. The radiative imprisonment may be computed using Holstein's model [18, 19] which gives the formula to the correction coefficient g ($A_{lf}^{(corr)} = g_{lf} A_{lf}$) depending on the plasma geometry and absorbing concentration.

The two last terms in the right-hand side of equation (11.34) represent the destruction of other species in the boundaries which may produce new molecules of species N_j, i.e.

$$N_\kappa \xrightarrow{\text{Wall}} pN_j. \tag{11.41}$$

or the destruction of species N_j in the wall.

$$N_j \xrightarrow{\text{Wall}} pN_l. \tag{11.42}$$

The first term on the left-hand side of equation (11.34) vanishes in the steady regime and the second term must only be evaluated if the convective flow is comparable with the reaction and diffusion losses. In the special case of a flow in a cylindrical container with negligible convective speed the diffusion term may be expressed as $D_j[N_j]/\Lambda^2$, where the constant Λ is the vessel radius R divided by the first zero $\lambda_0 \approx 2.405$ of the Bessel function $J_0(x)$.

One important observation must be made regarding the present form of equation (11.34). In the rigorous multi-component transport theory [20], the mass flux vector contains the contribution due to the concentration gradients of all species, and also additional flux terms due to pressure and temperature gradients. In the present and simplified form of the theory, we assumed that all molecules have much lower densities than the background gas, so that all the D_j are actually the binary diffusion coefficients of species j in the background gas. The density of the background gas is simply obtained using the equation of state of an ideal gas, i.e. $p = [N]k_BT \approx [N_{bg}]k_BT$, where N_{bg} is the density of the background gas. If the background gas is a mixture with fixed composition, the D_j of the species with low concentrations may be evaluated as the weighted mean of the binary diffusion coefficients [21]. We also note that the flux terms due to pressure and temperature gradients were neglected in equation (11.34).

The binary diffusion coefficients may be evaluated with aid of the kinetic theory [20] or, in some cases, extracted from theoretical and experimental tabulated data. In general, they vary with gas temperature. This is also true for the reaction coefficients, which are usually expressed by the Arrhenius formula [22] and can often be found in the literature (e.g. [23]). The electron reaction rates, on the other hand, must be evaluated by means of the calculated electron distribution function, i.e. $k_{e,j} = 4\pi \int_0^\infty \sigma_j(v)F_0(v)v^3 dv$, although tabulated values or empirical formulas from literature may be found for many electronic reactions for the case of Maxwellian distributions. The parameter γ_k can also be found in the literature for some species and surfaces, although often with large uncertainties.

With regard to the boundary conditions in the case of electrodeless discharges, simpler expressions can be defined because there is no need to account for the charge emission from electrodes. In the surfaces representing the vessel boundaries it is enough to impose the condition $N_j = 0$ for species that are totally destroyed in the walls; for species which only bounce back the boundary condition is $\hat{\mathbf{n}} \cdot \mathbf{\Gamma}_j = 0$, for species which are partially destroyed it is $\hat{\mathbf{n}} \cdot \boldsymbol{\Gamma}_j = \gamma_j \frac{1}{4}\langle v \rangle n_j$, and for species that may be created from wall interaction it is $\hat{\mathbf{n}} \cdot \boldsymbol{\Gamma}_j = \sum_k (1 - \gamma_k)\frac{1}{4}p_k \langle v \rangle n_k$. In the case of a free-stream jet, for example, zero gradient or zero value conditions may be imposed in the far away boundaries $((\nabla \cdot N_j) \cdot \hat{\mathbf{n}} = 0$ or $N_j = 0)$ for charged and reactive species, and fixed values for the background gas.

A last observation concerns the *quasi*-neutrality hypothesis. If it is assumed that for every small volume of the plasma medium the negative charge density equals the positive one, then there is no need to solve the continuity equation for one of the charged species. The electron density is usually chosen as this species. Indeed, the quasi-neutrality hypothesis is valid in the plasma, except in the sheath region. In atmospheric pressure plasmas, the sheath dimensions are short compared to the bulk dimensions, except in the case of microplasmas. Unless the modeler's effort is directed towards sheath modeling, the application of the quasi-neutrality hypothesis to the entire space domain is valid.

11.1.3 Macroscopic equations for the weakly ionized gas flow

In the previous section, we dropped some comments on the convective term in equation (11.34), but did not give the pathway to compute it. In the simplest case, the convective velocity may be considered constant if the flow can be modeled as a uniform and constant field. This would be the case in a laminar flow of an inviscid fluid through a tube. In this case, the velocity could be dropped out from the divergence term and convective losses would only be relevant in the regions where density gradients are present in the flow direction. In the general case of a weakly ionized plasma jet, the flow may be turbulent and in principle the flow field has to be determined from the Navier–Stokes equation for the background gas which is assumed to be totally unperturbed by the plasma,

$$\frac{\partial \mathbf{u}}{\partial t} + \mathbf{u} \cdot \nabla \mathbf{u} = -\frac{1}{\rho}\nabla \cdot \boldsymbol{\Pi} + \mathbf{g}, \tag{11.43}$$

where $\boldsymbol{\Pi}$ is the *pressure tensor*, given by [20]

$$\boldsymbol{\Pi} = -p\mathbf{I} + 2\eta\mathbf{S}, \quad \text{where } \mathbf{S} = \frac{1}{2}\left[\nabla\mathbf{u} + (\nabla\mathbf{u})^T - \frac{2}{3}(\nabla \cdot \mathbf{u})\mathbf{I}\right]. \tag{11.44}$$

In this particular form of the Navier–Stokes equation (11.43) the fluid is assumed to be compressible. In the derivation of equation (11.43) we use the fact that the overall mass conservation equation holds even when chemical reactions occur $((\partial\rho)/(\partial t) + \nabla \cdot (\rho\mathbf{u}) = 0)$. It also includes the gravity acceleration term $(\mathbf{g})$, which can be neglected in most cases. The bulk viscosity was also neglected. Although the charged species are supposed to have no influence on $\mathbf{u}$, the dynamic viscosity, represented by the symbol η, varies with temperature, which is determined mainly by the plasma heating. Similarly to the case of the diffusion coefficient, the dynamic viscosity may be determined from kinetic theory [20] or from parametrized formulas obtained from fittings of different sets of experimental data.

The energy equation of the flow closes the set of fluid equations, giving the gas temperature. In a weakly ionized plasma flow it assumes the following form[1]:

$$\rho C_p\left[\frac{\partial T}{\partial t} + \mathbf{u} \cdot \nabla T\right] = \nabla \cdot (\lambda \nabla T) + e\mathbf{E} \cdot \mathbf{\Gamma}_e - \sum_i K_i^{\text{ch}} h_i - \sum_{i,j<i} A_{ij} g_{ij} h\nu_{ij}[N_i], \quad (11.45)$$

where C_p is the heat capacity of the mixture at constant pressure, λ is the thermal conductivity of the system, K_i^{ch} and h_i are the reaction rate and enthalpy per molecule of particle i, respectively. The energy losses due to friction are neglected, which is perfectly valid as the plasma input power is much larger. The second term on the right-hand side accounts for the energy transferred from the accelerated electrons to heavy particles. This is usually the most important source of energy in a chemical plasma. Note that energy transfer from ions to the medium was not considered, since their flux can be assumed to be very small in a weakly ionized plasma at atmospheric pressure. The third term accounts for the energy losses or gains in endothermic or exothermic chemical reactions. The last term is usually small and accounts for the radiative power losses due to electronic transitions of atoms (a similar but more complicated formula could be written in the molecular case).

The boundary conditions for $\mathbf{u}$ and T are strongly dependent on the particular problem. We could take, for example, a plasma enclosed in a tube with a background gas being fed at a constant mass flow rate. The velocity boundary condition at the inlet is determined from the mass flow rate, while the outlet velocity boundary condition may be set as zero gradient in the outlet surface normal direction. The temperature at the inlet is usually fixed to a known value, while the outlet temperature boundary condition may be also set as zero gradient. At the tube walls the velocity component normal to the surface should be set to zero in the case of an inviscid flow, or zero velocity in the case of a viscous flow. If the temperature is kept constant at the walls by some means or is known by measurement then the boundary temperature at the walls is given by this value. If the value is not known, the temperature boundary condition may be defined as zero gradient in the surface normal direction. If use is made of the axis of symmetry to reduce the space domain, the projection of the gradient of the temperature, particle fluxes and velocity in the symmetry surface should be set to zero.

11.1.4 Electrodynamics

Poisson equation and ambipolar diffusion
In accordance with the quasi-neutrality hypothesis, any volume of the plasma is electrically neutral as long as the sampled volume dimensions are not smaller than the Debye length. If a small departure from neutrality occurs by a small charge separation due to some kind of transport effect, then an electric field will build up

---

[1] The energy equation can be derived from the kinetic theory of dilute gases presented in [20] and the kinetic theory of electrons presented here.

pushing the particles with opposite charges back to neutrality, in accordance with the Poisson equation

$$\nabla \cdot \mathbf{E} = \frac{e}{\epsilon_0}\left(\sum_{j}^{N^{(+)}} Z_j n_j^{(+)} - \sum_{j}^{N^{(-)}} Z_j n_j^{(-)}\right), \tag{11.46}$$

where Z_j is the number of elementary charges of charge carrier j. For simplicity, let us consider a plasma with only electrons as negative charged particles and only one singly ionized positive species. Let us further suppose that a small departure from neutrality produces a local field $\mathbf{E}_l$, much smaller than the amplitude of an alternating microwave field. In this case, the electron energy distribution and transport parameters are determined by the alternating field and the local field influence on them can be neglected. The time averaged flow rates of ions and electrons are given by

$$\Gamma_i = -D_i \nabla N_i + N_i \mu_i \mathbf{E}_l \quad \text{and} \quad \Gamma_e = D_e \nabla N_e - N_e \mu_e \mathbf{E}_l. \tag{11.47}$$

From the continuity equations, it can be readily found that $\nabla \cdot (\Gamma_i - \Gamma_e) = 0$. Removing the electric field from the above equations, using the quasi-neutrality condition and the latter result, it can be shown that

$$\nabla \cdot \Gamma_i = \nabla \cdot \Gamma_e = -D_a \nabla^2 n, \tag{11.48}$$

where D_a is then a *ambipolar diffusion coefficient*

$$D_a = \frac{\mu_i D_e + \mu_e D_i}{\mu_i + \mu_e}. \tag{11.49}$$

Therefore, when the quasi-neutrality condition is assumed in a plasma—i.e. in a condition where space charges could have relevant effects—it is required that the diffusion coefficients of the charge carriers are replaced by the ambipolar diffusion coefficient D_a in the continuity equations. In the case of a plasma composed of many positive ions, the ambipolar diffusion coefficient must written as

$$D_{a,i} = D_i - \mu_i \frac{\sum_{j=1}^{n+1} n_j D_j}{\sum_{j=1}^{n+1} n_j \mu_j}, \quad i = 1, \ldots, n+1. \tag{11.50}$$

In this expression the index $n + 1$ represents the electrons. In the evaluation of the ambipolar diffusion coefficient the signal of the electronic density, n_{n+1}, and electron mobility, μ_{n+1}, must be negative in equation (11.50).

Note that the Poisson equation must be solved in the study of the intermediate case between a very faint electric discharge and a plasma fulfilling the quasi-neutrality condition. It is also required when the study of the details of the sheath structure is desired. It is clear that in these cases D_a should not be replaced in the continuity equations.

Sustaining plasma using high-frequency surface waves

Plasmas can be sustained by microwaves by several mechanisms. In principle, all of them can be well described by the solution of the Maxwell equations given the proper boundary and initial conditions and the plasma properties. The complexity of this problem lies in the fact that the electric field is in general strongly coupled to the plasma properties, which complicates the task of finding the solution.

In this section we treat the simple case of a plasma sustained by a propagating wave in magnetic transversal mode TM_{00}, which is the simplest model commonly applied to describe surface-wave discharges. These discharges are sustained by high-frequency waves which in turn have the plasma as their propagating medium.

In the case of a TM_{00} propagating wave, the electric and magnetic fields $\mathbf{E}$ and $\mathbf{B}$ can be written in cylindrical coordinates as $\mathbf{E} = (E_r, 0, E_z)$ and $\mathbf{B} = (0, B_\phi, 0)$. These components are assumed to be proportional to the factor $\exp(-j\omega t + j\int_z k(z')\,dz')$, where $k = \beta + j\alpha$, β is the propagation coefficient and α is the space damping factor (attenuation coefficient), and another factor which only depends on the radial coordinate. The relation between these components can be found using the Maxwell equations

$$\frac{\partial B_\phi}{\partial z} = j\frac{\omega}{c^2}\epsilon E_r; \quad \frac{1}{r}\left[\frac{\partial}{\partial r}(rB_\phi)\right] = -j\frac{\omega}{c^2}\epsilon E_z; \quad \frac{\partial E_r}{\partial z} - \frac{\partial E_z}{\partial r} = j\omega B_\phi, \tag{11.51}$$

where ϵ is the medium relative dielectric constant. In the case of a vacuum (or free space), the value is $\epsilon_0 = 1$; in the case of a dielectric, the value is ϵ_d (~4, if the medium material is quartz, for example), and in the case of a plasma the dielectric constant is given by [10]

$$\epsilon_p = 1 - \frac{\omega_p^2}{\omega^2 + \langle\hat{\Omega}\rangle^2} + j\frac{\langle\hat{\Omega}\rangle}{\omega}\frac{\omega_p^2}{\omega^2 + \langle\hat{\Omega}\rangle^2} = 1 - \xi + j\frac{\langle\hat{\Omega}\rangle}{\omega}\xi, \tag{11.52}$$

where $\omega_p = (\frac{Ne^2}{\epsilon_0 m_e})^{1/2}$ is the plasma oscillation frequency. After some algebraic manipulation of equation (11.51), we arrive at the following expression for the component E_z:

$$\frac{1}{r}\left[\frac{\partial}{\partial r}\left(r\frac{\partial E_z}{\partial r}\right)\right]\left[\frac{k^2}{\kappa_p^2} - 1\right] + \frac{\partial E_z}{\partial r}\frac{\partial}{\partial r}\left(\frac{k^2}{\kappa_p^2}\right) = \frac{\omega^2}{c^2}\epsilon E_z, \tag{11.53}$$

where $\kappa_p = k^2 - (\omega^2\epsilon)/c^2$. The radial component may be determined from E_z,

$$E_r = -j\frac{k}{\kappa_p^2}\frac{\partial E_z}{\partial r}. \tag{11.54}$$

We note that the solutions of the above equations are complex, so that the physical solution is obtained taking the real part. Note also that the factor $\exp(-j\omega t + j\int_z k(z')\,dz')$ may be canceled out from equations (11.53) and (11.54) so that only the radial dependence is retained.

Equation (11.53) assumes a simpler form if the plasma relative dielectric constant is assumed to be independent of the radial coordinate,

$$\frac{1}{r}\frac{\partial}{\partial r}\left(r\frac{\partial E_z}{\partial r}\right) + \zeta^2 E_z = 0, \tag{11.55}$$

where $\zeta^2 = -\kappa_p^2$. This equation may be written in the standard form as

$$r^2 E_z'' + rE_z' + \zeta^2 r^2 E_z = 0. \tag{11.56}$$

The general solution of this equation in a medium i with dielectric constant ϵ_i can be written with special complex functions [24]

$$E_{z,i}(r) = A_i J_0\left[\left(\frac{\omega^2}{c^2}\epsilon_i - k^2\right)^{1/2} r\right] + B_i H_0^{(1)}\left[\left(\frac{\omega^2}{c^2}\epsilon_i - k^2\right)^{1/2} r\right], \tag{11.57}$$

where J_0 is the Bessel function of first kind and order zero, and $H_0^{(1)}$ is the Hankel function of first kind and first order. The constants A_i and B_i may be determined from the boundary conditions in the interfaces between the different media and the imposed limits in $r = 0$ and $r \to \infty$. In a coaxial system plasma–dielectric–vacuum, we have $B_i = 0$ in the plasma and $A_i = 0$ in the vacuum from the requirement of finiteness in the limits $r = 0$ and $r \to \infty$. The signal of the arguments of the square-root functions were chosen so that the imaginary parts of the arguments of J_0 and $H_0^{(1)}$ were positive. This is necessary so that the function $H_0^{(1)}$ goes to zero when $r \to \infty$ [24].

The propagating and attenuation coefficients may be determined from the continuity condition applied to the components E_z and B_ϕ. The resulting values will also depend on the values of n_e, ω_p, and $\hat{\Omega}$. Dispersion curves for $\beta \times (\omega/\omega_p)$ and $\alpha \times (\omega/\omega_p)$ can be found in [24].

11.2 Application: contraction of an argon discharge

Let us consider the case of an argon discharge sustained by a surface wave at atmospheric pressure. The discharge is supposed to be ignited inside a cylindrical tube made of a dielectric material. It expands into free space, but we shall ignore the diffusion of species from the environment into the discharge. Suppose also that the flow field of the background gas is always laminar and sufficiently small to be neglected. The axial gradients of the physical quantities are supposed to be small, so that at each axial position we may restrict the analysis to the radial description of the physical quantities. The axial mean or maximum values of the physical quantities can be determined from experiments (i.e. mean electron density or absorbed power) or fixed to a reasonable value in order to proceed with a prospective study.

The continuity equation (11.34) can be written as

$$\frac{\partial[N_j]}{\partial t} = \Gamma_j^{(+)}\left(T_e,\ T_g,\ [N_j],\ ...,\ [N_M]\right) - \Gamma_j^{(-)}\left(T_e,\ T_g,\ [N_j],\ ...,\ [N_M]\right)$$
$$+ \frac{1}{r}\frac{\partial}{\partial r}\left(rD_j(r)\frac{\partial}{\partial r}[N_j]\right),$$

$$(11.58)$$

where $\Gamma^{(+,-)}$ are the gain (+) and loss terms (−) due to chemical reactions or spontaneous emission and M is the number of different species considered in the model. In this case, the reaction coefficients and transport parameters are functions of the space coordinates. The Boltzmann equation is solved for each point of the discretized space domain, so that the radial dependence of the transport parameters and electronic reaction coefficients can be determined. The local field approximation is assumed. As the EEDF depends on the electron density, which is given by the continuity equations in the quasi-neutrality approximation, an iterative procedure is required. Table 11.1 shows the electronic collisional processes which are included in the computation of the electron Boltzmann equation. The corresponding cross sections were extracted from the IST-Lisbon database, which can be accessed on-line on the LXCat project's webpage [25]. Some processes were not considered in the computation of the EEDF, but the reaction rates were computed *a posteriori* to obtain the reaction coefficients. These mechanisms are also listed in table 11.1.

Since quasi-neutrality is assumed, only the continuity equations for the positive ions have to be solved. In this model, only the ions Ar^+ and Ar_2^+ are considered. Although the electron continuity equation does not need to be computed, the electron diffusion coefficient and mobility which are used in the ambipolar diffusion formula, as well as the ionization rates, must be evaluated, since these quantities enter into the ion continuity equations.

In total, the model considers seven neutral heavy species: ($Ar(^1S_0)$, $Ar(^3P_2)$, $Ar(^1P_1)$, $Ar(^3P_0)$, $Ar(^3P_1)$, $Ar(4p)$, Ar_2^*). The 10 states that have the electronic configuration $4p$ are grouped together in a single state, indicated as $Ar(4p)$. The same applies to the excimer case Ar_2^*, which in reality has a more complex structure, including vibrational states [26].

Table 11.2 shows the reactions included in the model. The reaction coefficients were extracted from several sources and in order to abbreviate the extensive work of bibliographical survey we refer to previous compilations [27, 28] except for the cross sections which are not included in those works.

The solutions of the partial differential equation (11.58) are obtained using the boundary and initial conditions described next. In the walls, the concentration of the charged and reactive species were supposed to be zero, i.e. $[N_j]_{r = R_a} = 0$; from the requirement of symmetry, we have $(\partial[N_j]/\partial r)|_{r=0} = 0$, where R_a is the tube radius. The initial conditions are arbitrary, but one should choose conditions which lead to a feasible solution; we chose a Bessel function of the first kind and order zero to describe the initial profile of the charged particles, and a Gaussian profile for the

Table 11.1. Electronic cross sections.

Collision[a]	Use[b]	Ref.	Thr. (eV)
Elastic:			
$Ar(^1S_0) + e \rightarrow Ar(^1S_0) + e$	EB	IST	–
Excitation[c]:			
Excitation from $Ar(^1S_0)$:			
$Ar(^1S_0) + e \rightarrow Ar(^3P_1) + e$	EB, RC	IST	11.6
$Ar(^1S_0) + e \rightarrow Ar(^1P_1) + e$	EB, RC	IST	11.6
$Ar(^1S_0) + e \rightarrow Ar(^3P_2) + e$	EB, RC	IST	11.6
$Ar(^1S_0) + e \rightarrow Ar(^3P_0) + e$	EB, RC	IST	11.6
$Ar(^1S_0) + e \rightarrow Ar(4p) + e$	EB, RC	IST	13.0
$Ar(^1S_0) + e \rightarrow Ar^{**} + e$	EB, RC	IST	14.0
Excitation from 4s:			
$Ar(4s) + e \rightarrow Ar(4p) + e$	RC	IST	1.4
Excitation between states 4s:			
$Ar(4s_i) + e \rightarrow Ar(4s - j) + e, i < j$	RC	IST	—
Ionization:			
Ionization from $Ar(^1S_0)$:			
$Ar(^1S_0) + e \rightarrow Ar^+ + 2e$	EB, RC	IST	15.6
Ionization from states 4s:			
$Ar(4s) + e \rightarrow Ar^+ + 2e$	RC	[29]	4.0
Ionization from states 4p:			
$Ar(4p) + e \rightarrow Ar^+ + 2e$	RC	[29]	2.6
Ionization from excimer Ar_2^*:			
$Ar_2^* + e \rightarrow Ar_2^+ + 2e$	RC	[29]	3.8
Superelastic processes:			
$Ar(4s) + e \rightarrow Ar(^1S_0) + e$	EB, RC	Klein–Rosseland	—
$Ar(4p) + e \rightarrow Ar(^1S_0) + e$	EB, RC	Klein–Rosseland	—
$Ar(4p) + e \rightarrow Ar(4s) + e$	RC	Klein–Rosseland	—
$Ar(4s_i) + e \rightarrow Ar(4s_j) + e, i > j$	RC	IST	—

[a] The electronic states of argon are written in $L - S$ notation using capital letters. The lowercase letter, such as 4s and 4p, indicates the electronic configuration of the excited electron and it is used to group all the states with the same configuration.
[b] The cross sections included in the Boltzmann code are indicated with the tag 'EB'; the cross sections used in post-processing calculations are indicated with the tag 'RC'. Many cross sections were used both in the Boltzmann equation evaluation and electronic reaction rate calculations.
[c] The 4p states were grouped in one 'effective' state. The four states in the 4s group were discriminated.

neutral species. The initial density of the background gas (ground-state argon) is considered to be constant. The value was given by the state equation of an ideal gas at constant pressure. The temperature needed for the evaluation of the background gas density equation is determined from the heat equation

$$\frac{1}{r}\frac{d}{dr}\left(r\lambda\big(T_g(r)\big)\frac{dT_g}{dr}\right) = -\sigma_e(r)E_0(r)^2, \tag{11.59}$$

Table 11.2. Reaction between heavy species, spontaneous emission, and reaction rates.

Reaction[a]	Ref.	Reaction coefficient
Diffusion:		$[\mathrm{s}^{-1}]$
Free diffusion		
$Ar(^3P_2)$, $Ar(^3P_0)$	[28]	D_f/Λ^2
Ambipolar diffusion		
Ar^+, Ar_2^+	[30]	D_a/Λ^2
Spontaneous emmision		$[\mathrm{s}^{-1}]$
$Ar(^3P_1) \rightarrow Ar(^1S_0) + h\nu$	[31]	$g \times (1.19 \cdot 10^8)$
$Ar(^1P_1) \rightarrow Ar(^1S_0) + h\nu$	[31]	$g \times (5.1 \cdot 10^8)$
$Ar(4p) \rightarrow Ar(^3P_2) + h\nu$	[31]	$1.43 \cdot 10^7$
$Ar(4p) \rightarrow Ar(^3P_1) + h\nu$	[31]	$7.3 \cdot 10^6$
$Ar(4p) \rightarrow Ar(^3P_0) + h\nu$	[31]	$3.5 \cdot 10^6$
$Ar(4p) \rightarrow Ar(^1P_1) + h\nu$	[31]	$9.83 \cdot 10^6$
$Ar_2(*) \rightarrow 2Ar(^1S_0) + h\nu$	[26]	$3.13 \cdot 10^5$
Molecular conversion:		$[\mathrm{cm}^6\ \mathrm{s}^{-1}]$
$Ar^+ + Ar(^1S_0) + Ar(^1S_0)$ $\rightarrow Ar_2^+ + Ar(^1S_0)$	[32]	$2, 5 \cdot 10^{-31}$
Dissociative recombination[a]:		$[\mathrm{cm}^3\ \mathrm{s}^{-1}]$
$Ar_2^+ + e \rightarrow Ar(^1S_0) + Ar(^1S_0)$	[33]	$9.6 \cdot 10^{-7}\left(1 - \exp\left(-\frac{630}{T_g}\right)\right)\chi\left(\frac{300}{T_e}\right)^{0.67}$
Penning ionization		$[\mathrm{cm}^3\ \mathrm{s}^{-1}]$
$Ar(4s_i) + Ar(4s_j) \rightarrow Ar(^1S_0) + Ar^+ + e$	[28]	$3.69 \cdot 10^{-11}\ T_g^{0.5}$
$Ar(4p) + Ar(4p) \rightarrow Ar^+ + Ar(^1S_0) + e$	[34]	$5 \cdot 10^{-10}$
$Ar(4p) + Ar(4s) \rightarrow Ar^+ + Ar(^1S_0) + e$	[34]	$5 \cdot 10^{-10}$
$Ar(4s_i) + Ar(4s_j) \rightarrow Ar_2^+ + e$	[34]	$5 \cdot 10^{-10}$
Electronic impact dissociation		$[\mathrm{cm}^3\ \mathrm{s}^{-1}]$
$e + Ar_2^+ \rightarrow e + Ar(^1S_0) + Ar^+$	[35]	$1.11 \cdot 10^{-6}\exp\left(-\frac{(2.94 - 3(T_g[\mathrm{eV}] - 0.026))}{(T_e[\mathrm{eV}])}\right)$
Atomic conversion		$[\mathrm{cm}^3\ \mathrm{s}^{-1}]$
$Ar_2^+ + Ar(^1S_0) \rightarrow Ar^+ + 2Ar(^1S_0)$	[35]	$5.22 \cdot 10^{-10}\dfrac{\exp\left(-\frac{1.304}{T_g[\mathrm{eV}]}\right)}{T_g[\mathrm{eV}]}$
Three-body recombination		$[\mathrm{cm}^6\ \mathrm{s}^{-1}]$
$e + e + Ar^+ \rightarrow e + Ar(4p)$	[29]	$1.4303 \cdot 10^{-28}(T_e[\mathrm{eV}])^{-3}$
$e + e + Ar^+ \rightarrow e + Ar(^3P_2)$	[29]	$6.1437 \cdot 10^{-29}(T_e[\mathrm{eV}])^{-3}$
$e + e + Ar^+ \rightarrow e + Ar(^3P_1)$	[29]	$1.33027 \cdot 10^{-29}(T_e[\mathrm{eV}])^{-3}$
$e + e + Ar^+ \rightarrow e + Ar(^3P_0)$	[29]	$3.81289 \cdot 10^{-29}(T_e[\mathrm{eV}])^{-3}$
$e + e + Ar^+ \rightarrow e + Ar(^1P_1)$	[29]	$4.19207 \cdot 10^{-29}(T_e[\mathrm{eV}])^{-3}$
Excimer formation		$[\mathrm{cm}^6\ \mathrm{s}^{-1}]$
$Ar(4s) + 2Ar(^1S_0) \rightarrow Ar_2(*) + Ar$	[27]	10^{-32}

Excimer extinction $\qquad$ [cm^3 s^{-1}]

$$\text{Ar}_2(*) + \text{Ar}_2(*) \rightarrow \text{Ar}_2^+ + 2\text{Ar}(^1\text{S}_0) \qquad [26] \qquad 5 \cdot 10^{-10}$$

Atomic three-body recombination $\qquad$ [cm^6 s^{-1}]

$$\text{Ar}^+ + \text{Ar}(^1\text{S}_0) + e \rightarrow 2\text{Ar}(^1\text{S}_0) \qquad [27] \qquad 3.7 \cdot 10^{-29} T_e(\text{eV})^{-1.5} T_g(K)^{-1}$$

---

[a] Fast atoms produced after dissociative recombination usually lost their charge, so only a small percentage of the collisional events effectively results in charge destruction [36]. We considered that just one percent ($\chi = 0, 01$) of the collisional processes produce charge destruction. Other authors [35] adopted the value $\chi = 0, 05$.

where λ is the thermal conductivity of argon and σ is the electric conductivity ($\sigma_e = en_e\mu_e$). Note that the the energy lost or gained due to radiation and chemical reactions is neglected, as electron collisions may be considered as the strongest heating source in the present case. The terms containing axial gradients were also neglected, in accordance with our earlier assumptions.

The thermal conductivity in equation (11.59) depends on the temperature, so an empirical relation for $\lambda(T)$ [37] had to be used to obtain a realistic solution. The electron conductivity $\sigma(r)$ was obtained from the solution of the electron Boltzmann equation. Rigorously, the electric field should be obtained from equation (11.53), but we used the approximate solution given by equation (11.55), using a similar approach to that proposed in [38]. The general solution of the $E_z(r)$ component is given by the formula

$$F_p(r) = E_0 I_0\big(a_p r/R_a\big) = E_0 I_0\left[\left(\beta^2 - \frac{\omega^2}{c^2}\epsilon_p^2\right)^{1/2} r\right], \qquad (11.60)$$

where R_a is the tube radius or the frontier between the plasma and the environment and I_0 is the modified Bessel function of the first kind. The dispersion relation is calculated using the explicit formulas derived by Zhelyazkov and Benova [38]. Equation (11.60) assumes a constant electron density, which was estimated as the radially averaged electron density

$$\langle N_e \rangle = \frac{\int_0^{2\pi}\int_0^{R_a} N_e(r)r \, dr d\theta}{\int_0^{2\pi}\int_0^{R_a} r \, dr d\theta} = \frac{2\int_0^{R_a} N_e(r)r \, dr}{R_a^2}. \qquad (11.61)$$

This value was used to obtain the plasma permittivity, ϵ_p, and evaluate the dispersion relation [38].

The boundary conditions of the heat equation are defined similarly to the continuity equations. From the requirement of symmetry we have $(\partial T/\partial r)|_{r=0} = 0$. The borders are assumed to have a constant temperature, $T|_{r=R_a} = T_0$, where T_0 is the laboratory temperature.

All the equations are coupled, since the continuity equations need the transport parameters and reaction rates from the Boltzmann equation, which in turn needs the electric field and electric density profiles. All the equations must be solved in an iterative and self-consistent fashion. Figure 11.1 shows the work-flow diagram of the iterative procedure. One of the parameters, the electric amplitude, acts as an 'eigenvalue' of the problem, which should be adjusted in order to produce the electric density or another physical quantity (e.g. the attenuation coefficient) given by the experiment. In the present case, we set the electric density at the center of the discharge, N_{e0}, as the fixed physical quantity, which in general is given by

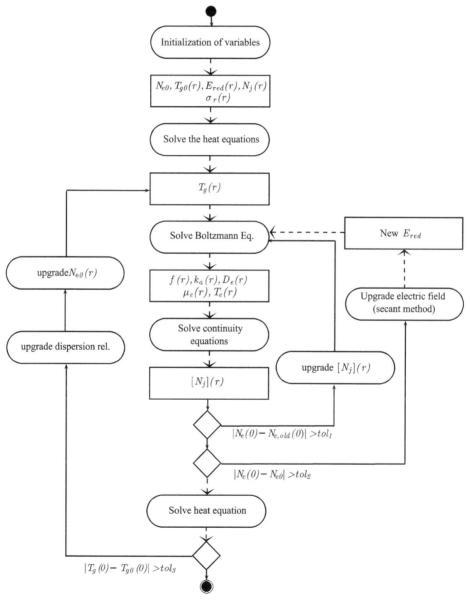

Figure 11.1. Work-flow graph representing the iterative and auto-consistent procedure.

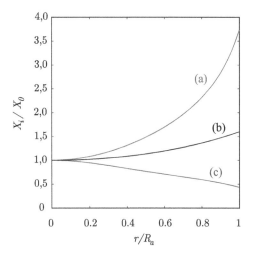

Figure 11.2. Normalized radial profile of particle density N, electric field $E_z(r)$, and reduced electric field $E_{red} = E_z(r)/N$. (a) $N(r)/N_0$, where $N_0 = 6.46 \cdot 10^{18}$ cm^{-3}; (b) $E_z(r)/E_0$, where $E_0 = 2505$ V m^{-1}; (c) $E_{red}(r)/E_{red0}$, where $E_{red0} = 0.39$ Td. The index 0 represents the value of a physical quantity in the column center.

experiment. Here we adopted $N_e = 3.5 \cdot 10^{14}$ cm^{-3}, a typical value which may be found in this kind of discharge operating at a flow rate of 2.5 slm, an applied power of 50 W, and a tube radius of 1.0 mm [39].

The solution converged to an electrical field amplitude (eigenvalue) of $E_0 = 2505$ V m^{-1}. Figure 11.2 shows the normalized profile of the field, obtained from equation (11.60). It also shows the reduced field $E_{red}(r)$ and the particle density, computed from $N(r) = p/(k_B T(r))$. The value of $N(r)$ grows faster than the electric field $E_z(r)$, resulting in a decreasing reduced electric field, which seems to be in contradiction with the experimental observation of a electron temperature increase towards the border [40]. Figure 11.4 shows two plots of the charged species profiles. From this graphic, it can be seen that the atomic ion in the discharge center is predominant. In the borders, where the gas temperature and electric density are smaller, the molecular ion density becomes larger. These results agree qualitatively with experiments [41].

In reality, there are many factors which may contribute to the discharge contraction. These factors may be interrelated. The diffusion process naturally causes the density decrease towards the borders. Because of the dependence of the EEDF and ionization coefficient on the electron density (see figure 11.3), an abrupt fall in the ionization rate occurs. At the same time, the metastable densities decrease, which reinforces the decrease of the effective ionization rate due to the reduction of Penning ionization. On the other hand, the large relative concentration of the dimer Ar_2^+ in the discharge border guarantees an elevated recombination rate, which effectively contributes to negative charge destruction and electron density collapse. Note that the dimer creation rate increases when the temperature decreases, due to the dependence of the background gas on T_g, explaining why the relative concentration of the dimer increases towards the border. Another important effect

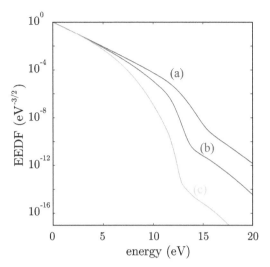

Figure 11.3. Comparison of the EEDFs at different positions in the plasma column, with $R_a = 1.0$ mm and $n_{e0} = 3.5 \cdot 10^{14}$ cm^{-3}. (a) $\rho = 0$, $T_e = 1.05$ eV; (b) $\rho = 0.1$, $T_e = 1.04$ eV; (c) $\rho = 0.2$, $T_e = 1.005$ eV.

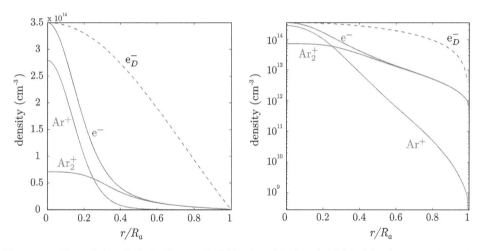

Figure 11.4. The radial profile in the linear scale (left) or logarithmic scale (right) of the charged particles with $n_{e0} = 3.5 \cdot 10^{14}$ cm^{-3} in the column center and $R_a = 1.0$ mm (e_D^- electron density in diffusion regime).

particular to this model is the reduced electric field fall, which is also due to the increase of the background gas density towards the border (see figure 11.2). We call attention to the fact that the electric field was obtained assuming a constant electron density in the plasma medium; the detailed description of the field could possibly explain the experimental observation of an electric temperature increase towards the border [40]. In order to give a better idea of the contraction effect, the plot in figure 11.4 also shows the Bessel profile which is expected in the diffusion regime characteristic of low pressure conditions.

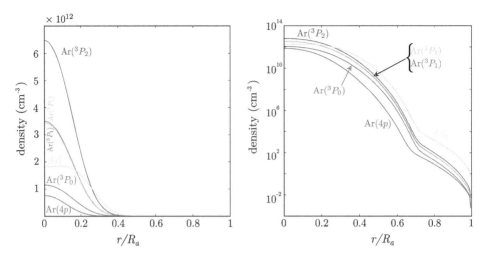

Figure 11.5. Radial profile in linear scale (left) or logarithmic scale (right) of the excited neutrals, with $n_e = 3.5 \cdot 10^{14}$ cm^{-3} and $R_a = 1.0$ mm.

The profile of the excited neutral species is shown in figure 11.5. The concentration of this species also decreases towards the border, except in the excimer case. One of the destruction channels of this species is ionization by electron impact, and the only creation channel has a rate which depends quadratically on Ar($^1S_0$), which explains why the excimer maximum is displaced from the column center.

11.3 Conclusion

In this chapter we presented the theoretical foundations of plasma modeling for problems in high-frequency fields and atmospheric pressure. We gave special emphasis to the classical electron kinetic theory and its connections with fluid modeling, since there is much evidence that non-equilibrium EEDF play a very important role in many phenomena. The treatment of heavy species, flow dynamics, heat transport, and electrodynamics was also discussed. In the last section, the theory was applied to the description of an argon discharge at atmospheric pressure, and the mechanisms leading to contraction were identified and discussed.

We hope that this chapter may serve as a starting point for those interested in the design of more elaborate models, which would eventually contain more ingredients. As a matter of fact, a truly comprehensive model is very difficult to implement. Efforts in this direction may be found in the recent literature [27, 42–44], and full understanding of the mechanisms of contraction clearly depends on the level of detail and completeness of the model. In addition, descriptions of plasma phenomena at atmospheric pressure in turbulent flows or plasmas sustained by transversal mode fields at higher-order modes, for instance, are poorly studied. In view of the many interesting applications of these discharges, further developments in this field would certainly be worthwhile.

Bibliography

[1] Crookes W 1879 The Bakerian lecture: on the illumination of lines of molecular pressure, and the trajectory of molecules *Phil. Trans. R. Soc. Lond.* **170** 135–64

[2] Crookes W 1819 Contributions to molecular physics in high vacua. Magnetic deflection of molecular trajectory. Laws of magnetic rotation in high and low vacua. Phosphorogenic properties of molecular discharge *Phil. Trans. R. Soc. Lond.* **170** 641–62

[3] Thomson J XIX 1899 On the theory of the conduction of electricity through gases by charged ions *Lond. Edin. Dublin Phil. Mag. J. Sci.* **47** 253–68

[4] Weisskopf V F 1965 In defence of high energy physics *Nature of Matter: Purposes of High Energy Physics* (Long Island, NY: Brookhaven National Laboratory) pp. 24–7

[5] Moisan M and Zakrzewski Z 1991 Plasma sources based on the propagation of electromagnetic surface waves *J. Phys. D: Appl. Phys.* **24** 1025

[6] Beenaker C 1977 Evaluation of a microwave-induced plasma in helium at atmospheric pressure as an element-selective detector for gas chromatography *Spectrochim Acta* B **32** 173–87

[7] Hopwood J, Hoskinson A R and Gregório J 2014 Microplasmas ignited and sustained by microwaves *Plasma Sources Sci. Technol.* **23** 064002

[8] Hnilica J, Kudrle V, Vašina P, Schäfer J and Aubrecht V 2012 Characterization of a periodic instability in filamentary surface wave discharge at atmospheric pressure in argon *J. Phys. D: Appl. Phys.* **45** 055201

[9] Blevin H A and Fletcher J 1984 Electron transport and rate coefficients in Townsend discharges *Austral. J. Phys.* **37** 593–600

[10] Bittencourt J 2003 *Fundamentals of Plasma Physics* (Berlin: Springer)

[11] Shkarofsky I, Johnston T and Bachynski M 1966 *The Particle Kinetics of Plasmas* (Reading, MA: Addison-Wesley)

[12] Lin S, Robson R and Mason E 1979 Moment theory of electron drift and diffusion in neutral gases in an electrostatic field *J. Chem. Phys.* **71** 3483–98

[13] Loureiro J and Amorim J 2016 *Kinetics and Spectroscopy of Low Temperature Plasmas* 1st edn (New York: Springer)

[14] Holstein T 1946 Energy distribution of electrons in high frequency gas discharges *Phys. Rev.* **70** 367–84

[15] Ridenti M A, Alves L L, Guerra V and Amorim J 2015 The role of rotational mechanisms in electron swarm parameters at low reduced electric field in N_2, O_2 and H_2 *Plasma Sources Sci. Technol.* **24** 035002

[16] Hagelaar G J M and Pitchford L C 2005 Solving the Boltzmann equation to obtain electron transport coefficients and rate coefficients for fluid models *Plasma Sources Sci. Technol.* **14** 722–33

[17] White R D, Robson R, Schmidt B and Morrison M A 2003 Is the classical two-term approximation of electron kinetic theory satisfactory for swarms and plasmas? *J. Phys. D: Appl. Phys.* **36** 3125

[18] Holstein T 1947 Imprisonment of resonance radiation in gases *Phys. Rev.* **72** 1212–33

[19] Holstein T 1951 Imprisonment of resonance radiation in gases. II *Phys. Rev.* **83** 1159–68

[20] Hirschfelder J O, Curtis C F and Bird R B 1954 *Molecular Theory of Gases and Liquids* (New York: Wiley)

[21] Drake G W 2006 *Springer Handbook of Atomic, Molecular, and Optical Physics* (New York: Springer)

[22] Laidler K 1987 *Chemical Kinetics* (New York: Harper and Row)

[23] Mallard W 1992 *NIST Chemical Kinetics Database* (Washington, DC: National Institute of Standards and Technology) http://kinetics.nist.gov/kinetics/index.jsp

[24] Glaude V M M, Moisan M, Pantel R, Leprince P and Marec J 1980 Axial electron density and wave power distributions along a plasma column sustained by the propagation of a surface microwave *J. Appl. Phys.* **51** 5693–8

[25] Alves L and Ferreira C 2014 *IST-Lisbon Database* www.lxcat.net

[26] Lam S K, Zheng C-E, Lo D, Dem'yanov A and Napartovich A P 2000 Kinetics of Ar_2^* in high-pressure pure argon *J. Phys. D: Appl. Phys.* **33** 242

[27] Petrov G M and Ferreira C M 2013 Numerical modelling of Ar glow discharge at intermediate and high pressures arXiv: 1308.2593

[28] Kutasi K, Guerra V and Sá P 2010 Theoretical insight into Ar-O_2 surface-wave microwave discharges *J. Phys. D: Appl. Phys.* **43** 175201

[29] Vriens L and Smeets A H M 1980 Cross-section and rate formulas for electron-impact ionization, excitation, deexcitation, and total depopulation of excited atoms *Phys. Rev.* A **22** 940–51

[30] Jonkers J, van de Sande M, Sola A, Gamero A and van der Mullen J 2003 On the differences between ionizing helium and argon plasmas at atmospheric pressure *Plasma Sources Sci. Technol.* **12** 30

[31] Wiese W L, Brault J W, Danzmann K, Helbig V and Kock M 1989 Unified set of atomic transition probabilities for neutral argon *Phys. Rev.* A **39** 2461–71

[32] Smith D, Dean A G and Plumb I C 1972 Three body conversion reactions in pure rare gases *J. Phys. B: At. Mol. Phys.* **5** 2134

[33] O'Malley T F, Cunningham A J and Hobson R M 1972 Dissociative recombination at elevated temperatures. II. Comparison between theory and experiment in neon and argon afterglows *J. Phys. B: At. Mol. Phys.* **5** 2126

[34] Brunet H, Lacour B, Rocca Serra J, Legentil M, Mizzi S, Pasquiers S and Puech V 1990 Theoretical and experimental studies of phototriggered discharges in argon and neon *J. Appl. Phys.* **68** 4474–80

[35] Jonkers J, van de Sande M, Sola A, Gamero A, Rodero A and van der Mullen J 2003 The role of molecular rare gas ions in plasmas operated at atmospheric pressure *Plasma Sources Sci. Technol.* **12** 464

[36] Ramos G, Schlamkowitz M, Sheldon J, Hardy K and Peterson J 1995 Observation of dissociative recombination of Ne_2^+ and Ar_2^+ directly to the ground state of the product atoms *Phys. Rev.* A **51** 2945

[37] Younglove B and Hanley H J 1986 The viscosity and thermal conductivity coefficients of gaseous and liquid argon *J. Phys. Chem. Ref. Data* **15** 1323–37

[38] Zhelyazkov I and Benova E 1989 Modeling of a plasma column produced and sustained by a traveling electromagnetic surface wave *J. Appl. Phys.* **66** 1641–50

[39] Ridenti M A, Souza-Corrêa J A and Amorim J 2014 Experimental study of unconfined surface wave discharges at atmospheric pressure by optical emission spectroscopy *J. Phys. D: Appl. Phys.* **47** 045204

[40] van Gessel A F H, Carbone E A D, Bruggeman P J and van der Mullen J J A M 2012 Laser scattering on an atmospheric pressure plasma jet: disentangling Rayleigh, Raman and Thomson scattering *Plasma Sources Sci. Technol.* **21** 015003

[41] Ridenti M, Spyrou N and Amorim J 2014 The crucial role of molecular ions in the radial contraction of argon microwave-sustained plasma jets at atmospheric pressure *Chem. Phys. Lett.* **595** 83–6

[42] Jimenez-Diaz M, Carbone E A D, van Dijk J and van der Mullen J J A M 2012 A two-dimensional Plasimo multiphysics model for the plasma–electromagnetic interaction in surface wave discharges: the surfatron source *J. Phys. D: Appl. Phys.* **45** 335204

[43] Castaños Martinez E, Kabouzi Y, Makasheva K and Moisan M 2004 Modeling of microwave-sustained plasmas at atmospheric pressure with application to discharge contraction *Phys. Rev.* E **70** 066405

[44] Dyatko N A, Ionikh Y Z, Kochetov I V, Marinov D L, Meshchanov A V, Napartovich A P, Petrov F B and Starostin S A 2008 Experimental and theoretical study of the transition between diffuse and contracted forms of the glow discharge in argon *J. Phys. D: Appl. Phys.* **41** 055204

IOP Publishing

Plasma Modeling
Methods and Applications
Sergey T Surzhikov

Chapter 12

High-enthalpy radiating flows in aerophysics

The expression 'high-enthalpy radiating flow' is often use to indicate gas flows heated up to high temperatures, with developed physicochemical processes, such as dissociation, ionization, and photon emission. Such high-enthalpy gas flows are very often encountered in nature (e.g. astrophysical plasmas [1–4]). The study of the aerothermodynamics of space vehicles and of hypersonic aircrafts, as well as of different high-energy devices, is directly related to the solution of the complex problem of the physics of high-enthalpy flows, which in turn is closely related to the problem of radiative heat transfer.

Unique features of hypersonic flight for space access are embedded in the phenomena of rarefied gas dynamics, viscous-inviscid interaction, surface ablation, non-equilibrium thermodynamics, and chemical reactions, as well as radiative heat transfer [5–9]. The traditional computational simulation capability has been developed on the framework of the kinetic theory of gases. In a high-temperature environment, the internal degrees of freedom of the gaseous medium become excited and knowledge of the internal structure of the gas components is required to understand the interactions at molecular and atomic scales [10, 11]. Therefore the relatively poorly understood but most challenging topics for high-enthalpy flow simulation can be identified as the physics-based models for the non-equilibrium thermodynamic state of the medium.

Flows in chemical and thermal non-equilibrium are often investigated by using conservation laws of interdisciplinary computational fluid dynamics (CFD) for high-enthalpy gases. In the framework of continuum mechanics, the bridge linking the Boltzmann equation and the conservative laws of aerodynamics is the kinetic theory of gas through Chapman–Enskog expansion [10]. Since the classical kinetic theory does not include the internal structure of atoms and molecules, the higher degrees of excitation beyond the molecular translational motion must be solved by approximations. Traditionally the connection between molecular and atomic structure and the thermodynamic behavior of high-temperature gas is described by statistical thermodynamics

doi:10.1088/978-0-7503-1200-4ch12
© IOP Publishing Ltd 2016

and augmented by knowledge from quantum physics. The evaluation of finite-rate chemical reactions usually adopts chemical kinetic models through the law of mass action [12–15]. Some attempts have recently been undertaken dedicated to the application of quantum chemistry and quantum statistics in modern aerophysics [16–18].

In a hypersonic atmospheric entry, the radiation heat transfer contributes a substantial amount of energy transfer in addition to the conductive and convective contributions. For example, at the peak heating condition of the Stardust re-entry [19], the radiative heat transfer rate is 120 W cm^{-2} versus the convective–conductive heat transfer rate of 839 W cm^{-2}. Therefore it is critically important to understand the basic heat transfer mechanisms when developing accurate and efficient predictive methods.

All transitions involving radiation emission or absorption can be divided into bound–bound, bound–free, and free–free processes according to the initial and final quantum states of the atom or molecule, resulting in a continuum or discrete spectrum [20, 21]. In general, the thermal radiative heat transfer has a frequency spectrum from the far infrared to the vacuum ultraviolet bands. High-enthalpy gas mixtures in the shock layers and in the wakes of re-entering space vehicles, also with species from the ablative heat shield, usually contain optically active components such as CO_2, H_2O, CH_4, N_2, O_2, NO, N_2^+, C_2, CO, etc. At a re-entry speed around 10 km s^{-1} strong absorption by C, N, O, NO, CO, and O_2 has been detected in the vacuum ultraviolet (80–200 nm) and ultraviolet spectra. Some of these transition processes are not fully understood, but are a critical mechanism for radiative heat transfer [22, 23].

The computation of radiative heat transfer requires the development of effective numerical simulation methods not only for the prediction of spectral radiation fluxes inside the radiating control volume, but also for the evaluation of optical properties. The energy exchange mechanism of radiative heat transfer is fundamentally different from the convective or conductive mechanisms. It is a phenomenon associated with quantum transitions of molecules and atoms. Therefore the electronic excitation is not ignorable in radiative calculation, as it is in convective–conductive heat transfer simulations in most Earth re-entry conditions.

The development of computational models and codes for the description of the aerothermodynamics of re-entry space vehicles—which are intended to investigate the planets of the Solar System then return to Earth—is an important part of the scientific and engineering programs currently being realized in different space agencies in the world. It is also well known that the general method for the verification and validation of such models and codes is comparison with available experimental and flight data, which are still very scanty. Existing well-known flight experiments, such as RAM-C-II [24–26] and Fire-II [27], which are still referred to for computational works [27–31], probed radiation properties in the shock layer, under different conditions, and from full non-equilibrium to local thermodynamic equilibrium (LTE). Ionization processes in air compressed layers at altitudes of 61–81 km in the vicinity of the spherically blunted cone at orbital velocity were studied in the RAM-C-II flight experiments, while the density of the convective and radiative heat fluxes to the surface of the superorbital spacecraft Fire-II were detected, conditions under which non-equilibrium dissociation and ionization play an important role.

The highest entry velocity of a spacecraft into Earth's atmosphere (12.4 km s^{-1}) was achieved for the spacecraft Stardust. In the first series of works focused on the

Stardust re-entry [19, 22, 23], predictions can be found on the aerothermodynamics and ablation of the thermal protection system, based on flight data analysis after the return of the spacecraft [31] and remote observation [32].

In 2010, a special issue of the *Journal of Spacecraft and Rockets* published a number of computational results performed using the new models created during the preceding decade. It was shown that the predicted data of 1999 [19] were confirmed with good accuracy. It is worth noting that the data from the flight experiments of Fire-II allowed the verification of the computer models used to describe the conditions of Stardust's entry with good reliability.

Non-equilibrium is a critical aspect for spacecraft entering the atmosphere with super-orbital velocity, where strong gas dynamics–radiation interaction is expected and a large contribution to the radiative heat transfer comes from atomic and ionic transitions [28–30].

Particularly relevant is the problem of thermal non-equilibrium, which actually means that there are no equilibrium conditions between the Boltzmann distributions of the populations of the vibrational and electronic excited states, which determine the intensity of the luminescence of the thermal radiation. To take into account the disequilibrium a multi-temperature approach can be used, where the energy stored in each vibrational or electronically excited state can be estimated by the Boltzmann distribution, with the introduction of a characteristic vibrational and electron temperatures. This last set is equal to that of the electronically excited states.

An alternative approach, called the *collisional–radiative* model, consists of solving the master equation for each excited state level of the atoms and molecules [33] (see also chapter 8).

Another challenging task which can be solved using models developed in in the framework of high-enthalpy flows is the physics of meteoroids. It is well known that the determination of the concentration of atoms and molecules, and relative ions, is the main problem in the quantitative analysis of meteor spectra [34]. The models should allow the determination of the chemical composition and the types of materials making up meteoroids and, consequently, the quantitative abundances of the constituent elements, thus we can establish to what degree the minor bodies of the Solar System are similar. This is a fundamental problem, as it is now widely accepted that there exists a genetic interrelation between asteroids, comets, and meteoroids. Different methods for the quantitative analysis of meteor spectra have been developed and used by researchers, classified as the quantitative physical approach. However, other research groups are developing methodologies based on the computational physical gas dynamics of high-enthalpy flows. The emission and ablation of large meteoroids in the course of their motion through the Earth's atmosphere are considered in [35]. The radiative gas dynamics and emission of small meteoroids are considered in [36]. It was shown that the non-equilibrium effects of the vibrational excitation of diatomic molecules N_2, O_2, and NO behind the leading shock wave play a significant role in forming the spectral signatures of bolides, as well as in ionization processes in the shock layer and wake.

The radiative gas dynamic model of high-enthalpy flows, which can be applied to the investigation of most of the mentioned problems, will be described in this chapter. The theory is based on the equations of viscous, heat conducting, chemically non-equilibrium, emitting and absorbing gas, and includes databases of thermodynamic, kinetic, and radiation properties.

12.1 Fluid dynamic model

Most general radiative gas dynamic models should contain Navier–Stokes or Euler continuity equations with mass conservation of chemical species, vibrational energy conservation, and the radiation heat transfer equation (in multi-group approximation). These are

$$\frac{\partial \rho}{\partial t} + \nabla \cdot \rho \mathbf{v} = 0 \qquad\qquad (a)$$

$$\frac{\partial \rho v_k}{\partial t} + \nabla \cdot \rho v_k \mathbf{v} = -\frac{\partial p}{\partial x_k} + F_{\mu,k} \qquad\qquad (b) \;\; k = x, y, z$$

$$\frac{\partial \rho_i}{\partial t} + \nabla \cdot \rho_i \mathbf{v} = -\nabla \cdot \mathbf{J}_i + \dot{\omega} \qquad\qquad (c) \;\; i = 1...N_s$$

$$\rho c_p \frac{\partial T}{\partial t} + \rho c_p \mathbf{v} \cdot \nabla T = \nabla \cdot \left(\lambda \nabla T \right) + \frac{\partial p}{\partial t} + \mathbf{v} \cdot \nabla p \qquad\qquad (12.1)$$

$$+ \Phi_\mu + Q_{\text{vib}} - \nabla \cdot \mathbf{q}_R - \sum_{i=1}^{N_s} \left[h_i \dot{\omega}_i + c_{p,i} \mathbf{J}_i \cdot \nabla T \right] \qquad (d)$$

$$\frac{\partial \rho e_{v,m}}{\partial t} + \nabla \cdot \rho \mathbf{v} e_{v,m} = \dot{e}_{v,m} \qquad\qquad (e) \; i = 1...N_v$$

$$\mathbf{\Omega} \cdot \nabla J_\omega \left(\mathbf{r}, \mathbf{\Omega} \right) + k_\omega(\mathbf{r}) J_\omega \left(\mathbf{r}, \mathbf{\Omega} \right) = j_\omega(\mathbf{r}) \qquad\qquad (f),$$

where for the Navier–Stokes equations

$$F_{\mu,x} = -\frac{2}{3} \frac{\partial}{\partial x} \left(\mu \nabla \cdot \mathbf{v} \right) + 2 \frac{\partial}{\partial x} \mu \frac{\partial v_x}{\partial x} + \frac{\partial}{\partial y} \left[\mu \left(\frac{\partial v_y}{\partial x} + \frac{\partial v_x}{\partial y} \right) \right] + \frac{\partial}{\partial z} \left[\mu \left(\frac{\partial v_z}{\partial x} + \frac{\partial v_x}{\partial z} \right) \right]$$

$$F_{\mu,y} = -\frac{2}{3} \frac{\partial}{\partial y} \left(\mu \nabla \cdot \mathbf{v} \right) + 2 \frac{\partial}{\partial y} \mu \frac{\partial v_y}{\partial y} + \frac{\partial}{\partial x} \left[\mu \left(\frac{\partial v_x}{\partial y} + \frac{\partial v_y}{\partial x} \right) \right] + \frac{\partial}{\partial z} \left[\mu \left(\frac{\partial v_z}{\partial y} + \frac{\partial v_y}{\partial z} \right) \right]$$

$$F_{\mu,z} = -\frac{2}{3} \frac{\partial}{\partial z} \left(\mu \nabla \cdot \mathbf{v} \right) + 2 \frac{\partial}{\partial z} \mu \frac{\partial v_z}{\partial z} + \frac{\partial}{\partial x} \left[\mu \left(\frac{\partial v_x}{\partial z} + \frac{\partial v_z}{\partial x} \right) \right] + \frac{\partial}{\partial y} \left[\mu \left(\frac{\partial v_y}{\partial z} + \frac{\partial v_z}{\partial y} \right) \right]$$

$$\Phi_\mu = \mu \left[2 \left(\frac{\partial v_x}{\partial x} \right)^2 + 2 \left(\frac{\partial v_y}{\partial y} \right)^2 + 2 \left(\frac{\partial v_z}{\partial z} \right)^2 \right. \qquad\qquad (12.2)$$

$$\left. + \left(\frac{\partial v_y}{\partial x} + \frac{\partial v_x}{\partial y} \right)^2 + \left(\frac{\partial v_z}{\partial x} + \frac{\partial v_x}{\partial z} \right)^2 + \left(\frac{\partial v_y}{\partial z} + \frac{\partial v_z}{\partial y} \right)^2 - \frac{2}{3} \left(\frac{\partial v_x}{\partial x} + \frac{\partial v_y}{\partial y} + \frac{\partial v_z}{\partial z} \right)^2 \right]$$

$$\mathbf{J}_i = -\rho D_i \nabla Y_i$$

$$c_p = \sum_{i=1}^{N_s} Y_i c_{p,i},$$

and for vibrational energy conservation equations

$$\dot{e}_{v,i} = \rho_{i(m)} \frac{e_{v,m}^0 - e_{v,m}}{\tau_m} - e_{v,m}\dot{\omega}_{im}$$

$$e_{v,m} = \frac{R_0 \theta_m}{M_{i(m)}[e^{\theta_m/T_{v,m}} - 1]}.$$

(12.3)

This system is closed by the ideal gas equation of state

$$p = \rho \frac{R_0}{M} T = \rho R_0 T \sum_{i=1}^{N_s} \frac{Y_i}{M_i}$$

(12.4)

and the total enthalpy equation

$$h = \int_{T_0}^{T} c_p \mathrm{d}T + h_0,$$

(12.5)

where t is the time, v_i are the projections of velocity $\mathbf{v}$ on the x, y, and z axes, p is the pressure, ρ and ρ_i are the total and single species mass densities, T is the translational temperature, M is the mean molar mass, μ and λ are the viscosity and heat conductivity coefficients, and c_p is the specific heat capacity of gas mixture. The symbol i is the index running over the N_s number of species with Y_i the mass fraction, $c_{p,i}$ the constant pressure specific heat, h_i the specific enthalpy, $\dot{\omega}_i$ the production rate, D_i the effective diffusion coefficient, $\mathbf{J}_i$ the mass diffusion flux, and M_i the molar mass. The index m runs over N_v vibrational modes where $i(m)$ is the index of species with the mth vibrational mode and $e_{v,m}(T_{v,m})$ is the specific vibrational energy corresponding to vibrational temperature $T_{v,m}$ and $e_{v,m}^0 = e_{v,m}(T_{v,m} = T)$ is the corresponding equilibrium energy. $R_0 = 8.314 \ 10^7$ erg/(K mole) is the universal gas constant. Other variables are defined below.

Equation (12.1(e)) describes the vibrational relaxation processes in non-equilibrium gas flows. The term Q_{vib} in equation (12.1(d)) is the energy production due to vibrational relaxation in a multi-component gas mixture. For air $N_v = 3$ assigning $m = 1$ for N_2, $m = 2$ for O_2, and $m = 3$ for the NO.

The radiation heat transfer equation (see equation (12.1(f))) is formulated in a general form neglecting light scattering. This equation being solved allows the calculation of the total radiation heat flux

$$\mathbf{q}_R = \int_{\Delta\omega_{tot}} \mathbf{q}_{R,\omega} \mathrm{d}\omega$$

$$q_R = \mathbf{q}_R \cdot \mathbf{n}$$

$$\mathbf{q}_{R,\omega} = \int_{4\pi} J_\omega\big(\mathbf{R}, \mathbf{\Omega}\big)\mathbf{\Omega} \ \mathrm{d}\mathbf{\Omega}$$

$$q_{R,\omega} = \mathbf{q}_{R,\omega} \cdot \mathbf{n},$$

(12.6)

where J_ω is the spectral intensity of heat radiation, $k_\omega(\mathbf{r})$ is the spectral absorption coefficient, and $j_\omega(\mathbf{r})$ is the spectral emission coefficient, which is calculated in the case of LTE by the Kirchhoff law

$$j_\omega(\mathbf{r}) = k_\omega(\mathbf{r})J_{b,\omega}(\mathbf{r}), \tag{12.7}$$

where $J_{b,\omega}(\mathbf{r})$ is the black body spectral intensity (the Planck function). The spectral intensity $J_\omega(\mathbf{r}, \mathbf{\Omega})$ is a function of the wave number ω, of the space coordinate $\mathbf{r}$ and of the direction of propagation $\mathbf{\Omega}$. This means that equation (12.1(f)) is a differential equation in a space with six dimensions, considering that the direction vector is completely determined by two angles. The total spectral region of heat radiation $\Delta\omega_{tot}$ in the given case is 1000–200 000 cm^{-1}.

A cumulative function, which will be used below, is defined as

$$Q_{rad}(\omega) = \int_{\omega_{min}}^{\omega} q_{R,\omega'}\,d\omega'. \tag{12.8}$$

The cumulative function Q_{rad} gives a representation concerning part of the total radiation heat flux for the given spectral region $[\omega_{min}, \omega]$.

In the case under consideration, the thermodynamic properties of the gas mixture cannot be described with the use of a single temperature because there are relaxation zones with non-equilibrium (i.e. non-Boltzmann, see section 8.1) populations of internal degrees of freedom of molecules and atoms. The most simple and most popular non-equilibrium thermodynamic models are based on the introduction of various translational and vibrational temperatures. Such models are used for the description of the relaxation processes behind shock wave fronts and in the expansion parts of gas flows. In such cases the Kirchhoff law is applied to the averaged vibrational temperature or to individual temperatures.

Using a symbolic formulation of the nth chemical reaction

$$\sum_{j=1}^{N_s} a_{j,n}[X_j] \underset{k_n^r}{\overset{k_n^f}{\rightleftharpoons}} \sum_{j=1}^{N_s} b_{j,n}[X_j] \tag{12.9}$$

the rate of formation of the ith species due to the nth chemical reaction can be written as

$$\left(\frac{dX_i}{dt}\right)_n = \left(b_{i,n} - a_{i,n}\right)\left(k_n^f \prod_{j=1}^{N_s} X_j^{a_{j,n}} - k_n^r \prod_{j=1}^{N_s} X_j^{b_{j,n}}\right)$$

$$= \left(b_{i,n} - a_{i,n}\right)\left(S_n^f - S_n^r\right) \tag{12.10}$$

and for N_r reactions the equation for the time evolution is given by

$$\frac{dX_i}{dt} = \sum_{n=1}^{N_r}\left(\frac{dX_i}{dt}\right)_n = \sum_{n=1}^{N_r}\left(b_{i,n} - a_{i,n}\right)\left(S_n^f - S_n^r\right), \tag{12.11}$$

where $a_{i,n}$ and $b_{i,n}$ are the stoichiometric coefficients of the nth reaction; $[X_i]$ is the chemical symbol of the ith species and X_i is its molar density; and k_n^f and k_n^r are the rate constants of the forward and reverse reactions, respectively. Equation (12.11) is written for molar density. To have the equation for the mass density ρ_i, equation (12.11) must be multiplied for the molar mass M_i, i.e.

$$\dot{\omega}_i = M_i \frac{dX_i}{dt} = M_i \sum_{n=1}^{N_r} \left(b_{i,n} - a_{i,n}\right)\left(S_n^f - S_n^r\right), \tag{12.12}$$

where $\dot{\omega}_i$ is the chemical contribution in equation (12.1). In the framework of the Arrhenius theory of reactions, the forward and reverse constants for the nth reaction are given by

$$k_n^{f/r} = A_n^{f/r} T^{p_n^{f/r}} \exp\left(-\frac{E_n^{f/r}}{kT}\right), \tag{12.13}$$

where $A_n^{f/r}$, $p_n^{f/r}$, and $E_n^{f/r}$ are the approximation coefficients for the forward (f) and reverse (r) reaction rates. Note that the equilibrium constants for each nth chemical reaction is determined as

$$K_n^{eq} = \frac{k_n^f}{k_n^r}. \tag{12.14}$$

In turn, the equilibrium constant can be calculated directly from thermodynamic data of individual species. Different databases (see for example [37, 38]) give approximate expressions for the thermodynamic properties of individual chemical species in the temperature range of 100–20 000 K. For example the Gibbs function Φ_i and its derivatives are given by the following expression as a function of $x = T \cdot 10^{-4}$:

$$\Phi_i = \phi_{1,i} + \phi_{2,i} \ln x + \phi_{3,i} x^{-2} + \phi_{4,i} x^{-1} + \phi_{5,i} x + \phi_{6,i} x^2 + \phi_{7,i} x^3$$

$$\frac{d\Phi_i}{dx} = \frac{1}{x}\left(\phi_{2,i} - 2\phi_{3,i} x^{-2} - \phi_{4,i} x^{-1} + \phi_{5,i} x + 2\phi_{6,i} x^2 + 3\phi_{7,i} x^3\right) \tag{12.15}$$

$$\frac{d^2\Phi_i}{dx^2} = \frac{1}{x^2}\left(-\phi_{2,i} + 6\phi_{3,i} x^{-2} + 2\phi_{4,i} x^{-1} + 2\phi_{6,i} x^2 + 6\phi_{7,i} x^3\right),$$

from which it is possible to determine for each species the molar enthalpy H_i (in J/mole) and the constant pressure heat capacity $C_{p,i}$ (in J/mole/K)

$$H_i = xT\frac{d\Phi_i}{dx} + \phi_{8,i} \tag{12.16}$$

$$C_{p,i} = 2x\frac{d\Phi_i}{dx} + x^2\frac{d^2\Phi_i}{dx^2}. \tag{12.17}$$

The equilibrium constants for partial pressures (with respect to the reference value $p_0 = 101325$ Pa $= 1$ atm) are given by

$$\ln K_n^{eq} = \frac{1}{R_0 T} \sum_{j=0}^{N_s} \left(a_{j,n} - b_{j,n}\right)\left(-T\Phi_j + \phi_{8,j} \cdot 10^3\right). \tag{12.18}$$

Transport coefficients such as the viscosity (μ) and heat conductivity (λ) of gas mixtures are calculated by the following mixing rules [39, 40]:

$$\mu = \left(\sum_{i=1}^{N_s} \frac{Y_i}{\mu_i} \right)^{-1} \tag{12.19}$$

$$\lambda = \frac{1}{2} \left[\sum_{i=1}^{N_s} x_i \lambda_i + \left(\sum_{i=1}^{N_s} \frac{x_i}{\lambda_i} \right)^{-1} \right], \tag{12.20}$$

where the single species quantities are calculated as

$$\mu_i \left(\mathrm{g\,cm^{-1}\,s^{-1}} \right) = 2.67 \cdot 10^{-5} \frac{\sqrt{M_i T}}{\sigma_i^2 \Omega_i^{(2,2)\star}} \tag{12.21}$$

$$\lambda_i \left(\mathrm{erg\,cm^{-1}\,K^{-1}} \right) = 8330 \cdot 10^{-5} \sqrt{\frac{T}{M_i}} \frac{1}{\sigma_i^2 \Omega_i^{(2,2)\star}}, \tag{12.22}$$

where x_i are the molar fractions, σ_i is the diameter of the collision, and $\Omega_i^{(2,2)\star}$ is the viscosity type reduced collision integral. The effective diffusion coefficient D_i for the ith species is calculated from the binary diffusion coefficients $D_{i,j}$ by the Wilke formula [41]

$$D_i \left(\mathrm{cm^2\,s^{-1}} \right) = (1 - x_i) \left(\sum_{\substack{j \neq i}}^{N_s} \frac{x_j}{D_{i,j}} \right)^{-1}, \tag{12.23}$$

where

$$D_{i,j} \left(\mathrm{cm^2\,s^{-1}} \right) = 1.858 \cdot 10^{-3} \sqrt{T^3 \frac{M_i + M_j}{M_i M_j}} \frac{1}{p \sigma_{i,j}^2 \Omega_{i,j}^{(1,1)\star}}, \tag{12.24}$$

where $\Omega_{i,j}^{(1,1)\star}$ are the diffusion type reduced collision integrals. The reduced collision integrals are calculated using the approximations suggested by N Anfimov [42]

$$\Omega_i^{(2,2)\star} = 1.157 T_i^{-0.1472}$$

$$\Omega_{i,j}^{(1,1)\star} = 1.074 T_{i,j}^{-0.1604}, \tag{12.25}$$

where $T_i = kT/\varepsilon_i$, with ε_i the well depth of the attractive interaction potential and

$$T_{i,j} = \frac{kT}{\varepsilon_{i,j}}$$
$$\varepsilon_{i,j} = \sqrt{\varepsilon_i \varepsilon_j} \tag{12.26}$$
$$\sigma_{i,j} = \frac{\sigma_i + \sigma_j}{2}.$$

The last aspect to consider is the interaction with the entry vehicle. A pseudo-catalytic surface is presumed, i.e. $Y_i(\text{wall}) = Y_i(\infty)$ where $Y_i(\infty)$ are the mass fractions in the oncoming flow. The surface temperature is calculated with the assumption of an equilibrium emitting surface

$$Q_w = \varepsilon \sigma T_w^4, \tag{12.27}$$

where ε is the surface emissivity factor, $\sigma = 5.67 \cdot 10^{-12}$ W cm^2 K^4 is the Stephan–Boltzmann constant, and Q_w is the total (convective plus radiative) heat flux density on the surface.

12.2 Radiative gas dynamics of re-entry space vehicles

Predictions of the shock-layer radiative heating of entry vehicles and the coupled radiative gas dynamic processes in shock layers under orbital and super-orbital return conditions have received renewed interest due to current national plans for a return to the exploration of the Moon and our neighboring planets. Despite the fact that radiation heat transfer processes applied to re-entry space vehicles have been considered for more than 40 years, unanswered questions still remain in existing planetary data. The scientific unknowns following from these questions have to be resolved to provide acceptable aerothermodynamic predictions for creating new-generation space vehicles.

One of the specific distinguishing features of the new generation of space vehicles is their large size in comparison to those of the previous generation. Space vehicles with midship diameters of about 5 m are being considered in the current plans. For example, the ORION project is considering the possibility of using spacecrafts with a nose radius of about 400 cm [43]. This concept has significant advantages, as it decreases the convective heat load. However, it is also necessary to take into account a significant disadvantage, namely the increase of the radiative heating load. It can be estimated that typical standoff distances for large spacecraft of similar size are about 30–50 cm.

To determine the radiative heating, atomic lines in the spectral range $\Delta\omega \approx 1000$–150 000 cm^{-1} ($\approx 10$ μm–0.067 μm) must be considered, modeling with good accuracy the line shape, whose typical half-widths are in the range $\gamma \approx 0.01$–1.0 cm^{-1}. A significant challenge is to manage the large number of atomic lines (for air ≈ 2000–3000), which should be considered individually or by using special spectral methods.

The radiative gas dynamics of different kinds of space vehicles will be considered below. These are: Fire-II [27–30, 33], Stardust [22, 23, 31, 32], RAM-C-II [24–26],

ORION [43], the Prospective Transport Vehicle (PTV) [44], and Mars Science Laboratory (MSL) [45, 46]. Their re-entry trajectories are shown in figure 12.1, and some of their trajectory parameters are presented in tables 12.1–12.6.

12.2.1 Fire-II

The flight data from the Fire-II experimental program have been used for the validation of physical-chemical models and computer codes for the aerophysics of re-entry vehicles for more than 40 years [27–30]. Nevertheless, it is well known that

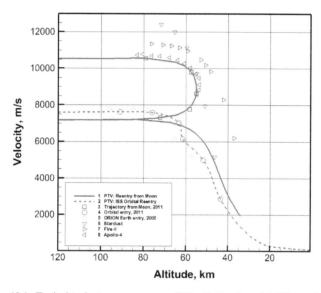

Figure 12.1. Typical trajectory parameters of Fire-II, Stardust, ORION, and PTV.

Table 12.1. Trajectory parameters for Fire-II [27–30]. H is the altitude; t the trajectory time; ρ_∞, V_∞, and T_∞ are the gas density, velocity, and temperature in the free stream; T_w is the wall temperature; p_0 is the stagnation pressure; and R_n is the blunting radius.

t s	H km	$\rho_\infty \cdot 10^7$ g cm^{-3}	p_∞ erg cm^{-3}	V_∞ km s^{-1}	T_∞ K	T_w K	$p_0 \cdot 10^{-5}$ erg cm^{-3}	R_n cm
1634	76.42	0.372	20.8	11.36	195	615	0.48	
1636	71.04	0.857	51.7	11.31	210	810	1.10	
1637	67.05	1.47	96.2	11.25	228	1030	1.86	93.5
1639	63.11	2.41	167.4	11.14	242	1325	2.99	
1640	59.26	3.86	281.4	10.97	254	1560	4.65	
1643	53.04	7.80	617.9	10.48	276	640	8.57	
1644	50.67	10.2	822.6	10.19	281	1100	10.5	80.5
1645	48.37	13.2	1079.7	9.83	285	1520	12.8	
1648	42.14	30.0	2298.9	8.30	267	1560	20.7	70.2
1651	37.19	60.5	4393.0	6.19	253	1060	23.2	

there are several basic issues in the aerophysics of re-entry probe simulation that need further development:

- kinetic models for non-equilibrium dissociation, ionization, relaxation, and radiation processes;
- efficient CFD tools for solving the governing equations on both structured and unstructured two- and three-dimensional grids; and
- adequate models for turbulent flow and laminar–turbulent transition.

The uncertainty in the numerical predictions for the convective and radiative heating of the re-entering capsule, using different models of chemical and physical kinetics, have been analyzed in several studies [27–30]. These calculations were performed with the Dunn and Kung [47] and Park [13, 14] models of chemical kinetics. Several models of dissociation have also been used, among them the one-temperature model assuming LTE, and different models of non-equilibrium dissociation, such as the studies by Treanor and Marrone [48, 49]. In this chapter, all the calculations are performed with and without coupling effects between radiation and gas dynamics. The spectral optical properties of high-temperature air are calculated using the *ab initio*, quasi-classical, and quantum mechanics approaches. The line-by-line spectral data calculations are performed with high accuracy for about 80000 nodes of spectral points.

All computational results are compared with experiments and other calculations. For each trajectory point the following data have been analyzed: velocity components, pressure, density, translational and vibrational temperatures, as well as mass fractions along the stagnation streamline. Finally, the one-sided integral radiation heat fluxes, spectral distributions of radiation heat flux density, and the cumulative function are presented and delineated.

Two-dimensional fields of translational temperature and longitudinal velocity are shown in figure 12.2(a) for trajectory point $t = 1639$ s. The axial distributions of translational and vibrational temperatures for molecules N_2, O_2, and NO along the stagnation line are shown in figure 12.2(b). These results were obtained using two kinetic models [13, 47]. The two-dimensional fields of temperature and velocity were obtained with the Park kinetic model [13].

The density of spectral radiation heat fluxes and the corresponding cumulative functions are shown in figure 12.2(c). These data allow us to demonstrate the contribution of the spectral lines of atoms and ions into total radiative heating. One can see that the atomic lines give a $\approx 100\%$–200% contribution to the radiative heating of the frontal radiation heat shield.

12.2.2 Stardust

The calculation data for the radiation gas dynamics of Stardust will be considered in this section. The coupled radiative gas dynamic problems are solved for the trajectory point $t = 54$ s (table 12.2).

Figure 12.3(a) shows the longitudinal velocity and translational temperature in the calculation domain, including the leading shock layer and flow-field in the wake.

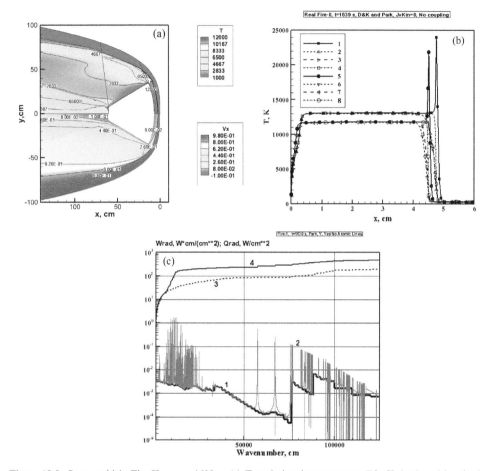

Figure 12.2. Space vehicle Fire-II at $t = 1639$ s. (a) Translational temperature T in K (top) and longitude velocity normalized to V_∞ (bottom). (b) Axial distribution of translational (1,5) and vibrational (2,6 – $T_v(N_2)$, 3,7 – $T_v(O_2)$, 4,8 – $T_v(NO)$) temperatures along the critical line. The Dunn–Kang (1–4) and the Park (5–8) kinetic models with the second model of Marrone–Treanor [49] at $U = 6$ are used. (c) Spectral radiation heat fluxes without (1) and with (2) atomic lines and corresponding cumulative functions (3,4) of radiation heat fluxes at the stagnation point. Note that dimensionless parameter U is used in model [49] to characterize the probability of dissociation from vibrationally excited states of diatomic molecules.

Table 12.2. Trajectory parameters for Stardust. See table 12.1 for parameter definitions.

t s	H km	$\rho_\infty \cdot 10^8$ g cm^{-3}	p_∞ erg cm^{-3}	V_∞ km s^{-1}	T_∞ K	$p_0 \cdot 10^{-5}$ erg cm^{-3}
42	71.9	4.16	26.40	12.4	221.4	0.972
48	65.4	10.6	103.0	12.0	229.0	2.23
54	59.8	23.4	160.0	11.1	238.5	3.96
66	51.2	72.1	688.0	7.96	253.5	5.36
76	46.5	135	1230.0	5.18	256.9	5.61

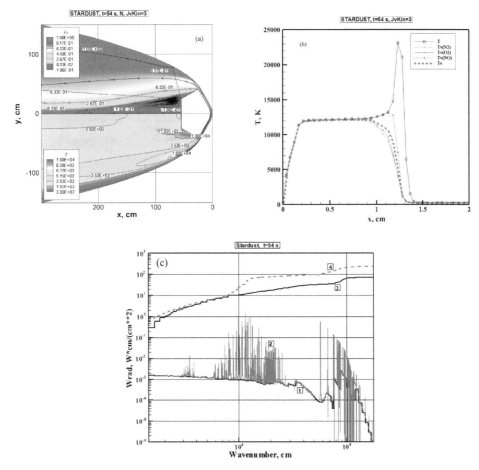

Figure 12.3. Space vehicle Stardust at $t = 54$ s. (a) Translational temperature T in K (top) and longitude velocity normalized to V_∞ (bottom). (b) Axial distribution of translational and vibrational ($T_v(N_2)$, $T_v(O_2)$, $T_v(NO)$) temperatures along the stagnation line of the flow-field. Park's kinetic model with the second model of Marrone–Treanor [49] at $U = 6$ is used. (c) Spectral radiative heat fluxes without (1) and with (2) atomic lines and the corresponding cumulative functions (3,4) of the radiation heat fluxes at the stagnation point.

These data are also important for the analysis of the spectral signature of the space vehicle entering the atmosphere, which can be compared with experimental measures [32].

A representation of the distributions of the translational and vibrational temperatures in the leading shock layer can be obtained from figure 12.3(b). It can be observed that translational temperatures achieve values larger than 20 000 K, while vibrational temperatures are located below 12 000 K. As the altitude and velocity decrease, the thermal equilibrium in the shock layer becomes more obvious. The depth of the non-equilibrium zone at $t = 54$ s is about 0.3 cm and, obviously, the temperature distributions in this region play a significant role in the radiative heating. The spectral distributions of the radiation heat fluxes achieved and corresponding cumulative functions are presented in figure 12.3(c). Comparing the spectral distribution of the heat fluxes with the cumulative functions shows that the two significant jumps of the total heat fluxes in the regions 10 000–20 000 cm^{-1} and 78 000–105 000 cm^{-1} are connected to free–free (radiative-recombination) and bound–bound transitions in atomic and ionic species. The main contribution in the spectral region 20 000–80 000 cm^{-1} is due to neutral and ionized diatomic molecules. One can see that molecular emission, located in the region between near-infrared and near-ultraviolet, gives a small contribution to the integral radiation heat flux. However, the photon emission of electronic bands is used for spectral analysis, therefore several electronic bands of heated air have attracted much attention from specialists in the aerophysics of return space vehicles, these are: the Vegard–Kaplan bands, the Birge–Hopfield 1 and 2 bands; the first, second, and fourth positive systems of N_2; the γ^-, β^-, δ^-, and ε^- systems of NO, namely Meinel's auroral system; and the first and fourth negative systems of N_2^+. The population of highly excited electronic states just behind the shock wave front reflects in full measure the differences in translational and electronic temperatures.

12.2.3 RAM-C-II

Examples of the two-dimensional and three-dimensional numerical simulation results for the RAM-C-II space vehicle are presented in this section. A kinetic model of air ionization behind the shock front at speeds of $V_\infty < 9$ km s^{-1} was analyzed in [26, 50]. The seven components of partially ionized air (N_2, O_2, N, O, NO, NO$^+$, e^-) and the following six chemical reactions were analyzed:

$$
\begin{array}{ll}
(R1) & O_2 + M \underset{k_r}{\overset{k_f}{\rightleftharpoons}} N + N + M \\[2mm]
(R2) & O_2 + M \underset{k_r}{\overset{k_f}{\rightleftharpoons}} O + O + M \\[2mm]
(R3) & NO + M \underset{k_r}{\overset{k_f}{\rightleftharpoons}} N + O + M \\[2mm]
(R4) & N_2 + O \underset{k_r}{\overset{k_f}{\rightleftharpoons}} NO + N \\[2mm]
(R5) & O_2 + N \underset{k_r}{\overset{k_f}{\rightleftharpoons}} NO + O \\[2mm]
(R6) & N + O \underset{k_r}{\overset{k_f}{\rightleftharpoons}} NO^+ + e^-.
\end{array}
\tag{12.28}
$$

The first three reactions of this model are the reactions of dissociation and recombination in collisions with a third particle M. Bimolecular reactions ($R4$) and ($R5$) are important in the kinetics of the formation of NO. In the model under consideration, the basic process with participation of the charged particles is considered the reaction of associative ionization ($R6$) (and the reverse reaction of dissociative recombination). In [51] the role of electron impact ionization is discussed. Based on these results [26, 51], the conclusion has been made that the main reaction is the associative ionization. At higher speeds, and correspondingly at higher temperatures, collisional ionization and ion–molecule reactions become more significant.

Park's kinetic model [13] was chosen for the calculations in the approximation of LTE, together with various models of non-equilibrium dissociation [12, 48, 49]. It is well known that assuming an Arrhenius rate for the associative ionization reaction (R6) introduces some errors, because it proceeds at least in two steps. First N and O atoms form $NO^\star$ which then decays into a NO^+ and an electron. This is the reason for the large differences between the rate constants used by different authors. A detailed description of such a reaction is also given in [52]. The free stream conditions for the calculation presented here have been taken from the experimental values reported in table 12.3. For reference, characteristic Mach numbers and Reynolds numbers are also reported in the table. The distributions of the translational temperature and the longitudinal velocity at altitude $H = 71$ km are shown in figure 12.4(a).

Table 12.3. Trajectory parameters for RAM-C-II. Re is the Reynolds number. For the other parameters see table 12.1.

H km	V_∞ km s^{-1}	p_∞ erg cm^{-3}	$\rho_\infty \cdot 10^8$ g cm^{-3}	T_∞ K	Mach	Re $\cdot 10^{-3}$
61	7.50	192.	27.3	244	23.9	19.5
71	7.64	44.8	7.20	217	25.9	6.28
81	7.95	8.89	1.57	197	28.3	1.59

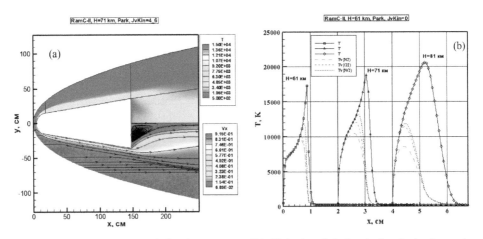

Figure 12.4. RAM-C-II: (a) Translational temperature T in K (top) and the longitude velocity (bottom) at $H = 71$ km and $V_\infty = 7.64$ km s^{-1}. (b) Axial distribution of translational and vibrational temperatures, along the front critical line for points of the trajectory $H = 61$ km and $V_\infty = 7.5$ km s^{-1}, $H = 71$ km and $V_\infty = 7.64$ km s^{-1}, $H = 81$ km and $V_\infty = 7.95$ km s^{-1}. Park's kinetic model with the second model of Marrone–Treanor [49] at $U = 6$ is used. The temperature distributions for different heights are shifted to the right by 2 cm.

With increasing altitude, the degree of non-equilibrium in the stream grows. Figure 12.4(b) shows the axial distributions of translational and vibrational temperatures along the stagnation line. It can be observed that relevant differences between the temperatures are achieved at a relatively large height ($H = 81$ km). In this case, the calculation of the rate constants of dissociation is performed in LTE, i.e. the translational temperature is used in equation (12.13).

Figure 12.5 shows the maximum electronic density along the surface of the blunted cone at three different altitudes, calculated using different models of non-equilibrium dissociation as reported in table 12.4 and compared with the experimental data [24, 25]. Increasing U from 3 to 6 results in faster dissociation from

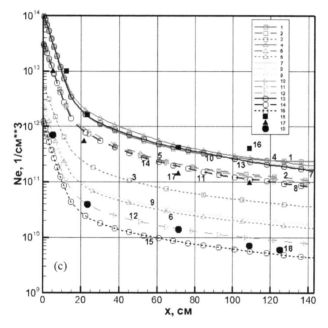

Figure 12.5. Distribution of the maximum electron density along the x-axis of the experimental spacecraft. See table 12.4 for curve descriptions.

Table 12.4. Conditions for the curve numbers in figure 12.5.

H [km]	61	71	81
V_∞ [km s^{-1}]	7.5	7.64	7.95
LTE	1	2	3
Park dissociation model [12]	4	5	6
First dissociation model in [48]	7	8	9
Second dissociation model in [49] with $U = 6$	10	11	12
Second dissociation model in [49] with $U = 3$	13	14	15
Experimental data [24, 25]	16	17	18

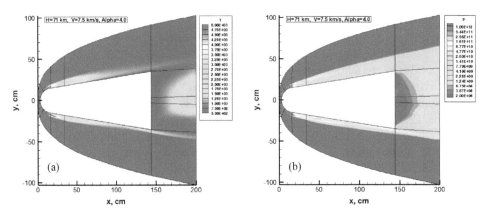

Figure 12.6. RAM-C-II: three-dimensional numerical simulation results for translational temperature T (in K) and electron number density N_e (cm^{-3}) in the plane of symmetry at $H = 61$ km and $V_\infty = 7.5$ km s^{-1} at an angle of attack $\alpha = 4°$.

vibrational states with a small vibrational quantum number. All the dissociation models show a good agreement with experimental data for low altitudes (61 and 71 km), however differences larger than one order of magnitude are obtained. A better agreement is obtained using the second Marrone–Treanor model [49], while the LTE model overestimates the electron density by more than a factor 10.

Let us analyze the behaviors of the flow around hypersonic vehicle RAM-C-II by changing the angle of attack. A windward surface of the cone is exposed to higher pressure than in the axisymmetric case. This is clearly seen when comparing the temperature fields shown in figure 12.6. Even if a good accuracy was achieved in reproducing the flight test data of RAM-C-II in a wide range of altitudes, it should be pointed out that this result was actually possible only by selecting suitable physical and chemical models. Constructing realistic models for non-equilibrium conditions that are capable of predicting flight conditions still remains an open problem in modern aerophysics.

12.2.4 ORION

In this section we consider the general elements of the radiative gas dynamics of the large-scale space vehicle ORION at its entry into Earth's atmosphere with orbital velocity. Two-dimensional fields of translational temperature and longitudinal velocity are shown in figure 12.7(a) while temperature distributions along the stagnation line for trajectory point $t = 200$ s are shown in figure 12.7(b). The dotted lines correspond to calculations without coupling gas dynamics with radiation, while the solid lines correspond to translational and vibrational temperatures including radiation–gas dynamic coupling. These temperature profiles indicate the non-equilibrium relaxation zone behind the shock wave front (larger than 2 cm). The total stand-off distance of the shock wave reaches ≈ 32 cm. This shock layer is a noticeable source of heat radiation. It is well known that increasing the blunt radius of a space vehicle leads to a decrease of convective heat flux, as well as increasing the shock wave stand-off distance and, as a consequence, radiative heat fluxes also

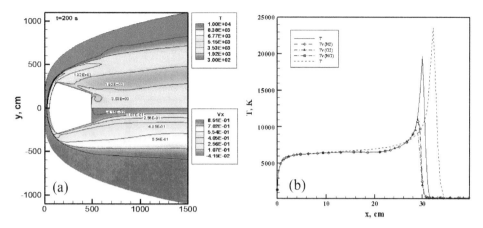

Figure 12.7. Space vehicle ORION. (a) Field of longitudinal velocity (normalized to V_∞) and translational temperature (in K) at $t = 200$ s. (b) Translational and vibrational temperatures along the forward stagnation line. The solid and dotted lines without markers are the translational temperatures with and without radiation–gas dynamic coupling, respectively.

Table 12.5. Trajectory parameters for ORION. See table 12.1 for parameter definitions.

trajectory points	t s	H km	$\rho_\infty \cdot 10^8$ g cm^{-3}	p_∞ erg cm^{-3}	V_∞ km s^{-1}	T_∞ K	$p_0 \cdot 10^{-4}$ erg cm^{-3}
1	150	83.0	0.630	3.37	7.7	187	0.374
2	200	78.2	2.45	14.2	7.7	202	1.45
3	300	65.6	15.1	100.	7.0	232	7.40
4	400	65.6	15.1	100.	6.2	232	5.82
5	500	57.1	44.3	324.	5.2	255	12.0
6	600	42.8	267	1980.	3.0	258	24.0

increases. Therefore it is expected that for some trajectory points one can find conditions where radiative heating overcomes the convective contribution. Such a situation dictates that more attention needs to be paid as to which relaxation kinetics should be used to model the shock formation (see table 12.5).

More realistic cases for these kinds of space vehicles must consider entering conditions with given angles of attack. Therefore three-dimensional radiative gas dynamic models must be used for these purposes.

Numerical simulation results obtained for the ORION spacecraft at an angle of attack $\alpha = 15°$ at the first trajectory points ($t = 200$ s) are shown in figures 12.8 and 12.9. It should be noted that for the analysis of the aerothermodynamics of large-sized space vehicles it is necessary to consider not only the flow-field above the windward surface, but also above the leeward surface and in the wake because radiative heating has a volumetric genesis. Let us focus on the translational and vibrational temperatures in the near and far wake reported in figure 12.8. These figures indicate the degree of non-equilibrium between the translational and vibrational degrees of freedom in N_2 and O_2.

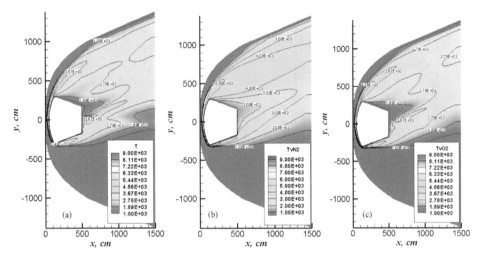

Figure 12.8. Space vehicle ORION: translational and vibrational temperatures of N_2 and O_2 at $t = 200$ s. Angle of attack $\alpha = 15°$.

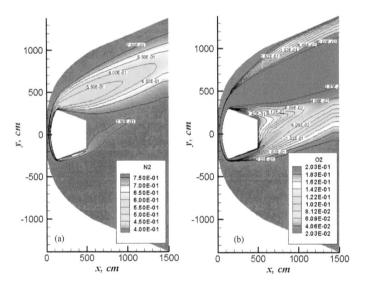

Figure 12.9. Space vehicle ORION: mass fractions of N_2 and O_2 at $t = 200$ s. Angle of attack $\alpha = 15°$.

The effective temperature of the vibrational excitation of molecules of N_2 is significantly higher than other temperatures in a large part of the flow field. Noticeably, not only should temperature profiles be considered correctly to predict the radiative heating, but also the mass fractions of optically active species (figure 12.9), and in particular N_2, O_2, and NO, the main emitters of heat radiation in the case under consideration. The distribution of species concentrations is strongly inhomogeneous, not only along the leeward surface but also above the windward surface, which is accompanied by visible variation of the shock layer depth, having a direct impact on the intensity of the heat radiation.

12.2.5 PTV

The radiating aerothermodynamics of PTV in a two-dimensional formulation was investigated at three points of trajectory of super-orbital entry, which are presented in table 12.6. These points correspond to non-equilibrium, quasi-equilibrium, and actual full-equilibrium conditions in the shock layer. The calculations of flow fields around the space vehicle were carried out in three series of multi-block calculation grids of different minuteness. All calculations were performed for a pseudo-catalytic surface.

Figure 12.10 shows a comparison between the convective and radiative heating of the super-orbital PTV versus time of flight. In the first part of the trajectory the radiative heating is dominant and the contribution of atomic lines is significant. Therefore, the conclusions concerning the influence of atomic lines on radiative

Table 12.6. PVT trajectory parameters (see tables 12.1 and 12.3 for definitions).

H km	$\rho_\infty \cdot 10^8$ g cm^{-3}	p_∞ erg cm^{-3}	T_∞ K	V_∞ km s^{-1}	$p_0 \cdot 10^{-4}$ erg cm^{-3}
80	1.85	10.5	199	8.5	1.33
60	31	220	247	9.5	27.7
50	103	798	247	9.9	100

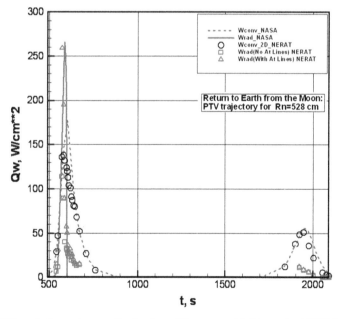

Figure 12.10. PVT: radiative and convective heating of the super-orbital large-sized PTV for two trajectory parts. The solid red line and dotted blue line correspond to correlation [53]. The black circles show the convective heating, predicted by code NERAT [44]. The red squares and triangles show radiative heating without and with atomic lines, predicted by codes NERAT + ASTEROID [44].

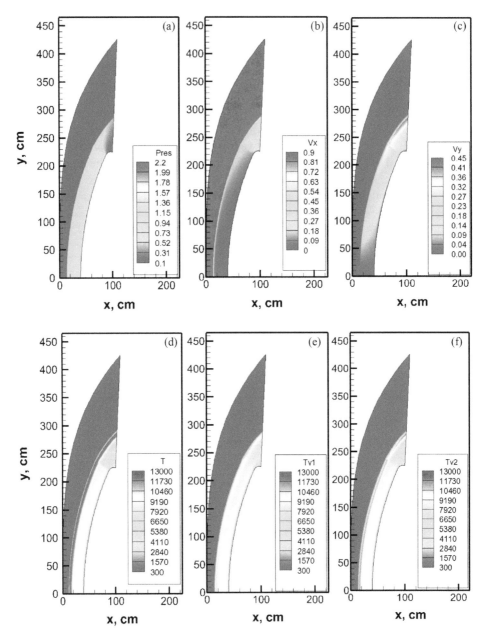

Figure 12.11. PTV: (a) p/p_0 (p_0 is the break pressure); (b) longitudinal and (c) radial velocities normalized to V_∞; (d) translational and vibrational temperature of (e) N_2 ($Tv1$) and (f) O_2 ($Tv2$) in K.

heating, already discussed for Fire-II flight data, holds also for super-orbital large-sized space vehicles.

Let us discuss the basic laws of radiative and convective heat transfer in a shock layer on the surface of a super-orbital spacecraft of large size, considering the example of the first trajectory point in table 12.6. Figure 12.11 shows flow quantities

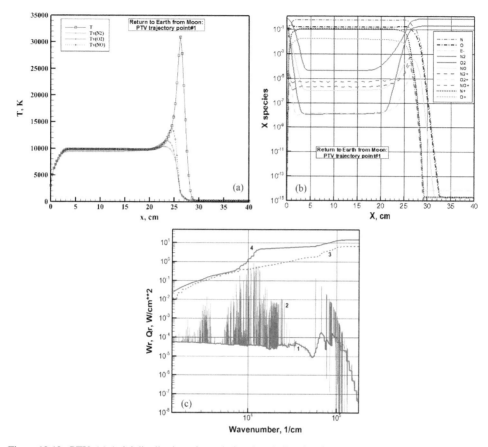

Figure 12.12. PTV. (a) Axial distribution of translational and vibrational temperatures in K. (b) Species molar fractions along the stagnation line. (c) Spectral radiation heat fluxes (1,2) and cumulative functions of radiation heat fluxes with (2,4) and without (1,3) atomic lines at the stagnation point.

in the shock layer and figure 12.12(a) reports the profiles of the translational and vibrational temperatures along the stagnation line. Significant differences between the translational and vibrational temperatures can be observed, evidence of thermodynamic non-equilibrium behind the shock wave front. Distributions of mole fractions along the stagnation line are presented in figure 12.12(b). These profiles allow us to estimate the degree of dissociation and ionization in the shock layer. The spectral structure of the radiative heat fluxes at PTV's stagnation point is illustrated in figure 12.12(c), reporting the cumulative functions and corresponding spectral heat fluxes. Note that the value of the cumulative function at the maximal value of the wave numbers $\omega_{max} = 120\,000$ cm^{-1} corresponds to the total integral flux to the stagnation point while the cumulative function as a function of ω is calculated according to equation (12.8).

These results allow us to determine the characteristics of the radiation gas dynamics for the super orbital re-entry of PTV with a radius for the blunting of the aerodynamic heat shield of 528 cm:

1. The thickness of the compressed layers at the front surface varies in the range $\delta = 22$–35 cm. The thickness of the relaxation zone behind the shock front varies in the range $h = 1$–10 cm;

2. At the first entry of the spacecraft into Earth's atmosphere at velocities $V_\infty > 9.5$ km s^{-1}, the temperature in the compressed layer reaches $T = 10\,000$ K. The detachment of shocks from the surface and the thickness of the relaxation zone are $\delta_1 = 23$ cm and $h_1 = 4$ cm.

3. The distributions of radiation fluxes towards the surface are of special practical interest. In the first trajectory point the radiative heat flux reaches its maximum in the non-equilibrium relaxation zone behind the shock front. In compressed layers are regions where weakening and strengthening of radiation fluxes are observed. In the second trajectory point, corresponding to the maximum of radiative heating, radiation fluxes to the surface grow over the whole compressed layer. Only in the boundary layer near the surface is there a slight weakening. In the third point of the trajectory, corresponding to the beginning of PTV's re-entry into the rarefied layers of the atmosphere, there is a sharp increase in the radiative flux in the relaxation zone behind the shock front, monotonically increasing in the compressed layer. Thus, we can conclude that a significant portion of radiative flux to the surface is initiated in the non-equilibrium relaxation zone.

4. Analysis of the spectral distribution of the radiation fluxes and cumulative functions shows that the contribution of the atomic lines may exceed 100% in the conditions under investigation. By decreasing the velocity below 10 km/s, the contribution of atomic lines is reduced.

5. Distributions of the gas dynamic functions in the compressed layer show a sufficiently high uniformity along the front surface of the aerodynamic shield. This justifies the use of the approximation of the locally one-dimensional layer to calculate radiative transfer. The increase of viscosity and thermal conductivity in the region of the shock wave is notable. The degree of ionization decreases quite rapidly, departing from the critical line of flow to the periphery of the aerodynamic shield.

The three-dimensional gas dynamics fields around a landing module of large size under an angle of attack of $\alpha = 25°$ are shown in figures 12.13 and 12.14. Figure 12.13 presents the translational and vibrational temperatures of N_2. These figures are provided in the same color scale in order to allow direct comparison.

The degree of non-equilibrium in the flow field can be appreciated, noticing that the vibrational temperature is much higher than the translational one. The high temperature in the shock layer and the strong asymmetry of the flow field, consequences of the angle of attack, are clearly seen here.

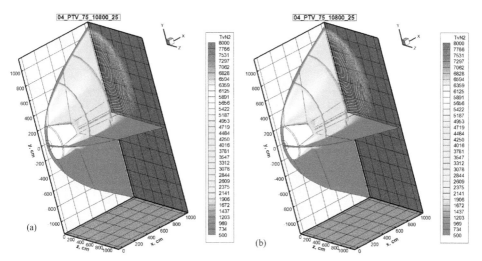

Figure 12.13. PTV: three-dimensional field of (a) T and (b) $T_v(N_2)$ at $H = 75$ km, $V_\infty = 10.8$ km s^{-2}, $\alpha = 25°$.

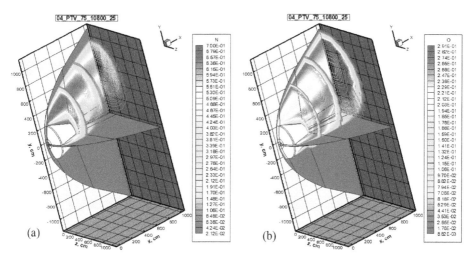

Figure 12.14. PTV: three-dimensional field of molar fraction of (a) N and (b) O at $H = 75$ km, $V_\infty = 10.8$ km s^{-1}, $\alpha = 25°$.

    The molar fractions of atoms N and O, as well as of diatomic molecules N_2 and O_2, are presented in figure 12.14. These figures give a general representation concerning the degree of dissociation of diatomic molecules, which was calculated using models of non-equilibrium dissociation.

    The axial distributions of the translational and vibrational temperatures, as well as the molar fractions of the gas components along the stagnation line, are shown in figure 12.15. These data give a full representation concerning the thermal conditions in the most high-temperature region and the degree of ionization and dissociation.

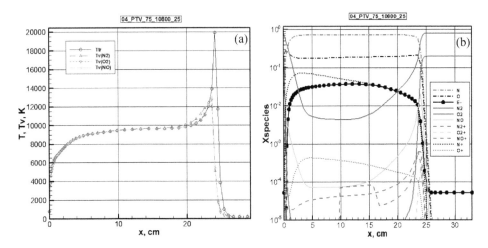

Figure 12.15. PTV: axial distributions of translational and vibrational temperatures (a) and molar fractions (b) along the stagnation line at $H = 75$ km, $V_\infty = 10.8$ km s^{-1}, $\alpha = 25°$.

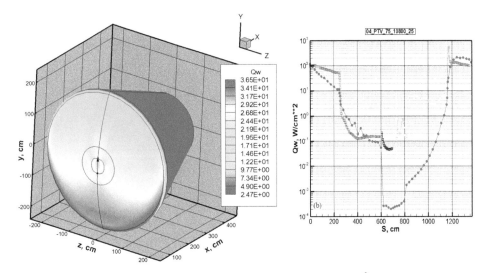

Figure 12.16. PTV: (a) distribution of convective heat flux density (in W cm^{-2}) on the front shield and (b) convective (squares) and radiative (circles) heating of the PTV surface at angle of attack $\alpha = 25°$ and $H = 75$ km, $V_\infty = 10.8$ km s^{-1}. Coordinate S along the surface is counted off from the black point in the direction marked by an arrow in the left side of the figure.

The general results of the numerical simulations are presented in figure 12.16, where the convective heat flux on the front aerodynamic shield, the distributions of the convective heat flux, and the integral radiation heat fluxes along the surface in meridional cross sections, are shown. In the case under consideration, the radiation heat flux exceeds the convective heat flux on the front shield surface, in particular in the region of the leeward part.

12.2.6 MSL

In the previous sections we considered the radiative gas dynamic problems of the Earth return of hypersonic flights. The radiative gas dynamics of Martian space vehicles are also of practical interest. Recently obtained experimental data on the heating of MSL during a real descent trajectory [45] provide the possibility to validate computational data.

Figure 12.17 shows the flow field around the space vehicle MSL with the catalytic surface under angle of attack $\alpha = 16°$. The distinctive features of the flow field are:

- The predicted asymmetry of the flow field (in particular, of the translational temperature and longitudinal velocity) is formed due to the angle of attack $\alpha = 16°$.
- An expansion of the shock layer above the windward surface from the direction of incident flow is observed, as well as the vortex structure above the top part of the leeward surface.
- High-mass fractions of CO_2 and low mass fractions of CO are observed for the catalytic surface.

The distributions of convective and radiative heat fluxes along the meridional surface of the MSL entry module are presented in figure 12.18(b). The starting position of the S coordinate is counted from the point shown in figure 12.18(a). It should be noted that radiative heating is significant for the back of the landing module.

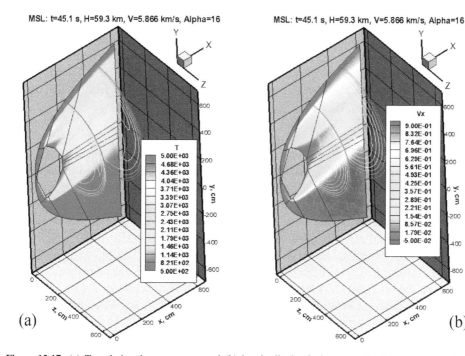

Figure 12.17. (a) Translational temperature and (b) longitudinal velocity normalized to V_∞ at $t = 45.1$ s, $H = 59.3$ km, $\rho_\infty = 2.67 \cdot 10^{-8}$ g cm^{-3}, $p_\infty = 4.29$ erg cm^{-3}, $V_\infty = 5.866$ km s^{-1}, $T = 85.7$ K, $\alpha = 16°$.

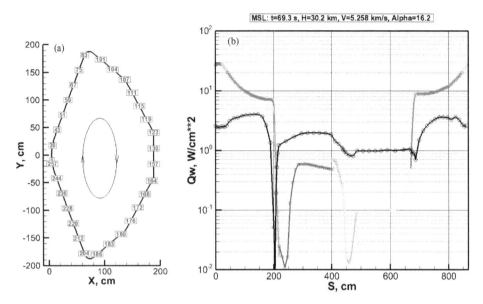

Figure 12.18. (a) Location of grid nodes in equatorial cross section and (b) convective heat flux (color curve) and integral radiation heat flux (black line) along surface coordinate S in the meridional plane. $t = 69.3$ s, $H = 30.2$ km, $\rho_\infty = 6.96 \cdot 10^{-7}$ g cm^{-3}, $p_\infty = 222$ erg cm^{-3}, $V_\infty = 5.258$ km s^{-1}, $T = 168.5$ K, $\alpha = 16.2°$.

The maximum value of radiative heat fluxes reaches 1.7 W cm^{-2}. Spectral radiation heat fluxes for separate points on the landing module surface are shown in figure 12.19, reporting the spectral radiation heat fluxes at six different points on the surface, corresponding to positions 1, 36, 95, 146, 209, 256 in figure 12.18(a). Two bands of CO_2 and CO contribute to surface illumination for all trajectory points. These are 2.7 μ and 4.3 μ bands, corresponding to wave numbers $\omega \approx 3703$ cm^{-1} and $\omega \approx 2325$ cm^{-1}. Some electronic bands are also observed on the front shield of the lander, the contribution of which is practically small.

12.3 Conclusions

The modern theory of high-enthalpy radiating flows has been realized in computational models of the aerophysics of spacecraft. Numerical analysis of the radiative aerothermodynamics of Earth and Martian space vehicles has been presented and analyzed. The distinguishing feature of the presented data is the comparison of convective and radiative heating of the landing module surfaces, as well as the comparison of obtained calculation data on convective heating with measured experimental data. Three-dimensional CFD and Radiative Gas Dynamic (CFD/RadGD) code NERAT(3D) has been used for the investigation of the radiative gas dynamics of different landing modules under angles of attack. Three-dimensional calculations of convective and radiative heating have been performed. It has been shown that for some trajectory points the radiative heating of the leeward surface exceeds the convective heating. The data obtained on convective heating are in good agreement with real space mission data.

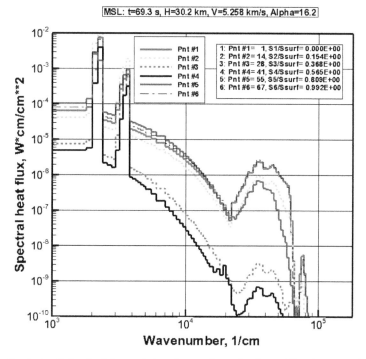

Figure 12.19. Spectral radiation heat fluxes at six points on the surface in meridional cross section. $t = 69.3$ s, $H = 30.2$ km, $\rho_\infty = 6.96 \cdot 10^{-7}$ g cm^{-3}, $p_\infty = 222$ erg cm^{-3}, $V_\infty = 5.258$ km s^{-1}, $T = 168.5$ K, $\alpha = 16.2°$.

Bibliography

[1] Pringle G and King A 2007 *Astrophysical Flows* (Cambridge: Cambridge University Press) p 206

[2] LeVeque R J, Mihalas D, Dorfi E and Müller E 1998 *Computational Methods for Astrophysical Fluid Flow* (*SAAS-FEE Advanced Course 27 Lecture Notes 1997 Swiss Society for Astrophysics and Astronomy* vol 27) (Berlin: Springer) p 508

[3] Shu F 1992 *Gas Dynamics* (*Physics of Astrophysics* vol 2) (Mill Valley, CA: University Science) p 476

[4] Rybicki G B and Lightman A P 2008 *Radiative Processes in Astrophysics* (Weinheim: Wiley) p 382

[5] Pai S-I 1966 *Radiation Gas Dynamics* (New York: Springer) p 228

[6] Martin J 1966 *Atmospheric Entry—An Introduction to its Science and Engineering* (Englewood Cliffs, NJ: Prentice-Hall) p 264

[7] Bond J W, Watson K M and Welch J A 1965 *Atomic Theory of Gas Dynamics* (Reading, MA: Addison-Wesley) p 518

[8] Siegel R and Howell J 1992 *Thermal Radiation Heat Transfer* (Washington DC: Hemisphere Publishing Company) p 1072

[9] Özisi M N 1973 *Radiative Transfer and Interactions with Conduction and Convection* (New York: Wiley) p 575

[10] Chapman S and Cowling T G 1970 *The Mathematical Theory of Non-uniform Gases* (Cambridge: Cambridge University Press) p 427

[11] Clarke J F and McChesney M 1964 *The Dynamics of Real Gases* (Washington, DC: Butterworth) p 419

[12] Park C 1990 *Nonequilibrium Hypersonic Aerothermodynamics* (New York: Wiley)

[13] Park C 1993 Review of chemical–kinetic problems of future NASA missions. I. Earth entries *J. Thermophys. Heat Transfer* **7** 385–98

[14] Park C, La Jaffe R and Partridge H 2001 Chemical–kinetic parameters of hyperbolic Earth entry *J. Thermophys. Heat Transfer* **15** 76–90

[15] Tchuen G and Zeitoun D E 2009 Effects of chemistry in nonequilibrium hypersonic flow around blunt bodies *J. Thermophys. Heat Transfer* **23** 433–42

[16] Bird G 2011 The Q–K model for gas-phase chemical reaction rates *Phys. Fluids* **23** 106101

[17] Wagnild R and Gallis M A 2013 Continuum simulation of hypersonic flows using the quantum-kinetic chemical reaction model *44th AIAA Thermophysics Conf.* AIAA 2013-3144

[18] Liechty D S 2013 State-to-state internal energy relaxation following the quantum-kinetic model in DSMC *44th AIAA Thermophysics Conf.* AIAA 2013-2901

[19] Olynick D, Chen Y K and Tauber M E 1999 Aerothermodynamics of the Stardust sample return capsule *J. Spacecraft Rockets* **36** 442–62

[20] Zel'dovich Y B and Raizer Y P 2002 *Physics of Shock Waves and High-temperature Hydrodynamics Phenomena* (Mineola, NY: Dover) pp 246–348

[21] Mihalas D and Weibel-Mihalas B 1984 *Foundations of Radiation Hydrodynamics* (New York: Oxford University Press) pp 309–402

[22] Shang J S and Surzhikov S T 2011 Simulating Stardust Earth reentry with radiation heat transfer *J. Spacecraft Rockets* **48** 385–96

[23] Liu Y, Prabhu D, Trumble K A, Saunders D and Jenniskens P 2010 Radiation modeling for the reentry of the Stardust sample return capsule *J. Spacecraft Rockets* **47** 741–52

[24] Akey N D and Cross A E 1970 Radio blackout alleviation and plasma diagnostic results from a 25000 foot per second blunt-body reentry *NASA Technical Report* TN D-5615

[25] Grantham W L 1970 Flight results of a 25000-foot-per-second reentry experiment using microwave reflectometers to measure plasma electron density and standoff distance *NASA Technical Report* TN D-6062

[26] Candler G V and MacCormack R W 1991 Computation of weakly ionized hypersonic flows in thermochemical nonequilibrium *J. Thermophys. Heat Transfer* **5** 266–73

[27] Olynick D, Henline W, Hartung L and Candler G 1994 Comparison of coupled radiative Navier–Stokes flow solutions with the project Fire–II flight data. *6th Joint Thermophysics and Heat Transfer Conf.* AIAA 94-1955

[28] Johnston C O, Hollis B R and Sutton K 2008 Spectrum modeling for air shock-layer radiation at lunar-return conditions *J. Spacecr. Rockets* **45** 865–78

[29] Johnston C O, Hollis B R and Sutton K 2008 Nonequilibrium stagnation-line radiative heating for Fire-II *J. Spacecraft Rockets* **45** 1185–95

[30] Surzhikov S and Shang J 2015 Fire–II flight data simulations with different physical–chemical kinetics and radiation models *Front. Aerospace Eng.* **4** 70–92

[31] Desai P N, Lyons D T, Tooley J and Kangas J 2008 Entry, descent, and landing operations analysis for the Stardust entry capsule *J. Spacecraft Rockets* **45** 1262–8

[32] Jenniskens P 2010 Observations of the Stardust sample return capsule entry with a slitless echelle spectrograph *J. Spacecraft Rockets* **47** 718–35

[33] Panesi M, Magin T, Bourdon A, Bultel A and Chazot O 2011 Electronic excitation of atoms and molecules for the Fire–II flight experiment *J. Thermophys. Heat Transfer* **25** 361–74

[34] Trigo-Rodriguez J M, Rietmeijer F J, Llorca J, Llorca J and Janches D 2008 *Advances in Meteoroid and Meteor Science* (New York: Springer) p 561

[35] Kosarev I 2009 The optical properties of vapors of matter of cosmic bodies invading the Earth atmosphere *High Temp* **47** 777–87

[36] Surzhikov S T 2014 Non-equilibrium radiative gas dynamics of small meteor *44th AIAA Fluid Dynamics Conf.* AIAA 2014–2636

[37] Gurvich L, Veitc I and Medvedev V *et al* 1978 *Thermodynamic Properties of Individual Substances, Handbook* (Moscow: Nauka)

[38] Chase M, Davies J R J and Downey C A *et al* 1985 JANAF thermochemical tables *J. Phys. Chem. Ref. Data* **14** suppl. 1 p 1856

[39] Bird R, Stewart W and Lightfoot E 2002 *Transport Phenomena* (New York: Wiley) p 884

[40] Hirschfelder J O, Curtiss C F, Bird R B and Mayer M G 1954 *Molecular Theory of Gases and Liquids* vol 26 (New York: Wiley) p 1219

[41] Wilke C R 1950 Diffusional properties of multicomponent gases *Chem. Eng. Prog* **46** 95–104

[42] Anfimov N 1962 The laminar boundary layer above chemically active surface *Proc. of the Academy of Sciences, Mechanics and Machine-building* **3** 46–52

[43] Stanley D *et al* 2005 NASA's exploration systems architecture study *NASA Technical Report* TM-2005-214062

[44] Surzhikov S 2016 Radiative gas dynamics of large superorbital space vehicle at angle of attack *54th AIAA Aerospace Sciences Meeting* AIAA 2016-0741

[45] Edquist K T, Hollis B R, Johnston C O, Bose D, White T R and Mahzari M 2014 Mars science laboratory heat shield aerothermodynamics: design and reconstruction *J. Spacecraft Rockets* **51** 1106–24

[46] Edquist K T 2005 Afterbody heating predictions for a Mars science laboratory entry vehicle *38th AIAA Thermophysics Conf.* AIAA 2005–4817

[47] Dunn M G and Kang S 1873 Theoretical and experimental studies of reentry plasmas *NASA Technical Report* CR-2232

[48] Treanor C E and Marrone P V 1962 Effect of dissociation on the rate of vibrational relaxation *Phys. Fluids* **5** 1022–6

[49] Marrone P V and Treanor C E 1963 Chemical relaxation with preferential dissociation from excited vibrational levels *Phys. Fluids* **6** 1215–21

[50] Park C, Howe J T, Jaffe R L and Candler G V 1994 Review of chemical-kinetic problems of future NASA missions. II. Mars entries *J. Thermophys. Heat Transfer* **8** 9–23

[51] Josyula E, Bailey W F and Ruffin S M 2003 Reactive and nonreactive vibrational energy exchanges in nonequilibrium hypersonic flows *Phys. Fluids* **15** 3223–35

[52] Boyd I D 2007 Modeling of associative ionization reactions in hypersonic rarefied flows *Phys. Fluids* **19** 096102

[53] Brandis A M, Johnston C. O. 2014 Characterization of stagnation-point heat flux for Earth entry *45th AIAA Plasmadynamics and Lasers Conf.* AIAA 2014–2374

IOP Publishing

Plasma Modeling

Methods and Applications

Gian Luca Delzanno and Xian-Zhu Tang

Chapter 13

Dust–plasma interaction: a review of dust charging theory and simulation

13.1 Introduction

The interaction of material objects with plasmas has been a major research area since the beginning of plasma physics, spanning a variety of applications in the laboratory and in space and astrophysics. In laboratory experiments, this includes the development of diagnostic techniques to probe the plasma properties [1], while experiments on solid particulates (dust) affecting streamer discharges were reported by Langmuir *et al* as early as 1924 [2]. Almost in parallel, remote observations indicated the presence of dust as a medium responsible for the absorption, scattering, and emission of light in astrophysical environments [3].

With the advent of the space age and the beginning of *in situ* observations, it became clear that dust is also a ubiquitous component of the Solar System [4, 5]. The Voyager observations of the mysterious Saturn spokes [6] indicated that the dust grains were charged by the plasma and gave impetus to the development of some of the theories of dust–plasma interaction. In the early 1990s further interest was driven by the microelectronics industry, particularly with regard to plasma deposition and etching techniques, where the negative impact of the presence of dust particulates in the processing chamber was recognized [7, 8]. Finally, tremendous growth of the field of dusty plasmas was driven by the discovery of ordered structures (i.e. plasma crystals) [9, 10] and the possibility to investigate phase transitions in these systems. Indeed, the fact that dust particles can be diagnosed using simple optical techniques (and in fact can usually be seen with the naked eye) is one of the main advantages of dusty plasmas, as it allows measurements at the kinetic level [11]. Today the field continues to grow with new emerging applications. For instance, dust created by plasma–material interaction at the chamber walls of tokamak devices poses important safety and operational concerns [12]. This has led to the development of sophisticated dust transport codes [13–18] and experimental campaigns [19–22] aimed at understanding the impact of dust in magnetic fusion energy applications. Furthermore, plasma medicine, exploiting the interaction of plasmas with biological matter, is a new research frontier that can have many aspects in common with dust–plasma interaction [23] and is showing much promise for the treatment of diseases for which traditional medicine has yet to find a definitive answer [24]. New laboratory experiments with high magnetic fields are also leading the development of the field of magnetized dusty plasmas [25].

A dusty plasma is loosely defined as a system containing electrons, ions, neutrals, and dust particulates [26]. The dust is macroscopic relative to the size of the plasma particles, atoms, and molecules, but still microscopic relative to the system size. There is indeed a great variety of dust sizes in the Universe. In astrophysical environments one can find macromolecules, micron-sized grains, planetesimals all the way to boulders [5]. In the laboratory the dust size can also vary widely. There are two characteristic regimes that depend on the relative ratios of the dust characteristic radius r_d: the plasma Debye length λ_D and the dust interparticle distance d [26]. Typically, the case $r_d < \lambda_D < d$ is referred to as 'dust-in-a-plasma' since the dust grains act individually as screened charged particles, while the case $r_d < d < \lambda_D$ is that of a 'dusty plasma' since the dust particles can act collectively. Note that in space and astrophysics $r_d/\lambda_D \ll 1$ (see for instance [3, 27] for some representative parameters), while in laboratory experiments parameters can vary widely and $r_d/\lambda_D \gg 1$ is also possible.

Dusty plasmas are quite different from conventional plasmas [28–30]. The dust particles are charged by absorption of plasma particles (and possibly also by other processes like electron emission) and are coupled to the plasma through the electromagnetic force. Since the dust charge-to-mass ratio is much smaller than that of the plasma, the dust particles introduce new characteristic scales and dynamics into the system. In addition, the dust can significantly affect the plasma through the quasi-neutrality condition, $q_i n_i = e n_e - q_d n_d$, where n_d, n_e, and n_i are the dust, electron, and ion densities, respectively, q_d is the dust charge (assumed here to be negative), e is the elementary charge, and q_i is the ion charge. Note that the dust charge is a dynamical variable that depends on the local dust and plasma conditions. The dust particles can often accumulate 10^3–10^5 electron charges, implying that even a small (relative to the plasma density) dust density can contribute appreciably to quasi-neutrality. Furthermore, dust particles can also act as a source of plasma through electron emission processes.

A dust charging theory is the building block of any study in dusty plasma physics and the main objective of the present chapter is to review this theory. In section 13.2 we review the basics of the interaction of a single dust grain with a collisionless, unmagnetized plasma in the simplest case of primary charging. We derive the equations of the orbital-motion-limited (OML) theory, the most widely used charging theory in the dusty plasma community. In section 13.3 we briefly discuss the numerical approaches that can be used in the context of dust charging. In section 13.4 we extend the charging theory to the case of dust electron emission. Electron emission can significantly alter the heating of a dust grain in a plasma, playing a major role in dust survivability. With electron emission, we show that OML can become inaccurate in the regime where the dust grain is positively charged and the power collected by the dust from the background electrons can be significantly reduced relative to the OML value. We conclude by highlighting some emerging applications and related important effects where further developments of dust–plasma interaction theories are necessary.

13.2 The basics of dust–plasma interaction

Let us begin by considering a spherical, absorbing dust grain of radius r_d at rest in an unmagnetized, collisionless plasma. The grain becomes charged by absorbing plasma particles, creating a sheath electric field near the grain. If the grain is at floating potential with respect to the plasma, the net current collected by the grain from the plasma vanishes as the electron and ion currents cancel each other. In this simple case, often denoted as primary charging, the grain is normally negatively charged due to the higher mobility of the electron relative to the ions.

We assume spherical symmetry, introduce a spherical reference system centered on the grain, and seek the steady-state solution of this system. Such a state is determined by the laws of Hamiltonian mechanics in a central force field (known as orbital motion (OM) theory), where the particle's motion is confined to a plane determined by its initial radial (v_r) and tangential (v_θ) velocities and the conservation of energy

$$\frac{m_\alpha}{2}\left(v_r^2 + v_\theta^2\right) + q_\alpha\phi(r) = E \tag{13.1}$$

and angular momentum

$$m_\alpha r v_\theta = J \tag{13.2}$$

apply to each particle. In equations (13.1) and (13.2), m_α (q_α) is the mass (charge) of particles of species α ($\alpha = e, i$ labels electrons or ions), $E(J)$ is the particle energy (angular momentum), and $\phi(r)$ is the radially symmetric electrostatic potential, while r is the radial distance. From classical mechanics, the motion of a particle can be reduced to a one-dimensional (radial) motion in an effective potential field given by

$$U(r, J) = q_\alpha\phi(r) + \frac{1}{2}\frac{J^2}{m_\alpha r^2}. \tag{13.3}$$

Clearly the first term in equation (13.3) corresponds to the electrostatic force while the second is the centrifugal force.

In terms of boundary conditions, at infinity $\lim_{r \to +\infty}\phi(r) = 0$, the plasma is unperturbed and we will assume the plasma distribution function to be Maxwellian:

$$f_\alpha(\mathbf{v}) = n_\infty\left(\frac{m_\alpha}{2\pi T_\alpha}\right)^{3/2}\exp\left[-\frac{m_\alpha}{2T_\alpha}\left(v_r^2 + v_\theta^2\right)\right], \tag{13.4}$$

where n_∞ is the unperturbed plasma density and T_α is the temperature of species α expressed in electronvolts. This implies that in equation (13.1), $E > 0$, a condition that must be taken into account when the distribution function is integrated to obtain plasma densities and currents. Note also that, by virtue of Liouville's theorem, the distribution function at any point in space remains Maxwellian and is given by

$$f_\alpha(r, \mathbf{v}) = n_\infty \left(\frac{m_\alpha}{2\pi T_\alpha}\right)^{3/2} \exp\left[-\frac{m_\alpha}{2T_\alpha}\left(v_r^2 + v_\theta^2\right) - \frac{q_\alpha \phi(r)}{T_\alpha}\right]. \tag{13.5}$$

It is important to emphasize that we are only considering *unbounded* particles, i.e. particles whose orbits connect to infinity.

Furthermore, one must take into account the absorbing nature of the grain, namely the fact that particles on inbound orbits ($v_r < 0$) hitting the grain will not be present in the outbound orbit ($v_r > 0$). Again from equations (13.1) and (13.2), it follows that the condition for absorption is

$$v_r^2 + \left[1 - \left(\frac{r}{r_d}\right)^2\right]v_\theta^2 + 2\frac{q_\alpha}{m_\alpha}\left[\phi(r) - \phi_d\right] \geq 0, \quad v_r \leq 0, \tag{13.6}$$

where $\phi_d = \phi(r_d)$ is the dust potential. This implies that, in order to calculate the plasma density around the grain, the distribution function must be integrated with the condition

$$v_r^2 + \left[1 - \left(\frac{r}{r_d}\right)^2\right]v_\theta^2 + 2\frac{q_\alpha}{m_\alpha}\left[\phi(r) - \phi_d\right] \leq 0, \quad v_r \geq 0. \tag{13.7}$$

The density of the plasma can then be computed by integrating equation (13.5) over the *accessible* part of phase space:

$$n_\alpha(r) = 2\pi \int dv_r \int dv_\theta v_\theta f_\alpha(r, \mathbf{v}). \tag{13.8}$$

Similarly, the current reaching the dust grain is obtained by

$$I_\alpha = -q_\alpha 4\pi r_d^2 2\pi \int dv_r v_r \int dv_\theta v_\theta f_\alpha(r_d, \mathbf{v}), \tag{13.9}$$

and the power collected by the dust grain from species α is

$$h_\alpha = -4\pi r_d^2 2\pi \int dv_r v_r \int dv_\theta v_\theta \frac{m_\alpha}{2}\left(v_r^2 + v_\theta^2\right) f_\alpha(r_d, \mathbf{v}), \tag{13.10}$$

where the domain of integration is now restricted to incoming particles (i.e. $v_r < 0$) and, as in equation (13.8), the limits of integration will be specified on a case-by-case basis. The factor $2\pi v_\theta$ in equations (13.8)–(13.10) comes from expressing the tangential velocity in terms of its magnitude v_θ and an isotropic angle that can be integrated away and leads to 2π. Consistently, $v_\theta > 0$.

13.2.1 Repelled species ($q_\alpha \phi_d > 0$)

Let us now proceed to treat the species that is repelled by the dust grain, i.e. electrons in the case of a negatively charged grain. The effective potential $U(r)$ is always

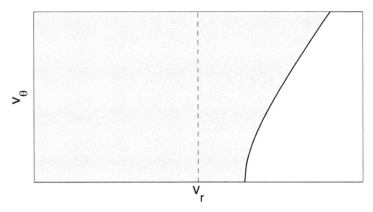

Figure 13.1. Diagram of the phase-space integration domain to obtain the density of the repelled species.

positive and monotonically decreasing with r, implying that all the particles are on unbounded orbits and that $E > U(r_d, J)$ is necessary to reach the dust grain. The distribution function must be integrated with the absorption condition (13.7), yielding

$$\frac{n_\alpha(z)}{n_\infty} = \frac{1}{2}\left\{1 + \text{Erf}\left(\sqrt{\varphi_d - \varphi}\right) + \sqrt{1 - z^{-2}}\left[1 - \text{Erf}\left(\sqrt{\frac{\varphi_d - \varphi}{1 - z^{-2}}}\right)\right]\right.$$

$$\left.\exp\left[\frac{\varphi_d - \varphi}{z^2 - 1}\right]\right\}\exp(-\varphi), \tag{13.11}$$

where we have introduced $z = r/r_d$ and $\varphi = q_\alpha\phi/T_\alpha$. Figure 13.1 shows the phase-space domain for the repelled species. One can compute the current reaching the dust grain by integrating over the whole $v_r < 0$ domain. It yields

$$I_\alpha = q_\alpha 4\pi r_d^2 n_\infty \sqrt{\frac{T_\alpha}{2\pi m_\alpha}}\exp\left(-\frac{q_\alpha\phi_d}{T_\alpha}\right). \tag{13.12}$$

Similarly, the dust collected power is

$$h_\alpha = 2T_\alpha\frac{I_\alpha}{q_\alpha}. \tag{13.13}$$

13.2.2 Attracted species ($q_\alpha\phi_d < 0$)

For the attracted species, the situation is complicated by the fact that the electro-static and centrifugal forces now have opposite signs and therefore the effective potential can be negative and/or non-monotonic. To illustrate this point, let us consider a Debye–Huckel potential (normalized to the temperature T_α) around the grain

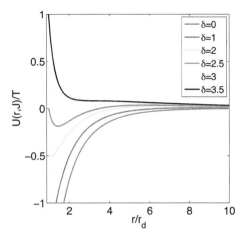

Figure 13.2. Profiles of the effective potential varying J via $\delta = J^2/(2m_a r_d^2 T_a)$ for the Debye–Huckel potential given in equation (13.14), illustrating the non-monotonic nature of the effective potential for particles of the attracted species in a certain range of J.

$$\varphi(r) = \frac{q_a \phi(r)}{T_\alpha} = \varphi_d \frac{r_d}{r} \exp\left(-\frac{r - r_d}{\lambda}\right), \tag{13.14}$$

where λ is the screening length. We choose $\varphi_d = -2.5$ and $\lambda/r_d = 2$ and show the effective potential for different values of the dimensionless parameter $\delta = J^2/(2m_a r_d^2 T_a)$ in figure 13.2. For $\delta = 0$, the effective potential is monotonically increasing with r. For small values of δ, $\delta < \delta_1$ ($\delta_1 = -3\varphi_d/4 = 1.875$ for the parameters considered here) the effective potential is non-monotonic, with a maximum at a certain $r_m(J)$. For $\delta_1 \leqslant \delta \leqslant \delta_2$ ($\delta_2 \simeq 3.46$), the effective potential is still non-monotonic, but with a minimum close to r_d and a maximum away from r_d. For $\delta \geqslant \delta_2$, the effective potential is again monotonic, but decreasing with r. The fact that the effective potential can be non-monotonic has two important consequences:

1. There could be particles performing bounded motion, i.e. motion confined between $r_A \leqslant r \leqslant r_B$, where r_A and r_B depend on the particle energy and angular momentum and $r_A > r_d$. Bounded particles are typically neglected in theories of dust–plasma interaction and we will follow the same path here, but the reader should keep in mind that this might not always be the case.

2. Particles might be reflected before reaching the dust grain if the condition

$$U^{max} = U(r_m(J), J) > E \tag{13.15}$$

holds, where r_m is the position of the maximum of U and, most importantly, is dependent on the particle's angular momentum J. This is known as *the absorption radius* effect [31], which might reflect away some particles of the attractive species even if $E > U(r_d, J)$ is satisfied. One would then need to integrate equation (13.5) with the condition (13.15) (instead of equation (13.7)) when $v_r > 0$. It yields an expression for the density in integral form and the interested reader is referred to [32, 33] for the exact expressions.

Here we take an alternative path, known as the OML theory [34]. The simplifying assumption is to assume that the effective potential is monotonic or, in other words, that all the particles such that $E > U(r_d)$ (which of course is equivalent to equation (13.7)) can reach the dust grain. From figure 13.2, one expects that the OML theory would overestimate the current and the density near the grain since it neglects that some particles can be reflected by (effective) potential barriers.

Let us now proceed to calculate the density of the attracted species in the OML framework. First, the distribution function has to be integrated with the condition $E > 0$, whose boundary is a circle in phase space

$$v_r^2 + v_\theta^2 \geqslant -\frac{2q_\alpha \phi(r)}{m_\alpha}, \tag{13.16}$$

which intersects the $v_r = 0$ axis at

$$v_\theta^b = \sqrt{-2q_\alpha \phi(r)/m_\alpha}. \tag{13.17}$$

For particles moving towards the grain, $v_r < 0$, equation (13.16) is the only condition necessary to calculate the density. For $v_r > 0$, on the other hand, one must also consider equation (13.7), which corresponds to a parabola in phase space whose intersection with the $v_r = 0$ axis is

$$v_\theta^r = \sqrt{\frac{2q_\alpha}{m_\alpha} \frac{\phi - \phi_d}{z^2 - 1}}. \tag{13.18}$$

Thus, the domain of integration in the $v_r > 0$ space depends on the value of v_θ^b and v_θ^r. For the case $v_\theta^b > v_\theta^r$, which corresponds to $\phi(r) < \phi_d/z^2$, the domain of integration is

$$v_r < 0, \quad v_r^2 + v_\theta^2 \geqslant -2q_\alpha \phi(r)/m_\alpha,$$

$$v_r \geqslant 0, \quad v_r \in \left[0, v_r^c\right], \quad v_\theta \in \left[\sqrt{-(2q_\alpha/m_\alpha)\phi - v_r^2}, \infty\right],$$

$$v_r \geqslant 0, \quad v_r \in \left[v_r^c, \infty\right], \quad v_\theta \in \left[\sqrt{\frac{v_r^2 + \frac{2q_\alpha}{m_\alpha}(\phi - \phi_d)}{z^2 - 1}}, \infty\right], \tag{13.19}$$

where v_r^c and v_θ^c correspond to $v_\theta^b = v_\theta^r$, i.e.

$$v_r^c = \sqrt{-\frac{2q_\alpha}{m_\alpha}\left(\phi - \frac{\phi_d}{z^2}\right)}; \quad v_\theta^c = \sqrt{-\frac{2q_\alpha}{m_\alpha}\frac{\phi_d}{z^2}}, \tag{13.20}$$

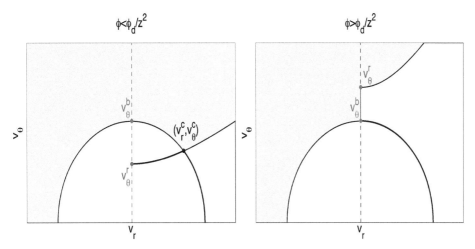

Figure 13.3. Illustration of the phase-space integration domain for the attracted species. The intersection of the two curves, $E > 0$ and equation (13.7), gives rise to two different integration limits depending on the behavior of $\phi(r)$ versus ϕ_d/z^2. Adapted from [35].

see figure 13.3 (left). This yields

$$
\frac{n_\alpha(z)}{n_\infty} = \sqrt{-\frac{\varphi}{\pi}}\left[1 + \sqrt{1 - \frac{\varphi_d}{z^2\varphi}}\right] + \frac{e^{-\varphi}}{2}\left[1 - \mathrm{Erf}\left(\sqrt{-\varphi}\right)\right]
$$
$$
+ \frac{\sqrt{1 - z^{-2}}}{2}e^{-\tilde{\varphi}}\left[1 - \mathrm{Erf}\left(\sqrt{-\tilde{\varphi}}\right)\right], \quad \phi \leqslant \phi_d/z^2,
\tag{13.21}
$$

with

$$
\tilde{\varphi} = \frac{z^2\varphi - \varphi_d}{z^2 - 1}.
\tag{13.22}
$$

For $v_\theta^b < v_\theta^r$, which corresponds to $\phi(r) > \phi_d/z^2$, the domain of integration becomes

$$
v_r < 0, \quad v_r^2 + v_\theta^2 \geqslant -2q_\alpha\phi(r)/m_\alpha,
$$
$$
v_r \geqslant 0, \quad v_r \in [0, +\infty], \quad v_\theta \in \left[\sqrt{\frac{v_r^2 + \dfrac{2q_\alpha}{m_\alpha}(\phi - \phi_d)}{z^2 - 1}}, \infty\right],
\tag{13.23}
$$

see figure 13.3 (right). It yields

$$
\frac{n_\alpha(z)}{n_\infty} = \sqrt{-\frac{\varphi}{\pi}} + \frac{e^{-\varphi}}{2}\left[1 - \mathrm{Erf}\left(\sqrt{-\varphi}\right)\right] + \frac{\sqrt{1 - z^{-2}}}{2}e^{-\tilde{\varphi}}, \quad \phi \geqslant \phi_d/z^2.
\tag{13.24}
$$

We note that the different expressions for the density depending on the behavior of ϕ relative to ϕ_d/z^2 have only been derived fairly recently [35, 36]. The distinction between ϕ and ϕ_d/z^2 is not necessary to compute the current, which is obtained from equation (13.9) by integrating over $v_r < 0$ and $v_r^2 + v_\theta^2 \geqslant -2q_\alpha\phi_d/m_\alpha$. This leads to

$$I_\alpha = q_\alpha 4\pi r_d^2 n_\infty \sqrt{\frac{T_\alpha}{2\pi m_\alpha}}\left(1 - \frac{q_\alpha\phi_d}{T_\alpha}\right).$$ (13.25)

The dust collected power is

$$h_\alpha = T_\alpha\left(\frac{2 - \varphi_d}{1 - \varphi_d} - \varphi_d\right)\frac{I_\alpha}{q_\alpha}.$$ (13.26)

13.2.3 Summary of OML theory

In summary, the steady-state solution of the interaction between a spherical dust grain and a plasma in the OML framework can be obtained by the solution of the non-linear Poisson equation

$$\nabla^2\phi = \frac{1}{r^2}\frac{d}{dr}\left(r^2\frac{d\phi}{dr}\right) = -\frac{\sum_\alpha q_\alpha n_\alpha(\phi)}{\varepsilon_0},$$ (13.27)

where ε_0 is vacuum permittivity and the densities are given by equation (13.11), equation (13.21) for $\phi \leqslant \phi_d/z^2$ and equation (13.24) for $\phi \geqslant \phi_d/z^2$. Equation (13.27) is solved with the following boundary conditions:

$$\text{floating potential condition:} \quad \sum_\alpha I_\alpha(\phi_d) = 0$$ (13.28)

and

$$\lim_{r \to +\infty}\phi(r) = 0.$$ (13.29)

From a numerical point of view, equation (13.29) is often substituted by $\phi(R) = 0$, where $R \gg r_d, \lambda_D$ is the boundary of the computational domain, or by $d\phi/dr|_{r=R} + \frac{\alpha}{r}\phi(R)$ if one wants to impose an asymptotic form of the potential given by $\phi = C/r^\alpha$ (with C a constant). For a probe biased at ϕ_{bias} equation (13.28) is substituted by the condition $\phi(r_d) = \phi_{\text{bias}}$.

The theory is based on the following assumptions:

1. Spherical symmetry.
2. Unbounded particles: the trajectory of all the plasma particles (either collected by the dust grain or reflected at some radius) connects to infinity.
3. Collisionless plasma: collisions between plasma particles or between particles and neutrals are neglected, implying that the theory is valid when $r_d, \lambda_D \ll \lambda_{\text{mfp}}$, where λ_D is the plasma Debye length and λ_{mfp} is the collisional mean-free-path.
4. Unmagnetized plasma: the effect of the magnetic field on the trajectory of the plasma particles is neglected, implying that the theory is valid when $r_d, \lambda_D \ll \rho_e$, where ρ_e is the electron gyroradius.

5. Potential barriers to the motion of the particles of the attracted species are neglected, namely all the attracted particles satisfying $E > U(r_d, J)$ are assumed to be collected by the dust grain. Relative to the assumptions listed above, this is the only one that differentiates the OML theory from the more complete (but rarely used) OM theory.

13.2.4 Some important considerations

We have shown that the OML approximation consists in neglecting that particles (of the attracted species) with angular momentum in a certain range will be reflected before reaching the dust grain if their absorption radius is larger than the dust radius. Thus, OML can/will overestimate the collected current and the density near the grain. This quite innocent approximation has, in practice, remarkable consequences: one can obtain the dust potential by solving equation (13.28) and *without* solving Poisson's equation! Given its simplicity and considering that the dust potential is the quantity that determines currents, forces, and heat fluxes to the grain, all necessary ingredients to study dust transport in a plasma, one can understand why OML is the most widely used dust charging theory.

One might ask the question of the validity of OML theory. The answer, of course, has to do with the particular form of the shielding potential and whether or not potential barriers to the particles' motion are present. By studying the behavior of $r^3 dU/dr$ one can conclude that if $\phi(r)$ varies more slowly than $1/r^2$ (say like a Coulomb potential $\propto 1/r$), then the effective potential can only have a minimum or be monotonic, see [32, 37]. For these conditions OML theory would be exact. Unfortunately, these are not the conditions that one encounters for dust–plasma interaction, where the potential decreases more slowly than $1/r^2$ (i.e. Coulomb) close to the grain, faster than $1/r^2$ in the Debye shielding region $r \sim \lambda_D$ and exactly as $1/r^2$ asymptotically for $r \gg \lambda_D$. Thus, OML can only be an approximation.

What is then the accuracy of the OML theory? To answer this question one must compare OML with an exact solution obtained either by solving the OM theory or through simulations. The fact that OML reproduces OM in the limit of a small dust radius relative to the Debye length has been investigated by several authors [33, 38–41] and is well established. Furthermore particle-in-cell (PIC) simulations of dust–plasma interaction were performed in [40, 41] to investigate the limit of the large dust radius. It was concluded that OML is a very good approximation for r_d/λ_{De} of order unity (with a dependence on the electron-to-ion temperature ratio [40]), with λ_{De} the electron Debye length. For $r_d \gg \lambda_{De}$, the sheath near the dust grain approaches a planar sheath (with the ion density increasing monotonically with r) and the transition from a spherical to a planar sheath is not captured by OML. Nevertheless, from [41] we find that for $T_i/T_e = 1$, $m_i/m_e = 1836$ and up to $r_d/\lambda_{De} = 10$, OML is still reasonable with $\sim 20\%$ accuracy on dust charge, currents, and heat fluxes.

An important issue relates to the dust charge Q_d, a quantity critically important for dust transport since it is directly related to the Lorentz force. The dust charge is given by the electric field at the dust surface as dictated by Gauss's law

$$Q_d = -4\pi\varepsilon_0 r_d^2 \frac{d\phi}{dr}\bigg|_{r=r_d}, \tag{13.30}$$

where we are still taking advantage of spherical symmetry. For a Debye–Huckel potential (13.14), equation (13.30) becomes

$$Q_d = 4\pi\varepsilon_0 r_d \left(1 + \frac{r_d}{\lambda}\right)\phi_d, \tag{13.31}$$

where one recognizes the capacitance of a spherical conductor in vacuum, $C_d = 4\pi\varepsilon_0 r_d$. In the limit $r_d/\lambda \ll 1$, expression (13.31) simplifies to

$$Q_d = 4\pi\varepsilon_0 r_d \phi_d, \tag{13.32}$$

namely the dust charge is completely determined by the dust potential ϕ_d. However, equation (13.31) reveals that in general the dust charge depends on the space-charge distribution outside the grain, i.e. by the screening λ, and this requires the solution of the non-linear Poisson equation (13.27). Note also that the screening contribution to Q_d can in fact be dominant in regimes where $r_d/\lambda > 1$. For instance, [41] shows that for $T_i/T_e = 1$, $m_i/m_e = 1836$ and $r_d/\lambda_{De} = 10$, the dust charge is six times higher than that given by equation (13.32).

13.3 A note on the numerical solution of dust–plasma interaction problems

Before we proceed with investigating further aspects of dust–plasma interaction, let us discuss how the problem can be treated numerically. There are essentially two approaches. Under certain assumptions such as those of the OML or OM theories, the problem is reduced to the numerical solution of a non-linear Poisson equation. In order to do so, let us introduce a computational mesh for the radial coordinate, where mesh points are given by

$$r_i = r_d + (i - 1)\Delta r, \quad i = 1, \ldots, N+1, \tag{13.33}$$

with N the number of mesh points and $\Delta r = (R - r_d)/N$ the mesh spacing (assumed as uniform for simplicity). The non-linear Poisson equation can be discretized using, for example, finite differences techniques and cast in residual form as

$$R[\phi_i] = R_i = \frac{r_{i+1/2}^2}{r_i^2}\frac{\phi_{i+1} - \phi_i}{\Delta r^2} - \frac{r_{i-1/2}^2}{r_i^2}\frac{\phi_i - \phi_{i-1}}{\Delta r^2}$$
$$+ \sum_\alpha \frac{q_\alpha}{\varepsilon_0} n_\alpha(\phi_i), \quad i = 2, \ldots, N, \tag{13.34}$$

where $\phi_i = \phi(r_i)$, together with the boundary conditions

$$R[\phi_1] = R_1 = \sum_\alpha I_\alpha(\phi_1),$$
$$R[\phi_{N+1}] = R_{N+1} = \phi_{N+1}. \tag{13.35}$$

In matrix form, we write equations (13.34) and (13.35) as $\mathbf{R}(\boldsymbol{\phi}) = 0$, with $\boldsymbol{\phi}$ the column vector containing all the unknowns. The system of discrete equations can be solved numerically with standard techniques of numerical analysis, such as Newton's method [42]. The idea is to solve iteratively successive linear systems of the form

$$\mathbf{J}(\boldsymbol{\phi}^n)\Delta\boldsymbol{\phi}^n = -\mathbf{R}(\boldsymbol{\phi}^n) \tag{13.36}$$

with the Jacobian matrix $\mathbf{J}$ defined as $J_{ij} = \frac{\partial R_i(\boldsymbol{\phi})}{\partial \phi_j}$ (where $\mathbf{R}_i$ labels the ith row of $\mathbf{R}$), n labels the Newton iteration, and $\boldsymbol{\phi}^{n+1} = \boldsymbol{\phi}^n + \Delta\boldsymbol{\phi}^n$. Non-linear convergence is reached when

$$\left\|\mathbf{R}(\boldsymbol{\phi}^n)\right\|_2 \leqslant \tau_a + \tau_r\left\|\mathbf{R}(\boldsymbol{\phi}^0)\right\|_2, \tag{13.37}$$

with $\|\cdot\|$ the L_2-norm and τ_r and τ_a are relative and absolute tolerances that are provided by the user. A discussion of iterative methods for linear and non-linear equations can be found in [42] and a series of MATLAB codes that implement these solvers is available on the related website. We use a Jacobian-free Newton–Krylov (JFNK) method for the solution of $\mathbf{R}(\boldsymbol{\phi}) = 0$, which has the advantage that the Jacobian matrix is never formed explicitly and the user only has to provide the residual function.

An alternative approach is to solve directly the time-dependent Vlasov–Poisson equations. For plasma–material interaction problems, the most popular technique is the PIC method. In PIC the plasma is discretized with macroparticles, each carrying the statistical information (such as the charge) corresponding to many particles of the physical system, moving through a computational mesh. At each time step, the plasma density is obtained by interpolating the particles' information on the mesh, Poisson's equation is solved on the mesh to construct the electrostatic field there, and the electrostatic force is interpolated back to the particles' position. The time step is completed by moving the particles via Newton's equations. The reader interested in the details of the PIC algorithm is referred to chapter 4 and the classic monographs [43, 44]. Note that the presence of a dust grain requires two adjustments to the classic PIC algorithm. First, plasma particles absorbed by the grain are removed from the simulation and their charge is accumulated onto the dust grain, translating in a boundary condition on the electric field given by Gauss's law. Second, since the dust grain is a sink of plasma particles, a source of plasma must be injected in the system (typically at the outer boundaries) to be able to reach a steady state.

Each of the two techniques has advantages and disadvantages. The Newton solver can deliver the solution within minutes on a conventional laptop (although convergence can be difficult depending on the level of non-linearity of the densities and might require some user intervention to provide a good initial guess to the iterative procedure) but the non-linear Poisson's equation can only be derived under some quite restrictive assumptions. PIC, on the other hand, is a first-principle method that can be easily extended to more general cases, such as to include the effect of collisionality or of an external magnetic field, or can be used to study transients. The price to pay for this flexibility is that it might take a long time to obtain the steady-state solution of the problem and this might require significant computational resources depending on the system parameters. Nevertheless, several PIC codes are in use in the dusty plasma community [45–49].

In the following section we will use both techniques to explore the limitations of OML in the limit of electron-emitting dust grains.

13.4 Dust electron emission

Let us now investigate dust–plasma interaction for electron-emitting dust grains, a case of interest in many applications in nature and in the laboratory. Dust grains can emit electrons according to three main mechanisms: thermionic emission, photo-emission, and secondary emission. Thermionic emission occurs when a material is heated to a high enough temperature that some electrons have enough kinetic energy to overcome the attractive force that holds them in the material (i.e. the material work function). Thermionic emission is important for the dynamics of dust particles created in magnetic fusion energy devices [14] or for meteoroids frictionally heated by the interaction with the Earth's atmosphere [50]. Electron emission can also be induced by the photo-electric effect, when incident photons have enough energy to release electrons from the material. This occurs commonly in space because of the UV field of nearby stars. Examples include dust transport in the Solar System [51], planetary rings [52], cometary environments [53], the Earth's moon [54], and spacecraft–plasma interaction when the spacecraft is exposed to sunlight [55]. Secondary electron emission, on the other hand, occurs when incident energetic plasma particles eject electrons from materials [56]. Once again, secondary emission is important in many applications in the laboratory [14, 57], and space and astrophysical environments [55, 58, 59]. A recent review on secondary electron emission can be found in [60].

In what follows, we will consider Maxwellian emitted electrons

$$f_{\text{em}}(r_d, v_r, v_\theta) = \frac{J_{\text{em}}}{e 2 \pi v_{\text{th,em}}^4} \exp\left[-\frac{m_e}{2 T_{\text{em}}} \left(v_r^2 + v_\theta^2 \right) \right], \tag{13.38}$$

where J_{em} is total emitted current density and $v_{\text{th,em}} = \sqrt{T_{\text{em}}/m_e}$ is the thermal velocity of the emitted electrons with temperature T_{em}. A Maxwellian distribution function is representative of thermionic emission, where J_{em} is given by the Richardson–Dushman current [61]

$$J_{\text{em}} = e \frac{4 \pi T_{\text{em}}^2 m_e}{h^3} \exp\left(-\frac{W}{T_{\text{em}}} \right), \tag{13.39}$$

where h is Planck's constant and W is the thermionic work function of the dust material. A Maxwellian distribution function is also sometimes used to model photo-emission, with the current density J_{em} depending on properties of the material and of the incident light. For secondary electron emission, on the other hand, the current density depends non-linearly on the plasma response to the charging process (via the electron background current to the dust grain) and changes with the polarity of the grain.

In order to treat the emitted electrons we can again use OM. The procedure follows that of section 13.2, with the difference that now the particle energy is in the range $-e\phi_d < E < \infty$. The expressions for the emitted-electron densities, current, and emitted power are equivalent to equations (13.8), (13.9), and (13.10). The steady state of the system including electron emission amounts again to solving Poisson's equation with the emitted-electron density n_{em}:

$$\nabla^2\phi = -\frac{\sum_\alpha q_\alpha n_\alpha(\phi) - en_{em}(\phi)}{\varepsilon_0}. \tag{13.40}$$

Similarly, the floating condition (13.28) is modified to include the emitted-electron current, I_{em}:

$$I_e(\phi_d) + I_i(\phi_d) + I_{em}(\phi_d) = 0. \tag{13.41}$$

13.4.1 The OML approach

Let us begin our investigation of electron emission effects by reviewing the OML approach, which assumes a monotonic shielding potential around the grain. We will treat the negatively charged and positively charged cases separately, depending on the sign of ϕ_d.

The negatively charged case with monotonically increasing ϕ
Let us consider the case where the grain is negatively charged and the shielding potential is *negative and monotonically increasing*. Under these assumptions, the effective potential is always positive and monotonically decreasing. All the emitted electrons are accelerated away from the grain, since $E > U(r_d, J)$ holds. There are no potential barriers (as described above for the attracted species) to the motion of the emitted electrons and the expressions that are derived below are exact. The emitted-electron density is obtained by integrating the distribution function over the domain

$$v_r \in \left[+\sqrt{\frac{2e}{m_e}(\phi - \phi_d)}, +\infty \right], \quad v_\theta \in \left[0, +\sqrt{\frac{v_r^2 - \frac{2e}{m_e}(\phi - \phi_d)}{z^2 - 1}} \right] \tag{13.42}$$

leading to

$$
n_{\text{em}}(r) = \sqrt{\frac{\pi}{2}} \frac{J_{\text{em}}}{e v_{\text{th,em}}} \exp\left(e\frac{\phi - \phi_d}{T_{\text{em}}}\right)\left[1 - \text{erf}\sqrt{e\frac{\phi - \phi_d}{T_{\text{em}}}}\right.
$$
$$
\left. - \exp\left[\frac{e(\phi - \phi_d)}{T_{\text{em}}(z^2 - 1)}\right]\frac{\sqrt{z^2 - 1}}{z}\left(1 - \text{erf}\sqrt{\frac{e(\phi - \phi_d)z^2}{T_{\text{em}}(z^2 - 1)}}\right)\right], \tag{13.43}
$$

while the emitted-electron current is

$$
I_{\text{em}} = 4\pi r_d^2 J_{\text{em}} \tag{13.44}
$$

and the emitted-electron power is

$$
h_{\text{em}} = -2T_{\text{em}}\frac{I_{\text{em}}}{e}. \tag{13.45}
$$

Since in this regime the only approximation in OML is to neglect potential barriers to the motion of the particles of the attracted species, as discussed in section 13.2, but the treatment of the emitted electrons is exact, one presumes that OML remains a good approximation for the negatively charged case for $r_d/\lambda_{\text{De}} \sim 1$. Simulations confirm that this is indeed true [62].

The positively charged case with monotonically decreasing ϕ
As the emitted current is increased, the grain becomes positively charged. In the spirit of section 13.2, let us now assume a *positive, monotonically decreasing* shielding potential. All the considerations performed in section 13.2 for the background plasma remain the same, with the obvious difference that now the background electrons are the attracted species.

Let us then focus on the emitted electrons. An inspection of the effective potential quickly reveals that the contribution of the electrostatic force is negative while that from the centrifugal force is positive, exactly as for the attracted species in section 13.2. Thus, potential barriers can appear. Depending on the angular momentum, the slow emitted electrons can be attracted back to the grain, while the conditions for emitted electrons to escape the grain attraction is given by

$$
E \geqslant U^{\text{max}}(r_m(J), J), \tag{13.46}
$$

where as before r_m is the position of the maximum of U, given by

$$
-r_m^3\phi'(r_m) = \frac{J^2}{em_e}, \tag{13.47}
$$

where $\phi' = \mathrm{d}\phi/\mathrm{d}r$. At this point the attentive reader will not be surprised to discover that, in the OML framework, a monotonic effective potential is assumed and potential barriers for the emitted electrons are neglected. Let us calculate the emitted-electron density. For particles moving away from the grain, the distribution function has to be integrated over the whole $v_r > 0$ space. For particles attracted back to the grain, i.e. moving with $v_r < 0$, one has to distinguish between negative and positive energy E. For $-e\phi_d \leqslant E \leqslant 0$, the effective potential is monotonically increasing and its maximum is at infinity, $U^{\max} = 0$. Thus, one has $E < U^{\max} = 0$, corresponding to

$$v_r^2 + v_\theta^2 \leqslant \frac{2e}{m_e}\phi. \tag{13.48}$$

For $0 \leqslant E \leqslant \infty$, the maximum of the effective potential is at $r = r_d$. The emitted electrons cannot reach infinity if $E < U^{\max}(r_d)$, which can be reformulated as

$$v_r^2 + v_\theta^2(1 - z^2) \leqslant \frac{2e}{me}(\phi - \phi_d), \tag{13.49}$$

whose intersection with the $v_r = 0$ axis corresponds to v_θ^r given by equation (13.18) (in this case $q_\alpha = -e$). Equation (13.49) is valid when $v_r^2 + v_\theta^2 \geqslant \frac{2e}{me}\phi$, whose intersection with the $v_r = 0$ axis corresponds to v_θ^b in equation (13.17). As for the attracted species in section 13.2, the limits of phase-space integration depend on the relative magnitude of v_θ^r and v_θ^b. For $v_\theta^r > v_\theta^b$, namely $\phi < \phi_d/z^2$, the phase-space domain of integration is

$$v_r > 0, \quad v_r \in [0, +\infty), \quad v_\theta \in [0, +\infty),$$

$$v_r \leqslant 0, \quad v_r^2 + v_\theta^2 \leqslant \frac{2e}{m_e}\phi,$$

$$v_r \leqslant 0, \quad v_r \in (-\infty, 0], \quad v_\theta \in \left[\sqrt{\frac{v_r^2 - \frac{2e}{m_e}(\phi - \phi_d)}{z^2 - 1}}, \infty\right], \tag{13.50}$$

leading to

$$n_{\mathrm{em}}(r) = \sqrt{\frac{\pi}{2}} \frac{J_{\mathrm{em}}}{e v_{\mathrm{th,em}}} \exp\left(e\frac{\phi - \phi_d}{T_{\mathrm{em}}}\right)\left[1 + \mathrm{erf}\sqrt{\frac{e\phi}{T_{\mathrm{em}}}} - \frac{2}{\sqrt{\pi}}\sqrt{\frac{e\phi}{T_{\mathrm{em}}}}\exp\left(-\frac{e\phi}{T_{\mathrm{em}}}\right)\right.$$

$$\left. - \frac{\sqrt{z^2 - 1}}{z}\exp\left[\frac{e(\phi - \phi_d)}{T_{\mathrm{em}}(z^2 - 1)}\right]\right], \quad \phi \leqslant \phi_d/z^2. \tag{13.51}$$

For $v_\theta^r < v_\theta^b$, namely $\phi > \phi_d/z^2$, the phase-space domain of integration is

$$v_r > 0, \quad v_r \in [0, +\infty), \quad v_\theta \in [0, +\infty),$$

$$v_r \leqslant 0, \quad v_r \in \left[-\sqrt{\frac{2e}{m_e}\left(\phi - \frac{\phi_d}{z^2}\right)}, 0 \right], \quad v_\theta \in [0, +\infty],$$

$$v_r \leqslant 0, \quad v_r \in \left[-\sqrt{\frac{2e}{m_e}\phi}, -\sqrt{\frac{2e}{m_e}\left(\phi - \frac{\phi_d}{z^2}\right)} \right], \quad v_r^2 + v_\theta^2 \leqslant \frac{2e}{m_e}\phi, \tag{13.52}$$

$$v_r \leqslant 0, \quad v_r \in \left(-\infty, -\sqrt{\frac{2e}{m_e}\left(\phi - \frac{\phi_d}{z^2}\right)} \right], \quad v_\theta \in \left[\sqrt{\frac{v_r^2 - \frac{2e}{m_e}(\phi - \phi_d)}{z^2 - 1}}, \infty \right),$$

leading to

$$
\begin{aligned}
n_{em}(r) = \sqrt{\frac{\pi}{2}} \frac{J_{em}}{e v_{th,em}} \exp\left(e \frac{\phi - \phi_d}{T_{em}} \right) & \left[1 + \mathrm{erf}\sqrt{\frac{e\phi}{T_{em}}} \right. \\
& + \frac{2}{\sqrt{\pi}} \left[\sqrt{\frac{e}{T_{em}}\left(\phi - \frac{\phi_d}{z^2}\right)} - \sqrt{\frac{e\phi}{T_{em}}} \right] \exp\left(-\frac{e\phi}{T_{em}} \right) \\
& \left. - \frac{\sqrt{z^2 - 1}}{z} \exp\left[\frac{e(\phi - \phi_d)}{T_{em}(z^2 - 1)} \right] \left[1 + \mathrm{erf}\sqrt{\frac{e}{T_{em}}\frac{z^2\phi - \phi_d}{z^2 - 1}} \right] \right],
\end{aligned}
\tag{13.53}
$$

$$\phi \geqslant \phi_d/z^2.$$

The emitted-electron current is obtained by integrating over $v_r > 0$ and

$$v_r^2 + v_\theta^2 \geqslant 2e\phi_d/m_e. \tag{13.54}$$

It leads to

$$I_{em} = 4\pi r_d^2 J_{em}\left(1 + \frac{e\phi_d}{T_{em}} \right) \exp\left(-\frac{e\phi_d}{T_{em}} \right). \tag{13.55}$$

Once again, one can solve for the dust potential via the floating condition (13.41) without solving Poisson's equation. The emitted-electron power is

$$h_{em} = -T_{em}\left(\frac{2 + \dfrac{e\phi_d}{T_{em}}}{1 + \dfrac{e\phi_d}{T_{em}}} + \frac{e\phi_d}{T_{em}} \right) \frac{I_{em}}{e}. \tag{13.56}$$

13.4.2 Transition from negatively to positively charged states

It was shown in equation (13.31) that in the conventional OML theory for non-emitting dust, the dust charge is linearly proportional to the dust potential. This implies that the dust charge vanishes when $\phi_d = 0$. For a dust that emits electrons, the transition from negative to positive dust charge occurs at a dust potential that is less than zero, which we denote as $\phi_d^* < 0$. In this section, we contrast the prediction of a modified OML theory, which includes the contribution of emitted electrons, with the first-principles kinetic simulations to elucidate the physics underlying ϕ_d^*.

PIC simulations of dust–plasma interaction with electron emission
We present some PIC simulations (which are equivalent to the OM theory) to contrast with the OML approximation described above [62]. We consider a system with a spherical dust grain of radius $r_d/\lambda_{\mathrm{De}} = 1$ ($\lambda_{\mathrm{De}} = \sqrt{\varepsilon_0 T_e/(n_\infty e^2)}$) interacting with a collisionless, unmagnetized hydrogen plasma ($m_i/m_e = 1836$). At time $t = 0$ the plasma is Maxwellian with $T_i/T_e = 1$ and the outer domain is at $R/\lambda_{\mathrm{De}} = 10$. The dust emits electrons thermionically (see equation (13.38)). We fix $T_{\mathrm{em}}/T_e = 0.03$ and vary $\hat{J}_{\mathrm{em}} = \dfrac{J_{\mathrm{em}}}{e n_\infty \sqrt{T_e/m_e}}$ parametrically to investigate negatively and positively charged regimes. The smallest time step in the simulations is $\Delta t \omega_{pe} = 0.0125$ ($\omega_{pe} = \sqrt{e^2 n_\infty/(\varepsilon_0 m_e)}$ is the electron plasma frequency).

Figure 13.4 (left) shows the steady-state dust potential from OML (i.e. solving equation (13.41)) and from PIC. For PIC, the dust potential is averaged over the last quarter of the simulation, once the system has reached a steady state, to reduce statistical noise. For small values of $\hat{J}_{\mathrm{em}}$ the grain is negatively charged and the potential asymptotes to the value obtained from the expressions derived in section 13.2, $e\phi_d/T_e = -2.5$. As the emitted current is increased, the dust potential starts to decrease

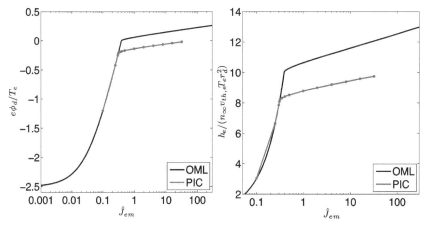

Figure 13.4. The dust potential (left) and power collected from the background electrons (right) from OML theory and PIC simulations as a function of the emitted-electron current $\hat{J}_{\mathrm{em}}$. Adapted from [62].

(in magnitude) significantly since the grain is less negatively charged. At a critical current the two curves bend, signaling that the dust is now positively charged and part of the emitted electron population is attracted back to the grain. One can see that OML and PIC agree very well in the negatively charged regime. However, OML misses the transition to the positively charged regime (which in OML corresponds to $\phi_d = 0$). For PIC, this transition occurs at a negative dust potential, $e\phi_d^*/T_e = -0.22$. Along the same lines, figure 13.4 shows the power collected by the dust grain from the background electrons, indicating that OML overestimates this quantity by ∼30% for the parameters considered. This is a particularly important result, since it is well known that in the positively charged regime the heating of the dust grain is dominated by the background electrons. This can quickly lead to dust destruction via a positive feedback [14], and might affect dust survival in applications such as magnetic fusion energy. The reason for the discrepancy between OML and PIC resides in the presence of a non-monotonic shielding potential (a potential well, also known as a virtual cathode) [62, 63], which invalidates the assumption of OML. This can be seen in figure 13.5 which shows the steady-state shielding potential obtained by PIC for various values of $\hat{J}_{\mathrm{em}}$. We note that the importance of potential-well effects increases if, for instance, one increases the dust radius (while the other parameters remain fixed). This can be seen in figure 13.6, again showing the dust potential (left) and the power collected from the background electrons (right) varying the dust radius for the case $\hat{J}_{\mathrm{em}} = 2$. While the OML prediction is independent of the dust radius, the discrepancy between PIC and OML widens as the dust radius increases. For instance, for $r_d/\lambda_{\mathrm{De}} = 4$ OML overestimates the power collected from the background electrons by ∼50%. Finally, the relation between the dust charge and potential given by the vacuum capacitance no longer holds, as the screening length is dominated by the emitted electrons.

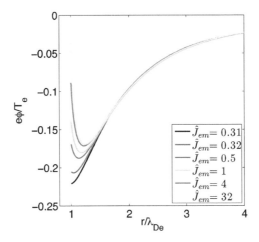

Figure 13.5. The shielding potential varying the emitted-electron current. Adapted from [62].

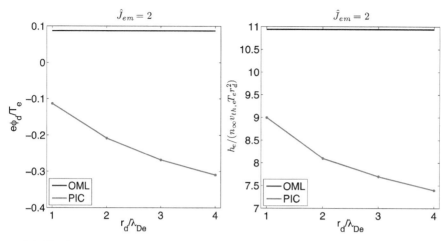

Figure 13.6. The dust potential (left) and power collected from the background electrons (right) from OML theory and PIC simulations as a function of dust radius ($\hat{J}_{em} = 2$).

Critical ϕ_d^* for transition from negatively to positively charged states
In the positively charged regime the slowest emitted electrons can be attracted back to the grain, creating an electron population confined near the grain and the development of a non-monotonic shielding potential. Let us examine this case from the point of view of the OM theory, assuming that the shielding potential has a minimum at $r = r_{\min}$ which is labeled as $\phi_{\min}$. The emitted electrons experience potential barriers to their motion. The position of the extremum of U is still given by equation (13.47). Clearly, $\phi' < 0$ for $r_d \leqslant r_m \leqslant r_{\min}$ and there is only one solution to equation (13.47) giving $r_m = r_m(J)$ corresponding to a maximum of $U(r)$. Also, $J = 0$ corresponds to $r_m = r_{\min}$. No solution exists for $r > r_{\min}$, since $\phi' > 0$, corresponding to a monotonically decreasing effective potential. One can define a *critical angular momentum* above which all the emitted electrons leave the dust grain because of the strong centrifugal force:

$$J^* = \sqrt{-er_d^3 m_e \phi'(r_d)}. \tag{13.57}$$

Again, the emitted electrons that are able to escape the grain attraction and contribute to the emitted current are those with enough energy to satisfy $E \geqslant U^{\max}(r_m(J), J)$, equation (13.46).

The implications of equation (13.46) can be understood if one considers the limit where $J^* \gg m_e r_d v_{\text{th,em}}$. In this case, since the emitted electrons are Maxwellian, most of them are characterized by $J \leqslant m_e r_d v_{\text{th,em}}$ and see the potential barrier located at $r_m \sim r_{\min}$. Thus one can reduce expression (13.46) to a somewhat more manageable expression

$$E \geqslant U(r_{\min}, J) \longrightarrow v_r^2 + \left[1 - \left(\frac{r_d}{r_{\min}}\right)^2\right]v_\theta^2 \geqslant \frac{2e}{m_e}(\phi_d - \phi_{\min}), \tag{13.58}$$

corresponding to an ellipsoid in phase space with aspect ratio given by $1 - (r_d/r_{min})^2$. When compared to the OML condition for the emitted-electron current (13.54), it is clear that if the shielding potential has a deep $(|\phi_d - \phi_{min}| \gg |\phi_d|)$ and localized (near the grain, $r_{min} \sim r_d$) potential well, then the emitted-electron current can be significantly different from that predicted by OML. Note that a full OM theory for non-monotonic shielding potentials (with $\phi_d > 0$) has been developed in [64].

Furthermore, a consequence of the presence of non-monotonic potentials is that the negatively-to-positively charged transition $Q_d = 0$ does not occur for $\phi_d = 0$. In fact, in section 13.2 we have discussed that, in general, the dust charge depends on the space-charge distribution outside the grain, equations (13.30) and (13.31), and that this requires the solution of the non-linear Poisson equation. Motivated by the good agreement between PIC and OML in the negatively charged regime, we solve the OML Poisson's equation in this regime to calculate the value of the critical current (or dust potential) corresponding to $Q_d = 0$. Thus, we solve equation (13.40) with the JFNK method discussed in section 13.3, with the emitted-electron density given by equation (13.43). The boundary conditions are given by the floating condition (13.41) at $r = r_d$ and by the condition $\phi'(R) = 0$. An inspection of the plasma densities, equations (13.11), (13.21), (13.24), and (13.43), reveals that the system depends on the following dimensionless parameters: $\frac{m_e}{m_i}$, $\frac{T_i}{T_e}$, $\frac{T_{em}}{T_e}$, $\hat{J}_{em}$, while the dependence on the dust radius comes through the floating condition (13.41). We fix $m_i/m_e = 1836$, consider four representative values of T_i/T_e (=0.001, 0.1, 1, 10), and vary parametrically T_{em}/T_e and r_d/λ_{De}. The emitted-electron current density is obtained by imposing the additional condition $Q_d = 0$ and we label quantities corresponding to the negatively to positively charged transition with the superscript *. Once $\hat{J}_{em}^*$ is obtained, ϕ_d^* follows from the floating condition (13.41).

The results are presented in figure 13.7, showing the normalized current density $\hat{J}_{em}^*$, the dust potential ϕ_d^* and γ^*, the ratio between the emitted current and the background electron current $\gamma^* = \frac{I_{em}^*}{I_e(\phi_d^*)}$, a quantity that is often used in the literature to characterize electron emission. One can see in figure 13.7 that $\hat{J}_{em}^*$ varies between $0.2 - 0.4$, quite insensitive to T_i/T_e. Interestingly, γ^* has a rather weak dependence on r_d/λ_{De} and T_{em}/T_e: it varies between $0.98 - 0.995$, $0.96 - 0.99$, $0.94 - 0.98$, and $0.86 - 0.93$ for $T_i/T_e = 0.01, 0.1, 1, 10$. Most importantly, the negatively to positively charged grain transition occurs with a negative dust potential. Only for $r_d/\lambda_{De} \ll 1$ and for smaller values of T_i/T_e, the dust potential remains quite small and the OML approximation is reasonable. On the other hand, for large values of r_d/λ_{De} and small T_{em}/T_e, figure 13.7 shows that $e\phi_d^*/T_e$ can be sizable, with values in the range -0.3 to -0.6 for $T_i/T_e \geqslant 0.1$. The dust potential is more negative when T_i/T_e is higher, consistent with the fact that higher T_i/T_e implies a higher ion current and therefore less emitted current is necessary to reach $Q_d = 0$. Note also that for the parameters of section 13.4.2, $r_d/\lambda_{De} = 1$ and $T_{em}/T_e = 0.03$, this procedure gives $e\phi_d^*/T_e = -0.22$ and $\hat{J}_{em}^* = 0.31$, in excellent agreement with figure 13.4. These results

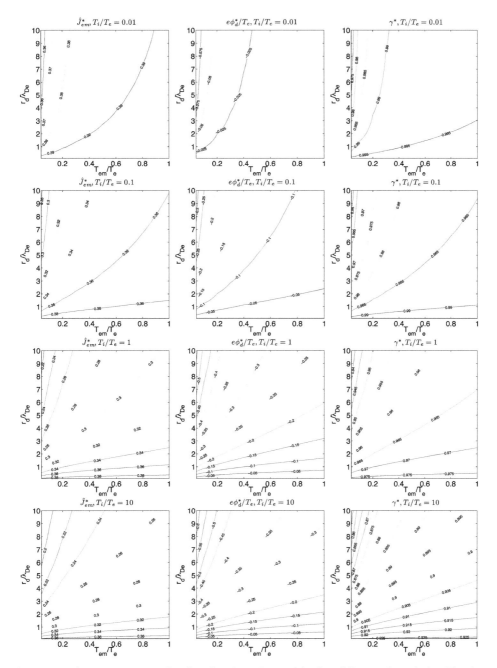

Figure 13.7. The case $Q_d = 0$ as a function of T_{em}/T_e, r_d/λ_{De} and for four different values of T_i/T_e. The plots show the value of the emitted current J_{em}^*, the dust potential ϕ_d^*, and the ratio between the emitted current and the background electron current defined as γ^*.

confirm that, depending on the regime of parameters, the OML approximation can be inaccurate in capturing the negatively to positively charged transition. Most importantly, ϕ_d^* can be used as a proxy to signal when potential-well effects are important and the OML theory becomes inadequate.

13.5 Final remarks

In this chapter we have reviewed the basics of the theory of charging and shielding of a spherical object in a collisionless, unmagnetized plasma, one of the building blocks for studies of plasma–material interaction including dusty-plasma physics. We have derived the basic equations in the OML approximation and contrasted them with first-principles PIC simulations. While PIC simulations of dust–plasma interaction can be computationally expensive depending on the scale separation in the system, the beauty of OML is that it delivers a simple and fast solution where Poisson's equation does not even need to be solved. For this reason, the OML theory is the basis of all existing codes to study dust transport in a plasma. Although OML is traditionally accepted in the limit of $r_d/\lambda_D \ll 1$ (a regime which is very common in space and astrophysics), PIC simulations show that in the negatively charged regime it can be quite accurate for $r_d/\lambda_D \gtrsim 1$ (a regime of interest for laboratory applications). In the positively charged regime driven by dust electron-emission processes, on the other hand, OML can become very inaccurate when $r_d/\lambda_D \gtrsim 1$ because of the development of a non-monotonic shielding potential. This also implies that the dust collected power from the background electrons can be significantly reduced relative to OML and that electron emission can play a major role in determining the dust survivability in a plasma.

As new emerging applications continue to drive the development of the field, it is clear that further developments in dust–plasma interaction theories are necessary. This is particularly true if one considers the necessity of bridging the microscopic scales associated with dust charging to the macroscopic scales of dust transport, which makes OML-like models very popular. For instance, in magnetic fusion energy applications and recent laboratory experiments with high magnetic fields, the electron gyroradius can become comparable or smaller than the characteristic dust size, thus affecting the charging currents (particularly in the positively charged regime). Furthermore, in many applications the dust grains are far from spherical: a variety of shapes arises depending on the dust formation processes. Whether OML-like models can be formulated to capture these non-OML effects (including the anisotropy induced by the presence of a directed ion flow) accurately is at this point an open question. Furthermore, the interaction of plasmas with biological matter, at the hearth of the infant field of plasma medicine, is already showing great promise for the treatment of diseases such as cancer, opening new perspectives and a host of opportunities at the intersection of plasma physics, atomic physics, and chemistry.

In summary, this is an exciting time for researchers interested in plasma–material interaction problems, and in particular for dust–plasma interaction.

Bibliography

[1] Hutchinson I H 2005 *Principles of Plasma Diagnostics* (Cambridge: Cambridge University Press)

[2] Irving Langmuir A F D and Found C G 1924 A new type of electric discharge: the streamer discharge *Science* **60** 392–4

[3] Mendis D A and Rosenberg M 1994 Cosmic dusty plasma *Ann. Rev. Astron. Astrophys* **32** 419–63

[4] Goertz C K 1989 Dusty plasmas in the solar system *Rev. Geophys.* **27** 271–92

[5] Verheest F 2000 *Waves in Dusty Space Plasmas* (Dordrecht: Kluwer)

[6] Smith B A *et al* 1982 A new look at the Saturn system—the Voyager 2 images *Science* **215** 504–37

[7] Selwyn G S, Haller K L and Patterson E F 1993 Trapping and behavior of particulates in a radio frequency magnetron plasma etching tool *J. Vacuum Sci. Technol.* A **11** 1132–5

[8] Bouchoule A 1999 *Dusty Plasmas: Physics, Chemistry, and Technological Impacts in Plasma Processing* (New York: Wiley)

[9] Chu, L I J H 1994 Direct observation of coulomb crystals and liquids in strongly coupled RF dusty plasmas *Phys. Rev. Lett.* **72** 4009–12

[10] Thomas H, Morfill G E, Demmel V, Goree J, Feuerbacher B and Möhlmann D 1994 Plasma crystal: coulomb crystallization in a dusty plasma *Phys. Rev. Lett.* **73** 652–5

[11] Fortov V E, Khrapak A G, Khrapak S A, Molotkov V I and Petrov O F 2004 Dusty plasmas *Phys. Usp.* **47** 447–92

[12] Krasheninnikov S I, Smirnov R D and Rudakov D L 2011 Dust in magnetic fusion devices *Plasma Phys. Control. Fusion* **53** 083001

[13] Ticos C M, Wang Z, Delzanno G L and Lapenta G 2006 Plasma dragged microparticles as a method to measure plasma flows *Phys. Plasmas* **13** 103501

[14] Smirnov R D, Pigarov A Y, Rosenberg M, Krasheninnikov S I and Mendis D A 2007 Modelling of dynamics and transport of carbon dust particles in tokamaks *Plasma Phys. Control. Fusion* **49** 347–71

[15] Bacharis M, Coppins M and Allen J E 2010 Critical issues for modeling dust transport in tokamaks *Phys. Rev.* E **82** 026403

[16] Ratynskaia S, Vignitchouk L, Tolias P, Bykov I, Bergsåker H, Litnovsky A, den Harder N and Lazzaro E 2013 Migration of tungsten dust in tokamaks: role of dust–wall collisions *Nucl. Fusion* **53** 123002

[17] Delzanno G L and Tang X Z 2014 Survivability of dust in tokamaks: dust transport in the divertor sheath *Phys. Plasmas* **21** 022502

[18] Vignitchouk L, Tolias P and Ratynskaia S 2014 Dust–wall and dust–plasma interaction in the MIGRAINe code *Plasma Phys. Control. Fusion* **56** 095005

[19] Rudakov D *et al* 2009 Dust studies in DIII-D and TEXTOR *Nucl. Fusion* **49** 085022

[20] Litnovsky A *et al* 2013 Dust investigations in TEXTOR: impact of dust on plasma–wall interactions and on plasma performance *J. Nucl. Mater.* **438** S126–32

[21] Ratynskaia S *et al* 2015 Elastic–plastic adhesive impacts of tungsten dust with metal surfaces in plasma environments *J. Nucl. Mater.* **463** 877–80

[22] Tolias P *et al* 2016 Dust remobilization in fusion plasmas under steady state conditions *Plasma Phys. Control. Fusion* **58** 025009

[23] Fridman A and Friedman G 2013 *Plasma Medicine* (New York: Wiley)

[24] Keidar M 2015 Plasma for cancer treatment *Plasma Sources Sci. Technol.* **24** 033001

[25] Thomas E Jr., Merlino R L and Rosenberg M 2012 Magnetized dusty plasmas: the next frontier for complex plasma research *Plasma Phys. Control. Fusion* **54** 124034

[26] Shukla P K and Mamun A A 2011 *Introduction to Dusty Plasma Physics* (Philadelphia, PA: Institute of Physics)

[27] de Angelis U 1992 The physics of dusty plasmas *Phys. Scr.* **45** 465

[28] Tsytovich V N and de Angelis U 1999 Kinetic theory of dusty plasmas. I. General approach *Phys. Plasmas* **6** 1093–106

[29] Schram P P J M, Trigger S A and Zagorodny A G 2003 New microscopic and macroscopic variables in dusty plasmas *New J. Phys.* **5** 27

[30] Tolias P, Ratynskaia S and de Angelis U 2011 Kinetic models of partially ionized complex plasmas in the low frequency regime *Phys. Plasmas* **18** 073705

[31] Allen J 1992 Probe theory—the orbital motion approach *Phys. Scr.* **45** 497–503

[32] Al'pert Y L, Gurevich A V and Pitaevskii L P 1965 *Space Physics with Artificial Satellites* (New York: Plenum)

[33] Kennedy R and Allen J 2003 The floating potential of spherical probes and dust grains. II: orbital motion theory *J. Plasma Phys.* **69** 485–506

[34] Mott-Smth H and Langmuir I 1926 The theory of collectors in gaseous discharges *Phys. Rev.* **28** 727–63

[35] Tang X-Z and Delzanno G L 2014 Orbital-motion-limited theory of dust charging and plasma response *Phys. Plasmas* **21** 123708

[36] Bystrenko T and Zagorodny A 2002 Effects of bound states in the screening of dust particles in plasmas *Phys. Lett.* A **299** 383–91

[37] Allen J, Annaratone B and de Angelis U 2000 On the orbital motion limited theory for a small body at floating potential in a Maxwellian plasma *J. Plasma Phys.* **63** 299–309

[38] Daugherty J, Porteous R, Kilgore M D and Graves D 1992 Sheath structure around particles in low-pressure discharges *J. Appl. Phys.* **72** 3934–42

[39] Lampe M 2001 Limits of validity for orbital-motion-limited theory for a small floating collector *J. Plasma Phys.* **65** 171–80

[40] Willis C T N, Coppins M, Bacharis M and Allen J E 2011 The effect of dust grain size on the floating potential of dust in a collisionless plasma *Plasma Sources Sci. Technol.* **19** 065022

[41] Delzanno G L and Tang X-Z 2015 Comparison of dust charging between orbital-motion-limited theory and particle-in-cell simulations *Phys. Plasmas* **22** 113703

[42] Kelley C T 1995 *Iterative Methods for Linear and Nonlinear equations* (Philadelphia, PA: SIAM)

[43] Birdsall C K and Langdon A B 2004 *Plasma Physics Via Computer Simulation* (London: Taylor and Francis)

[44] Hockney R and Eastwood J 1988 *Computer Simulation Using Particles* (London: Taylor and Francis)

[45] Koziel S, Leifsson L, Lees M, Krzhizhanovskaya V V, Dongarra J, Sloot P M and Miloch W 2015 Simulations of several finite-sized objects in plasma *Proc. Comput. Sci.* **51** 1282–91

[46] Patacchini L and Hutchinson I H 2007 Ion-collecting sphere in a stationary, weakly magnetized plasma with finite shielding length *Plasma Phys. Control. Fusion* **49** 1719

[47] Lapenta G 2011 DEMOCRITUS: an adaptive particle in cell (PIC) code for object-plasma interactions *J. Comput. Phys.* **230** 4679–95

[48] Taccogna F 2012 Dust in plasma I. Particle size and ion-neutral collision effects *Contrib. Plasma Phys.* **52** 744–55

[49] Delzanno G L, Camporeale E, Moulton J D, Borovsky J E, MacDonald E A and Thomsen M 2013 CPIC: a curvilinear particle-in-cell code for plasma–material interaction studies *IEEE Transa. Plasma Sci.* **41** 3577

[50] Sorasio G, Mendis D and Rosenberg M 2001 The role of thermionic emission in meteor physics *Planet. Space Sci.* **49** 1257–64

[51] Kimura H and Mann I 1998 The electric charging of interstellar dust in the solar system and consequences for its dynamics *Astrophys. J.* **499** 454

[52] Horányi M and Cravens T E 1996 The structure and dynamics of Jupiter's ring *Nature* **381** 293–5

[53] Klumov B, Vladimirov S and Morfill G 2007 On the role of dust in the cometary plasma *JETP Lett* **85** 478–82

[54] Poppe A, Halekas J S and Horányi M 2011 Negative potentials above the day-side lunar surface in the terrestrial plasma sheet: evidence of non-monotonic potentials *Geophys. Res. Lett.* **38** L02103

[55] Hastings D and Garrett H 2004 *Spacecraft–Environment Interactions* (Cambridge: Cambridge University Press)

[56] Bruining H 1962 *Physics and Applications of Secondary Electron Emission* (London: Pergamon)

[57] Wang X, Pilewskie J, Hsu H-W and Horanyi M 2016 Plasma potential in the sheaths of electron-emitting surfaces in space *Geophys. Res. Lett.* **43** 525–31

[58] MacDonald E A, Lynch K A, Widholm M, Arnoldy R, Kintner P M, Klatt E M, Samara M, LaBelle J and Lapenta G 2006 *In situ* measurement of thermal electrons on the SIERRA nightside auroral sounding rocket *J. Geophys. Res.: Space Phys.* **111** A12310

[59] Meyer-Vernet N 1982 Flip-flop of electric potential of dust grains in space *Astron. Astrophys.* **105** 98–106

[60] Tolias P 2014 On secondary electron emission and its semi-empirical description *Plasma Phys. Control. Fusion* **56** 123002

[61] Ashcroft N and Mermin N 1976 *Solid State Physics* (Philadelphia, PA: Saunders College)

[62] Delzanno G L and Tang X Z 2014 Charging and heat collection by a positively charged dust grain in a plasma *Phys. Rev. Lett.* **113** 035002

[63] Delzanno G L, Lapenta G and Rosenberg M 2004 Attractive potential around a thermionically emitting microparticle *Phys. Rev. Lett.* **92** 350021

[64] Delzanno G L, Bruno A, Sorasio G and Lapenta G 2005 Exact orbital motion theory of the shielding potential around an emitting, spherical body *Phys. Plasmas* **12** 062102

IOP Publishing

Plasma Modeling
Methods and Applications
Ivo Furno, Paolo Ricci, Ambrogio Fasoli, Fabio Riva and Christian Theiler

Chapter 14

Verification and validation in plasma physics

The goal of the verification and validation (V&V) procedure is to assess that the physics models are correctly implemented in numerical codes (code verification), to estimate the numerical errors affecting the simulation results (solution verification), and to establish the consistency of the code results, and thus of the physics model, with experimental data (validation). This chapter discusses the methodology to carry out a rigorous V&V exercise and shows, through a practical example, how this can contribute to advancing our physics understanding of plasma turbulence. The method of manufactured solutions (MMS) is used to assess that the model equations are correctly solved within the order of accuracy of the numerical scheme. The technique to carry out a solution verification is described to estimate the uncertainty affecting the numerical results. A methodology for plasma turbulence code validation is also discussed, focusing on quantitative assessment of the agreement between experiments and simulations. The V&V methodology is applied to the study of plasma turbulence in the basic plasma physics experiment the 'Toroidal Plasma Experiment' (TORPEX) [19], considering both two-dimensional and three-dimensional simulations carried out with the GBS code [21, 132]. The validation procedure allows progress in the understanding of the turbulent dynamics in TORPEX, by pinpointing the presence of a turbulent regime transition due to the competition between the resistive and ideal interchange instabilities.

14.1 Introduction

Errors affecting the simulations used to improve our understanding and to describe the dynamics of plasmas can have far-reaching consequences. To limit these errors, which can be due both to mistakes in the code and to an incomplete or inaccurate physics model, there is an increasing motivation in plasma physics to use V&V procedures [1–3]. V&V is composed of the *code verification* process, the *solution verification* procedure, and the *validation*.

doi:10.1088/978-0-7503-1200-4ch14 © IOP Publishing Ltd 2016

The code verification procedure is a mathematical and numerical procedure aimed at assessing that the physics model is correctly implemented in a numerical simulation code, with the accuracy expected by the numerical algorithm. The goal of the code verification procedure is therefore to assess that no mistakes are present in the simulation code.

Due to finite computational power and the related finite precision achievable, the simulation results are always affected by numerical errors, even if the model equations are implemented correctly. Estimating the amplitude of the numerical errors is crucial to ensure the reliability of the numerical results. The estimate of the numerical error affecting the simulations constitutes the solution verification procedure [2, 3, 14, 15].

Finally, validation is used to assess the consistency of the code results, and therefore of the physics model, with experiments. Validation requires the comparison of the simulation results with experimental measurements. The V&V procedure is visualized in figure 14.1.

The purpose of the present chapter is two-fold. First, we summarize the methodology for code verification, simulation verification, and validation (section 14.2). Second, in section 14.3, we describe the V&V effort that we have carried out in recent years to study the plasma dynamics in the basic plasma physics experiment TORPEX [19, 20] by using the GBS code [21, 132]. This exemplifies the application of the proposed methodology. Owing to its detailed diagnostics, the possibility of parameter scans, and relative simple configuration, TORPEX is an ideal testbed to perform experiment/simulation comparisons and to investigate the corresponding methodological framework. Three models have been considered for the simulation, all based on the fluid equations developed by Braginskii for collisional plasmas and simplified by neglecting the fast particle gyration, considering therefore the drift limit of those equations: (a) a three-dimensional two-fluid model, that is able to describe the global evolution of the TORPEX plasma [22, 23]; (b) the same three-dimensional model completed by an appropriate first-principles set of boundary conditions that has been recently derived [24]; and (c) a reduced two-dimensional two-fluid model [25, 26], that is able to describe only the evolution of $k_{\parallel} = 0$ modes.

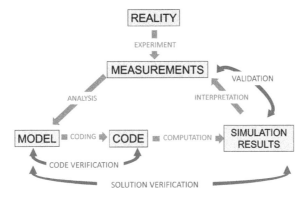

Figure 14.1. Schematic representation of the V&V procedure.

The three models are implemented in the GBS code. We therefore discuss the verification of the GBS code and of its results. We then compare the simulation results with the experimental measurements, showing that the validation metric is able to point out that the agreement between the two-dimensional model and the experiment is no longer satisfactory when $k_{\parallel} \neq 0$ modes are present in the experiment [27–29, 133]. The conclusions follow in section 14.4.

14.2 The V&V methodology

14.2.1 Code verification methodology

To perform a code verification, one can [2] (a) perform simple tests (e.g. verify that the code conserves energy), (b) compare codes with each other (this is known as code-to-code benchmarking), (c) quantify the discretization error with respect to a known solution, (d) check the convergence of the numerical result to a known solution, and (e) prove that the numerical solution converges to the analytical one at the rate expected for the numerical scheme (order-of-accuracy tests). As they do not require an analytical solution, the first two techniques ((a) and (b)) are the simplest to use and, in fact, code-to-code benchmarks are routinely used in plasma physics (see, e.g. [4–10]). While valuable, this exercise requires that at least one code is fully verified and, generally, it is very difficult to understand if the difference in the code results is due to discretization errors or to an incorrect implementation of the model. Only the three other approaches ((c)–(e)) are rigorous and, in particular, the order-of-accuracy test is the only one that can quantitatively assess the correct implementation of the model equations and of the numerical scheme, verifying that the discretization errors reduce at the rate expected for the numerical scheme as the spatial mesh and the time step are refined. Order-of-accuracy tests are performed in other scientific domains, such as computational fluid dynamics. The MMS [11–13] allows order-of-accuracy tests even when an analytical solution is not available for the physics model considered.

Given a theoretical model M with an analytical solution s, such that $M(s) = 0$, and the numerically discretized model of M, M_h, with a numerical solution s_h that satisfies $M_h(s_h) = 0$ (h is a parameter representing the degree of refinement of the mesh), the error affecting the numerical results is expressed as $\epsilon_h = \|s_h - s\|$, where $\|\cdot\|$ denotes a designed norm. The theoretical order of accuracy, p, associated with the numerically discretized operator M_h, represents the rate at which the numerical solution converges to the analytical solution as the mesh is refined. The numerical error satisfies the relation $\epsilon_h = C_p h^p + O(h^{p+1})$, where C_p is independent of h, and p is the order of accuracy of the numerical scheme, typically evaluated through its Taylor expansion [2, 3, 13]. With the two numerical solutions of M_h and M_{rh}, i.e. s_h and s_{rh}, where rh indicates coarsening the h mesh by a factor r, we evaluate the observed order of accuracy

$$\hat{p} = \frac{\ln(\epsilon_{rh}/\epsilon_h)}{\ln(r)}. \tag{14.1}$$

If $\hat{p}$ converges to p for $h \to 0$, i.e. when the discretization error is dominated by the lowest-order term in the expansion (the so-called asymptotic regime), we can state that the code is verified and the equations are correctly solved, with the order of accuracy expected for the numerical scheme.

The main issue related to the systematic evaluation of $\hat{p}$ is the need for the analytical solution s to compute the numerical error ϵ_h; in fact, the analytical solution is unknown in most cases. The MMS, developed to overcome this problem [11–13], instead of solving analytically a theoretical model, imposes a solution to the model, the so-called manufactured solution, and modifies the model equations to accommodate the imposed solution. The obtained modified model is then solved to compute the discretized error. More precisely, for a given model M, we choose an analytical function u and we compute a source term, $S = M(u)$, which is subtracted from M to obtain a new analytical model $G = M - S$, whose analytical solution is u ($G(u) = M(u) - S = 0$). The discretization of the new model is straightforward, i.e. $G_h = M_h - S$. Since the source term S is computed analytically, we do not add any new discretization errors to the numerical model considered. The behavior of the numerical error is preserved. This can be expressed as: $\epsilon_h = \|u_h - u\| = D_p h^p + O(h^{p+1})$. In practice, using the MMS for an order-of-accuracy test implies adding source terms to the discretized equations, performing a simulation scan to obtain the observed order of accuracy, and comparing the observed order of accuracy to the theoretical one to verify the code.

Although the idea behind the MMS is simple, its implementation requires some subtleties. First, the manufactured solution should: (a) be smooth enough and not singular, (b) be general enough to excite all the terms present in the equations, (c) satisfy the code constraints (e.g. positivity for the density or the temperature), and (d) ensure that the magnitude of the different terms composing the equations are of the same order of magnitude. The manufactured solutions are usually built as a combination of trigonometric and/or hyperbolic functions. We note that the MMS cannot be applied to codes used to model singularities, shocks, or discontinuities; the verification of these codes is still an open issue [2].

14.2.2 Solution verification methodology

Even when a model is correctly implemented in a code, we note that simulations are affected by numerical errors, which should be evaluated through a solution verification procedure [2, 3, 14, 15]. The sources of numerical error are: (a) round off errors due to the finite number of digits carried over by a computer, (b) statistical sampling errors, (c) errors associated with iterative numerical methods stopped before they reach full convergence, and (d) discretization errors due to the finite grid spacing. While all these errors have to be evaluated, if iterative methods are applied using a sufficiently large number of iterations, and simulations are run long enough to decrease the statistical errors, discretization errors typically dominate over the other sources. Focusing on these errors, for grid-based algorithms the Richardson extrapolation can be used as higher-order estimator of the solution, as well as the Roache grid convergence index as a relative numerical uncertainty estimate [16].

In the early 20th century, Richardson developed a method [30, 31], later extended [32, 33], to accelerate the rate of convergence of a numerical sequence. This method is based upon the use of two numerical solutions obtained using two different meshes, s_h and s_{rh}, to compute an estimate of the analytical solution that presents a convergence rate one order higher than the original numerical solution. Concretely, the Richardson extrapolation is defined as

$$\bar{s} = s_h + \frac{s_h - s_{rh}}{r^p - 1}, \tag{14.2}$$

where p is the formal order of accuracy defined in section 14.2.1. Noting that $\|s_h - s\| = C_p h^p + O(h^{p+1})$, it follows that the extrapolated solution $\bar{s}$ satisfies $\|\bar{s} - s\| = D_p h^{p+1} + O(h^{p+2})$, where D_p is independent of h. Therefore, for $h \to 0$, $\bar{s} \to s$ faster than the numerical solutions obtained from the simulations. Consequently, we can use $\bar{s}$ as an estimate of the exact solution s, and approximate the numerical error with

$$\epsilon_h \simeq \|s_h - \bar{s}\| = \left\| \frac{s_{rh} - s_h}{r^p - 1} \right\|. \tag{14.3}$$

The relative discretization error, RDE, is therefore approximated as

$$RDE = \frac{s_h - s}{s} \simeq \frac{s_h - \bar{s}}{\bar{s}} = \frac{s_{rh} - s_h}{s_h r^p - s_{rh}}. \tag{14.4}$$

For $\bar{s}$ to be a reasonable estimate of s, however, several assumptions should be satisfied. First, the Richardson extrapolation method requires the use of uniform mesh spacing, meaning that the degree of the refinement of the meshes can be represented solely by the parameter h discussed before. Therefore, s_h and s_{rh} should be computed over two meshes that are the uniform systematic refinement of the other. Consequently, the application of Richardson extrapolation to computations involving local mesh refinement or mesh adaptation is not allowed. Second, the simulations used to evaluate $\bar{s}$ should be in the asymptotic regime, meaning that the discretization error is dominated by its lower-order term, $C_p h^p$. This requirement could result in computationally very expensive simulations, due to the potential need of very fine meshes. Third, to apply the method presented above for the estimate of the numerical error, the solutions should be smooth enough, with no singularities and/or discontinuities. To allow the expansion of the numerical error in powers of the parameter h, the derivatives of the analytical solution should exist and be continuous. Moreover, we should note that we do not have any guarantee that the Richardson extrapolated solution will meet the same governing equations satisfied by either the numerical solution or the analytical one. Consequently, we use this extrapolation for the computation of the numerical error only.

Usually, it is problematic to satisfy the requirement of being in the asymptotic regime, due to the high computational cost of the simulations. Moreover, it has to be demonstrated that the numerical solutions are in the asymptotic regime, by showing that the observed order of accuracy matches the formal one. This requires at least three simulations, resulting from two subsequent refinements of the coarser mesh of a factor r, from which the observed order of accuracy can be evaluated as

$$\hat{p} = \frac{\ln\left[(s_{r^2 h} - s_{rh})/(s_{rh} - s_h)\right]}{\ln(r)}. \tag{14.5}$$

If only two simulations are available, or if the observed order of accuracy does not match the formal one, we should substitute the numerical error estimates in equations (14.3) and (14.4) with a numerical uncertainty quantification. In general, the error estimate in equations (14.3) and (14.4) may depend strongly on the refinement factor r and on the precision of the numerical scheme. It is therefore difficult to rely on such an error estimate. To overcome these issues, [16] introduces the grid convergence index, GCI, defined as

$$GCI = \frac{F_s}{r^{\tilde{p}}-1}\left|\frac{s_{rh} - s_h}{s_h}\right|, \qquad (14.6)$$

which represents an estimate of the relative discretization error affecting the simulation results. The GCI value is obtained by approximating in equation (14.4) $s_h r^p - s_{rh} \simeq (r^{\tilde{p}}-1)s_h$. The factor of safety F_s and $\tilde{p}$ ensure that GCI is larger than the numerical discretization error in 95% of the cases. Oberkampf and Roy [2] propose the following: if $|p - \hat{p}| < 0.1p$, we can assume that the simulation is in the asymptotic regime and we use $F_s = 1.25$, as well as $\tilde{p} = p$. If $|p - \hat{p}| > 0.1p$, a more conservative factor of safety, $F_s = 3$, has to be used and $\tilde{p} = \min[\max(0.5, \hat{p}), p]$. If $\hat{p}$ is not evaluated (for example, if only two solutions are available), $F_s = 3$ and $\tilde{p} = p$ are used. We remark that, although these definitions are reasonable, there still is an ongoing discussion in the verification community about their generality.

To conclude our presentation of the error estimate methodology, we have to discuss a few details. First, we draw attention to the fact that the present procedure can be applied not only to point-by-point solution values, but also to solution functionals. This is important to estimate the numerical error affecting the observables used in the validation of the physical model [3]. Second, as s_h and s_{rh} are in general computed on different meshes, the results on the coarser mesh have to be interpolated on the finer grid, using an interpolation scheme whose order is equal or higher than the order of the numerical scheme used by the code. A complete discussion of this topic is found in [32]. Third, using GCI as an evaluation of the numerical error requires a non-oscillatory convergence of the numerical solution. If oscillatory convergence is observed, the numerical error has to be evaluated from the difference between the obtained numerical solutions. Finally, we illustrate a useful propriety of GCI, that is the possibility of computing the overall GCI analyzing each coordinate of the problem independently. As it can be numerically very expensive to perform a uniform refinement of the grid along all the coordinates at the same time, it is possible to refine each coordinate of the mesh separately by a factor r_i, where the index i refers to the coordinate under investigation. This allows us to compute GCI_i and a $\tilde{p}_i$ for the i coordinate, and obtain the overall GCI as $GCI = \sum_i GCI_i$.

14.2.3 Validation methodology

The guidelines for carrying out the validation between experiments and simulations have been imported from other domains, such as computational fluid dynamics, to plasma physics, and are described in [17, 18]. Simulations and experiments have to be compared considering a number of physical quantities, common to the experimental measurements and simulation results, analyzed using the same techniques. These physical quantities, denoted as validation observables, should be identified and organized into a hierarchy. This hierarchy is based on the number of model assumptions and the combinations of measurements necessary to obtain the observable, i.e. how stringent each observable is for comparison purposes. By combining the results of the comparison of all the observables, while taking into account the position in the hierarchy and precision, the agreement between simulations and experiments needs to be quantified by using an appropriate composite metric, χ. Such a metric χ should be complemented by an index, Q, which assesses the quality of the comparison. Practically, Q is related to the number of observables that are used for the validation and the strength of the constraints they impose. We remark that the validation should take into account the experimental and simulation uncertainties, therefore these have to be accurately evaluated. Sources of experimental uncertainties are the approximations of the models used to interpret the experimental results, the difficulties in the evaluation of the properties of the measuring devices, and the imperfect reproducibility of the experiments. For the simulations, uncertainties are due to numerical errors (evaluated through the solution verification procedure) and to the uncertainty associated with the not-well-known input parameters. The latter can be estimated through a sensitivity analysis.

We point out that the validation procedures should remain simple. The goal is a useful tool that can be easily applied in order to compare different models with experimental results. Through this comparison, it is possible to assess the physics elements that play a role in the dynamics of the system and that should be taken into account for its description. On the other hand, it is very delicate to judge a single model in absolute terms in view of testing its predictive capabilities and this subject is not addressed in this chapter.

Simulations and experiments have to be compared considering a number of physical quantities, common to the experimental measurements and simulation results, and analyzed using the same techniques. These validation observables should be independent of each others, and their resolution is sufficient to describe their variation well.

Once the observables are defined and evaluated, the agreement between experiments and simulations relative to each observable has to be quantified. We denote with e_j and s_j the values of the jth observable used in the comparison, as coming from the experimental measurement or the simulation results, respectively. Most observables depend on space and time, and are given on a discrete number of points, denoted as N_j. We denote with $e_{j,i}$ and $s_{j,i}$ the values of the jth observable at points $i = 1, 2, ..., N_j$ (this notation can therefore be used for zero-, one-, two-, etc, dimensional observables). For the jth observable, we normalize the distance d_j between experiments and simulations to the uncertainty related to these quantities:

$$d_j = \sqrt{\frac{1}{N_j} \sum_{i=1}^{N_j} \frac{(e_{j,i} - s_{j,i})^2}{\Delta e_{j,i}^2 + \Delta s_{j,i}^2}}, \qquad (14.7)$$

where $\Delta e_{j,i}$ and $\Delta s_{j,i}$ are the uncertainties related to the evaluation of $e_{j,i}$ and $s_{j,i}$. Since simulations and experiments can be considered to agree if their difference is smaller than their uncertainties, we define the level of agreement between experiments and simulations with respect to observable j as

$$R_j = \frac{\tanh\left[\left(d_j - 1/d_j - d_0\right)/\lambda\right] + 1}{2}. \qquad (14.8)$$

with $R_j \lesssim 0.5$ corresponding to agreement and $R_j \gtrsim 0.5$ denoting disagreement. The function in equation (14.8) is chosen as it describes a transition, whose sharpness can be changed with λ, from agreement to disagreement when $d_j \simeq d_0$. Here, we choose $d_0 = 1$ and $\lambda = 0.5$; our tests show that the conclusions of a validation exercise are not affected by the specific choices of the parameters d_0 and λ, if these parameters are within the reasonable range that point out agreement between experiments and simulations when they fall within their uncertainties. Some authors prefer to normalize the distance between experimental and simulation results to the actual value of the observables, rather than to their uncertainty [2]. We believe that the normalization to the uncertainty is the most appropriate choice in the present case, as we are interested in understanding if the basic physics mechanisms at play in the system are well captured by the model under consideration. The normalization to the actual value of the observable is instead preferable in the case that the predictive capabilities of the code are tested.

In the case of the experiments, we can identify three main uncertainty sources. First, the model of a measuring device provides predictions through which one can infer the physical quantities of interest (e.g. from the I–V curve of a Langmuir probe (LP) one can infer n and T_e). Experimental measurements typically do not follow perfectly the model predictions: thus, a fit has to be made in order to evaluate the relevant physical parameters, introducing an uncertainty that we denote with $\Delta e_{j,i}^{\text{fit}}$. Second, a source of uncertainty is due to properties of the measuring device that are often difficult to evaluate accurately (e.g. the geometry and surface conditions of an LP). Thus, measurements should be performed with different tools (e.g. LPs which differ in dimension, surface condition, and electronics). The quantity $\Delta e_{j,i}^{\text{prb}}$ denotes the uncertainty related to the probe properties. Finally, the plasmas are not perfectly reproducible due to control parameters that are difficult to set or know precisely (e.g. the vacuum pressure). Experiments should be repeated in order to check the reproducibility of the plasma, while measurements are taken with different measurement devices. The quantity $\Delta e_{j,i}^{\text{rep}}$ is the uncertainty due to the plasma reproducibility, averaged over the different measuring devices. The total experimental uncertainty is given by $\Delta e_{j,i}^2 = (\Delta e_{j,i}^{\text{fit}})^2 + (\Delta e_{j,i}^{\text{prb}})^2 + (\Delta e_{j,i}^{\text{rep}})^2$.

Simulations are also affected by uncertainties resulting from two sources: (i) errors due to the numerics and (ii) errors due to unknown or imprecise input parameters. While errors due to the numerics, $\Delta s_{j,i}^{\text{num}}$, can be estimated through the methodology described in section 14.2.1, the evaluation of the error related to not perfectly known input parameters, $\Delta s_{j,i}^{\text{inp}}$, requires an uncertainty propagation study, i.e. an investigation of how the model results are affected by the input parameter variations. The number of input parameters of a turbulence simulation code is usually quite large and a complete study of the model response is prohibitive. However, the theory can indicate to which input parameters the results are particularly sensitive. The analysis has then to focus on those. We remark that in the literature, a number of useful techniques have been proposed to predict the response of the model to variation of simulation parameters using the smallest possible number of simulations (see, e.g. [34]). As in the case of the experimental error bars, the two sources of error should be added, such that $\Delta s_{j,i}^2 = (\Delta s_{j,i}^{\text{num}})^2 + (\Delta s_{j,i}^{\text{inp}})^2$.

We note that the error bars should not take into account the uncertainties related to model assumptions and/or to combinations of measurements, which are often needed to deduce the comparison observables from the simulation results and the raw experimental data [27]. Evaluating these uncertainties rigorously is usually very challenging. The idea is to take them into account approximately through the observables' primacy hierarchy. More specifically, the higher the hierarchy level of an observable is, the lower the importance of the observable in the comparison metric.

The overall level of agreement between simulations and experiments can be measured by considering a composite metric, which should take into account the level of agreement of each observable, R_j, and weight it according to how constraining each observable is for comparison purposes. This means that the hierarchy level of each observable and the level of confidence characterizing the measurement or the simulation of each observable have to be considered. The higher the level in the primacy hierarchy and the larger the error affecting the observable measurement, the smaller the weight of the observable. We thus define the metric χ as

$$\chi = \frac{\sum_j R_j H_j S_j}{\sum_j H_j S_j},$$

(14.9)

where H_j and S_j are functions defining the weight of each observable according to its hierarchy level and the precision of the measurement, respectively. Thanks to the definition of R_j, χ is normalized in such a way that perfect agreement is observed for $\chi = 0$, while simulation and experiment disagree completely for $\chi = 1$.

The definition of H_j and S_j is somewhat arbitrary. H_j should be a decreasing function of the hierarchy level. The definition we adopt is $H_j = 1/h_j$, where h_j is the combined experimental/simulation primacy hierarchy level, which takes into account the number of assumptions or combinations of measurements used in evaluating the observables both from the experiments and from the simulations. In practice, if no assumptions or combinations of measurements are used for obtaining an observable, $h_j = 1$, any assumption or combination of measurement leads h_j to increase of a unity (see [27] for more details on the h_j definition). The quantity S_j

should be a decreasing function of the experimental and simulation uncertainty. We introduce the following definition:

$$S_j = \exp\left(-\frac{\sum_i \Delta e_{j,i} + \sum_i \Delta s_{j,i}}{\sum_i |e_{j,i}| + \sum_i |s_{j,i}|}\right)$$

(14.10)

such that $S_j = 1$ in the case of zero uncertainty.

The validation metric should be complemented by an index, Q, that assesses the 'quality' of the comparison. The idea is that a validation is more reliable with a larger number of independent observables, particularly if they occupy a low level in the primacy hierarchy and the measurement and simulation uncertainties are low. The quality of the comparison Q can thus be defined as

$$Q = \sum_j H_j S_j.$$

(14.11)

14.3 A practical example of using the V&V methodology

We present an example of the V&V procedure, to which we apply the methodology described in section 14.2. We discuss the simulations of plasma turbulence in the basic plasma physics experiment TORPEX [19, 20] that have been carried out using the GBS code. The GBS code has been developed in the last few years to simulate plasma turbulence in the open field region of magnetic confinement devices, evolving the drift-reduced Braginskii two-fluid equations [35], without any separation between equilibrium and perturbation quantities [21, 132]. In this development, increasingly complex magnetic configurations have been considered: first, the code was developed to describe basic plasma physics devices, in particular, linear devices such as LAPD [36] and simple magnetized toroidal devices such as TORPEX [23], whose simulations are the focus of the present chapter. GBS was then extended to the tokamak geometry, and it is now able to model the tokamak scrape-off layer (SOL) region in limited plasmas (see, e.g. [37–40]).

We first introduce the TORPEX device (section 14.3.1) and then the models used for the simulations of TORPEX and the GBS code (section 14.3.2). GBS has been the subject of a rigorous code verification procedure, described fully in [29, 132], which we briefly summarize in section 14.3.3. The application of the solution verification procedure is the subject of section 14.3.4. Finally, the validation is described in section 14.3.5. We remark that the present example of using the V&V methodology is reported in [133].

14.3.1 The TORPEX device, its diagnostics, and ancillary systems

TORPEX is a highly flexible toroidal device that allows for full diagnostic access and a variety of magnetic configurations. Plasmas of different gases are produced and sustained by microwaves at 2.45 GHz with typical densities of $n_e \approx 10^{15}-10^{17}$ m^{-3} and temperatures $T_e \approx 5-10$ eV. The main TORPEX elements are reviewed in this section.

Figure 14.2. View of the TORPEX device with the main elements. A helical magnetic field line of an SMT configuration is shown in violet. Simulated plasma potential profiles at two toroidal positions are shown. These are obtained from numerical simulations using the GBS code.

Figure 14.2 shows a drawing of the TORPEX device with the main elements. The circular cross-section toroidal vacuum chamber with major radius $R = 1$ m and minor radius $a = 0.2$ m is made of stainless-steel and is divided into twelve sectors, each spanning over 30 degrees. The four-fold symmetrical vessel is divided into four electrically insulated sets, of three sectors each. Four sectors of the vessel, toroidally separated from each other by 90 degrees, are extractable, to facilitate the installation of diagnostics. Spare sectors are available and can be modified for new experiments without stopping the experimental activity. A total of 48 port-holes provide access for diagnostics and pumping. Four turbo-pumps (a nominal pumping speed of 345 l s^{-1} for nitrogen), backed by four primary pumps, maintain a base pressure $\lesssim 5 \times 10^{-5}$ mbar inside the vessel. Gas or a mixture of different gases (argon, hydrogen, deuterium, helium, and neon) are injected at one toroidal position; the injection rate is controlled remotely.

The magnetic field coil system comprises a toroidal, a poloidal, and an ohmic system, allowing for a variety of magnetic configurations (see [88] for details). The toroidal system (figure 14.2) consists of 28 water-cooled coils for the generation of a toroidal magnetic field of up to ~100 mT on-axis. The poloidal coil system includes 10 water-cooled coils and allows for a variety of different vertical magnetic field configurations with a maximum vertical field of 5 mT. The design and arrangement of the poloidal coils are optimized to also generate a cusp field up to 100 mT about 10 cm from the center of the vessel, with a magnetic X-point at the center of the cross section. A subset of the poloidal coils can be used to induce in the vessel a loop voltage of the order of 10 V for approximately 30 ms, to drive a plasma current in a tokamak-like configuration [57]. The combination of a toroidal magnetic field B_φ with a vertical magnetic field B_z results in helical open field lines, terminating on the vacuum vessel, as shown in figure 14.2. This magnetic geometry is known as a *simple magnetized torus* (SMT) [89], and incorporates the main ingredients for drift and interchange instabilities and turbulence, namely pressure gradients, ∇B and magnetic field curvature. Although it does not possess a rotational transform of the magnetic field lines, the SMT mimics the SOL of magnetic confinement devices for fusion, where magnetic field lines are open and terminate either on the vessel wall or on limiters/divertors.

TORPEX plasmas are produced and sustained by microwaves at 2.45 GHz, corresponding to the electron–cyclotron (EC) range of frequencies. The choice of the frequency is based on the constraints imposed by the maximum available toroidal magnetic field, $B_\varphi \approx 0.1$ T, and by the large availability of sources in this frequency range for industrial applications. The microwave system installed on TORPEX consists of a commercial microwave source and a transmission line, which connects the source to the vacuum chamber. The source can deliver up to 5 kW of microwave power in continuous mode, or up to 50 kW during 100 ms in pulsed mode. The microwave power can be modulated with frequencies up to 100 kHz for a sinusoidal waveform, or 20 kHz for a square waveform. The modulation amplitude is arbitrary, with a lower limit for the injected power of 200 W. The injected and reflected power are measured on the transmission line and provide the value of the power absorbed by the plasma. The microwaves are injected perpendicularly to the magnetic field from the low-field side (LFS). A truncated waveguide is used as antenna, which introduces finite $k_\perp$ components in the injected spectrum. The ordinary (O-mode) polarization has been chosen to avoid the cut-off for the extraordinary mode at relatively low densities. A complete description of the TORPEX microwave system can be found in [100].

The plasma production mechanism associated with the injection of microwaves in the EC range of frequencies in STM plasmas has been investigated in great detail [101, 103, 104]. The main contribution to the particle source comes from the electrons accelerated at the EC and upper-hybrid (UH) resonant layers. The electrons in the tail of the thermal distribution give only a negligible contribution to the total ionization rate. The dependence of the plasma source upon experimental parameters, such as the absorbed microwave power, the magnetic field configuration, and the neutral gas pressure, was also investigated and a expression for the spatial profile of the particle source is derived from the experimental results and a simple numerical code. These studies are of paramount importance for the code validation program.

TORPEX possesses an extensive set of diagnostics, comprising approximately 200 probes, together with a fast optical imaging system [75–77], which allow for a full reconstruction of plasma profiles and turbulent structures across the entire device. Measurements of time-averaged plasma parameters and electrostatic fluctuations are performed using combinations of arrays of LPs in single, double, and triple configurations [122], and Mach probes [83]. The LPs can be operated in ion saturation, floating potential, or swept mode and their signals are acquired at a rate of 250 kHz. The LPs are disposed at different toroidal locations and are installed on remotely controlled two-dimensional positioning systems, which allow one to cover almost the entire plasma poloidal cross section.

An array of 85 LPs at one toroidal location, the Hexagonal Turbulence Imaging Probe (HEXTIP), allows the characterization of fluctuations and turbulent structures in terms of wavelength, coherence lengths, frequency, speed, and statistical properties, in a single plasma discharge [91]. The original HEXTIP design with 86 probes was recently modified to be compatible with the new toroidal conductor system. HEXTIP covers the entire poloidal cross section with a spatial resolution of

3.5 cm and constitutes the workhorse for high temporal resolution (250 kHz) real space imaging of the ion saturation current and floating potential dynamical evolution. HEXTIP allows the identification of observables related to turbulent structures, such as size, mass, shape, and trajectories from two-dimensional plasma imaging data. The probability distributions of these observables from measurements of many turbulent realizations can then be used to characterize the nature of turbulence and the related transport. A new version of the HEXTIP array with improved (and variable) spatial resolution has recently been developed. Two new systems are under construction and will be installed in TORPEX at different toroidal locations, thus extending to three dimensions the measurements available with the previous HEXTIP.

Special arrangements of single and multiple single-sided LPs are used to measure plasma currents collected on a poloidal limiter, in particular for blob studies [59]. The simplest version of these probes consists of a conductor plate (usually made out of tungsten) with a collecting area A_{LP}, which is exposed to the plasma only from one side a few cm in front of the limiter. The plate is oriented perpendicularly to the magnetic field and is kept at the limiter potential, such that the current to the limiter, I_0, is now measured by the probe. The current density is then computed as $J_\parallel = I_0/A_{LP}$. A good agreement is found between single-sided probe and magnetic probe measurements [61].

Magnetic and current probes complement electrostatic probes and provide three-dimensional fluctuating components of the magnetic field with nano-Tesla resolution. A specially designed magnetic probe, developed in collaboration with Consorzio RFX in Padova (IT), consists of an L-shaped array of three miniaturized three-axial pick up coils (3.5 cm spacing, an effective area of 2.3×10^{-3} m^{-2}) [59, 61]. The probe is installed on a two-dimensional movable system and allows mapping the magnetic field and current density fluctuations associated with turbulent structures [59, 61].

Together with magnetic probes, direct measurements of the current density are obtained using a dedicated miniaturized current probe diagnostic [53]. The probe consists of a Rogowski coil wound around a ferrite core to amplify the magnetic flux. Electrical and thermal insulation from the plasma is obtained by placing the detecting coil inside a vacuum safe cylindrical housing, machined as two halves of a container. The toroidal detecting coil consists of a single layer helical winding with an inside return loop to cancel the net single turn formed by the solenoid windings. The probe is operated in current mode, which provides a direct measurement of the total current flowing inside the detection area.

14.3.2 The simulation model

Owing to the low TORPEX plasma temperature, the drift-reduced Braginskii equations (see, e.g. [35]) can be used to model the TORPEX plasma dynamics. In the limit of $T_i \ll T_e$ and $\beta \ll 1$, and assuming that $B_z \ll B_\varphi$ so that $B \simeq B_0 R/(R + r)$, since the magnetic curvature is constant along a field line and equal to $R + r$, these equations can be written as

$$\frac{\partial n}{\partial t} = \frac{c}{B}[\phi, n] + \frac{2c}{eRB}\left(\frac{\partial p_e}{\partial z} - en\frac{\partial \phi}{\partial z}\right) - \frac{\partial\left(nV_{\|e}\right)}{\partial x_\|} + S_n \qquad (14.12)$$

$$\frac{\partial \nabla_\perp^2 \phi}{\partial t} = \frac{c}{B}[\phi, \nabla_\perp^2 \phi] - V_{\|i}\frac{\partial \nabla_\perp^2 \phi}{\partial x_\|} + \frac{2B}{cm_i Rn}\frac{\partial p_e}{\partial z} + \frac{m_i \Omega_{ci}^2}{e^2 n}\frac{\partial j_\|}{\partial x_\|} \qquad (14.13)$$

$$\frac{\partial T_e}{\partial t} = \frac{c}{B}[\phi, T_e] + \frac{4c}{3eRB}\left(\frac{7}{2}T_e\frac{\partial T_e}{\partial z} + \frac{T_e^2}{n}\frac{\partial n}{\partial z} - eT_e\frac{\partial \phi}{\partial z}\right)$$
$$+ \frac{2}{3}\frac{T_e}{en}0.71\frac{\partial j_\|}{\partial x_\|} - \frac{2}{3}T_e\frac{\partial V_{\|e}}{\partial x_\|} - V_{\|e}\frac{\partial T_e}{\partial x_\|} + S_T \qquad (14.14)$$

$$\frac{\partial V_{\|i}}{\partial t} = \frac{c}{B}\left[\phi, V_{\|i}\right] - V_{\|i}\frac{\partial V_{\|i}}{\partial x_\|} - \frac{1}{nm_i}\frac{\partial p_e}{\partial x_\|} \qquad (14.15)$$

$$m_e n\frac{\partial V_{\|e}}{\partial t} = m_e n\frac{c}{B}\left[\phi, V_{\|e}\right] - m_e n V_{\|e}\frac{\partial V_{\|e}}{\partial x_\|}$$
$$- T_e\frac{\partial n}{\partial x_\|} + en\frac{\partial \phi}{\partial x_\|} - 1.71n\frac{\partial T_e}{\partial x_\|} + \frac{en}{\sigma_\|}j_\|, \qquad (14.16)$$

where $p_e = nT_e$, $[a, b] = \partial_r a\partial_z b - \partial_z a\partial_r b$, $j_\| = en(V_{\|i} - V_{\|e})$, and S_n and S_T are the density and temperature sources. The r-coordinate denotes the radial direction, $x_\|$ is parallel to B, and z is the direction perpendicular to r and $x_\|$ (for $B_v \ll B_\varphi$ the vertical and z-directions are approximately the same).

The computational domain has an annular shape with a cross section $r = -L_v/2$ to $r = L_v/2$ and $z = 0$ to $z = L_v$. At $r = -L_v/2$ and $r = L_v/2$, Dirichlet boundary conditions are used for n, T_e, ϕ, and $\nabla_\perp^2\phi$, and Neumann boundary conditions for $V_{\|e}$ and $V_{\|i}$.

At the upper and lower walls, at $z = 0$ and $z = L_v$, we consider two sets of boundary conditions. First, we implement an ad hoc set of boundary condition, where we impose $V_{\|i} = \pm c_s$ and $V_{\|e} = \pm c_s \exp(\Lambda - e\phi/T_e)$ with $\Lambda = \log\sqrt{m_i/(2\pi m_e)}$ and we explore both Dirichlet and Robin boundary conditions for n and T_e, while for ϕ we use both Dirichlet boundary conditions $e\phi = \Lambda T_e$ (implying $V_{\|e} = V_{\|i}$) and a boundary condition of the form $\partial_z\phi \propto (e\phi - \Lambda T_e)$. Second, we use the first-principles set of boundary conditions derived in [24], which is $V_{\|i} = \pm c_s$, $V_{\|e} = \pm c_s\exp(\Lambda - e\phi/T_e)$, $\partial_z T_e = k_T\partial_z\phi$, $\partial_z n = \mp(n/c_s)\partial_z V_{\|i}$, $\Omega = \nabla_\perp^2\phi = -\cos^2\alpha[(\partial_z V_{\|i})^2 \pm c_s\partial_z^2 V_{\|i})]$, and $\partial_z\phi = \mp c_s\partial_z V_{\|i}$.

We have used source profiles that mimic the EC and UH resonance layers in TORPEX, i.e. $S_{n,T} = S_{0;n,T}\{S_{UH}\exp[-(r - r_{UH})^2/\lambda_{UH}^2] + S_{EC}\exp[-(r - r_{EC})^2/\lambda_{EC}^2]\}$, with $S_{UH} = 1.5$, $S_{EC} = 1$, $\lambda_{UH} = 1$ cm, $\lambda_{EC} = 0.5$ cm, $r_{UH} = -2$ cm, $r_{EC} = -6$ cm, and values of the source strength ($S_{0n} = 1.5 \cdot 10^{20}$ m^{-3} s^{-1}, $S_{0T} = 3.5 \cdot 10^4$ eV s^{-1}) estimated experimentally through a global balance of the TORPEX plasma. We remark that the dependence of the UH resonance position on n is neglected in the present model. The other values used are: $R = 1$ m, $L_v = 40$ cm, $m_i/m_e = 200$, $\Lambda = 3$, the resistivity has been varied, ranging from $\nu = 0.01 c_s/R$ to $\nu = 1 c_s/R$, in order to check the influence of this parameter ($\nu = e^2 nR/(m_i \sigma_\parallel c_s)$.

GBS solves equations (14.12)–(14.16) on a field-aligned grid using a second-order finite difference scheme with Runge–Kutta time-stepping and small diffusion terms. The Poisson bracket are evaluated using the Arakawa scheme [134].

If only $k_\parallel \simeq 0$ modes are considered, simple two-dimensional fluid equations that describe the plasma turbulence can be considered. The Braginskii equations are integrated in the parallel direction in order to evolve the line-integrated density, $n(r, z) = \int n(r, z, x_\parallel) \mathrm{d}x_\parallel/L_c$, potential, $\phi(r, z) = \int \phi(r, z, x_\parallel) \mathrm{d}x_\parallel/L_c$, and temperature, $T_e(r, z) = \int T_e(r, z, x_\parallel) \mathrm{d}x_\parallel/L_c$, $L_c = 2\pi NR$ being the magnetic field line length. The magnetic field lines are characterized by the parameter $N = L_v B_\varphi/(2\pi R B_z)$, which is the number of field line turns in the device. We use Bohm's boundary conditions to take into account the ion and electron parallel flow at the sheath edge: by assuming that the density at the edge is equal to $n(r,z)/2$, it is possible to approximate the ion and electron flows as $\Gamma_{\parallel,i} = nc_s/2$ and $\Gamma_{\parallel,e} = nc_s \exp(-e\phi/T_e + \Lambda)/2$. The evolution equations for n, ϕ, and T_e thus become

$$
\frac{\partial n}{\partial t} = \frac{c}{B}[\phi, n] + \frac{2c}{eRB}\left(n\frac{\partial T_e}{\partial z} + T_e\frac{\partial n}{\partial z} - en\frac{\partial \phi}{\partial z}\right)
$$
$$
- \frac{\sigma n c_s}{R}\exp(\Lambda - e\phi/T_e) + S_n,
$$
(14.17)

$$
\frac{\partial \nabla^2 \phi}{\partial t} = \frac{c}{B}[\phi, \nabla^2 \phi] + \frac{2B}{cm_iR}\left(\frac{T_e}{n}\frac{\partial n}{\partial z} + \frac{\partial T_e}{\partial z}\right)
$$
$$
+ \frac{\sigma c_s m_i \Omega_i^2}{eR}\left[1 - \exp(\Lambda - e\phi/T_e)\right],
$$
(14.18)

$$
\frac{\partial T_e}{\partial t} = \frac{c}{B}[\phi, T_e] + \frac{4c}{3eRB}\left(\frac{7}{2}T_e\frac{\partial T_e}{\partial z} + \frac{T_e^2}{n}\frac{\partial n}{\partial z} - eT_e\frac{\partial \phi}{\partial z}\right)
$$
$$
- \frac{2}{3}\frac{\sigma T_e c_s}{R}\left[1.71 \exp(\Lambda - e\phi/T_e) - 0.71\right] + S_T,
$$
(14.19)

where $\sigma = R/L_c = \Delta/(2\pi L_v)$. We note that a similar system of equations has been used in [129]. The system of equations (14.17)–(14.19) has been solved numerically, using the earliest version of GBS that has been developed from the ESEL code [130]. The algorithm used is described in [131].

For the two-dimensional simulations, we consider a domain with extension L_r in the radial direction and Δ along z. The boundary conditions are periodic along the vertical direction (due to the flute property of the interchange mode) and we use Dirichlet boundary conditions in the radial direction. In order to study the sensitivity of the results to the parallel boundary conditions, a scan of the σ parameter has been performed.

Both for the three-dimensional and the two-dimensional case, the simulation is started from random noise. Then, the sources introduce plasma and heat, increasing the plasma pressure and triggering the interchange instability. The interchange instability leads to density and particle transport in the radial direction from the source region to the LFS of the machine. At the same time, plasma is removed from the system by parallel losses. The results discussed in this chapter focus on the quasi-steady-state period, established after the initial simulation transient, as a result of a balance between parallel losses, perpendicular transport, and sources.

A detailed analysis of the plasma dynamics described by the three-dimensional model (14.12)–(14.16) has been presented in [22, 23], while the simulation results obtained from the two-dimensional model (14.17)–(14.19) have been discussed in [25].

For the comparison of simulations and experiments, we consider a set of TORPEX configurations with different values of vertical magnetic field, characterized by different properties of plasma turbulence. We analyze four scenarios, characterized by different windings of the magnetic field lines, i.e. $N = 2, 4, 8$, and 16. By using GBS, we carry out simulations of TORPEX plasma turbulence by considering three different models: (a) a global three-dimensional two-fluid model that describes the evolution of the plasma dynamics in the full TORPEX volume, (b) a global three-dimensional two-fluid model that describes the evolution of the plasma dynamics in the full TORPEX volume provided with recently derived first-principles boundary conditions [24], and (c) a two-dimensional model that is able to represent only $k_\parallel = 0$ modes.

14.3.3 Code verification

GBS has been subject to the code verification procedure described in [29, 132]. Here we recall the main elements of the procedure. To verify the implementation of the drift-reduced Braginskii equations into GBS and to satisfy the requirements given in section 14.2.1, we choose to manufacture the model solution as the combination of trigonometric functions

$$s(z, r, \varphi, t) = A_s\left\{B_s + C_s \sin[D_s x_\parallel]\sin(E_s\varphi + F_s)\sin[G_s r + H_s t]\right\}, \quad (14.20)$$

where A_s, B_s, C_s, D_s, E_s, F_s, G_s, and H_s are arbitrary constants, φ is the toroidal angle, and s represents the scalar fields $\{n, V_{\parallel i}, V_{\parallel e}, T_e, \phi\}$ present in the GBS equations. The constant B_s is used to ensure $n > 0$, $T_e > 0$, while the other factors are used to calibrate the numerical error, ensuring that no term dominates the numerical error in the equations. The dependencies imposed through these constants manifest the physical problem of interest, with a term being perfectly aligned to the magnetic field lines, and a perturbation along the vertical z-coordinate. Variation along r and a temporal dependence of the manufactured solution are also included.

Computation of the source terms S simply consists of plugging the analytical manufactured solutions into the drift-reduced Braginskii equations. This process is particularly tedious, but it involves only straightforward algebraic manipulations with no conceptual difficulties and can be carried out by using the symbolic manipulation software *Mathematica* [42], which allows the direct translation into the *Fortran* language in which GBS is written.

In figure 14.3 we report the results of the code verification. Six simulations are carried out, with mesh refinement parameters $h = 1, 2, 4, 8, 16, 32$ in each direction. The time step is varied with $\sqrt{h}$, as the Runge–Kutta time advance is fourth-order accurate compared to the second-order accurate spatial derivatives. The numerical error, evaluated as the norm of the difference between the numerical and the manufactured solutions at each grid point, decreases with the grid spacing linearly on a logarithmic scale, when refining the mesh, with the slope expected from the order of accuracy of the numerical method. The scan we perform shows a reduction of the numerical error by at least three orders of magnitude, giving confidence that there are no sub-dominant errors decreasing at a rate different than the expected one.

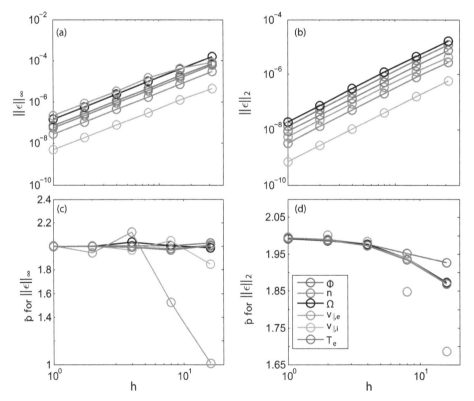

Figure 14.3. $\|\epsilon\|_\infty$ and $\|\epsilon\|_2$ norms of the discretization error, (a) and (b), and their respective order of accuracy estimates, (c) and (d), for GBS simulations where the refinement parameter, defined as $h = x/x_0 = y/y_0 = z/z_0 = (t/t_0)^2$, is varied by a factor of 32.

14.3.4 Solution verification

A previous solution verification procedure applied to the GBS code [29] shows that the numerical error affecting the simulations is mostly due to the space discretization, while the time discretization leads to a negligible numerical error. We therefore perform three simulations of TORPEX turbulence for $N = 2$ by increasing the spatial grid resolution in all directions, focusing our attention on the numerical error that affects the 10 validation observables considered in section 14.3.5. In particular, we consider the equilibrium radial profiles at the vertical mid-plane of density, n, electron temperature, T_e, electric potential, ϕ, ion saturation current, I_{sat}, normalized I_{sat} fluctuations, $\delta I_{\text{sat}}/I_{\text{sat}}$, and I_{sat} skewness and kurtosis. We also use as validation observables the vertical wavenumber, k_z, the probability distribution function (PDF), and the power spectrum density (PSD) of I_{sat} at the vertical mid-plane, at the radial point where I_{sat} is equal to 3/4 of its peak value (this is the location where we expect to identify more clearly the turbulence properties). We plot the results of the three simulations in figure 14.4. Convergence in some cases shows an oscillatory character: as the necessary condition to use the *GCI* estimate are missing, we estimate the numerical uncertainty as the spatial average of the maximum of the difference between the different simulations. The error bar associated with this uncertainty is plotted in figure 14.4.

14.3.5 Validation

To compare experiments and simulations, we use LP measurements and the observables [27, 43] listed in table 14.1. Here, the 10 observables are all at the second level of the comparison hierarchy ($H = 0.5$ in all cases). For each observable, the experimental uncertainty Δe is evaluated by repeating the experiments a number of times, and comparing the measurements of different probes. The simulation uncertainty, Δs, is evaluated by performing a number of simulations where we vary the input parameters. A complete sensitivity scan would require the analysis of all input parameters. However, due to the high computational cost of the present simulations, large parameter scans are prohibitive at the moment, and we have focused our attention on the parameters that are expected to most significantly affect the simulations and that are not well known. For the three-dimensional simulations, these are the plasma resistivity ν and the boundary conditions (in the case of simulations with ad hoc boundary conditions). An example of uncertainty evaluation is shown in figure 14.5, where we show three simulations carried out with different values of ν and the error bar we deduce. For the two-dimensional simulations, instead, we study the sensitivity to the parameter describing the parallel losses, σ.

The values of χ are plotted in figure 14.6, which describes the dependence of the simulation/experiment agreement as a function of the number of field line turns, N, for both the two-dimensional and three-dimensional simulations, with and without the first-principles set of boundary conditions. For the three-dimensional simulations, it is $\chi \simeq 0.5$, showing that the three-dimensional simulations are able to represent equally well low- and high-N scenarios. A clear trend is instead observed in

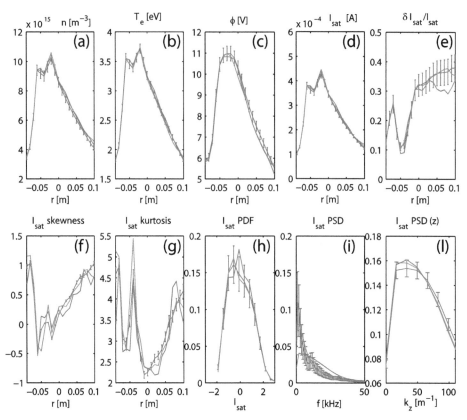

Figure 14.4. Evaluation of the numerical uncertainty due to the numerical discretization. The 10 observables considered for the validation procedure are plotted for TORPEX simulations with $N = 2$, considering an increasing refinement of the mesh: $N_r \times N_z \times N_\varphi = 128 \times 128 \times 32$ (blue), $N_r \times N_z \times N_\varphi = 192 \times 192 \times 48$ (green), and $N_r \times N_z \times N_\varphi = 288 \times 288 \times 72$ (red). The resulting estimated uncertainty is represented by the errorbars. The I_{sat} PDFs are rescaled to the same values of average, variance, and area. The PSDs are rescaled to the same area.

Table 14.1. Primacy hierarchy for some observables obtained from LP measurements.

	Experimental hierarchy	Simulation hierarchy	Comparison hierarchy
Moments, PDF, and PSD of I_{sat} and V_{fl}	1	2	2
$\bar{T}_e, \bar{n}, \bar{\phi}$	2	1	2
Results from spectral analysis (e.g. k_z)	2	1	2
Statistical analysis of turbulent structures	2	3	4
T_e	3	1	3
Particle flux	3	2	4

The experimental hierarchy counts the number of assumptions or combinations of measurements used to obtain the observables from experimental data (the first level in the hierarchy denotes no assumptions or combinations of measurements; then, each assumption or combination of measurements increases the hierarchy level of unity). The same definition is used for the simulation hierarchy. The comparison hierarchy sums the experimental and simulation assumptions.

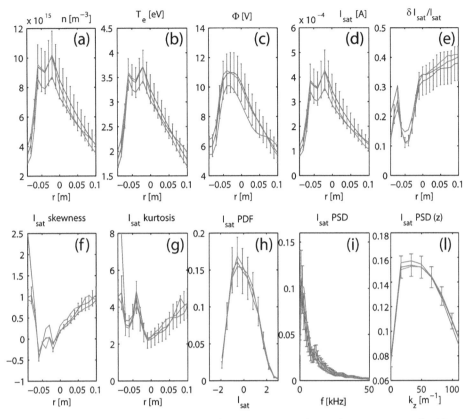

Figure 14.5. Evaluation of the numerical uncertainty due to poorly known input parameters (in this case plasma resistivity). The 10 observables considered for the validation procedure are plotted for TORPEX simulations with $N = 2$, considering $\nu = 0.01$ (green), $\nu = 0.1$ (blue), and $\nu = 1$ (red). The resulting estimated uncertainty is represented by the error bars. The I_{sat} PDFs are rescaled to the same values of average, variance, and area. The PSDs are rescaled to the same area.

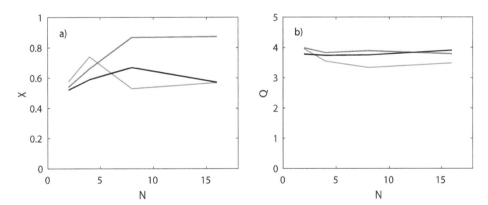

Figure 14.6. Global metric, χ, and quality of the comparison, Q, for the TORPEX simulations as a function of N. The three-dimensional simulations with ad hoc boundary conditions (blue), the three-dimensional simulations with first-principles boundary conditions (black), and two-dimensional simulations (red) are considered.

the case of the two-dimensional simulations, where the agreement decreases with N, passing from $\chi \simeq 0.6$ for $N = 2$ to $\chi \simeq 0.9$ (i.e. almost complete disagreement) for $N = 16$. With respect to the observables considered here, no clear improvement is observed in the simulations carried out with the first-principles set of boundary conditions.

If we look at the origin of the discrepancy between simulations and experiment in more detail, we observe that the three-dimensional simulations generally describe the equilibrium profiles of n and I_{sat} reasonably well. The agreement is worse in the case of the T_e and ϕ profiles. A significant discrepancy between simulations and experiments is revealed by the comparison of the turbulence amplitude levels, as the simulated turbulence amplitude is about a factor of two smaller than the experimental one. Analysis of the PSDs reveals that the difference between the experimental and simulated turbulence levels is present across all the frequencies.

For two-dimensional simulations, the agreement degrades with increasing N for the majority of observables. In fact, for $N = 2$ and $N = 4$ the agreement between the two-dimensional simulation and experiment is comparable to the one observed in the case of three-dimensional simulations. At $N = 16$, all the observables reveal complete disagreement.

The quality of the comparison is also plotted in figure 14.6. Since the uncertainties for all the observables are relatively small and all the observables are at the second level of the validation hierarchy, Q is almost independent of N, and in particular $Q \simeq 4$, which is close to the maximum value, $Q = 5$, that can be obtained using 10 observables at the second level of hierarchy. The Q values reported in the present validation project can be compared with the Q that would be obtained in a validation carried out by exclusively comparing the agreement of the experimental and simulation particle fluxes, i.e. $Q \leqslant 0.25$.

14.4 Conclusions

In this chapter, we propose a methodology for carrying out a V&V procedure to make progress in the understanding of plasma turbulence. Rigorous techniques have been introduced to assess the correct implementation of the model equations in a simulation code (code verification), and estimate the numerical error affecting the simulation results (solution verification). The assessment of the agreement between experiments and simulations (validation) makes use of a number of validation observables to obtain a global metric χ, normalized in order to be equal to 0 in the case of perfect agreement and 1 in the case of complete disagreement. A validation is not concluded until the quality of the comparison, Q, is also provided. The parameter Q is an index that can be used to compare validations; it reveals how well a comparison has been made, indicating the number of observables used for the comparison and how constraining they are.

The proposed methodology has been tested on the simulation of the basic plasma physics experiment TORPEX, focusing on measurements from LPs. We have considered simulations carried out with the GBS code, focusing on two-dimensional and three-dimensional models. The values of χ and Q are displayed in figure 14.6.

What progress has our V&V exercise allowed? First, by carrying out the verification procedure we have largely increased our confidence on the GBS simulation results. In fact, we have rigorously shown that the drift-reduced model is correctly implemented in GBS and, by establishing a methodology to evaluate the magnitude of the numerical error affecting the simulation results, we have quantified the numerical uncertainty of the TORPEX simulations. Second, by comparing different models, we have made progress in the understanding of plasma turbulence in TORPEX. As discussed in [23], there are two turbulent regimes in TORPEX, each primarily driven by a distinct plasma instability: the ideal and the resistive interchange modes. The most obvious difference between the two regimes is the wave number along the magnetic field: $k_\parallel = 0$ in the ideal interchange case, while $k_\parallel \neq 0$ in the resistive case. The main parameter that controls the transition from one instability to the other is the pitch of the field line, expressed in terms of the number of field line turns, N. At low values of N, TORPEX dynamics is dominated by the ideal interchange regime. Theoretical investigations [23] show that the transition to the resistive regime occurs at a value of N that depends on the plasma resistivity and on the vertical size of the device. For TORPEX resistivity and size, the transition is expected to take place at $N \simeq 10$, as has been confirmed experimentally [44] and by the three-dimensional simulations, which show a transition between the two regimes while passing from the $N = 4$ to the $N = 8$ simulations. The validation exercise points out that it is essential to correctly model the transition from the ideal to the resistive interchange mode regime that has $k_\parallel \neq 0$ (not allowed in the two-dimensional simulations), in order to describe the plasma dynamics at high N.

The validation has also pointed out that no significant improvement in the description of the experimental results is made by the implementation of the first-principles set of boundary conditions. However, its implementation eliminates the need of a sensitivity analysis to the boundary conditions, and the possibility of a fortuitous agreement between simulations and experiments. On the other hand, it has been pointed out that the effect of the boundary conditions is particularly important in the parallel velocity profile [45]. We therefore expect that it would be possible to discriminate better the physics of the boundary conditions by considering observables that provide insights on the parallel flow and are based on measurements taken in the proximity of the wall. As a matter of fact, the missing elements in the description of the T_e scale length, which are responsible for the underestimation of the fluctuation amplitude, should be searched for elsewhere (e.g. plasma source fluctuations, Boussinesq approximation, the presence of fast electrons). The importance of these elements can still be assessed through the use of the validation methodology. The example we provide shows that it is relatively easy to use this methodology to discriminate among models and assess whether or not they follow the right trend, pinpointing that the correct physics is described by them. On the other hand, it is much more delicate to judge a single model in absolute terms and to assess its predictive capabilities.

In TORPEX, several advances have been achieved in the basic comprehension of waves, instabilities, and turbulence in SMT configurations. To increase the relevance of TORPEX for fusion research and better reproduce the SOL-edge magnetic geometry in tokamaks, a new system that creates twisted field line configurations [46]

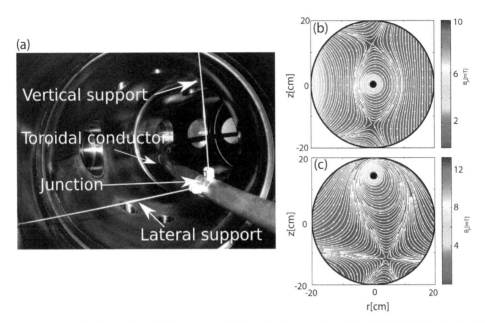

Figure 14.7. (a) The in-vessel toroidal conductor with the main elements installed on TORPEX. Simulation of a (b) double-null configuration and (c) a magnetic second-order null-point (magnetic snowflake) on TORPEX.

was developed, shown in figure 14.7. A toroidal copper conductor (1 cm radius) is suspended inside the vacuum vessel by an electrical coaxial feedthrough, figure 14.7(a), and by three vertical stainless steel wires (1 mm diameter). Four horizontal supports are also used to stabilize the conductor during a TORPEX discharge. This set of supports enables different vertical positions of the toroidal conductor, so that the SMT configuration can be recovered by pulling it to the top of the vacuum vessel. A dedicated external power supply drives a current in the toroidal conductor with a maximum flat top current of ≈1 kA. For simplicity, only the feedthrough is water-cooled. The flat top current duration is limited by the Ohmic heating of the conductor with almost pure radiative cooling in the vacuum. This new system opens the possibility of investigating magnetic configurations with both a confined region with closed magnetic flux surfaces and SOL regions with open field lines. Several new magnetic field configurations are possible, see figure 14.7(b) and (c), from simple plasmas limited on the LFS by the vessel, to more complex magnetic geometries, such as single- or double-null X-points, as well as advanced divertor concepts such as snowflakes [98, 116]. These more complex magnetic configurations will provide a basis for the validation of the numerical simulations, which is necessary to determine the complexity of the numerical model needed for a realistic description not only of TORPEX data, but in particular for fusion devices.

Acknowledgments

The simulations presented here were carried out at the Swiss National Supercomputing Centre (CSCS), under project ID s346, and at the Computational

Simulation Centre of the International Fusion Energy Research Centre (IFERCCSC), Aomori, Japan, under the Broader Approach collaboration between Euratom and Japan, implemented by Fusion for Energy and JAEA. This work was carried out within the framework of the EUROfusion Consortium. It was supported by the Swiss National Science Foundation and received funding from the European Union Horizon 2020 Research and Innovation Programme under grant agreement number 633053. The views and opinions expressed herein do not necessarily reflect those of the European Commission.

Bibliography

[1] Oberkampf W L and Trucano T G 2002 *Prog. Aerosp. Sci.* **38** 209

[2] Oberkampf W L and Roy C J 2010 *Verification and Validation in Scientific Computing* (New York: Cambridge University Press)

[3] Roache P J 1998 *Verification and Validation in Computational Science and Engineering* (Hermosa: Albuquerque, NM)

[4] Dimits A M *et al* 2000 *Phys. Plasmas* **3** 969

[5] Birn J *et al* 2001 *J. Geophys. Res.* **106** 3715

[6] Ricci P, Brackbill J U, Daughton W and Lapenta G 2004 *Phys. Plasmas.* **8** 4102

[7] Umansky V, Day M S and Rognlien T D Num 2005 *Hear Transfer* B **47** 533

[8] Myra J R, D'Ippolito D A, Russel D A and Umansky M 2007 *Bull. Am. Phys. Soc.* **52** 533

[9] Falchetto G L *et al* 2008 *Plasma Phys. Control. Fusion* **50** 124015

[10] Bravenec R V, Chen Y, Candy J, Wan W and Parker S 2013 *Phys. Plasmas* **20** 104506

[11] Seinberg S and Roache P J 1985 *J. Comput. Phys.* **57** 251

[12] Roache P J 1994 *J Fluids Eng.* **116** 405

[13] Roy C J 2005 *J. Comput. Phys.* **205** 131

[14] Roache P J 1997 *Annu. Rev. Fluid Mech.* **29** 123–60

[15] Stern F, Wilson R V, Coleman H W and Paterson E G 2001 *J. Fluids Eng.* **123** 793

[16] Roache P J 1994 *J. Fluids Eng.* **116** 405

[17] Terry P W, Greenwald M, Leboeuf J-N, McKee G R, Mikkelsen D R, Nevins W M, Newman D E and Stotler D P 2008 Task group on verification and validation, us burning plasma organization, and us transport task force *Phys. Plasmas* **15** 062503

[18] Greenwald M 2010 *Phys. Plasmas* **17** 058101

[19] Fasoli A, Labit B, McGrath M, Müller S H, Plyushchev G, Podestà M and Poli F M 2006 *Phys. Plasmas* **13** 055902

[20] Fasoli A *et al* 2010 *Plasma Phys. Control. Fusion* **52** 124020

[21] Ricci P, Halpern F D, Jolliet S, Loizu J, Mosetto A, Fasoli A, Furno I and Theiler C 2012 *Plasma Phys. Control. Fusion* **54** 124047

[22] Ricci P and Rogers B N 2009 *Phys. Plasmas* **16** 092307

[23] Ricci P and Rogers B N 2010 *Phys. Rev. Lett.* **104** 145001

[24] Loizu J, Ricci P, Halpern F D and Jolliet S 2012 *Phys. Plasmas* **19** 122307

[25] Ricci P, Rogers B N and Brunner S 2008 *Phys. Rev. Lett.* **100** 225002

[26] Ricci P and Rogers B N 2009 *Phys. Plasmas* **16** 062303

[27] Ricci P, Theiler C, Fasoli A, Furno I, Labit B, Müller S H, Podestà M and Poli F M 2009 *Phys. Plasmas* **16** 055703

[28] Ricci P, Theiler C, Fasoli A, Furno I, Labit B, Gustafson K, Iraji D and Loizu J 2011 *Phys. Plasmas* **18** 032109

[29] Riva F, Ricci P, Halpern F D, Jolliet S, Loizu J and Mosetto A 2014 *Phys. Plasmas* **21** 062301

[30] Richardson L F 1911 *Phil. Trans. R. Soc.* A **210** 307

[31] Richardson L F and Gaunt J A 1927 *Phil. Trans. R. Soc.* A **226** 299

[32] Roache P J and Knupp P M 1993 *Commun. Num. Methods Eng.* **9** 365

[33] Richards S A 1997 *Commun. Num. Methods Eng.* **13** 573

[34] Myers R H, Montgomery D C and Anderson-Cook C M 2009 *Response Surface Methodology: Process and Product Methodology, Using Designed Experiments* 3rd edn (New York: Wiley)

[35] Zeiler A, Drake J F and Rogers B 1997 *Phys. Plasmas* **4** 2134

[36] Rogers B N and Ricci P 2010 *Phys. Rev. Lett.* **104** 225002

[37] Ricci P and Rogers B N 2013 *Phys. Plasmas* **20** 010702

[38] Mosetto A, Halpern F D, Jolliet S, Loizu J and Ricci P 2013 *Phys. Plasmas* **20** 092308

[39] Halpern F D *et al* 2013 *Nucl. Fusion* **53** 122001

[40] Loizu J, Ricci P, Halpern F D, Jolliet S and Mosetto A 2013 *Plasma Phys. Control. Fusion* **55** 124019

[41] Theiler C, Diallo A, Fasoli A, Labit B, Podestà M, Poli F M and Ricci P 2008 *Phys. Plasmas* **15** 042303

[42] Wolfram Research Inc. 2010 Mathematica Version 8.0 (Champaign, IL: Wolfram Research Inc.)

[43] Müller S H, Diallo A, Fasoli A, Furno I, Labit B, Plyushchev G, Podestà M and Poli F M 2006 *Phys. Plasmas* **13** 100701

[44] Poli F M, Ricci P, Fasoli A and Podestà M 2008 *Phys. Plasmas* **15** 032104

[45] Loizu J, Ricci P, Halpern F D, Jolliet S and Mosetto A 2014 *Phys. Plasmas* **21** 062309

[46] Avino F, Fasoli A and Furno I 2014 The new TORPEX in-vessel toroidal conductor for the generation of a poloidal magnetic field *Rev. Sci. Instrum.* **85** 033506

[47] Bovet A, Furno I, Fasoli A, Gustafson K and Ricci P 2012 Investigation of fast ion transport in TORPEX *Nucl. Fusion* **52** 094017

[48] Bovet A, Furno I, Fasoli A, Gustafson K and Ricci P 2013 Three-dimensional measurements of non-diffusive fast ion transport in TORPEX *Plasma Phys. Control. Fusion* **55** 124021

[49] Bovet A, Gamarino M, Furno I, Ricci P, Fasoli A, Gustafson K, Newman D E and Sanchez R 2014 Transport equation describing fractional Lévy motion of suprathermal ions in TORPEX *Nucl. Fusion* **54** 104009

[50] Diallo A, Fasoli A, Furno I, Labit B, Podestà M and Theiler C 2008 Dynamics of plasma blobs in a shear flow *Phys. Rev. Lett.* **101** 115005

[51] Diallo A, Ricci P, Fasoli A, Furno I, Labit B, Müller S H, Podestáa M, Poli F M and Skiff F 2007 Antenna excitation of drift wave in a toroidal plasma *Phys. Plasmas* **14** 102101

[52] D'Ippolito D A, Myra J R and Zweben S J 2011 Convective transport by intermittent blob-filaments: comparison of theory and experiment *Phys. Plasmas* **18** 060501

[53] Fasoli A *et al* 2013 Basic investigations of electrostatic turbulence and its interaction with plasma and suprathermal ions in a simple magnetized toroidal plasma *Nucl. Fusion* **53** 063013

[54] Fasoli A *et al* 2010 Electrostatic instabilities, turbulence and fast ion interactions in the TORPEX device *Plasma Phys. Control. Fusion* **52** 124020

[55] Fasoli A *et al* 2003 Electrostatic turbulence and transport in a simple magnetized plasma *Bull. Am. Phys. Soc.* **13** 119

[56] Federspiel L, Labit B, Ricci P, Fasoli A, Furno I and Theiler C 2009 Observation of a critical pressure gradient for the stabilization of interchange modes in simple magnetized toroidal plasmas *Phys. Plasmas* **16** 092501

[57] Furno I, Diallo A, Fasoli A, Labit B, Mueller S H, Plyushchev G, Podestàa G and Poli F M 2006 Effect of the magnetic configuration on fluctuations and turbulence in a magnetized toroidal plasma *11th EU-US Transport Task Force Meeting*

[58] Furno I *et al* 2008 Mechanism for blob generation in the TORPEX toroidal plasma *Phys. Plasmas* **15** 055903

[59] Furno I, Spolaore M, Theiler C, Vianello N, Cavazzana R and Fasoli A 2011 Direct two-dimensional measurements of the field-aligned current associated with plasma blobs *Phys. Rev. Lett.* **106** 245001

[60] Furno I, Theiler C, Chabloz V, Fasoli A and Loizu J 2014 Pre-sheath density drop induced by ion-neutral friction along plasma blobs and implications for blob velocities *Phys. Plasmas* **21** 012305

[61] Furno I, Theiler C, Lançon D, Fasoli A, Iraji D, Ricci P, Spolaore M and Vianello N 2011 Blob current structures in TORPEX plasmas: experimental measurements and numerical simulations *Plasma Phys. Control. Fusion* **53** 124016

[62] Furno I *et al* 2008 Experimental observation of the blob-generation mechanism from interchange waves in a plasma *Phys. Rev. Lett.* **100** 055004

[63] Garcia O E, Naulin V, Nielsen A H and Juul Rasmussen J 2005 Turbulence and intermittent transport at the boundary of magnetized plasmas *Phys. Plasmas* **12** 062309

[64] Garcia O E 2012 Stochastic modeling of intermittent scrape-off layer plasma fluctuations *Phys. Rev. Lett.* **108** 265001

[65] Goodman T P *et al* 2003 An overview of results from the TCV tokamak *Nucl. Fusion* **43** 1619

[66] Greenwald M 2010 Verification and validation for magnetic fusion *Phys. Plasmas* **17** 058101

[67] Gustafson K and Ricci P 2012 Lévy walk description of suprathermal ion transport *Phys. Plasmas* **19** 032304

[68] Gustafson K, Ricci P, Bovet A, Furno I and Fasoli A 2012 Suprathermal ion transport in simple magnetized torus configurations *Phys. Plasmas* **19** 062306

[69] Gustafson K, Ricci P, Furno I and Fasoli A 2012 Nondiffusive suprathermal ion transport in simple magnetized toroidal plasmas *Phys. Rev. Lett.* **108** 035006

[70] Guszejnov D, Lazanyi N, Bencze A and Zoletnik S 2013 On the effect of intermittency of turbulence on the parabolic relation between skewness and kurtosis in magnetized plasmas *Phys. Plasmas* **20** 112305

[71] Halpern F D, Ricci P, Jolliet S, Loizu J and Mosetto A 2014 Theory of the scrape-off layer width in inner-wall limited tokamak plasmas *Nucl. Fusion* **54** 043003

[72] Halpern F D *et al* 2013 Theory-based scaling of the SOL width in circular limited tokamak plasmas *Nucl. Fusion* **53** 122001

[73] Heidbrink W W *et al* 2012 Measurements of interactions between waves and energetic ions in basic plasma experiments *Plasma Phys. Control. Fusion* **54** 124007

[74] Henderson M A *et al* 2003 Recent results from the electron cyclotron heated plasmas in Tokamak à configuration variable (TCV) *Phys. Plasmas* **10** 1796

[75] Iraji D, Diallo A, Fasoli A, Furno I and Shibaev S 2008 Fast visible imaging of turbulent plasma in TORPEX *Rev. Sci. Instrum.* **79** 10F508

[76] Iraji D, Furno I and Fasoli A 2011 Fast imaging of blob motion in TORPEX plasmas *IEEE Trans. Plasma Sci.* **39** 3010–1

[77] Iraji D *et al* 2010 Imaging of turbulent structures and tomographic reconstruction of TORPEX plasma emissivity *Phys. Plasmas* **17** 122304

[78] Katz N, Egedal J, Fox W, Le A and Porkolab M 2008 Experiments on the propagation of plasma filaments *Phys. Rev. Lett.* **101** 015003

[79] Krasheninnikov S I 2001 On scrape off layer plasma transport *Phys. Lett. A* **283** 368–70

[80] Krommes J A 2008 The remarkable similarity between the scaling of kurtosis with squared skewness for TORPEX density fluctuations and sea-surface temperature fluctuations *Phys. Plasmas* **15** 030703

[81] Labit B *et al* 2007 Statistical properties of electrostatic turbulence in toroidal magnetized plasmas *Plasma Phys. Control. Fusion* **49** B281

[82] Labit B, Furno I, Fasoli A, Diallo A, Müller S H, Plyushchev G, Podestà M and Poli F M 2007 Universal statistical properties of drift-interchange turbulence in TORPEX plasmas *Phys. Rev. Lett.* **98** 255002

[83] Labit B, Theiler C, Fasoli A, Furno I and Ricci P 2011 Blob-induced toroidal momentum transport in simple magnetized plasmas *Phys. Plasmas* **18** 032308

[84] Loizu J, Ricci P, Halpern F D, Jolliet S and Mosetto A 2013 On the electrostatic potential in the scrape-off layer of magnetic confinement devices *Plasma Phys. Control. Fusion* **55** 124019

[85] Loizu J, Ricci P, Halpern F D, Jolliet S and Mosetto A 2014 Intrinsic toroidal rotation in the scrape-off layer of tokamaks *Phys. Plasmas* **21** 062309

[86] Mekkaoui A 2013 Derivation of stochastic differential equations for scrape-off layer plasma fluctuations from experimentally measured statistics *Phys. Plasmas* **20** 010701

[87] Mosetto A, Halpern F D, Jolliet S, Loizu J and Ricci P 2013 Turbulent regimes in the tokamak scrape-off layer *Phys. Plasmas* **20** 092308

[88] Müller S 2007 Turbulence in basic toroidal plasmas *PhD Thesis* SB, Lausanne

[89] Müller S H, Fasoli A, Labit B, McGrath M, Podestà M and Poli F M 2004 Effects of a vertical magnetic field on particle confinement in a magnetized plasma torus *Phys. Rev. Lett.* **93** 165003

[90] Müller S H, Theiler C, Fasoli A, Furno I, Labit B, Tynan G R, Xu M, Yan Z and Yu J H 2009 Studies of blob formation, propagation and transport mechanisms in basic experimental plasmas (TOPEX and CSDX) *Plasma Phys. Control. Fusion* **51** 055020

[91] Müller S H *et al* 2006 Probabilistic analysis of turbulent structures from two-dimensional plasma imaging *Phys. Plasmas* **13** 100701

[92] Müller S H *et al* 2008 Plasma blobs in a basic toroidal experiment: origin, dynamics, and induced transport *Phys. Plasmas* **14** 110704

[93] Myra J R and D'Ippolito D A 2005 Edge instability regimes with applications to blob transport and the quasicoherent mode *Phys. Plasmas* **12** 092511

[94] Otsuji K *et al* 2007 Small-scale magnetic-flux emergence observed with Hinode solar optical telescope *Publ. Astron. Soc. Japan* **59** S649–54

[95] Park J, Lühr H, Stolle C, Rother M, Min K W and Michaelis I 2010 Field-aligned current associated with low-latitude plasma blobs as observed by the champ satellite *Ann. Geophys.* **28** 697–703

[96] Perri S and Zimbardo G 2012 Superdiffusive shock acceleration *Astrophys. J.* **750** 87

[97] Perrone D *et al* 2013 Nonclassical transport and particle-field coupling: from laboratory plasmas to the solar wind *Space Sci. Rev.* **178** 233–70

[98] Piras F *et al* 2009 Snowflake divertor plasmas on TCV *Plasma Phys. Control. Fusion* **51** 055009

[99] Plyushchev G *et al* 2006 Fast ion source and detector for investigating the interaction of turbulence with suprathermal ions in a low temperature toroidal plasma *Rev. Sci. Instrum.* **77** 10F503

[100] Podestà M 2007 Plasma production and transport in a simple magnetised toroidal plasma *PhD Thesis* SB, Lausanne

[101] Podestà M, Diallo A, Fasoli A, Furno I, Labit B, Müller S H and Poli F M 2007 Characterization of the electron distribution function in an electron-cyclotron driven toroidal plasma *Plasma Phys. Control. Fusion* **49** 175

[102] Podestà M, Fasoli A, Labit B, Furno I, Ricci P, Poli F M, Diallo A, Müller S H and Theiler C 2008 Cross-field transport by instabilities and blobs in a magnetized toroidal plasma *Phys. Rev. Lett.* **101** 045001

[103] Podestà M, Fasoli A, Labit B, McGrath M, Müller S H and Poli F M 2006 Experimental characterization and modelling of the particle source in an electron-cyclotron wave driven toroidal plasma *Plasma Phys. Control. Fusion* **48** 1053

[104] Podestà M *et al* 2005 Plasma production by low-field side injection of electron cyclotron waves in a simple magnetized torus *Plasma Phys. Control. Fusion* **47** 1989–2002

[105] Poli F M, Brunner S, Diallo A, Fasoli A, Furno I, Labit B, Mueller S H, Plyushchev G and Podestà M 2006 Experimental characterization of drift-interchange instabilities in a simple toroidal plasma *Phys. Plasmas* **13** 102104

[106] Poli F M, Podestàa M and Fasoli A 2007 Development of electrostatic turbulence from drift-interchange instabilities in a toroidal plasma *Phys. Plasmas* **14** 052311

[107] Poli F M, Ricci P, Fasoli A and Podestà M 2008 Transition from drift to interchange instabilities in an open magnetic field line configuration *Phys. Plasmas* **15** 032104

[108] Reames D V 1999 Particle acceleration at the Sun and in the heliosphere *Space Sci. Rev.* **90** 413–91

[109] Ricci P, Halpern F D, Jolliet S, Loizu J, Mosetto A, Fasoli A, Furno I and Theiler C 2012 Simulation of plasma turbulence in scrape-off layer conditions: the GBS code, simulation results and code validation *Plasma Phys. Control. Fusion* **54** 124047

[110] Ricci P and Rogers B N 2009 Three-dimensional fluid simulations of a simple magnetized toroidal plasma *Phys. Plasmas* **16** 092307

[111] Ricci P and Rogers B N 2010 Turbulence phase space in simple magnetized toroidal plasmas *Phys. Rev. Lett.* **104** 145001

[112] Ricci P and Rogers B N 2013 Plasma turbulence in the scrape-off layer of tokamak devices *Phys. Plasmas* **20** 010702

[113] Ricci P, Rogers B N and Brunner S 2008 High- and low-confinement modes in simple magnetized toroidal plasmas *Phys. Rev. Lett.* **100** 225002

[114] Ricci P, Theiler C, Fasoli A, Furno I, Gustafson K, Iraji D and Loizu J 2011 Methodology for turbulence code validation: quantification of simulation-experiment agreement and application to the torpex experiment *Phys. Plasmas* **18** 032109

[115] Ricci P, Theiler C, Fasoli A, Furno I, Labit B, Mueller S H, Podestà M and Poli F M 2009 Langmuir probe-based observables for plasma-turbulence code validation and application to the torpex basic plasma physics experiments *Phys. Plasmas* **16** 055703

[116] Ryutov D D 2007 Geometrical properties of a 'snowflake' divertor *Phys. Plasmas* **14** 064502

[117] Sandberg I, Benkadda S, Garbet X, Ropokis G, Hizanidis K and del-Castillo-Negrete D 2012 Universal probability distribution function for bursty transport in plasma turbulence *Phys. Rev. Lett.* **103** 165001

[118] Schaffner D A, Carter T A, Rossi G D, Guice D S, Maggs J E, Vincena S and Friedman B 2012 Modification of turbulent transport with continuous variation of flow shear in the large plasma device *Phys. Rev. Lett.* **109** 135002

[119] Terry P W *et al* 2008 Validation in fusion research: towards guidelines and best practices *Phys. Plasmas* **15** 62503

[120] Theiler C, Diallo A, Fasoli A, Furno I, Labit B, Podestà M, Poli F M and Ricci P 2008 The role of the density gradient on intermittent cross-field transport events in a simple magnetized toroidal plasma *Phys. Plasmas* **15** 042303

[121] Theiler C, Furno I, Fasoli A, Ricci P, Labit B and Iraji D 2011 Blob motion and control in simple magnetized plasmas *Phys. Plasmas* **18** 055901

[122] Theiler C, Furno I, Kuenlin A, Marmillod P and Fasoli A 2011 Practical solutions for reliable triple probe measurements in magnetized plasmas *Rev. Sci. Instrum* **82** 013504

[123] Theiler C, Furno I, Loizu J and Fasoli A 2012 Convective cells and blob control in a simple magnetized plasma *Phys. Rev. Lett.* **108** 065005

[124] Theiler C, Loizu J, Furno I, Fasoli A and Ricci P 2012 Properties of convective cells generated in magnetized toroidal plasmas *Phys. Plasmas* **19** 082304

[125] Theiler C *et al* 2009 Cross-field motion of plasma blobs in an open magnetic field line configuration *Phys. Rev. Lett.* **103** 065001

[126] Podestà M, Fasoli A, Labit B, McGrath M, Müller S H and Poli F M 2006 *Plasma Phys. Control. Fusion* **48** 1053

[127] Müller S H, Fasoli A, Labit B, McGrath M, Pisaturo O, Plyushchev G, Podestà M and Poli F M 2005 *Phys. Plasmas* **12** 090906

[128] Theiler C, Furno I, Kuenlin A, Marmillod P and Fasoli A 2011 Practical solutions for reliable triple probe measurements in magnetized plasmas *Rev. Sci. Instrum.* **82** 013504

[129] Bisai N, Das A, Deshpande S, Jha R, Kaw P, Sen A and Singh R 2004 *Phys. Plasmas* **11** 4018

[130] Naulin V, Nycander J and Rasmussen J J 1998 *Phys. Rev. Lett.* **81** 4148
Garcia O E, Naulin V, Nielsen A H and Rasmussen J J 2004 *Phys. Rev. Lett.* **92** 165003

[131] Naulin V and Nielsen A H 2003 *SIAM J. Sci. Comput.* **25** 104

[132] Halpern F D, Ricci P, Jolliet S, Loizu J, Morales J, Mosetto A, Musil F, Riva F, Tran T M and Wersal C 2016 *J. Comput. Phys.* **315** 388

[133] Ricci P, Riva F, Theiler C, Fasoli A, Furno I, Halpern F D and Loizu J 2015 *Phys. Plasmas* **22** 055704

[134] Arakawa A 1966 *J. Comput. Phys.* **1** 119

Lightning Source UK Ltd.
Milton Keynes UK
UKHW051627131021
392129UK00003B/87